Discovering
the Essential Universe

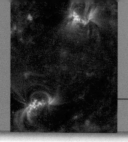

CONTENTS OVERVIEW

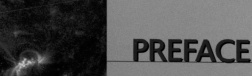

Teaching and learning introductory astronomy have evolved dramatically in recent years. Students enter the classroom with varying degrees of preparation, interest, and misconceptions. Instructors, considering the diverse audience before them, often adjust their methods to address the variety of student misconceptions and learning styles.

Discovering the Essential Universe, Fourth Edition, has been written specifically to address the issues that astronomy educators and students have been facing. This textbook is one of the briefest, least expensive introductory astronomy texts available, and is built on a learning method designed to help students overcome their misconceptions about astronomy. Accompanied by an impressive collection of multimedia resources designed and tested by astronomy education researchers, *Discovering the Essential Universe* provides comprehensive explanations of the core concepts in a flexible and student-friendly text.

Despite its brevity and low price, this book's topical coverage remains consistent with most introductory courses, and you'll find it at least as rich in celestial images and figures as most other textbooks for the same audience. Certain nonessential items have been removed, including detailed mathematical explanations, enrichment boxes, and some of the end-of-chapter material. The fourth edition of *Discovering the Essential Universe* continues the book's legacy of presenting up-to-date concepts clearly and accurately while providing all the pedagogical tools to make the learning process memorable. The pedagogy includes

- using both textual and graphical information to present concepts for students who learn in different ways.

- helping students compare their beliefs with the findings of modern science and understand why the scientific view is correct.

- using analogies from everyday life to make cosmic phenomena more concrete.

- presenting the observations and underlying physical concepts needed to connect astronomical observations to theories that explain them coherently and meaningfully.

NEW FEATURES BRING THE UNIVERSE INTO CLEARER FOCUS

Scientific American articles, chosen by the author, are included in the text to illuminate the core concepts in the text. These brief, recent, and relevant selections demonstrate the process of science and discovery, while also providing touchstones for classroom discussion.

A new chapter on astrobiology provides students with a thorough overview of this motivating and exciting area of astronomy, correcting common misconceptions and illustrating the ongoing development of our scientific knowledge.

Focus questions about important concepts are presented in most sections of the book. These questions encourage students to test themselves frequently on material presented in the preceding sections of the chapters and thereby correct their beliefs before errors accumulate. For example, after learning about Uranus's ring system in Section 5-29, students are asked why Uranus's rings remain in orbit. Answers to approximately one-third of these questions appear at the end of this text.

Why do Uranus's rings remain in orbit?

Star charts show the location in the sky of important astronomical objects cited in the text. Sufficient detail in the star charts allows students to locate the objects with either the unaided eye or a small telescope, as appropriate.

a M104: an Sa galaxy

New coverage of the planets Astronomers have created a new classification scheme for the objects in the solar system. These *planets, dwarf planets*, and *small solar-system objects*, along with new subclasses such as *plutoids* are explained and reconciled with the existing classes of objects that include planets, moons, asteroids, meteoroids, and comets. Also explained is how Pluto fits better with the dwarf planets than with the eight planets.

New dynamic art Summary figures appear throughout book to show either the interactions between importa concepts or the evolution of important objects introduced. For example, the location of the Sun in the sky, which varies over the seasons, as does the corresponding intensity of the light and the appropriate ground cover, is shown in a sequence of drawings combined into one figure.

PROVEN FEATURES SUPPORT LEARNING

What Do You Think? and **What Did You Think?** questions in each chapter ask students to consider their present beliefs and actively compare them with the correct science presented in the book. Margin numbers mark the places in the text where each concept is discussed. Encouraging students to think about what they believe is true and then work through the correct science step by step has proved to be an effective teaching technique, especially when time constraints prevent instructors from working with students directly to reconcile incorrect beliefs with proper science.

Learning objectives underscore the key chapter concepts.

Section headings are brief sentences that summarize section content and serve as a quick study guide to the chapter when reread.

Icons link the text to web material.

- *Starry Night Enthusiast*™ icons link the text to a specific interactivity in the *Starry Night Enthusiast*™ observing programs, which is an optional, free CD with the textbook. Also, *WorldWide Telecope* **questions** have been added to provide students who do not have *Starry Night* the opportunity to do observational projects. These questions guide the students through an investigative and inquiry-based process using the free *WorldWide Telecope* software now available from Microsoft®.

- Video icons link the text to relevant video clips available on the textbook's Web site.

- Animation icons link the text to animated figures available on the textbook's Web site.

- Active Integrated Media Modules (AIMM) and Interactive Exercise icons link the text to interactive exercises available on the textbook's Web site.

- Web link icons direct the reader to further information on a particular topic.

Insight into Science boxes are brief asides that relate topics to the nature of scientific inquiry and encourage critical thinking.

Wavelength tabs with photographic images show whether an image was made with radio waves (R), infrared radiation (I), visible light (V), ultraviolet light (U), X rays (X), or gamma rays (G).

R I V U X G

Review and practice material

- **Summary of Key Ideas** is a bulleted list of key concepts in the chapter.

- **What Did You Think?** questions at the end of each chapter answer the What Do You Think? questions posed at the beginning of the chapter.

- **Key Words, Review Questions,** and **Advanced Questions** help the students understand the chapter material.

- **Observing Projects,** featuring *Starry Night Enthusiast*™ and *WorldWide Telecope* activities, ask students to be astronomers themselves.

MEDIA AND SUPPLEMENTS PACKAGE

For Students
Online Study Center

W. H. Freeman's Online Study Center is a flexible resource for instructors and students, providing multimedia learning and assessment tools that build on the teaching strategies presented in this textbook. For consistency and convenience, the Online Study Center offers a **wide array of course resources in one location, with one login.**

- **Tutorials** engage students in the scientific process of discovery and interpretation. Dozens of these concept-driven, experiential walkthroughs allow students to make observations, draw conclusions, and apply their knowledge. Interwoven with multimedia, activities, and questions, students receive a deep, self-guided exploration of the concepts.

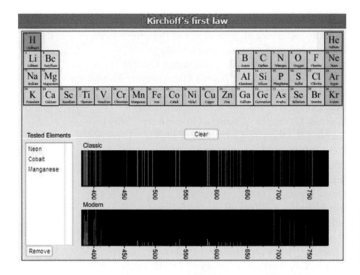

- **Automatically graded on-line quizzing with instructional feedback** corrects student misconceptions immediately and allows instructors to assess student comprehension.

- **Animations** bring to life key concepts and illustrations from the text.

- **Videos provided by NASA** are up-to-date renderings of astronomoical objects and phenomena.

- **Active integrated media modules** take students deeper into key topics that are presented in the text.

- **Interactive drag-and-drop exercises** based on text illustrations help students grasp the vocabulary in context.

If you would like more information or have any questions about the Online Study Center, please contact your local representative or visit www.whfreeman.com/astronomy.

Starry Night Enthusiast™

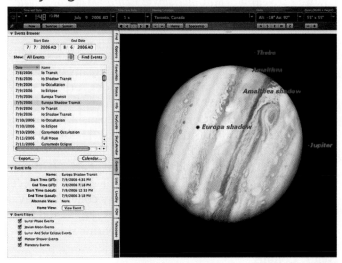

Starry Night Enthusiast™ 6.2 is a brilliantly realistic planetarium software package. It is designed for easy use by anyone with an interest in the night sky. See the sky from anywhere on Earth or lift off and visit any solar system body or any location up to 20,000 light-years away. View 2,500,000 stars along with more than 170 deep-space objects such as galaxies, star clusters, and nebulae. Travel 15,000 years in time, check out the view from the International Space Station, and see planets up close from any one of their moons. Included are stunning OpenGL graphics. Handy star charts can be printed to explore outdoors. The *Starry Night Enthusiast*™ CD is available at no extra charge with the text upon request. Use ISBN 1-4292-3038-X.

Observing Projects Using *Starry Night Enthusiast*™

T. Alan Clark and William J. F. Wilson, University of Calgary, and Marcel Bergman, ISBN 1-4292-1866-5

Available for packaging with the text and compatible with both PC and Mac, this workbook contains a variety of comprehensive lab activities for use with *Starry Night Enthusiast*™ 6.2. The Observing Projects workbook can also be packaged with the *Starry Night* software. Use ISBN 1-4292-2011-2.

Free Student Companion Web Site

The Companion Web site at www.whfreeman.com/deu4e features a variety of study and review resources designed to help students understand the concepts. The open-access Web site includes the following:

- **Online quizzing** offers questions and answers with instant feedback to help students review and prepare for exams. Instructors can access results.

- **Animations and videos,** both original and NASA-created, are keyed to specific chapters.

- **Web links** provide a wealth of online resources to the student.

FOR INSTRUCTORS

The Companion Web site at www.whfreeman.com/deu4e also includes password-protected Instructor Resources, including:

- All figures and photos from the text in both JPEG and PowerPoint formats

- Lecture PowerPoints

- Answers to the review questions from the text

ACKNOWLEDGMENTS

I am deeply grateful to the astronomers and teachers who reviewed chapters for this and previous editions.

William R. Alexander, *James Madison University*
Gordon Baird, *University of Mississippi*
Henry E. Bass, *University of Mississippi*
J. David Batchelor, *Community College of Southern Nevada*
Jill Bechtold, *University of Arizona*

Peter A. Becker, *George Mason University*
Michael Bennett, *DeAnza College*
John Bieging, *University of Arizona*
Greg Black, *University of Virginia*
Julie Bray-Ali, *Mt. San Antonio College*
John B. Bulman, *Loyola Marymount University*
John W. Burns, *Mt. San Antonio College*
Alison Byer, *Widener University*
Gene Byrd, *University of Alabama*
Eugene R. Capriotti, *Michigan State University*
Michael W. Castelaz, *Pisgah Astronomical Research Institute*
Gerald Cecil, *University of North Carolina*
David S. Chandler, *Porterville College*
David Chernoff, *Cornell University*
Tom Christensen, *University of Colorado, Colorado Springs*
Chris Clemens, *University of North Carolina*
Christine Clement, *University of Toronto*
Halden Cohn, *Indiana University*
John Cowan, *University of Oklahoma*
Antoinette Cowie, *University of Hawaii*
Volker Credé, *Florida State University*
Charles Curry, *University of Waterloo*
James J. D'Amario, *Harford Community College*
Purnas Das, *Purdue University*
Peter Dawson, *Trent University*
John M. Dickey, *University of Minnesota, Twin Cities*
Dan Durben, *Black Hills State University*
John D. Eggert, *Daytona Beach Community College*
Bernd Enders, *College of Marin*
Mark W. F. Fischer, *The College of Mt. St. Joseph*
Robert Frostick, *West Virginia State College*
Martin Gaskell, *University of Nebraska*
Bruce Gronich, *University of Texas–El Paso*
Siegbert Hagmann, *Kansas State University*
David Hedin, *Northern Illinois University*
Chuck Higgins, *Penn State University*
James L. Hunt, *University of Guelph*
Nathan Israeloff, *Northeastern College*
Kenneth Janes, *Boston University*
William C. Keel, *University of Alabama*
William Keller, *St. Petersburg Junior College*
Marvin D. Kemple, *Indiana University–Purdue University Indianapolis (IUPUI)*
Julia Kennefick, *University of Arkansas*
Pushpa Khare, *University of Illinois at Chicago*
F. W. Kleinhaus, *Indiana University–Purdue University Indianapolis (IUPUI)*
Rob Klinger, *Parkland College*
George F. Kraus, *College of Southern Maryland*
Patrick M. Len, *Cuesta College*
John Patrick Lestrade, *Mississippi State University*
C. L. Littler, *University of North Texas*
M. A. K. Lohdi, *Texas Tech University*
Michael C. LoPresto, *Henry Ford Community College*
Phyllis Lugger, *Indiana University*
R. M. MacQueen, *Rhodes College*
Robert Manning, *Davidson College*
Paul Mason, *University of Texas–El Paso*
P. L. Matheson, *Salt Lake Community College*
Rahul Mehta, *University of Central Arkansas*
J. Scott Miller, *University of Louisville*
L. D. Mitchell, *Cambria County Area Community College*

J. Ward Moody, *Brigham Young University*
Siobahn M. Morgan, *University of Northern Iowa*
David Morris, *Eastern Arizona College*
Steven Mutz, *Scottsdale Community College*
Paul J. Neinaber, *Saint Mary's University of Minnesota*
Gerald H. Newson, *Ohio State University*
Bob O'Connell, *College of the Redwoods*
William C. Oelfke, *Valencia Community College*
Richard P. Olenick, *University of Dallas*
John P. Oliver, *University of Florida*
Melvyn Jay Oremland, *Pace University*
Jerome A. Orosz, *San Diego State University*
David Patton, *Trent University*
Jon Pedicino, *College of the Redwoods*
Sidney Perkowitz, *Emory University*
David D. Reid, *Wayne State University*
Adam W. Rengstorf, *Indiana University*
James A. Roberts, *University of North Texas*
Henry Robinson, *Montgomery College*
Dwight P. Russell, *University of Texas–El Paso*
Barbara Ryden, *Ohio State University*
Larry Sessions, *Metropolitan State College*
C. Ian Short, *Florida Atlantic University*
John D. Silva, *University of Massachusetts at Dartmouth*
Michael L. Sitko, *University of Cincinnati*
Earl F. Skelton, *George Washington University*
George F. Smoot, *University of California at Berkeley*
Alex G. Smith, *University of Florida*
Michael Sterner, *University of Montevallo*
Brent W. Studer, *Kirkwood Community College*
David Sturm, *University of Maine, Orono*
Paula Szkody, *University of Washington*
Michael T. Vaughan, *Northeastern University*
Robert Vaughn, *Graceland University*
Andreas Veh, *Kenai Peninsula College*
John Wallin, *George Mason University*
William F. Welsh, *San Diego State University*
R. M. Williamon, *Emory University*
Edward L. (Ned) Wright, *University of California at Los Angeles*
Jeff S. Wright, *Elon College*
Nicolle E. B. Zellner, *Rensselaer Polytechnic Institute*

I would like to add my special thanks to the wonderfully supportive staff at W. H. Freeman and Company who make the revision process so enjoyable. Among others, these people include Anthony Palmiotto, Kharissia Pettus, Kerry O'Shaughnessy, Blake Logan, Ted Szczepanski, Bill Page, Lawrence Guerra, and Kathryn Treadway. Thanks also go to Louise B. Ketz, Anna Paganelli, and Black Dot Group. Warm thanks also to David Sturm, University of Maine, Orono, for his help in collecting current data on objects in the solar system; to my UM astronomy colleague David Batuski; and to my wife, Sue, and my sons, James and Josh, for their patience and support while preparing this book.

Neil F. Comins
neil_comins@umit.maine.edu

Professor **Neil F. Comins** is on the faculty of the University of Maine. Born in 1951 in New York City, he grew up in New York and New England. He earned a bachelor's degree in engineering physics at Cornell University, a master's degree in physics at the University of Maryland, and a Ph.D. in astrophysics from University College, Cardiff, Wales, under the guidance of Bernard F. Schutz. Dr. Comins's work for his doctorate, on general relativity, was cited in Subramanyan Chandrasekhar's Nobel laureate speech. He has done theoretical and experimental research in general relativity, observational astronomy, computer simulations of galaxy evolution, and science education. The fourth edition of *Discovering the Universe* was the first textbook that Dr. Comins wrote for W.H. Freeman and Company, having taken over following the death of Bill Kaufmann in 1994. This has been followed by 9 others. He is also the author of three trade books, *What If the Moon Didn't Exist?*, *Heavenly Errors,* and *The Hazards of Space Travel: A Tourist's Guide*. *What If the Moon Didn't Exist?* has been made into planetarium shows, been excerpted for television and radio, been translated into several languages, and was the theme for the Mitsubishi Pavilion at the World Expo 2005 in Aichi, Japan. *Heavenly Errors* explores misconceptions people have about astronomy, why such misconceptions are so common, and how to correct them. Dr. Comins has appeared on numerous television and radio shows and gives many public talks. Although he has jumped out of airplanes while in the military, today his activities are a little more sedate: He is a licensed pilot and avid sailor, having once competed against Prince Philip, Duke of Edinburgh.

Discovering the Night Sky

R | I | V | U | X | G

 Circumpolar Star Trails behind the Anglo-Australian Telescope, Siding Springs Mountain, New South Wales, Australia *(Anglo-Australian Observatory/David Malin Images)*

WHAT DO YOU THINK?

1 Is the North Star—Polaris—the brightest star in the night sky?

2 What causes the seasons?

3 When is Earth closest to the Sun?

4 How many zodiac constellations are there?

5 Does the Moon have a dark side that we never see from Earth?

6 Is the Moon ever visible during the daytime?

Answers to these questions appear in the text beside the corresponding numbers in the margins and at the end of the chapter.

You have picked an exciting time to study astronomy. Our knowledge about the cosmos (or the *universe*) is growing as never before. Current telescope technology makes it possible for astronomers to observe objects that were invisible to them just a few years ago. These new observations have deepened our understanding of virtually every aspect of the universe. We can now watch it expand and see stars explode in distant galaxies; we have discovered planets orbiting nearby stars; we have seen newly born stars enshrouded in clouds of gas and dust; and we have identified black holes and other remnants of stellar evolution. Many of these objects are so far away that the light we see from them began its journey to Earth millions and even billions of years ago. Thus, as we look farther and farther out into the universe—defined as everything that we can see or that can be seen—we also see farther and farther back into time.

Telescopes are not the only means by which we are deepening our understanding of the skies. We have also begun the process of physically exploring our neighborhood in space. In just the past half century, humans have walked on the Moon, and space probes have rolled over and dug into Martian soil. Other space missions have landed on an asteroid, brought back debris from a comet, discovered active volcanoes and barren ice fields on the moons of Jupiter, visited Saturn's shimmering rings and its murky moon Titan, and traveled beyond the realm of the planets in our solar system, to mention a few accomplishments.

As you proceed through this book, we hope that you will gain a new appreciation of the awesome power of the human mind to reach out, to observe, to explore, and to comprehend. One of the great lessons of modern astronomy is that by gaining, sharing, and passing on knowledge, we transcend the limitations of our bodies and the brevity of human life.

In this chapter you will discover

• how astronomers organize the night sky to help them locate objects in it

• that Earth's spin on its axis causes day and night

• how the tilt of Earth's axis of rotation and Earth's motion around the Sun combine to create the seasons

• that the Moon's orbit around Earth creates the phases of the Moon and lunar and solar eclipses

• how the year is defined and how the calendar was developed

SCALES OF THE UNIVERSE

While learning about a new field, it is often useful to see the "big picture" before exploring the details. For this reason, we begin by surveying the major types of objects in the universe, along with their ranges of size and the scale of distances between them.

1-1   Astronomical distances are, well, astronomical

One of the thrills and challenges of studying astronomy is becoming familiar and comfortable with the range of sizes that are used. In our everyday lives, we typically deal with distances ranging from millimeters to thousands of kilometers. (The metric system of units is standard in science and will be used throughout this book; however, we will often provide the equivalent in British units. Appendix C-9 lists conversion factors between the two sets of units.) A hundredth of a meter or a thousand kilometers are numbers that are easy to visualize and write. In astronomy, we deal

with particles as small as a millionth of a billionth of a meter and systems of stars as large as a thousand billion billion kilometers across. Similarly, the speeds of some things, such as light, are so large as to be cumbersome if you have to write them out in words each time. To deal with numbers much larger and much smaller than 1, we use a shorthand throughout this book called **scientific notation,** or "powers of ten." (Read Appendix A if you are unfamiliar with scientific notation.)

The size of the universe that we can observe and the range of sizes of the objects in it are truly staggering. Figure 1-1 summarizes the range of sizes from atomic particles up to the diameter of the entire universe visible to

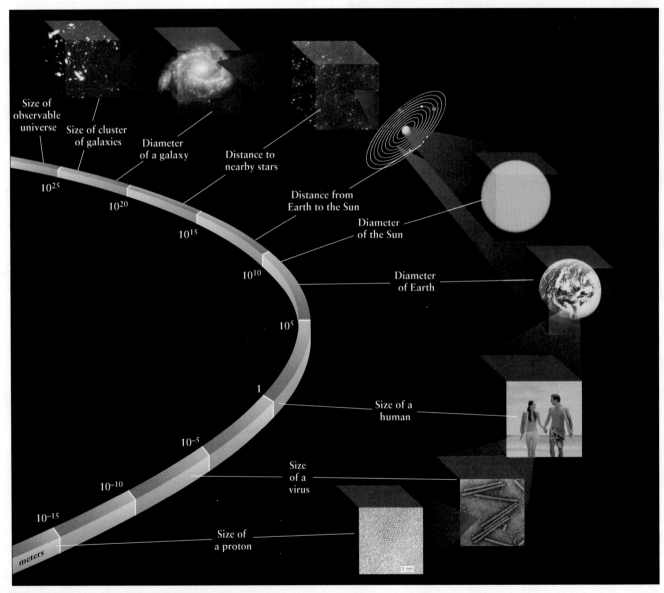

FIGURE 1-1 **The Scales of the Universe** This curve gives the sizes of objects in meters, ranging from subatomic particles at the bottom to the entire observable universe at the top. Every 0.5 cm up along the arc represents a factor of 10 larger. *(Top to bottom: R. Williams and the Hubble Deep Field Team [STScI] and NASA; AAT; L. Golub, Naval Observatory, IBM Research, NASA; Richard Bickel/Corbis; Scientific American Books; Jose Luis Pelaez/Getty Images; Rothamsted Research Centre for Bioimaging; Courtesy of Florian Banhart/University of Mainz)*

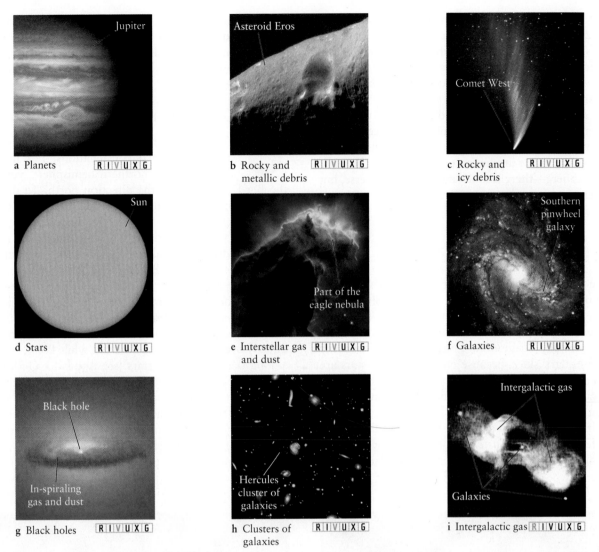

a Planters R I V U X G

b Rocky and   R I V U X G
metallic debris

c Rocky and   R I V U X G
icy debris

d Stars R I V U X G

e Interstellar gas R I V U X G
and dust

f Galaxies R I V U X G

g Black holes R I V U X G

h Clusters of   R I V U X G
galaxies

i Intergalactic gas R I V U X G

FIGURE 1-2 Inventory of the Universe Pictured here are examples of the major categories of objects that have been found throughout the universe. You will discover more about each type in the chapters that follow. *(a: NASA/Hubblesite; b: NASA; c: Peter Stättmayer/European Southern Observatory; d: Big Bear Observatory; e: NASA/Jeff Hester & Paul Scowen; f: Anglo-Australian Observatory; g: NOAO; h: NASA; i: N. F. Comins & F. N. Owen/NRAO)*

us. Unlike linear intervals measured on a ruler, moving up 0.5×10^{-2} m (0.5 cm) along the arc of this figure brings you to objects 10 times larger. Because of this, going from the size of a proton (roughly 10^{-15} m) up to the size of an atom (roughly 10^{-10} m) takes about the same space along the arc as going from the distance between Earth and the Sun to the distance between Earth and the nearby stars.

This wide range of sizes underscores the fact that astronomy *synthesizes* or brings together information from many other fields of science. We will need to understand what atoms are composed of and how they behave; the nature and properties of light; the response of matter and energy to the force of gravity; the generation of energy by fusing particles together in stars; the ability of carbon—and carbon alone—to serve as the foundation of life; as well as other information. These concepts will all be introduced as they are needed.

What, then, have astronomers seen of the universe? Figure 1-2 presents examples of the types of objects we will explore in this text. An increasing number of planets like Jupiter, rich in hydrogen and helium (Figure 1-2a), as well as rocky planets not much larger than Earth, have been discovered orbiting other stars. Much smaller pieces of space debris—some of rock and metal called **asteroids** or **meteoroids** (Figure 1-2b), and others of rock and ice called **comets** (Figure 1-2c)—orbit the Sun (Figure 1-2d) and other stars. Vast stores of interstellar gas and dust are found in many galaxies; these are often the incubators of new generations of stars (Figure 1-2e). Stars by the millions, billions, or even trillions are held together in galaxies by the force of gravity (Figure 1-2f). Galaxies like our own Milky Way often contain large amounts of that interstellar gas and dust, as well as regions of space where matter is so dense that it cannot radiate light; these regions are called **black holes** (Figure 1-2g).

Groups of galaxies are held together by gravity in clusters (Figure 1-2h), and clusters of galaxies are held together by gravity in superclusters. Huge quantities of intergalactic gas are often found between galaxies (Figure 1-2i).

Every object in astronomy is constantly changing—each has an origin, an active period you might consider as its "life," and each will have an end. We will study these processes along with the important physical concepts upon which they are based. You will also discover that all the matter astronomers see in stars and galaxies is but the tip of the cosmic iceberg—there is much more in the universe, but astronomers do not yet know its nature.

PATTERNS OF STARS

When you gaze at the sky on a clear, dark night where the air is free of pollution and there is not too much light, there seem to be millions of stars twinkling overhead. In reality, the unaided human eye can detect only about 6000 stars over the entire sky. At any one time, you can see roughly 3000 stars in dark skies, because only half of the stars are above the *horizon*—the boundary between Earth and the sky. In very smoggy or light-polluted cities, you may see only a tenth of that number, or less (Figure 1-3).

 In any event, you probably have noticed patterns, technically called *asterisms,* formed by bright stars, and you are familiar with some common names for these patterns, such as the ladle-shaped Big Dipper and broad-shouldered Orion. These recognizable patterns of stars are informally called *constellations* in everyday conversation and they have names derived from ancient legends (Figure 1-4a).

1-2   Constellations make locating stars easy

 You can orient yourself on Earth with the help of easily recognized constellations. For instance, if you live in the northern hemisphere, you can use the Big Dipper to find the direction north. To do this, locate the Big Dipper and imagine that its bowl is resting on a table (Figure 1-5). If you see the dipper upside down in the sky, as you frequently will, imagine the dipper resting on an upside-down table above it. Locate the two stars of the bowl farthest from the Big Dipper's handle. These are called the *pointer stars*. Draw a mental line through these stars in the direction away from the table, as shown in Figure 1-5. The first *moderately bright* star you then encounter is Polaris, also called the North Star because it is located almost directly over Earth's North Pole. So, while Polaris is not even among the 20 brightest stars (see Appendix C-5), it is easy to locate. Whenever you face Polaris, you are facing north. East is then on your right, south is behind you, and west is on your left. There is no equivalent star over the South Pole.

The Big Dipper example also illustrates the fact that being familiar with constellations makes it easy to locate other stars. The most effective way to do this is to use vivid visual

a

FIGURE 1-3 The Night Sky Without and With Light Pollution
(a) Sunlight is a curtain that hides virtually everything behind it. As the Sun sets, places with little smog or light pollution treat viewers to beautiful panoramas of stars that can inspire the artist or scientist

b   R I V U X G

in many of us. This photograph shows the night sky during a power outage seen from Goodwood, Ontario, Canada. (b) This photograph shows the same sky with normal city lighting. (© *Todd Carlson/ SkyNews Magazine*)

a

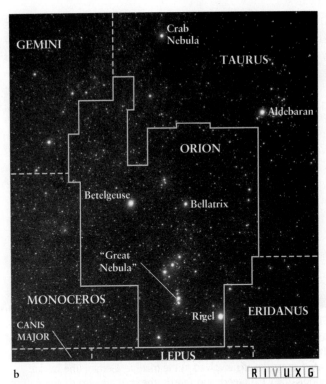

b   R I V U X G

FIGURE 1-4   The Constellation Orion (a) The pattern of stars (asterism) called Orion is prominent in the winter sky. From the northern hemisphere, it is easily seen high above the southern horizon from December through March. You can see in this photograph that the various stars have different colors, something to watch for when you observe the night sky.

(b) Technically, constellations are entire regions of the sky. The constellation called Orion and parts of other nearby constellations are depicted in this photograph. All the stars inside the boundary of Orion are members of that constellation. The celestial sphere is covered by 88 constellations of differing sizes and shapes. *(© 2004 Jerry Lodriguss/www.astropix.com)*

connections, especially those of your own devising. For example, imagine gripping the handle of the Big Dipper and slamming its bowl straight down onto the head of Leo (the Lion). Leo comprises the first group of bright stars your dipper encounters. As shown in Figure 1-5, the brightest star in this group is Regulus, the dot of the backward question mark that traces the lion's mane. As another example, follow the arc of the Big Dipper's handle away from its bowl. The first bright star you encounter along that arc beyond the handle is Arcturus in Boötes (the Herdsman). Follow the same arc farther to the prominent bluish star Spica in Virgo (the Virgin). Spotting these stars and remembering their names is easy if you remember the saying "Arc to Arcturus and speed on to Spica."

During the winter months in the northern hemisphere, you can see some of the brightest stars in the sky. Many of them are in the vicinity of the "winter triangle," which connects bright stars in the constellations of Orion (the Hunter), Canis Major (the Larger Dog), and Canis Minor (the Smaller Dog), as shown in Figure 1-6. The winter triangle passes high in the sky at night during the middle of winter. It is easy to find Sirius, the brightest star in the night sky, by locating the belt of Orion and following a straight mental line from it to the left (as you face Orion). The first bright star that you encounter is Sirius.

FIGURE 1-5   The Big Dipper As a Guide In the northern hemisphere, the Big Dipper is an easily recognized pattern of seven bright stars. This star chart shows how the Big Dipper can be used to locate the North Star as well as the brightest stars in three other constellations. While the Big Dipper appears right side up in this drawing of the sky shortly before sunrise, at other times of the night it appears upside down.

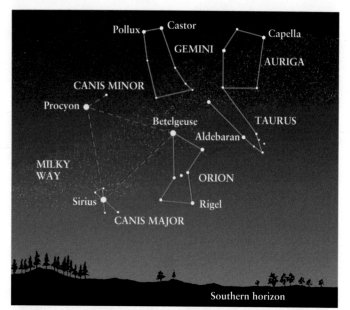

FIGURE 1-6 **The Winter Triangle** This star chart shows the southern sky as it appears during the evening in December. Three of the brightest stars in the sky make up the winter triangle. In addition to the constellations involved in the triangle, Gemini (the Twins), Auriga (the Charioteer), and Taurus (the Bull) are also shown.

FIGURE 1-7 **The Summer Triangle** This star chart shows the northeastern sky as it appears in the evening in June. In addition to the three constellations involved in the summer triangle, the faint stars in the constellations Sagitta (the Arrow) and Delphinus (the Dolphin) are also shown.

INSIGHT INTO SCIENCE

Flexible Thinking Part of learning science is learning to look at things from different perspectives. For example, when learning to identify the prominent constellations, be sure to view them from different orientations (that is, with the star chart rotated at different angles), so that you can find them at different times of the night and the year.

Create a story with which you can remember the connection between the constellations Orion and Taurus, which is up and to the right of Orion.

The "summer triangle," which graces the summer sky as shown in Figure 1-7, connects the bright stars Vega in Lyra (the Lyre), Deneb in Cygnus (the Swan), and Altair in Aquila (the Eagle). A conspicuous portion of the Milky Way forms a beautiful background for these constellations, which are nearly overhead during the middle of summer at midnight.

Astronomers require more accuracy in locating dim objects than is possible simply by moving from constellation to constellation. They have therefore created a sky map, called the **celestial sphere,** and applied a coordinate system to it, analogous to the coordinate system of north-south latitude and east-west longitude used to navigate on Earth. If you know a star's celestial coordinates, you can locate it quickly. For such a sky map to be useful in finding stars, the stars must be fixed on it, just as cities are fixed on maps of Earth.

1-3 The celestial sphere aids in navigating the sky

If you look at the night sky year after year, you will see that the stars do indeed appear fixed relative to one another. Furthermore, throughout each night the entire pattern of stars appears to rigidly orbit Earth. We employ this artificial, Earth-based view of the heavens to make celestial maps by pretending that the stars are attached to the inside of an enormous hollow shell, the celestial sphere, with Earth at its center (Figure 1-8). Visualized another way, you can imagine the half of the celestial sphere that is visible at night as a giant bowl covering Earth.

Whereas asterisms such as the Big Dipper are often called "constellations" in normal conversation, astronomers use the word **constellation** to describe an entire area of the sky and all the objects in it (see Figure 1-4b). The celestial sphere is divided into 88 constellations of differing sizes and shapes. (Keep in mind that most constellations and their asterisms—the recognizable star patterns—have the same name; for example, Orion, the asterism, and Orion, the constellation, which sometimes complicates conversation.) The boundaries of the constellations are straight lines that meet at right angles (see Figure 1-4b). Some constellations, like Ursa Major (the Large Bear), are very large, while others, like Sagitta (the Arrow), are relatively small. To describe a star's location, we might say "Albireo in the constellation Cygnus (the Swan)," much as we would refer to "Chicago in the state of Illinois," "Melbourne in the state of New South Wales," or "Ottawa in the province of Ontario."

The stars seem fixed on the celestial sphere only because of their remoteness. In reality, they are at widely varying distances from Earth, and they do move relative to one another. But we neither see their motion nor perceive their relative distances because the stars are so far from here. You can understand this by imagining a jet plane just 1 km overhead traveling at 1000 km (620 mi) per hour across the sky. Its motion is unmistakable. However, a plane moving at the same speed and altitude, but appearing along the horizon, seems to be moving about 100 times more slowly. And an object that is at the distance of the Sun, traveling at the same speed, would appear to be moving across the sky nearly 100 million times more slowly than the plane overhead.

The stars (other than the Sun) are all more than 40 trillion km (25 trillion mi) from us. Therefore, although the patterns of stars in the sky do change, their great distances prevent us from seeing those changes over the course of a human lifetime. Thus, as unrealistic as it is, the celestial sphere is so useful for navigating the heavens that it is used by astronomers even at the most sophisticated observatories around the world.

As shown in Figure 1-8, we can project key geographic features from Earth out into space to establish directions and bearings. If we expand Earth's equator onto the celestial sphere, we obtain the **celestial equator,** which divides the sky into northern and southern hemispheres, just as Earth's equator divides Earth into two hemispheres. We can also imagine projecting Earth's North Pole and South Pole out into space along Earth's axis of rotation. Doing so gives us the **north celestial pole** and the **south celestial pole,** also shown in Figure 1-8.

Using the celestial equator and poles as reference features, astronomers divide up the surface of the celestial sphere in precisely the same way that the latitude and longitude grid divides Earth. The equivalent to latitude on Earth is called **declination (dec)** on the celestial sphere. It is measured from 0° to 90° north or south of the celestial equator. The equivalent of longitude on Earth is called **right ascension (r.a.)** on the celestial sphere, measured from 0 h to 24 h around the celestial equator (see Figure 1-8). The boundaries of the constellations, introduced above, run along lines of constant right ascension and constant declination.

The planets move through constellations, while the stars do not. What two possible properties of the planets could allow them to do that?

Just as the location of Greenwich, England, defines the *prime meridian,* or zero of longitude on Earth, we need to establish a zero of right ascension. It is defined as one of the places where the Sun's annual path across the celestial sphere intersects the celestial equator. (We will explore later in this chapter why the Sun appears to move in a circle around the celestial sphere during the course of a year.) The celestial equator and the Sun's path intersect at two points. The equivalent on the celestial sphere of Earth's prime meridian is where the Sun crosses the celestial equator moving northward. Angles of right ascension are measured from this point, called the **vernal equinox** (see Figure 1-8).

In navigating on the celestial sphere, astronomers measure the distance between objects in terms of angles. Ancient mathematicians invented a system of angles and angular measure that is still used today to denote the relative positions and apparent sizes of objects in the sky. To locate stars on the celestial sphere, for example, we do not need to know their distances from Earth. All we need to know is the angle from one star to another in the sky, a property that remains fixed over our lifetimes because the stars are all so far away.

An **arc angle,** often just called an **angle,** is the opening between two lines that meet at a point. Angular measure is a method of describing the size of an angle. The basic unit of angular measure is the **degree,** designated by the symbol °. A full circle is divided into 360°. A right angle measures 90°. As shown in Figure 1-9, the angle between the two "pointer stars" in the Big Dipper is about 5°.

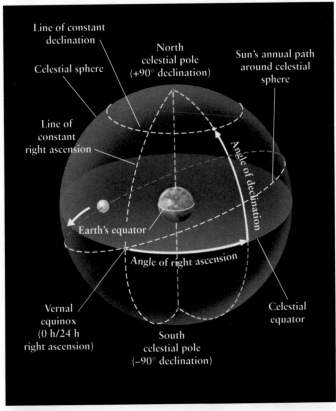

FIGURE 1-8 The Celestial Sphere The celestial sphere is the apparent "bowl" or hollow sphere of the sky. The celestial equator and poles are projections of Earth's equator and axis of rotation onto the celestial sphere. The north celestial pole is therefore located directly over Earth's North Pole, while the south celestial pole is directly above Earth's South Pole. Analogous to longitude and latitude, the coordinates in space are right ascension (r.a.) and declination (dec), respectively. The star in this figure has the indicated r.a. and dec.

FIGURE 1-9 **The Big Dipper** The angular distance between the two "pointer stars" at the front of the Big Dipper is about 5°. For comparison, the angular diameter of the Moon is about ½°.

Astronomers also use angular measure to describe the apparent sizes of celestial objects. For example, imagine the full Moon. As seen from Earth, the angle across the Moon's diameter is nearly ½°. We therefore say that the **angular diameter,** or **angular size,** of the Moon is ½°. Alternatively, astronomers say that the Moon "subtends" an angle of ½°. In this context, subtend means "to extend across."

To talk about smaller angles, we subdivide the degree into 60 arcminutes (abbreviated 60 arcmin or 60'). An arcminute is further subdivided into 60 arcseconds (abbreviated 60 arcsec or 60"). A dime viewed face-on from a distance of 1.6 km (1 mi) has an angular diameter of about 2 arcsec.

From everyday experience, we know that an object looks bigger when it is nearby than when it is far away. The angular size of an object therefore does not necessarily tell you anything about its actual physical size. For example, the fact that the Moon's angular diameter is ½° does not tell you how big the Moon really is.

EARTHLY CYCLES

The Sun systematically rises and sets at different times and in different places throughout the year. Similarly, the Moon rises and sets at different times each day and repeats its cycle roughly once every 29½ days. Furthermore, as noted in Section 1-2, we do not see the same constellations up in the sky every night of the year, but the cycle of constellations we can see at night repeats each year. The daily and annual rhythms of the sky, Earth, and all life on it arise from three celestial motions: Earth's spinning, which causes day and night, and also causes the apparent daily motion of the celestial sphere and all the objects on it; Earth's orbit around the Sun, which creates the seasons, the year, and the change

in times at which constellations are up at night; and the Moon's orbit around Earth, which creates the lunar phases, the cycle of tides, and the spectacular phenomena we call eclipses.

1-4 Earth's rotation creates the day-night cycle and its revolution defines a year

Earth spins on its axis. Such motion is called **rotation.** We do not feel Earth's rotation because our planet is so physically large compared to us that its gravitational attraction holds us firmly on its surface. Earth's rotation causes the stars—as well as the Sun, Moon, and planets—to appear to rise on the eastern horizon, move across the sky, and set on the western horizon. Earth's daily rotation, causing the Sun to rise and set, thereby creates day and night. The **diurnal motion,** or daily motion, of the celestial bodies is apparent in time-exposure photographs, such as that shown in Figure 1-10.

Take a friend outside on a clear, warm night to observe the diurnal motion of the stars for yourselves. Soon after dark, find a spot away from bright lights, and note the constellations in the sky relative to some prominent landmarks near you on Earth. A few hours later, check again from the same place. You will find that the entire pattern of stars (as well as the Moon, if it is visible) has shifted. New constellations will have risen above the eastern horizon, while other

R I V U X G

FIGURE 1-10 **Circumpolar Star Trails** This long exposure, taken from Australia's Siding Spring Mountain and aimed at the south celestial pole, shows the rotation of the sky. The stars that pass between the pole and the ground are all circumpolar stars. *(Anglo-Australian Observatory/David Malin Images)*

constellations will have disappeared below the western horizon. If you check again just before dawn, you will find the stars that were just rising in the east when the night began are now low in the western sky.

Different constellations are visible at night during different times of the year. This occurs because Earth orbits, or *revolves*, around the Sun. **Revolution** is the motion of any astronomical object around another astronomical object. Earth takes one year, or about 365¼ days, to go once around the Sun. A year on Earth is measured by the motion of our planet relative to the stars. For example, draw a straight line from the Sun through Earth to some star on the opposite side of Earth from the Sun. As Earth revolves, that line inscribes a straight path on the celestial sphere and returns to the original star 365¼ days later. The length of any cycle of motion, such as Earth's orbit around the Sun, that is measured with respect to the stars is called a **sidereal period.**

If Earth were spinning over a fixed place on the Sun, rather than revolving around it, then every star would rise and set at the same times throughout the year. As a result of Earth's motion around the Sun, however, the stars rise approximately 4 min earlier each day than they did the day (or night) before. This effect accumulates, bringing different constellations up at night throughout the year. Figure 1-11 summarizes this motion. When the Sun is within the boundaries of (colloquially, "in") Virgo (September 18–November 1), for example, the hemisphere containing

the Sun and the constellations around Virgo are in daylight (see Figure 1-11a). When the Sun is up, so are Virgo and the surrounding constellations, so we cannot see them. During that time of year, the constellations on the other side of the celestial sphere, centered on the constellation Pisces, are in darkness. Thus, when the Sun is "in" Virgo, Pisces and the constellations around it are high in our sky at night.

Six months later, when the Sun is "in" Pisces, that half of the sky is filled with daylight, while Virgo and the constellations around it are high in the night sky (see Figure 1-11b). These arguments apply everywhere on Earth at the same time because the Sun moves along its path very slowly as seen from Earth, taking a year to make one complete circuit.

We spoke earlier of stars rising on the eastern horizon and setting on the western horizon. Depending on your latitude, some of the stars and constellations never disappear below the horizon. Instead, they trace complete circles in the sky over the course of each night (see Figure 1-10). To understand why this happens, imagine that you are standing on Earth's North Pole at night. Earth is spinning around its axis directly under your feet. When you look forward, you see the stars sweeping from your left to your right in horizontal rings around you. Equivalently, when you look up, you see Polaris fixed directly overhead, and all the stars are moving in counterclockwise circles around it. Polaris is always directly overhead at the North Pole, that is, at the pole's **zenith.** (Every place has a different zenith, which is the point directly overhead anywhere on Earth.) As seen from the North Pole,

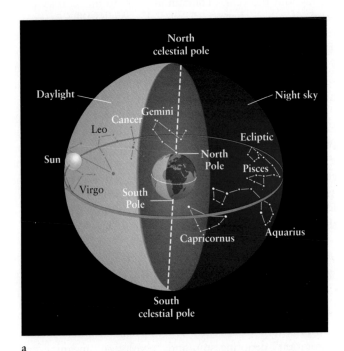

a

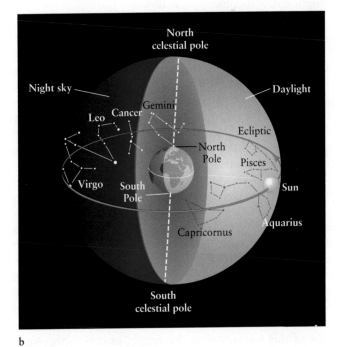

b

FIGURE 1-11 **Why Different Constellations Are Visible at Different Times of the Year** (a) On the autumnal equinox each year, the Sun is in the constellation Virgo. As seen from Earth, that part of the sky is then daylight and we see stars only on the other half of the sky, centered around the constellation Pisces. (b) Six months later, the Sun is in Pisces. This side of the sky is then bright, while the side centered on Virgo is in darkness.

FIGURE 1-12 **Motion of Stars at the Poles** Because Earth rotates around its poles, stars seen from these locations appear to move in huge, horizontal circles. This is the same effect you would get by standing up in a room and spinning around; everything would appear to move in circles around you. At the North Pole stars move left to right, while at the South Pole they move right to left.

FIGURE 1-13 **Rising and Setting of Stars at the Equator** Standing on the equator, you are perpendicular to the axis around which Earth rotates. As seen from there, the stars rise straight up on the eastern horizon and set straight down on the western horizon. This is the same effect you get when driving straight over the crest of a hill; the objects on the other side of the hill appear to move straight upward as you descend.

FIGURE 1-14

Rising and Setting of Stars at Middle North Latitudes Unlike the motion of the stars at the poles (see Figure 1-12), the stars at all other latitudes do change angle above the ground throughout the night. This time-lapse photograph shows stars setting. The latitude determines the angle at which the stars rise and set. *(David Miller/DMI)*

no stars rise or set (Figure 1-12). They just seem to revolve around Polaris in horizontal circles. Stars and constellations that never go below the horizon are called **circumpolar.** Although there is no bright South Pole star equivalent to Polaris, all stars seen from the South Pole are also circumpolar and move from right to left (clockwise).

Why are asterisms like the Big Dipper sometimes seen upright and sometimes upside down?

If you live in the northern hemisphere, Polaris is always located above your northern horizon at an angle equal to your latitude. Only the stars and constellations that pass between Polaris and the land directly below it are circumpolar. As you go farther south in the northern hemisphere, the number of stars and constellations that are circumpolar decreases. Likewise, as you go farther north in the southern hemisphere, the number of stars and constellations that are circumpolar decreases.

Now visualize yourself at the equator. All the stars appear to rise straight up in the eastern sky and set straight down in the western sky (Figure 1-13). Polaris is barely visible on the northern horizon. Although Polaris never sets, all the other stars do, and therefore none of the stars is circumpolar as seen from the equator.

As you can see from these last two mental exercises, the angle at which the stars rise and set depends on your viewing latitude. Figure 1-14 shows stars setting at 35° north latitude. Polaris is fixed at 35° above the horizon to the right of this figure, not at the zenith, as it is at the North Pole, nor on the horizon, as seen from the equator. As another example, except for those stars in the corners, all the stars whose paths are shown in Figure 1-10 are circumpolar because they are visible all night, every night.

1-5   The seasons result from the tilt of Earth's rotation axis combined with Earth's revolution around the Sun

Imagine that you could see the stars even during the day, so that you could follow the Sun's apparent motion against the background constellations throughout the year. (The Sun appears to move among the stars, of course, because Earth orbits around it.) From day to day, the Sun traces a straight path on the celestial sphere. This path is called the **ecliptic.** As you can see in Figure 1-15a, the ecliptic makes a closed circle bisecting the celestial sphere. The ecliptic is precisely the loop labeled "Sun's annual path around celestial sphere" in Figure 1-8.

The term "ecliptic" has a second use in astronomy. Earth orbits the Sun in a plane also called the ecliptic. The two ecliptics exactly coincide: Imagine yourself on the Sun watching Earth move day by day. The path of Earth on the celestial sphere as seen from the Sun is precisely the same as the path of the Sun as seen from Earth (Figure 1-15b). Recall also the discussion of the line running from the Sun through Earth to the celestial sphere in Section 1-4.

INSIGHT INTO SCIENCE

Define Your Terms As with interpersonal communication, using the correct words is very important in science. Usually, scientific words each have a specific meaning, such as "rotation" denoting spin and "revolution" meaning one object orbiting another. Be especially careful to understand the context in which words with more than one meaning, such as "ecliptic" and "constellation," are used.

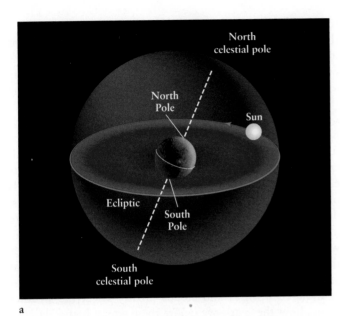

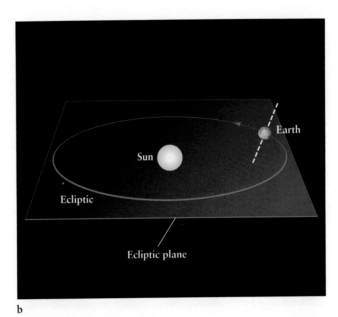

a

b

FIGURE 1-15 **The Ecliptic** (a) The ecliptic is the apparent annual path of the Sun on the celestial sphere. (b) The ecliptic is also the plane described by Earth's path around the Sun. The planes created by the two ecliptics exactly coincide. As in (a), the rotation axis of Earth is shown here tilted 23½° from being perpendicular to the ecliptic.

Equinoxes and Solstices The ecliptic and the celestial equator are different circles tilted 23½° with respect to each other on the celestial sphere. This occurs because Earth's rotation axis is tilted 23½° away from a line perpendicular to the ecliptic (Figure 1-15 and Figure 1-16). These two circles intersect at only two points, which are exactly opposite each other on the celestial sphere (Figure 1-17). Each of these two points is called an **equinox** (from the Latin words meaning "equal night"), because when the Sun appears at either point, it is directly over Earth's equator, resulting in 12 h of daytime and 12 h of nighttime everywhere on Earth on that day.

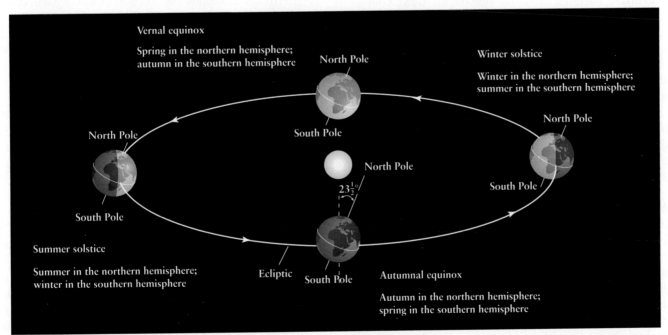

FIGURE 1-16 **The Tilt of Earth's Axis** Earth's axis of rotation is tilted 23½° from being perpendicular to the plane of Earth's orbit. Earth maintains this orientation (with its North Pole aimed at the north celestial pole near the star Polaris) throughout the year as it orbits the Sun. Consequently, the amount of solar illumination and the number of daylight hours at any location on Earth vary in a regular fashion with the seasons.

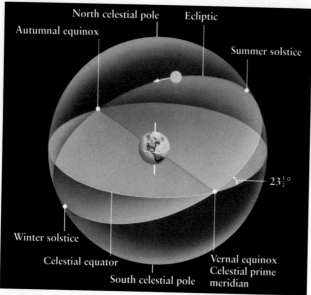

FIGURE 1-17 The Seasons Are Linked to Equinoxes and Solstices The ecliptic is inclined to the celestial equator by 23½° because of the tilt of Earth's axis of rotation. The ecliptic and the celestial equator intersect at two points called the equinoxes. The northernmost point on the ecliptic is called the summer solstice; the southernmost point is called the winter solstice.

Except for tiny changes each year, Earth maintains this tilted orientation as it orbits the Sun. Therefore, Polaris is above the North Pole throughout the year. For half the year, the northern hemisphere is tilted toward the Sun, and, as a result, the Sun rises higher in the northern hemisphere's sky than it does during the other half of the year (Figure 1-18). Equivalently, when the southern hemisphere is tilted toward the Sun, the Sun rises higher in the southern hemisphere's sky.

Consider the location of the Sun as seen from the northern hemisphere throughout the year. The day that the Sun rises farthest south of east is around December 22 each year (see Figure 1-18a) and is called the **winter solstice.** The winter solstice is the point on the ecliptic farthest south of the celestial equator (see Figure 1-17). It is also the day when the Sun rises to the lowest height at noon (Figure 1-18a) and it signals the day of the year in the northern hemisphere that has the fewest number of daylight hours.

As the Sun moves along the ecliptic after the winter solstice, it rises earlier, is more northerly on the eastern horizon, and it passes higher in the sky at midday than it did on preceding days. Three months later, around March 21, the Sun crosses the celestial equator heading northward. This is called the vernal equinox and is one of the two days on which the Sun rises due east and sets due west (Figure 1-18b). The vernal equinox is the "prime meridian" of the celestial sphere, as discussed earlier. Three months after the vernal equinox, around June 21, the Sun rises farthest north of east and passes highest in the sky (Figure 1-18c). This is the **summer solstice** (see Figure 1-17), the day of the year in the northern hemisphere with the most daylight.

From June 21 through December 21, the Sun rises farther south than it did the preceding day. Its highest point in the sky is lower each succeeding day—the cycle of the previous 6 months reverses. The **autumnal equinox** occurs around September 22 (Figure 1-18d), with the Sun heading southward across the celestial equator, as seen from Earth.

The higher the Sun rises during the day, the more daylight hours there are. During the days with longer periods of daylight, more light and heat from the Sun strike that hemisphere. Furthermore, when the Sun is higher in the sky, its energy is more concentrated on Earth's surface (see the ovals of sunlight in each of the illustrations in Figure 1-18). Thus, during these days more energy is deposited on each square meter of the surface—thereby warming them more—than when the Sun is lower in the sky. The temperature and hence the seasons are determined by the duration of daylight at any place and the height of the Sun in the sky there. (Bear in mind that winds and clouds greatly affect the weather throughout the year—we ignore these effects here.)

To summarize, the Sun is lowest in the northern sky on the winter solstice. This marks the beginning of winter in the northern hemisphere. As the Sun moves northward, the amount of daylight increases daily. The vernal equinox marks a midpoint in the amount of light and heat from the Sun onto the northern hemisphere and is the beginning of spring. When the Sun reaches the summer solstice, it is highest in the northern sky and is above the horizon for the most hours of any day of the year. This is the beginning of summer. Returning southward, the Sun crosses the celestial equator once again on the autumnal equinox, the beginning of fall.

Earth's orbit around the Sun is elliptical (we will discuss this oval shape in detail in Chapter 2). Although the distance between Earth and the Sun changes by 5 million km (3 million mi) throughout the year, this variation has only a minor effect on the seasons. If the seasons *were* caused by the changing distance from Earth to the Sun, all parts of Earth should have the same seasons at the same time. In fact, the northern and southern hemispheres have exactly opposite seasons. Furthermore, Earth is closest to the Sun on or around January 3 of each year—the dead of winter in the northern hemisphere!

Explain why Figure 1-14 must have been taken facing west. Hint: Examine Figure 1-18.

The variation in Earth's distance from the Sun over the year would have a greater effect if it were not for the fact that the southern hemisphere has more area covered by oceans than does the northern hemisphere. As a result, when Earth is closer to the Sun (and the Sun is high in the southern hemisphere's sky), the southern oceans scatter more light and heat directly back into space than occurs when the Sun is higher over the northern hemisphere during the other half of the year. Had the extra energy sent back into space when we are closer to the Sun been absorbed by

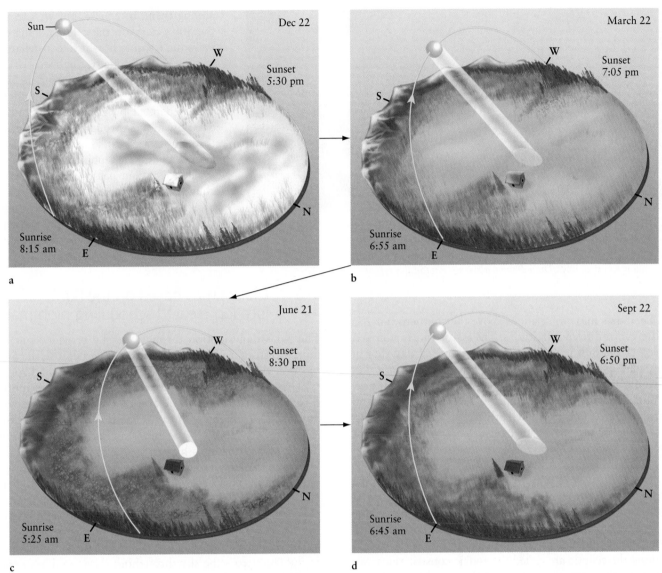

FIGURE 1-18 The Sun's Daily Path and the Energy It Deposits Here (a) On the winter solstice, the first day of winter, the Sun rises farthest south of east, is lowest in the noontime sky, stays up the shortest time, and its light and heat are least intense (most spread out) of any day of the year in the northern hemisphere. (b) On the vernal equinox, the first day of spring, the Sun rises precisely in the east and sets precisely in the west. Its light and heat have been growing more intense, as shown by the brighter oval of light than in (a). (c) On the summer solstice, first day of summer, the Sun rises farthest north of east of any day in the year, is highest in the noontime sky, stays up the longest time, and its light and heat are most intense of any day in the northern hemisphere. (d) On the autumnal equinox, the same astronomical conditions exist as on the vernal equinox.

our planet, Earth would, indeed, heat more during this time than when the Sun is over the northern hemisphere.

INSIGHT INTO SCIENCE

Expect the Unexpected Many phenomena in the universe defy commonsense explanations. The process of science requires that we question the obvious, that is, what we think we know. The fact that the changing distance from Earth to the Sun has a minimal effect on the seasons is an excellent example.

The Sun's Path across the Sky During the northern hemisphere's summer months, when the northern hemisphere is tilted toward the Sun, the Sun rises in the northeast and sets in the northwest. The Sun provides more than 12 h of daylight in the northern hemisphere and passes high in the sky. At the summer solstice, the Sun is as far north as it gets, giving the greatest number of daylight hours to the northern hemisphere.

During the northern hemisphere's winter months, when the northern hemisphere is tilted away from the Sun, the Sun rises in the southeast. Daylight lasts for fewer than 12 h, as the Sun skims low over the southern horizon and sets in the southwest. Night is longest in the northern hemisphere when the Sun is at the winter solstice.

11:40 P.M.   12:40 A.M.   1:40 A.M.   2:40 A.M.   3:40 A.M.

FIGURE 1-19 The Midnight Sun This time-lapse photograph was taken on July 19, 1985, at 69° north latitude in northeastern Alaska. At that latitude, the Sun is above the horizon continuously from mid-May until the end of July. *(Doug Plummer/Science Photo Library)*

R I V U X G

among them, of course, but we can plot the Sun's path on the celestial sphere to determine through which constellations it moves. Traditionally, there were 12 zodiac constellations whose borders were set in antiquity. In 1930, the boundaries were redefined by astronomers, and the Sun now moves through 13 constellations throughout the year. (The thirteenth zodiac constellation is Ophiuchus, the Serpent Holder. The Sun passes through Ophiuchus from December 1 to December 19 each year.) Table 1-1 lists all the zodiac constellations and the dates the Sun passes through them. You may not have the "sign" that you think you have.

The Sun's maximum angle above the southern horizon is different at different latitudes. The farther north you are, the lower the Sun is in the sky at any time of day than it is on that day at more equatorial locations. At latitudes above 66½° north latitude or below 66½° south latitude, the Sun does not rise at all during parts of their fall and winter months. During their spring and summer months, those same regions of Earth have continuous sunlight for weeks or months (Figure 1-19), hence the name "Land of the Midnight Sun."

4 The Sun takes 1 year to complete a trip around the ecliptic. Because there are about 365¼ days in a year and 360° in a circle, the Sun appears to move along the ecliptic at a rate of slightly less than 1° per day. The constellations through which the Sun moves throughout the year as it travels along the ecliptic are called the **zodiac** constellations. We cannot see the stars of these constellations when the Sun is

1-6 Clock times based on the Sun's location created scheduling nightmares

The Sun's daily motion through the sky provided our distant ancestors' earliest reference for time because the Sun's location determines whether it is day or night and roughly whether it is before or after midday. The Sun's motion through the sky came to determine the length of the **solar day,** upon which our 24-h day is based. This interval of time is ideally between when the Sun is highest in the sky on one day until the time it is highest in the sky on the next day. However, the length of the solar day actually varies throughout the year. This occurs because Earth's orbit around the Sun is not perfectly circular—our planet speeds up as it approaches the Sun and slows as it moves away—and because Earth's rotation axis is tilted 23½° from being perpendicular to the ecliptic. These two effects change the apparent speed of the Sun across the sky throughout the year. The *average* time interval between consecutive noontimes throughout the year is 24 h, which determines the time we use on our clocks. This is called the *mean* (or *average*) solar day.

What region of Earth has the smallest range of seasonal temperature changes and why?

Traditionally, noontime was taken to be the instant when the Sun is highest in the sky. But as we have just seen, the interval from one noontime to the next is not exactly 24 h, so the Sun is not always highest at noon. The difference between clock noontime and astronomical noontime (when the Sun is highest) is as much as 16 min. Because Earth is rotating eastward, the Sun is highest at different longitudes (measured east and west along lines running between Earth's poles) at different times. For example, astronomical noon in New York City occurs earlier than it does a little farther west, in Philadelphia. Before the advent of time zones, local time was based on astronomical noon. To travel west from New York to Philadelphia by train, for example, you had to know the departure time at New York, using New York time, as well as the arrival time in Philadel-

TABLE 1-1 The 13 Constellations of the Zodiac	
Constellation	Dates of the Sun's passage through
Pisces	March 13–April 20
Aries	April 20–May 13
Taurus	May 13–June 21
Gemini	June 21–July 20
Cancer	July 20–August 11
Leo	August 11–September 18
Virgo	September 18–November 1
Libra	November 1–November 22
Scorpius	November 22–December 1
Ophiuchus	December 1–December 19
Sagittarius	December 19–January 19
Capricorn	January 19–February 18
Aquarius	February 18–March 13

phia (say, if someone was going to meet you) in Philadelphia time. Such time considerations became very confusing and burdensome as society became more complex.

Time zones were established in the late nineteenth century to alleviate this problem. In a time zone, everyone agrees to set their clocks alike. Time zones are based on the time at 0° longitude in Greenwich, England, a location called the prime meridian, as mentioned earlier. With some variations due to geopolitical boundaries, every 15° of longitude around the globe begins a new time zone. The resulting 24 time zones are shown in Figure 1-20. Going from one time zone to the next usually requires you to change the time on your watch by exactly 1 h.

There is also a *sidereal day*, the length of time from when a star is in one place in the sky until it is next in the same place. The solar and sidereal days differ from each other in length because while Earth rotates, it revolves around the Sun. This motion of Earth in its orbit day by day changes the location of the stars, bringing them back to their original positions 4 min earlier each day. Therefore, the sidereal day is 23 h, 56 min long, while the solar day is 24 h long.

Why would a calendar based on sidereal days not be satisfactory?

1-7 Calendars based on equal-length years also created scheduling problems

Just as the day is caused by Earth's rotation, the year is the unit of time based on Earth's revolution about the Sun. Earth does not take exactly 365 days to orbit the Sun, so the year is not exactly 365 days long. Basing the year on a 365-day cycle led to important events occurring on the wrong day. To resolve this problem, Roman statesman Julius Caesar implemented a new calendar in the year 46 B.C. Because measurement revealed to ancient astronomers that the length of a year is approximately 365¼ days, this "Julian" calendar established a system of leap years to accommodate the extra quarter of a day. By adding an extra day to the calendar every 4 years, Caesar hoped to ensure that seasonal astronomical events, such as the beginning of spring, would occur on the same date year after year.

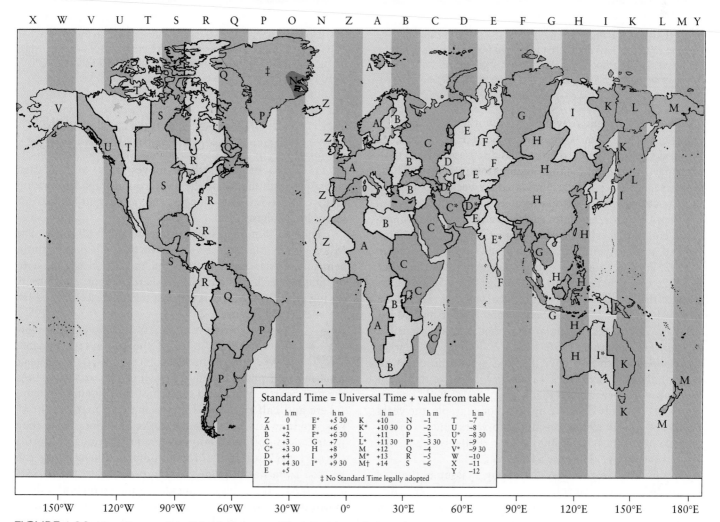

FIGURE 1-20 **Time Zones of the World** For convenience, Earth's 360° circumference is divided into 24 time zones. Ideally, each time zone would run due north–south. However, political considerations make many zones irregular. Indeed, there are even a few zones only a half-hour wide.

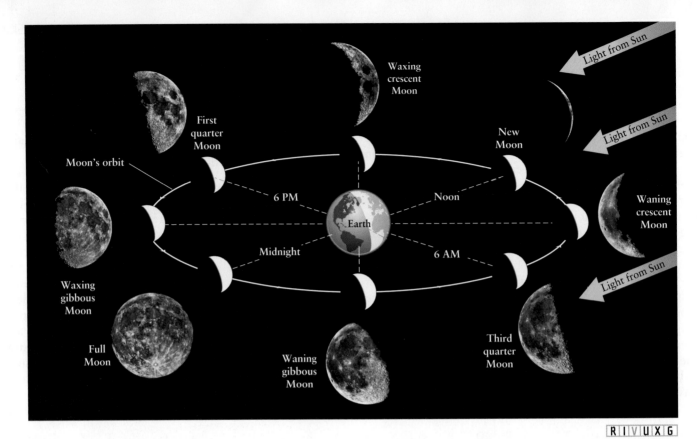

FIGURE 1-21 **The Phases of the Moon** The diagram shows the Moon at eight locations on its orbit as viewed from far above Earth's North Pole. The corresponding photographs show the resulting lunar phases *as seen from Earth.* The waning and third quarter photographs look backward, but they are correctly oriented as seen from Earth. Light from the Sun illuminates one-half of the Moon at all times, while the other half is dark. It takes about 29½ days for the Moon to go through all of its phases. *(Yerkes Observatory and Lick Observatory)*

The Julian calendar would have worked just fine if a year were exactly 365¼ days long and if Earth's rotation axis (now pointing toward Polaris, as discussed earlier) never changed direction. Neither assumption is correct. Thus, over time, a discrepancy accumulated between the calendar and the actual time—astronomical and cultural events began to fall on different dates each year. To straighten things out, a committee established by Pope Gregory XIII recommended a refinement, thus creating the Gregorian calendar in 1582. Pope Gregory began by dropping 10 days (October 5, 1582, was proclaimed to be October 15, 1582), which brought the first day of spring back to March 21. Next, he modified Caesar's system of leap years. Caesar had added February 29 to every calendar year that is evenly divisible by four. For example, 1996, 2000, 2004, and 2008 were all leap years with 366 days. But this system produces an error of about 3 days every 4 centuries. To solve the problem, Pope Gregory decreed that century years would be leap years only if evenly divisible by 400. For example, the years 1700, 1800, and 1900 were not leap years under the improved Gregorian system. But the year 2000—which can be divided evenly by 400—was a leap year.

We use the Gregorian system today. It assumes that the year is 365.2425 mean solar days long, which is very close to the length of the *tropical year,* defined as the time interval from one vernal equinox to the next. In fact, the error is only 1 day in every 3300 years. That won't cause any problems for a long time.

1-8 The phases of the Moon originally inspired the concept of the month

As the Moon orbits Earth, it moves from west to east, changing position among the background stars. Its position relative to the Sun also changes, and, as a result, we see different **lunar phases.**

The Sun illuminates half of the Moon at all times. The Moon's phase that we see depends on how much of its sunlit hemisphere is facing Earth. When the Moon is closest to the Sun in the sky, its dark hemisphere faces us. This phase, during which the Moon is at most a tiny crescent, is called the *new* Moon (Figure 1-21).

 During the 7 days following the new phase, more of the Moon's illuminated hemisphere becomes exposed to our view, resulting in a phase called the *waxing crescent* Moon. At the *first quarter* Moon, we see half of the illuminated hemisphere and half of the dark hemisphere. "Quarter Moon" refers to how far in its cycle the Moon has gone, rather than what fraction of the Moon appears lit by sunlight. During the next week still more of the illuminated hemisphere can be seen from Earth, giving us the phase called the *waxing gibbous* Moon. "Gibbous" means "rounded on both sides." When the Moon arrives on the opposite side of Earth from the Sun, we see virtually all of the fully illuminated hemisphere. This phase is the full Moon. Over the following 2 weeks, we see less and less of the illuminated hemisphere

R I V U X G

FIGURE 1-22 **The Moon during the Day** The Moon is visible at some time during daylight hours virtually every day. The time of day or night it is up in our sky depends on its phase. *(Richard Cummins/ SuperStock)*

as the Moon continues along its orbit. This movement produces the phases called the *waning gibbous* Moon, *third quarter* Moon, and the *waning crescent* Moon. The Moon completes a full cycle of phases in 29½ days.

5 Confusion often occurs over the terms "far side" and "dark side" of the Moon. The far side is the side of the Moon facing away from Earth. The dark side is the side of the Moon on which the Sun is not shining. By examining the photographs in Figure 1-21, you can see that the same side of the Moon faces Earth all the time. The half of the Moon that never faces Earth is the far side. However, the far side is not always the dark side, because we see part of the dark side whenever we see less than a full Moon.

Figure 1-21 shows the Moon at various positions in its orbit. Remember that the bright side of the Moon is on the right (west) side of the waxing Moon, while the bright side is on the left (east) side of the waning Moon. This information can tell you at a glance whether the Moon is waxing or waning. When looking at the Moon through a telescope, the best place to see details is where the shadows are longest. This occurs at

FIGURE 1-23 **The Sidereal and Synodic Months** The sidereal month is the time it takes the Moon to complete one revolution with respect to the background stars. However, because Earth is constantly moving in its orbit about the Sun, the Moon must travel through more than 360° to get from one new Moon to the next. The synodic month is the time between consecutive new Moons or consecutive full Moons. Thus, the synodic month is slightly longer than the sidereal month.

the boundary between the bright and dark regions, called the **terminator.**

Figure 1-21 also shows local time around the globe, **6** from noon, when the Sun is highest in the sky, to midnight, when it is on the opposite side of Earth. These time markings roughly indicate when the Moon is highest in the sky. For example, at first quarter, the Moon is 90° east of the Sun in the sky; hence, the Moon is highest at sunset. At full Moon, the Moon is opposite the Sun in the sky; thus, the Moon is highest at midnight. Using this information, you can see that the Moon is visible during the daytime (Figure 1-22) for a part of most days of the year.

Since the dawn of civilization, people have sought accurate timekeeping systems. Ancient Egyptians wanted to know when the Nile would flood, and farmers everywhere needed to know when to plant crops. Migratory tribes wanted to know when the weather would change. Religious leaders scheduled observances in accordance with celestial events. Thus, astronomers have traditionally been responsible for telling time. Indeed, of the four ways in which time cycles are set, three are astronomical in origin: Time is determined by the positions of the Moon, Sun, or stars or, in our own age, by technological means, such as atomic clocks.

The approximately 4 weeks that the Moon takes to complete one cycle of its phases inspired our ancestors to invent the concept of a month. Astronomers find it useful to define two types of months, depending on whether the Moon's motion is measured relative to the stars or to the Sun. Neither type corresponds exactly to the months of our usual calendar, which have different lengths.

The **sidereal month** is the time it takes the Moon to complete one full orbit of 360° around Earth (Figure 1-23). As with the sidereal day, the length of the sidereal month is determined by the location of the Moon in its orbit around Earth as measured with respect to the stars. Equivalently, this is the

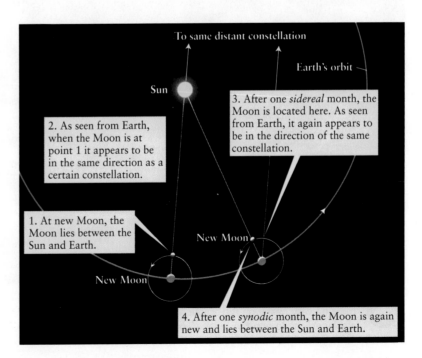

To same distant constellation

Earth's orbit

Sun

2. As seen from Earth, when the Moon is at point 1 it appears to be in the same direction as a certain constellation.

3. After one *sidereal* month, the Moon is located here. As seen from Earth, it again appears to be in the direction of the same constellation.

1. At new Moon, the Moon lies between the Sun and Earth.

New Moon

New Moon

4. After one *synodic* month, the Moon is again new and lies between the Sun and Earth.

time it takes the Moon to start at one place on the celestial sphere and return to exactly the same place again. The sidereal orbital period of the Moon takes approximately 27.3 days. The **synodic month,** or **lunar month,** is the time it takes the Moon to complete one 29½-day cycle of phases (that is, from new Moon to new Moon or from full Moon to full Moon) and thus is measured with respect to the Sun rather than the stars.

Is the Moon waning or waxing in Figure 1-22?

The synodic month is longer than the sidereal month because Earth is orbiting the Sun while the Moon goes through its phases. As shown in Figure 1-23, the Moon must travel *more* than 360° along its orbit to complete a cycle of phases (for example, from one new Moon to the next), which takes about 2.2 days longer than the sidereal month.

Both the sidereal month and synodic month vary somewhat, because the gravitational pull of the Sun on the Moon affects the Moon's speed as it orbits Earth. The sidereal month can vary by as much as 7 h, while the synodic month can change by as much as 12 h.

The terms *synodic* and *sidereal* are also used in discussing the motion of the other bodies in the solar system. The synodic period of a planet is the time between consecutive straight alignments of the Sun, Earth, and that planet (during which time interval the planet also goes through a cycle of phases, as seen from Earth). Recall that any orbit measured with respect to the stars is called "sidereal," including orbits

of the planets around the Sun, as well as orbits of moons around their planets. Earth's sidereal year is 365.2564 days. Our sidereal year differs from the time between consecutive vernal equinoxes (called a *tropical year*) primarily due to Earth's rotation axis slowly changing direction in space, an effect called *precession*.

ECLIPSES

Eclipses are among the most spectacular natural phenomena. During a **lunar eclipse,** the brilliant full Moon often darkens to a deep red. A lunar eclipse occurs when the Moon passes through Earth's shadow. This can happen only when the Sun, Earth, and Moon are in a straight line at full Moon. During a **solar eclipse,** broad daylight is transformed into an eerie twilight, as the Sun seems to be blotted from the sky. A solar eclipse occurs when the Moon's shadow moves across Earth's surface. As seen from Earth, the Moon moves in front of the Sun.

1-9   Eclipses occur only when the Moon crosses the ecliptic during the new or full phase

At first glance, it would seem that eclipses should happen at every new and full Moon, but, in fact, they occur much less often because the Moon's orbit is tilted 5° from the plane of the ecliptic (Figure 1-24). Consequently, the new Moon

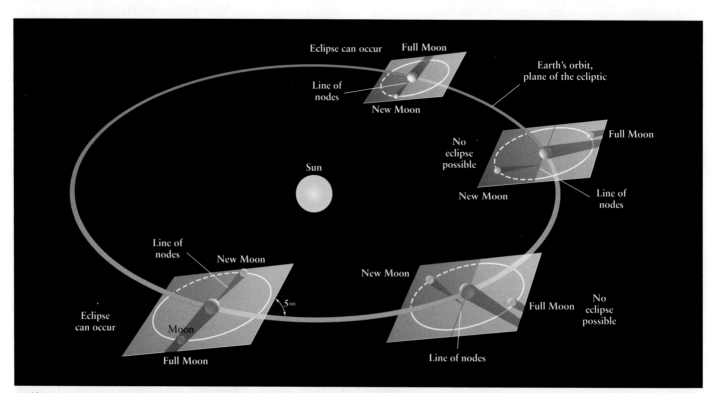

FIGURE 1-24 Conditions for Eclipses The Moon must be very nearly on the ecliptic at new Moon for a solar eclipse to occur. A lunar eclipse occurs only if the Moon is very nearly on the ecliptic at full Moon. When new Moon or full Moon phases occur away from the ecliptic, no eclipse is seen, because the Moon and Earth do not pass through each other's shadow.

and full Moon usually occur when the Moon is either above or below the plane of Earth's orbit. In such positions, a true alignment between the Sun, Moon, and Earth is not possible and so an eclipse cannot occur.

Indeed, because its orbit is tilted 5° from the ecliptic, the Moon is usually above or below the plane of our orbit around the Sun. The Moon crosses the ecliptic at what is called the **line of nodes** (see Figure 1-24). When the Moon crosses the plane of the ecliptic during its new or full phase, an eclipse takes place. By calculating the number of times a new Moon takes place on the line of nodes, we find that at least two and no more than five solar eclipses occur each year. Lunar eclipses occur just about as frequently as solar eclipses, with the maximum number of eclipses (solar plus lunar) possible in a year being seven.

1-10   Three types of lunar eclipse occur

Earth's shadow has two distinct parts, as shown in Figure 1-25a. The **umbra** is the part of the shadow where all direct sunlight is blocked by Earth. If you were in Earth's umbra looking at Earth, you would not see the Sun behind it at all. The **penumbra** of the shadow is where Earth blocks only some of the sunlight. If you were in Earth's penumbra looking at Earth, you would see a crescent Sun behind it. The Moon has an analogous umbra and penumbra.

Depending on how the Moon travels through Earth's shadow, three kinds of lunar eclipses may occur. A **penumbral eclipse,** when the Moon passes through only Earth's penumbra, is easy to miss. The Moon still looks full, just a little dimmer than usual and sometimes slightly reddish in color (path 1 in Figure 1-25a).

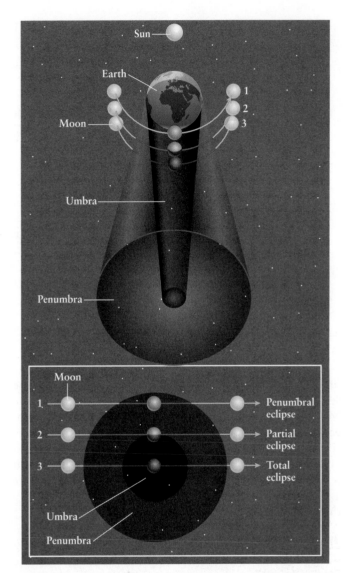

a

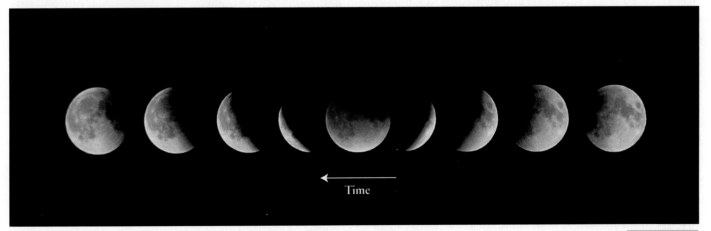

b

R I V U X G

  FIGURE 1-25   Three Types of Lunar Eclipse (a) People on the nighttime side of Earth see a lunar eclipse when the Moon moves through Earth's shadow. The umbra is the darkest part of the shadow. In the penumbra, only part of the Sun is covered by Earth. The inset shows the various lunar eclipses that occur, depending on the Moon's path through Earth's shadow.
(b) This sequence of nine photographs was taken over a 3-h period during the total lunar eclipse of January 20, 2000. During the total phase, the Moon has a distinctly reddish color. *(Fred Espenak, NASA/ Goddard Space Flight Center; © Fred Espenak, MrEclipse.com)*

TABLE 1-2 Lunar Eclipses, 2009–2012

Date	Visible from	Type	Duration of totality (h:min)
2009 February 9	Eastern Europe, Asia, Australia, Pacific, western North America	Penumbral	
2009 July 7	Australia, Pacific, Americas	Penumbral	
2009 August 6	Americas, Europe, Africa, western Asia	Penumbral	
2009 December 31	Europe, Africa, Asia, Australia	Partial	
2010 June 26	Eastern Asia, Australia, Pacific, western Americas	Partial	
2010 December 21	Eastern Asia, Australia, Pacific, Americas, Europe	Total	1:13
2011 June 15	South America, Europe, Africa, Asia, Australia	Total	1:41
2011 December 10	Europe, eastern Africa, Asia, Australia, North America	Total	:52
2012 June 4	Asia, Australia, Pacific, Americas	Partial	
2012 November 28	Europe, eastern Africa, Asia, Australia, North America	Penumbral	

When just part of the lunar surface passes through the umbra, a bite seems to be taken out of the Moon, and we see a **partial eclipse** (path 2 in Figure 1-25a). When the Moon travels completely into the umbra, we see a **total eclipse** of the Moon (path 3 in Figure 1-25a). Total lunar eclipses with the maximum duration, lasting for up to 1 h and 47 min, occur when the Moon is closest to Earth and it travels directly through the center of the umbra. Table 1-2 lists all the total and partial lunar eclipses from 2009 through 2012.

Even during a total eclipse, the Moon does not completely disappear. A small amount of sunlight passing through Earth's atmosphere is bent into Earth's umbra. The light deflected into the umbra is primarily red and orange, and thus the darkened Moon glows faintly in rust-colored hues (Figure 1-25b). At sunrise and sunset, the sky appears red or orange for the same reason, because at those times more of Earth's atmosphere deflects red and orange light from the Sun toward you.

Everyone on the side of Earth over which a lunar eclipse occurs can see it, provided that clouds don't obscure the event. Lunar eclipses are perfectly safe to watch with the naked eye.

Why does the new moon sometimes appear as a crescent?

1-11 Three types of solar eclipse also occur

Because of their different distances from Earth, the Sun and the Moon have nearly the same angular diameter as seen from Earth—about ½°. When the Moon completely covers the Sun, the result is a total solar eclipse. You must be at a location within the Moon's umbra to see a total solar eclipse. During those few precious moments, hot gases

(the **solar corona**) surrounding the Sun can be observed and photographed (Figure 1-26). In this way, astronomers have been able to learn more about the Sun's temperature, chemistry, and atmospheric activity.

R I V U X G

FIGURE 1-26 **A Total Eclipse of the Sun** During a total solar eclipse, the Moon completely covers the Sun's disk, and the solar corona can be photographed. This halo of hot gases extends for millions of kilometers into space. This gorgeous image is a composite of several taken in Chisamba, Zambia, during the June 21, 2001, solar eclipse. *(F. Espenak)*

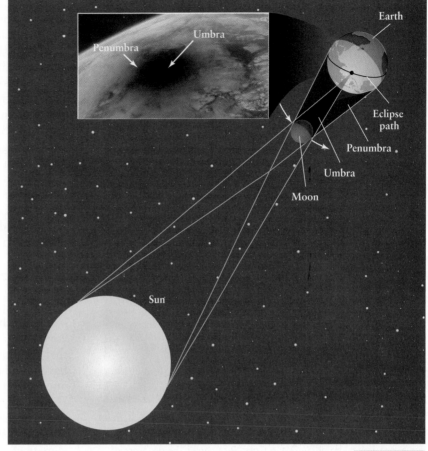

FIGURE 1-27 **The Geometry of a Total Solar Eclipse** During a total solar eclipse, the tip of the Moon's umbra traces an eclipse path across Earth's surface. People inside the eclipse path see a total solar eclipse, whereas people inside the penumbra see only a partial eclipse. The photograph in this figure shows the Moon's shadow on Earth. It was taken from the *Mir* space station during the August 11, 1999, total solar eclipse. The Moon's umbra appears as the very dark spot on the eastern coast of the United States. It is surrounded by the penumbra. *(Jean-Pierre Haigneré, Centre National d'Etudes Spatiales, France/GSFS)*

R I V U X G

You can see in Figure 1-27 that only the tip of the Moon's umbra ever reaches Earth's surface. As Earth turns and the Moon orbits, the tip traces an **eclipse path** across our planet. Only people within these areas are treated to the spectacle of a total solar eclipse. *Viewing the Sun directly for more than a moment at any time without an approved filter causes permanent eye damage.* It is only safe to look at a total solar eclipse without a filter during the brief time when the Moon completely blocks the Sun. At all other times during the eclipse, you must view it either through an approved filter or in the image made on a flat surface by a telescope or a pinhole camera.

Earth's rotation and the orbital motion of the Moon cause the umbra to race along the eclipse path at speeds in excess of 1700 km/h (1050 mph). For this reason, a total eclipse never lasts for more than 7½ min at any one location on the eclipse path, and it usually lasts for only a few moments.

The Moon's umbra is also surrounded by a penumbra (see Figure 1-27). The photograph in Figure 1-27 shows the dark spot produced by the Moon's umbra and penumbra on Earth's surface during a total solar eclipse. During a solar eclipse, the Moon's penumbra extends over a large portion of Earth's surface. When only the penumbra sweeps across Earth's surface, as happens in high latitude regions, the Sun is only partly covered by the Moon. This circumstance results in a *partial eclipse of the Sun.* Similarly, people in the penumbra of a total eclipse see a partial eclipse. In either case, the Sun looks crescent as seen from Earth.

The Moon's orbit around Earth is not quite a perfect circle. The distance between Earth and the Moon, which averages 384,400 km (238,900 mi), varies by a few percent as the Moon goes around Earth. The width of the eclipse path depends primarily on Earth-Moon distance during an eclipse. The eclipse path is widest—up to 270 km (170 mi)—when the new Moon happens to be at the point in its orbit nearest Earth. Usually, however, the path is much narrower.

If a solar eclipse occurs when the Moon is farthest from Earth, then the Moon's umbra falls short of Earth and no one sees a total eclipse. From Earth's surface, the Moon then appears too small to cover the Sun completely, and a thin ring or "annulus" of light is seen around the edge of the Moon at mid-eclipse. This type of eclipse is called an **annular eclipse** (Figure 1-28). The length of the Moon's umbra is nearly 5000 km (3100 mi) shorter than the average distance between the Moon and Earth's surface. Thus, the Moon's

R I V U X G

FIGURE 1-28 **An Annular Eclipse of the Sun** This composite of five exposures taken at sunrise in Costa Rica shows the progress of an annular eclipse of the Sun that occurred on December 24, 1974. Note that at mid-eclipse the edge of the Sun is visible around the Moon. *(Dennis Di Cicco)*

shadow often fails to reach Earth, making annular eclipses slightly more common than total eclipses. Table 1-3 lists all the total, partial, and annular solar eclipses from 2009 through 2012. It is *never* safe to look directly at a partial or annular eclipse.

A total solar eclipse is a dramatic event. The sky begins to darken, the air temperature decreases, and the winds increase as the Moon's umbra races toward you. All nature responds: Birds go to roost, flowers close their petals, and crickets begin to chirp as if evening had arrived. As the moment when the Sun becomes totally eclipsed approaches, the landscape is bathed in shimmering bands of light and dark as the last few rays of sunlight peek out from behind the edge of the Moon. Finally, the corona blazes forth in a star-studded daytime sky. It is an awesome sight, but always take care in viewing it.

1-12   Frontiers yet to be discovered

In this chapter we have examined several factors that explain the changing seasons on Earth. Geologists have discovered a variety of cycles of global temperature change that have occurred over the history of Earth. Indeed, they have found that much of Earth once suffered a global freezing. Conversely, we are now undergoing a global warming. Historically, these changes occur over tens of thousands of years, hundreds of thousands of years, and possibly longer cycles. Most scientists are now convinced that the present climate change is a result of a combination of human and other causes, some of which have yet to be discovered.

Why aren't there any annular lunar eclipses?

TABLE 1-3	Solar Eclipses, 2009–2012			
Date	Type	Visible from		Total/annular eclipse time (min:s)
2009 January 26	Annular	Africa, southeast Asia, Australia		7:54
2009 July 22	Total	Eastern Asia, Pacific, Hawaii		6:39
2010 January 15	Annular	Africa, Asia		11:08
2010 July 11	Total	Southern South America		5:20
2011 January 11	Partial	Europe, Africa, central Asia		
2011 June 1	Partial	Eastern Asia, northern North America, Iceland		
2011 July 1	Partial	Southern Indian Ocean		
2011 November 25	Partial	Southern Africa, Tasmania, New Zealand		
2012 May 20	Annular	Asia, Pacific, North America		5:46
2012 Nov 13	Total	Australia, New Zealand, southern South America		4:02

SUMMARY OF KEY IDEAS

Sizes in Astronomy

• Astronomy examines objects that range in size from the parts of an atom ($\sim 10^{-15}$ m) to the size of the observable universe ($\sim 10^{26}$ m).

• Scientific notation is a convenient shorthand for writing very large and very small numbers.

Patterns of Stars

• The surface of the celestial sphere is divided into 88 unequal areas called constellations.

• The boundaries of the constellations run along lines of constant right ascension or declination.

Earthly Cycles

• The celestial sphere appears to revolve around Earth once in each day–night cycle. In fact, it is Earth's rotation that causes this apparent motion.

• The poles and equator of the celestial sphere are determined by extending the axis of rotation and the equatorial plane of Earth out onto the celestial sphere.

• Earth's axis of rotation is tilted at an angle of 23½° from a line perpendicular to the plane of Earth's orbit (the plane of the ecliptic). This tilt causes the seasons.

• Equinoxes and solstices are significant points along Earth's orbit that are determined by the relationship between the Sun's path on the celestial sphere (the ecliptic) and the celestial equator.

• The length of the day is based upon Earth's rotation rate and the average motion of Earth around the Sun. These effects combine to produce the 24-h day upon which our clocks are based.

• The phases of the Moon are caused by the relative positions of Earth, Moon, and Sun. The Moon completes one cycle of phases in a synodic month, which averages 29½ days.

• The Moon completes one orbit around Earth with respect to the stars in a sidereal month, which averages 27.3 days.

Eclipses

• The shadow of an object has two parts: the umbra, where direct light from the source is completely blocked; and the penumbra, where the light source is only partially obscured.

• A lunar eclipse occurs when the Moon moves through Earth's shadow. During a lunar eclipse, the Sun, Earth, and Moon are in alignment with Earth between the Sun and the Moon, and the Moon is in the plane of the ecliptic.

• A solar eclipse occurs when a strip of Earth passes through the Moon's shadow. During a solar eclipse, the Sun, Earth, and Moon are in alignment with the Moon between Earth and the Sun, and the Moon is in the plane of the ecliptic.

• Depending on the relative positions of the Sun, Moon, and Earth, lunar eclipses may be penumbral, partial, or total, and solar eclipses may be annular, partial, or total.

WHAT DID YOU THINK?

1 *Is the North Star—Polaris—the brightest star in the night sky?* No. Polaris is a star of medium brightness compared with other stars visible to the naked eye.

2 *What causes the seasons?* The tilt of Earth's rotation axis with respect to the ecliptic causes the seasons. They are not caused by the changing distance from Earth to the Sun that results from the shape of Earth's orbit.

3 *When is Earth closest to the Sun?* On or around January 3 of each year.

4 *How many zodiac constellations are there?* There are 13 zodiac constellations, the lesser-known one being Ophiuchus.

5 *Does the Moon have a dark side that we never see from Earth?* Half of the Moon is always dark. Whenever we see less than a full Moon, we are seeing part of the Moon's dark side. So, the dark side of the Moon is not the same as the far side of the Moon, which we never see from Earth.

6 *Is the Moon ever visible during the daytime?* The Moon is visible at some time during daylight hours almost every day of the year. Different phases are visible during different times of the day.

Review Questions

1. Where is the horizon? **a.** directly overhead, **b.** along the celestial equator, **c.** the boundary between land and sky, **d.** along the path that the Sun follows throughout the day, **e.** the line running from due north, directly overhead, ending due south

2. How many constellations are there? **a.** 2, **b.** 12, **c.** 13, **d.** 56, **e.** 88

3. Which of the following lies on the celestial sphere directly over Earth's equator? **a.** ecliptic, **b.** celestial equator, **c.** north celestial pole, **d.** south celestial pole, **e.** horizon

4. The length of time it takes Earth to orbit the Sun is **a.** an hour, **b.** a day, **c.** a month, **d.** a year, **e.** a century.

5. In Figure 1-8, what is another name for the "Sun's annual path"?

6. How are constellations useful to astronomers?

7. What is the celestial sphere, and why is this ancient concept still useful today?

8. What is the celestial equator, and how is it related to Earth's equator? How are the north and south celestial poles related to Earth's axis of rotation?

9. What is the ecliptic, and why is it tilted with respect to the celestial equator?

10. By about how many degrees does the Sun move along the ecliptic each day?

11. Through how many constellations does the Sun move every day?

12. Through how many constellations does the Sun move every year?

13. Why does the tilt of Earth's axis relative to its orbit cause the seasons as Earth revolves around the Sun? Draw a diagram to illustrate your answer.

14. What are the vernal and autumnal equinoxes? What are the summer and winter solstices? How are these four points related to the ecliptic and the celestial equator?

15. How does the daily path of the Sun across the sky change with the seasons?

16. Why is it warmer in the summer than in the winter?

17. Why is it convenient to divide Earth into time zones?

18. Why does the Moon exhibit phases?

19. What is the difference between a sidereal month and a synodic month? Which is longer? Why?

20. What is the line of nodes, and how is it related to solar and lunar eclipses?

21. What is the difference between the umbra and the penumbra of a shadow?

22. What is a penumbral eclipse of the Moon? Why is it easy to overlook such an eclipse?

23. Which type of eclipse—lunar or solar—have most people seen? Why?

24. How is an annular eclipse of the Sun different from a total eclipse of the Sun? What causes this difference?

25. When is the next leap year?

26. At which phase(s) of the Moon does a solar eclipse occur? A lunar eclipse?

27. Is it safe to watch a solar eclipse without eye protection? A lunar eclipse?

28. During what phase is the Moon "up" least in the daytime?

Observing Projects

Many of the following projects are based on the planetarium program, *Starry Night Enthusiast*™. If assigned them, it is likely that your teacher had the software included with this text. Install this program on a suitable computer and run it.

You can learn to use the program with the help of the User's Guide, which can be accessed under **Help > User's Guide** on the main menu. A short introductory tutorial is available on the Web site associated with this textbook. You can also experiment on your own or with the help of a friend, or use the on-line help. **Start each Project by clicking the Home button in the toolbar to reset the program to the current time and to your home location.**

29. Use *Starry Night Enthusiast*™ to study the Moon's path in the sky and eclipses. Open the **Options** pane and turn off the *Local Horizon* and *Daylight* options in the **Local View** layer. Under the *Guides* layer, turn on *The Ecliptic* and *Celestial Grid*. Open the *Find* pane and double-click the entry for *The Moon*. Change the field of view of the observed sky to about 15° using the Zoom buttons on the upper right of the toolbar. Note whether the Moon lies on the ecliptic. Set the **Time Flow Rate** to *6 minutes* (adjust this if sky motion is too fast or too slow). Keeping the Moon locked in the center of your screen, set time in motion. **a.** In which direction does the Moon move against the background stars? Keep in mind that the celestial equator runs east-west (east is to the left) and the ecliptic intersects it at 23½°. Ignore the Moon's rocking motion, which is due to its rising and setting. Does it ever change direction relative to the stars? Why or why not? Describe its path relative to the ecliptic. **b.** Change the **Time Flow Rate** to *1 hour* (adjust as necessary). Determine how many days elapse between successive times that the Moon is on the ecliptic. **c.** Set the time and date to 9 AM on July 21, 2009. Running time forward at a rate of 1 min, stop time when the Moon is closest to the Sun. Sketch the Sun and Moon on a piece of paper. Even though the Moon may not be directly over the Sun on the screen, an eclipse is occurring. What type of eclipse is it? Why might the Moon *not* be directly over the Sun at that time? **d.** Click on the *Events* tab, expand the *Event Filters* layer and deselect all but the *Lunar and Solar Eclipse Events*. Expand the *Events Browser* layer, set the *Start* and *End* dates to cover the period from 1/21/2009 to 1/21/2010 and click *Find Events*. Right click on each solar eclipse in turn, choose *View Event*, zoom back out to about 15°, and *Run time forward* and *backward* at a rate of 1 min until the Moon is again closest to the Sun. Draw the Sun and Moon. Are they separated by the same amount as in part c? If not, why not? If there is a difference, what effect does it have on the eclipse?

Many of the following projects are based upon the *World-Wide Telescope* (WWT) program, which is freely available on the internet. You should download this program on a suitable computer and run it. A short introductory tutorial is available on the Web site associated with this textbook. You can access more complete operating instructions by clicking on the down arrow under the **Explore** tab in the menu at the top of the screen and selecting **Getting Started (Help)**. You can also get help by pressing the **F1** key on the keyboard.

30. **a.** Launch WWT and if you have not already done so, change the observing location to your home town or city using the **View** tab. Check the option to **View from this location** in the **Observing Location** panel to ensure that your horizon is visible. Drag the view so that this horizon is visible in the lower part of the screen and note the cardinal points of the compass along this horizon. Set the time flow rate in the **View** tab to ×1000 and observe the motion of the sky. From which half of the sky do the stars rise? In which half of the sky do the stars set? What causes this apparent motion of the stars? Drag the view so that the eastern horizon is at the center bottom of the screen. Do the stars rise straight from the horizon or at an angle? **b.** With time continuing to flow, click the **View** tab followed by the **Setup** button in the **Observing Location** panel. Change the latitude of the observing location to 0° and click **OK**. From this location, on the equator of Earth, do the stars rise straight up from the horizon or do they rise at an angle? **c.** Click the **View** tab followed by the **Setup** button in the **Observing Location** panel and set the latitude of the observing location to 45°, the mid-northern latitude. Click the **OK** button and observe the rising stars. Do the stars rise straight up from the horizon or at an angle? Change the observing location to the mid-southern latitude of −45°. Do the stars rise straight up from the horizon or at an angle? How does this compare with the rising of stars from the mid-northern latitude? **d.** Finally, change the observing location to a point very near to the North Pole by setting the latitude to 89.99°. Describe the motion of the stars in the sky as seen from this location.

31. You can use WWT to show eclipses of the Sun as seen from appropriate locations. **a.** Launch WWT and click the **View** tab. In the **Observing Location** panel, click the **Setup** button. Set the observing location to **Longitude** 0° and **Latitude** −10°. If you wish, set the **Name** parameter to **West Africa**. Then click the **OK** button to move to this position on Earth. Click the **Explore** tab followed by the **Solar System** thumbnail. Then double click the thumbnail for the **Sun** to center it in the view. Next, click the **View** tab and the **Pause** button in the **Observing Time** pane. Click the **UTC** checkbox to **on** to enter the time in Coordinated Universal Time. (UTC, used by astronomers throughout the world, is defined as the time at the prime meridian passing through Greenwich, England). Set the **Date** to September 22, 2006, and set the time to 10:00:00 UTC. Then click the **OK** button. In the **Overlays** pane, check the **Ecliptic** option. To the right of the word "Ecliptic" click on the black rectangle and, if necessary, change the color from black to one of your choice. Zoom out to a field of view of about 2°. In which constellation is the Sun? (Note that this information is shown on the sky diagram in the lower right corner of the screen). In what phase is the Moon in this view? Is the Moon near the ecliptic? Finally, use the time flow controls at ×1000 to make observations through to 14:00:00 UTC. What type of solar eclipse was seen from this location on this date? During your observations, was this location ever in the penumbra of the Moon's shadow? Was this location ever in the umbra of the Moon's shadow? If so, for how long? **b.** **Pause** time flow and reset the time to 10:00:00 UTC. Change the observing location to **Longitude** 0° and **Latitude** −31° and manipulate the time controls to observe the eclipse from this location on Earth. What type of eclipse was seen from this location? During your observations, was this location ever in the penumbra of the Moon's shadow? Was this location ever in the umbra of the Moon's shadow? If so, for how long? **c.** **Pause** time flow and change the observing location to Mexico City, Mexico. Change the date to July 11, 1991 and the time to 17:00:00 UTC. In which constellation is the Sun? What is the local time at this location? Hint: The longitude of Mexico City is displayed in the **Observing Location** pane. Use the time flow controls to make observations through to about 21:00:00 UTC. What type of eclipse was observed from this location on this date? During your observations, was this location ever in the penumbra of the Moon's shadow? Was this location ever in the umbra of the Moon's shadow? If so, for how long? Hint: Slow down the time rate to watch the appearance of the solar corona for a few minutes during this eclipse.

Gravitation and the Motion of the Planets

R I V U X G

Applying the Law of Gravitation and Other Laws of Physics Is Essential in Traveling Safely to the Moon and Back *(NASA)*

Answers to these questions appear in the text beside the corresponding numbers in the margins and at the end of the chapter.

Science provides explanations for activities and events that occur today and those that have occurred in the past. It also makes predictions about things that have yet to happen or that have yet to be observed. These tools are incredibly powerful and they enable us to understand what we see without having to accept events on faith or to fear that things, such as the force of gravity, will change unexpectedly. Science simplifies and takes some of the uncertainty out of the world. In this chapter we will explore the nature of science and use it to see how gravity forms planets, and how it keeps them, along with myriad other objects, orbiting our Sun and other stars.

In this chapter you will discover

- what makes a theory scientific

- the scientific revolution that dethroned Earth from its location at the center of the universe

- Copernicus's argument that the planets orbit the Sun

- why the direction of motion of the planets on the celestial sphere sometimes appears to change

- that Kepler's determination of the shapes of planetary orbits depended on the careful observations of his mentor Tycho Brahe

- how Isaac Newton formulated an equation to describe the force of gravity and how he thereby explained why the planets and moons remain in orbit

- how the solar system formed

- why the environment of the early solar system was much more violent than it is today

- how astronomers define the various types of objects in the solar system

- how the planets are grouped

- how the moons formed throughout the solar system

- what the debris of the solar system is made of

- that disks of gas and dust, as well as planets, have been observed around a growing number of stars

- that newly forming stars and planetary systems are being observed

WHAT DO YOU THINK?

1 What makes a theory scientific?

2 What is the shape of Earth's orbit around the Sun?

3 Do the planets orbit the Sun at constant speeds?

4 Do all of the planets orbit the Sun at the same speed?

5 How much force does it take to keep an object moving in a straight line at a constant speed?

6 How does an object's mass differ when measured on Earth and on the Moon?

7 Do astronauts orbiting Earth feel the force of gravity from our planet?

8 Were the Sun and planets among the first generation of objects created in the universe?

9 How long has Earth existed, and how do we know this?

10 What typical shape(s) do moons have, and why?

11 Have any Earthlike planets been discovered orbiting Sunlike stars?

SCIENCE: KEY TO COMPREHENDING THE COSMOS

Understanding how nature works enables us to manipulate the matter and energy that comprise our environment and thereby to create new things to make our lives better. Improvements in technology lead, in turn, to better research equipment, enabling us to make even deeper discoveries about space, time, matter, energy, and the relationships among them. This spiral of understanding and application began centuries ago. In this chapter we will explore the nature of science and use it to see how gravity keeps planets and other objects orbiting the Sun as well as the moons orbiting their respective planets.

2-1   Science is both a body of knowledge *and* a process of learning about nature

Science is actually two things. First, it is a body of knowledge that we acquire by observations and experiments. The details of the motions of the Moon, planets, and Sun on the celestial sphere, described in Chapter 1, are examples of that knowledge. While nature can be discussed descriptively, as it is for the most part in this book, science also provides mathematical equations that quantify the effects being studied.

Second, science is a process for gaining more knowledge in a way that ensures that the information can be tested and thereby accepted by everyone. Science as a process is also called the **scientific method,** and it describes how scientists ideally go about observing, explaining, and predicting physical reality. The scientific method (Figure 2-1) can begin in a variety of places, but most often it starts by people making observations or doing experiments. For example, the observation that some objects (such as planets) move along the celestial sphere, while others (stars) remained fixed on it,

demanded explanation. The results of observations or experiments are compared with the predictions of any preexisting theories that are supposed to explain them. If the new data and old ideas are not consistent, then a *hypothesis* that modifies or replaces the existing explanation is proposed. Hypotheses on related topics that make accurate predictions are incorporated together as a **scientific theory** (often just called a **theory**).

In everyday conversation, a theory is an idea based on common sense, intuition, or deep-seated personal beliefs. Such theories neither originate in equations nor lead to rigorous predictions. The word *theory* in science has a very different connotation. It is an explanation of observations or experimental results that can be described quantitatively and tested formally. The mathematical description of a scientific theory is considered a **model** of the real system. For example, Newton's *theory* (or, in earlier usage, *law*) of gravitation is written as an equation that predicts how bodies attract each other. The word *gravity* is often used as shorthand for *gravitation,* and both are used in this book.

As just noted, to be considered scientific, a theory must make *testable* predictions that can be verified using new observations and experiments. Testing is a crucial aspect of the scientific method, which requires that the theory accurately forecast the results of new observations in its realm of validity. Newton's law of gravitation predicts that the Sun's gravitational force makes the planets move in elliptical orbits, and it predicts how long it should take each planet to orbit the Sun. As we will see shortly, observations have confirmed most of these predictions. **1**

Scientists who develop new or more accurate models are going where no person has gone before. Many of them find this process of discovery as satisfying as an artist creating a masterpiece, an athlete breaking a world record, or an astronaut going into space. Scientists who make obser-

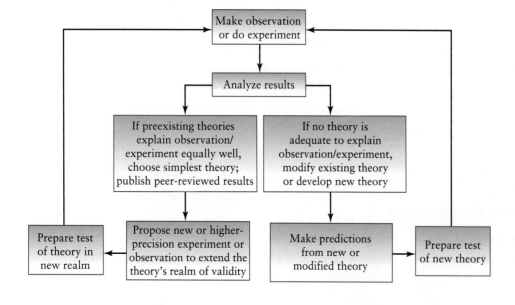

Figure 2-1 **The Scientific Method** This flow chart shows the basic processes by which scientists study nature and develop new scientific theories. Different scientists start at different places on this chart, including making observations or doing experiments; creating or modifying scientific theories; or making predictions from theories. Anyone interested in some aspect of science and willing to learn the tools of that science can participate in the adventure. (© *Neil F. Comins*)

vations or do experiments that reveal previously unknown facts about nature often have similar reactions to their discoveries.

INSIGHT INTO SCIENCE

Science Is Inclusive Science is intended as an inclusive endeavor. In principle, a scientific theory can be created, modified, or tested by anyone inclined to do so. In practice, however, being involved in the scientific enterprise requires that you understand the mathematical tools of science. Assuring that theories are written in terms of equations so that they can be carefully analyzed and tested by others is part of the process intended to prevent the scientific method from being derailed.

For a theory to be considered scientific, it must also be potentially possible to disprove it. For example, Newton's law of gravitation can be tested and potentially disproved by observations, and thus qualifies as a scientific theory. The idea that Earth was created in 6 days cannot be tested, much less disproved. It is not a scientific theory, but rather a matter of faith.

Give an example of one scientific hypothesis and one nonscientific hypothesis.

If the predictions of a theory are inconsistent with observations, the theory is modified, applied in more limited circumstances, or discarded in favor of a more accurate explanation. For example, Newton's law of gravitation is entirely adequate for describing the motion of an apple falling to Earth, the flight of a soccer ball, or the path of Earth orbiting the Sun; however, it is inaccurate in the vicinity of a black hole, where matter is especially dense. In this case, Newton's law of gravitation is replaced by Einstein's theory of general relativity, which describes gravitational behavior more accurately and over a much wider range of conditions than Newton's law, but at the cost of much greater mathematical complexity.

INSIGHT INTO SCIENCE

Theories and Beliefs New theories are personal creations, but science is not a personal belief system. As stated in the previous Insight into Science, scientific theories make predictions that can be tested independently. If everyone who performs tests of the theory's predictions gets results consistent with the theory, the theory is considered valid in that realm. In comparison, belief systems—such as which sports team or political system is best—are personal matters. People will always hold differing opinions about such issues.

Science also strives to explain as many things as possible using as few theories as possible. We see billions upon billions of objects in the universe. It would be virtually impossible to study all of them separately so that we could create detailed descriptions of each one. Fortunately, individual theories explaining each object are not necessary. Scientists overcome this problem by noting that many of the bodies in space appear similar to each other. By categorizing them suitably and then applying the scientific method to these groups of objects, we form a few theories that describe many objects and how they have evolved. These few theories can then be tested and refined as necessary. Such groupings of objects have proven invaluable, and they give us insights into the structure and organization of billions of stars and galaxies that are, indeed, very similar to one another.

INSIGHT INTO SCIENCE

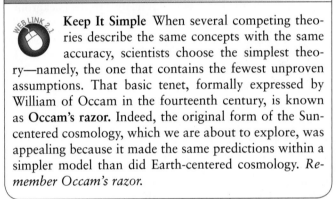 **Keep It Simple** When several competing theories describe the same concepts with the same accuracy, scientists choose the simplest theory—namely, the one that contains the fewest unproven assumptions. That basic tenet, formally expressed by William of Occam in the fourteenth century, is known as **Occam's razor**. Indeed, the original form of the Sun-centered cosmology, which we are about to explore, was appealing because it made the same predictions within a simpler model than did Earth-centered cosmology. *Remember Occam's razor.*

Although the vast majority of scientists carefully and scrupulously follow the rules of scientific research, we acknowledge that some experiments are run poorly. Some scientists have ignored experimental data or observations that do not mesh with cherished beliefs or have even fudged data or stolen data from others. Virtually all of these oversights and misdeeds are eventually discovered, because most theories and their predictions are tested by several independent researchers.

The scientific method can be summarized in six words: *observe, hypothesize, predict, test, modify, economize.* I urge you to watch for applications of the scientific method throughout this book. Our first encounter with it is the discovery that Earth orbits the Sun.

CHANGING OUR EARTH-CENTERED VIEW OF THE UNIVERSE

Early Greek astronomers tried to explain the motion of the five then-known planets: Mercury, Venus, Mars, Jupiter, and Saturn. Most people at that time held a *geocentric* view of the universe: Based on the observed motion of the celestial sphere, they believed that the Sun, the Moon, the stars, and the planets revolve around Earth. A theory of the overall structure and evolution of the universe is called a **cosmology,** so the prevailing Earth-centered cosmology was called *geocentric.* Geocentric cosmology, consistent as it is with casual observation of the sky, held sway for more than 2000 years (see Appendix G-1 for details).

2-2   The belief in a Sun-centered cosmology formed slowly

Explaining the motions of the five planets in a geocentric universe was one of the main challenges facing the astronomers of antiquity. The Greeks knew that the positions of the planets slowly shift relative to the "fixed" stars in the constellations. In fact, the word *planet* comes from a Greek term meaning "wanderer." They also observed that planets do not move at uniform rates through the constellations. From night to night, as viewed in the northern hemisphere, the planets usually move slowly to the left (eastward) relative to the background stars. This movement is called **direct motion.** Occasionally, however, a planet seems to stop and then back up for several weeks or months. This reverse movement (to the west relative to the background stars) is called **retrograde motion.** Both direct and retrograde motions are best observed by plotting or photographing the nightly position of a planet against the background stars over a long period (Figure 2-2).

All planetary motions on the celestial sphere are much slower than the apparent daily movement of the entire sky caused by Earth's rotation, and so they are superimposed on it. Therefore, the planets always rise in the eastern half of the sky and set in the western half, as the stars do.

Why do you think Mars is seen sometimes above the ecliptic and sometimes below it?

The effort to understand planetary motion—and especially to explain retrograde motion—using a geocentric cosmology resulted in an increasingly contrived and complex model. The ancient Greek astronomer Aristarchus proposed a more straightforward explanation of planetary motion, namely, that all of the planets, including Earth, revolve around the Sun. The retrograde motion of Mars in this **heliocentric (Sun-centered) cosmology** occurs just because the faster-moving Earth overtakes and passes the red planet (Figure 2-3). The occasional retrograde movement of a planet is merely the result of our changing viewpoint as we orbit the Sun—an idea that is beautifully simple compared to the geocentric system with all of its complex planetary motions. (The word *heliocentric* is misleading. Although the local planets, moons, and small pieces of space debris do orbit the Sun, the stars and innumerable other objects in space do not. In fact, the Sun and the bodies that orbit the Sun all orbit the center of our Milky Way Galaxy.)

Because simplicity and accuracy are hallmarks of science, the complex geocentric model eventually gave way to the simpler, more elegant heliocentric cosmology, but dethroning the geocentric model did not occur immediately. In the first place, Earth just does not seem to move! This observation, along with strict geocentric religious teachings and the human desire to be at the center of everything, outweighed the simpler plan proposed by Aristarchus. Not until 1300 years later did anyone seriously reconsider the advantages of a heliocentric cosmology.

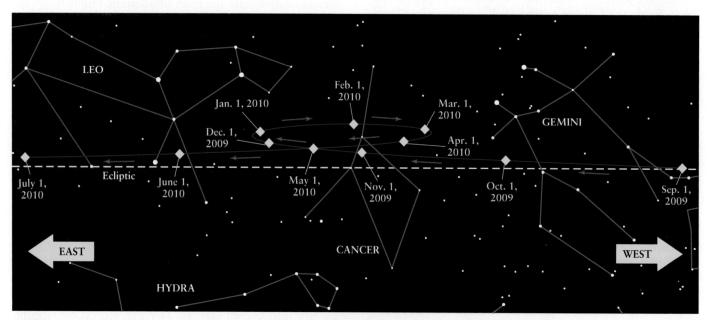

Figure 2-2 **Path of Mars** From September 2009 through June 2010, Mars moves through the constellations of Gemini, Cancer, and Leo. From December 23, 2009, through March 12, 2010, Mars is in retrograde motion. The retrograde loop is sometimes north of the normal path and sometimes south of it (see Figure 2-3).

Figure 2-3 **A Heliocentric Explanation of Planetary Motion** Earth travels around the Sun more rapidly than does Mars. Consequently, as Earth overtakes and passes this slower-moving planet, Mars appears (from points 4 through 6) to move backward among the background stars for a few months.

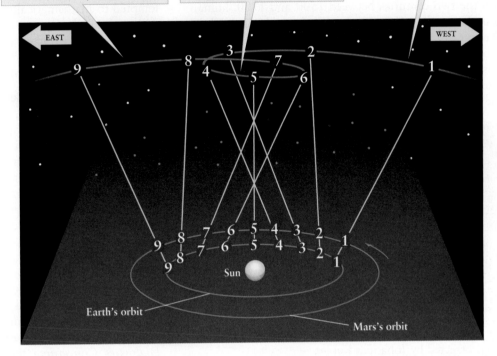

3. From point 6 to point 9, Mars again appears to move eastward against the background stars as seen from Earth (direct motion).

2. As Earth passes Mars in its orbit from point 4 to point 6, Mars appears to move westward against the background stars (retrograde motion).

1. From point 1 to point 4, Mars appears to move eastward against the background stars as seen from Earth (direct motion).

2-3 Copernicus devised the first comprehensive heliocentric cosmology

Over the centuries, increasingly accurate observations of the planets' locations revealed errors in the predictions of geocentric cosmology. To reconcile that cosmology with the data, more and more complex motions were attributed to the planets. By the mid-1500s, the geocentric cosmology had become truly unwieldy in its efforts to predict the motions of the planets accurately. It was then that the Polish mathematician, lawyer, physician, economist, cleric, and artist Nicolaus Copernicus resurrected Aristarchus's theory. Copernicus (see Guided Discovery: Astronomy's Foundation Builders) was motivated by an effort to simplify the celestial scheme.

 After assuming that the planets orbit the Sun rather than Earth, Copernicus, through observations, determined which planets are closer to the Sun than Earth and which are farther away. Because Mercury and Venus are always observed fairly near the Sun, he correctly concluded that their orbits must lie inside Earth's. The other planets visible to Copernicus—Mars, Jupiter, and Saturn—can sometimes be seen high in the sky in the middle of the night, when the Sun is far below the horizon. This can occur only if Earth comes between the Sun and a planet. Copernicus therefore concluded (correctly) that the orbits of Mars, Jupiter, and Saturn lie outside Earth's orbit.

The geometric arrangements among Earth, another planet, and the Sun are called **configurations**. For example, when Mercury or Venus is directly between Earth and the Sun (Figure 2-4), we say the planet is in a configuration called an **inferior conjunction;** when either of these planets is on the opposite side of the Sun from Earth, its configuration is called a **superior conjunction.**

The angle between the Sun and a planet as viewed from Earth is called the planet's **elongation.** A planet's elongation varies from 0° to a maximum value, depending upon where we see it in its orbit around the Sun. At *greatest eastern* or *greatest western elongation,* Mercury and Venus are as far from the Sun in angle as they can be. This is about 28° for Mercury and about 47° for Venus. When either Mercury or Venus rises before the Sun, it is visible in the eastern sky as a bright "star" and is often called the "morning star." Similarly, when either of these two planets sets after the Sun, it is visible in the western sky and is then called the "evening star." Because these two planets are not always at their greatest elongations, they are often very close in angle to the Sun. This is especially true of Mercury, often making it hard to see from Earth. Venus is often nearly halfway up the sky at sunrise or sunset and therefore quite noticeable during much of its orbit. Because they are so bright and sometimes appear to change color due to the motion of Earth's atmosphere, Venus and Mercury are often mistaken for UFOs. (The same motion of the air causes the road in front of your car to shimmer on a hot day.)

GUIDED DISCOVERY
Astronomy's Foundation Builders

In the two centuries between 1500 and 1700, human understanding of the motion of celestial bodies and the nature of the gravitational force that keeps them in orbit surged forward as never before. Theories related to this subject were developed by brilliant thinkers, whose work established and verified the heliocentric model of the solar system and the role of gravity.

(E. Lessing/Art Resource)

Nicolaus Copernicus (1473–1543) Copernicus, the youngest of four children, was born in Torun, Poland. He pursued his higher education in Italy, where he received a doctorate in canon law and studied medicine. Copernicus developed a heliocentric theory of the known universe and just before his death in 1543 published this work under the title *De Revolutionibus Orbium Coelestium*. His revolutionary theory was flawed in that he assumed that the planets had circular orbits around the Sun. This was corrected by Johannes Kepler.

(Painting by Jean-Leon Huens, courtesy of National Geographic Society)

Tycho Brahe (1546–1601) and **Johannes Kepler (1571–1630)** Tycho (depicted within the portrait of Kepler) was born to nobility in the Danish city of Knudstrup, which is now part of Sweden. At age 20 he lost part of his nose in a duel and wore a metal replacement thereafter. In 1576 the Danish king Frederick II built Tycho an astronomical observatory that Tycho named Uraniborg (after Urania, Greek muse of astronomy). Tycho rejected both Copernicus's heliocentric theory and the Ptolemaic geocentric system. He devised a halfway theory called the *Tychonic system*. According to Tycho's theory, Earth is stationary, with the Sun and Moon revolving around it, while all the other planets revolve around the Sun. Tycho died in 1601.

Kepler was educated in Germany, where he spent 3 years studying mathematics, philosophy, and theology. In 1596, Kepler published a booklet in which he attempted to mathematically predict the planetary orbits. Although his theory was altogether wrong, its boldness and originality attracted the attention of Tycho Brahe, whose staff Kepler joined in 1600. Kepler deduced his three laws from Tycho's observations.

(Art Resource)

Galileo Galilei (1564–1642) Born in Pisa, Italy, Galileo studied medicine and philosophy at the University of Pisa. He abandoned medicine in favor of mathematics. He held the chair of mathematics at the University of Padua, and eventually returned to the University of Pisa as a professor of mathematics. There Galileo formulated his famous law of falling bodies: All objects fall with the same acceleration regardless of their weight. In 1609 he constructed a telescope and made a host of discoveries that contradicted the teachings of Aristotle and the Roman Catholic Church. He summed up his life's work on motion, acceleration, and gravity in the book *Dialogues Concerning the Two Chief World Systems*, published in 1632.

(National Portrait Gallery, London)

Isaac Newton (1642–1727) Newton delighted in constructing mechanical devices, such as sundials, model windmills, a water clock, and a mechanical carriage. He received a bachelor's degree in 1665 from the University of Cambridge. While there, he began developing the mathematics that later became calculus (developed independently by the German Gottfried Leibniz). While pursuing experiments in optics, Newton constructed a reflecting telescope and also discovered that white light is actually a mixture of all colors. His major work on forces and gravitation was the tome *Philosophiae Naturalis Principia Mathematica*, which appeared in 1687. In 1704, Newton published his second great treatise, *Opticks*, in which he described his experiments and theories about light and color. Upon his death in 1727, Newton was buried in Westminster Abbey, the first scientist to be so honored.

Planets farther from the Sun than Earth have different configurations. When one of them is located behind the Sun, as seen from Earth, it is said to be in **conjunction**. When it is opposite the Sun in the sky, the planet is at **opposition**. It is not difficult to determine when a planet happens to be located at one of the key positions in Figure 2-4. For example, when Mars is at opposition, it appears high and bright in the sky at midnight.

Figure 2-4 **Planetary Configurations** Key points along a planet's orbit have names, as shown. These points identify specific geometric arrangements between Earth, another planet, and the Sun. Knowing where a planet is with respect to the Sun helps astronomers know when and where to look for the planet.

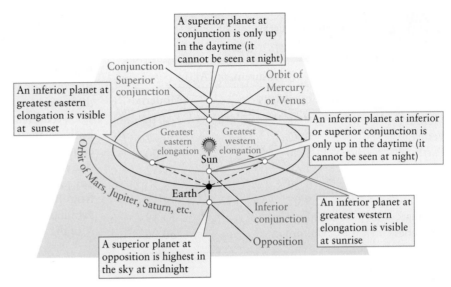

It is relatively easy to follow a planet as it moves from one configuration to another. However, these observations alone do not tell us the planet's actual orbit, because Earth, from which we make the observations, is also moving. Copernicus was therefore careful to distinguish between two characteristic time intervals, or *periods,* of each planet.

Recall from your study of the Moon in Chapter 1 that the sidereal period of orbit is the true orbital period of any astronomical body. A planet's **sidereal period** is the time it takes that body to make one complete orbit of the Sun. The sidereal period is the length of a year for each planet.

The other useful time interval that Copernicus used is the **synodic period.** The synodic period is the time that elapses between two successive identical configurations as seen from Earth. It can be from one opposition to the next, for example, or from one conjunction to the next (Figure 2-5). It tells us, among other things, when to expect a planet to be closest to Earth and, therefore, most easily studied.

INSIGHT INTO SCIENCE

Take a Fresh Look When a scientific concept is hard to visualize, try another perspective. For example, a planet's sidereal period of orbit is easy to understand when viewed from the Sun but more complicated as seen from Earth. The synodic period of each planet, on the other hand, is easily determined from Earth. As we will see, especially when we study Einstein's theories of relativity, each of these perspectives is called a *frame of reference.*

Thus, nearly 500 years ago, Copernicus was able to obtain the first six entries shown in Table 2-1 (the others are contemporary results included for completeness). Copernicus was then able to devise a straightforward geometric method for determining the distances of the planets from the Sun. His answers turned out to be remarkably close to the modern values, as shown in Table 2-2. From these two tables it is apparent that the farther a planet is from the Sun, the longer it takes to complete its orbit.

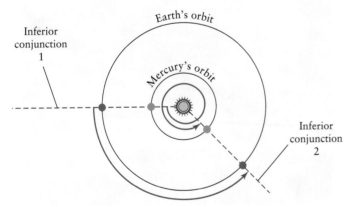

Figure 2-5 **Synodic Period** The time between consecutive conjunctions of Earth and Mercury is 116 days. Typical of synodic periods for all planets, the location of Earth is different at the beginning and end of the period. You can visualize the synodic periods of the exterior planets by putting Earth in Mercury's place in this figure and putting one of the outer planets in Earth's place.

TABLE 2-1	Synodic and Sidereal Periods of the Planets (in Earth Years)	
	Synodic (year)	Sidereal (year)
Mercury	0.318	0.241
Venus	1.599	0.616
Earth	—	1.0
Mars	2.136	1.9
Jupiter	1.092	11.9
Saturn	1.035	29.5
Uranus	1.013	84.0
Neptune	1.008	164.8

TABLE 2-2 Average Distances of the Planets from the Sun		
	Measurement (AU)	
	By Copernicus	Modern
Mercury	0.38	0.39
Venus	0.72	0.72
Earth	1.00	1.00
Mars	1.52	1.52
Jupiter	5.22	5.20
Saturn	9.07	9.54
Uranus	Unknown	19.19
Neptune	Unknown	30.06

Copernicus presented his heliocentric cosmology, including supporting observations and calculations, in a book entitled *De Revolutionibus Orbium Coelestium* (*On the Revolutions of the Celestial Spheres*), which was published in 1543, the year of his death. His great insight was the conceptual simplicity of a heliocentric cosmology compared to geocentric views, especially in explaining retrograde motion. However, Copernicus incorrectly assumed that the planets travel along circular paths around the Sun. Without using epicycles similar to those used in geocentric theory (see Appendix G-1), many of his predictions were no more accurate than those of the earlier theory! As we will see shortly, by changing the shape of the orbits, Kepler was able to do away with epicycles and make even more accurate predictions than either the geocentric or original Copernican theory.

2-4 Tycho Brahe made astronomical observations that disproved ancient ideas about the heavens

In November 1572, a bright star suddenly appeared in the constellation Cassiopeia. At first, it was even brighter than Venus, but then it began to grow dim. After 18 months, it faded from view. Modern astronomers recognize this event as a supernova explosion, the violent death of a certain type of star (see Chapter 10). In the sixteenth century, however, the prevailing opinion was quite different. Teachings dating back to Aristotle and Plato argued that the heavens were permanent and unalterable. From that perspective, the "new star" of 1572 could not really be a star at all, because the heavens do not change. Many astronomers and theologians of the day argued that the sighting must be some sort of bright object quite near Earth, perhaps not much farther away than the clouds overhead. A 25-year-old Danish astronomer named Tycho Brahe (see Guided Discovery: Astronomy's Foundation Builders) realized that straightforward observations might reveal the distance to this object.

Consider what happens when two people look at a nearby object from different places—they see it in different positions relative to the things behind it. Furthermore, their heads face at different angles when looking at it. This variation in angle that occurs when viewing a nearby object from different locations is called **parallax** (Figure 2-6).

Tycho reasoned as follows: If the new star is nearby, its position should shift against the background stars over the course of a night (Figure 2-7a). His careful observations, done in the spirit of the scientific method, failed to disclose any parallax, and so the new star had to be far away, farther from Earth than anyone had imagined (Figure 2-7b). Tycho summarized his findings in a small book, *De Stella Nova* (*On the New Star*), published in 1573.

Figure 2-6 **Parallax** Nearby objects are viewed at different angles from different places. These objects also appear to be in different places with respect to more distant objects when viewed at the same time by observers located at different positions. Both effects are called parallax, and they are used by astronomers, surveyors, and sailors to determine distances. (*Tobi Zausner*)

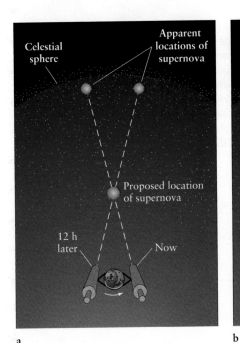
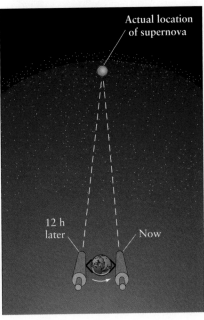

a   b

Figure 2-7 **The Parallax of a Nearby Object in Space** Tycho thought that Earth does not rotate and that the stars revolve around it. From our modern perspective, the changing position of the supernova would be due to Earth's rotation as shown. **(a)** Tycho argued that if an object is near Earth, its position relative to the background stars should change over the course of a night. **(b)** Tycho failed to measure such changes for the supernova in 1572. This is illustrated in (b) by the two telescopes being parallel to each other. He therefore concluded that the object was far from Earth.

Tycho's astronomical records were soon to play an important role in the development of a heliocentric cosmology. From 1576 to 1597, Tycho made comprehensive observations, measuring planetary positions with an accuracy of 1', about as precise as is possible with the naked eye. Upon Tycho's death in 1601, most of these invaluable records were given to his gifted assistant, Johannes Kepler (see Guided Discovery: Astronomy's Foundation Builders).

Describe a simple experiment to demonstrate that your eyes (and, implicitly, your brain) use parallax to determine distances.

KEPLER'S AND NEWTON'S LAWS

Until Kepler's time, astronomers had assumed that heavenly objects move in circles. For philosophical and aesthetic reasons, circles were considered the most perfect and most harmonious of all geometric shapes. However, using circular orbits failed to yield accurate predictions for the positions of the planets. For years, Kepler tried to find a shape for orbits that would fit Tycho's observations of the planets' positions against the background of distant stars. Finally, he began working with a geometric form called an ellipse.

2-5   Kepler's laws describe orbital shapes, changing speeds, and the lengths of planetary years

You can draw an ellipse as shown in Figure 2-8a. Each thumbtack is at a **focus** (plural: **foci**). The longest diameter (major axis) across an ellipse passes through both foci. Half of that distance is called the **semimajor axis.** In astronomy, the length of the semimajor axis is also the average distance between a planet and the Sun.

To Kepler's delight, the ellipse turned out to be the curve for which he had been searching. Predictions of the locations of planets based on elliptical paths were in very close agreement with where the planets actually were. He published this discovery in 1609 in a book known today as *New Astronomy.* This important discovery is now considered the first of **Kepler's laws:**

Kepler's first law: *The orbit of a planet around the Sun is an ellipse, with the Sun at one focus.*

The shapes of ellipses have two extremes. The roundest ellipse, occurring when the two foci merge, is a circle. The most elongated ellipse approaches being a straight line. The shape of a planet's orbit around the Sun is described by its *orbital eccentricity,* designated by the letter *e,* which ranges from 0 (circular orbit) to just under 1.0 (nearly a straight line). Figure 2-8b shows a sequence of ellipses and their associated eccentricities. Observations have revealed that there is no object at the second focus of each elliptical planetary orbit.

Tycho's observations also showed Kepler that planets do not move at uniform speeds along their orbits. Rather, a planet moves most rapidly when it is nearest the Sun, a point on its orbit called **perihelion.** Conversely, a planet moves most slowly when it is farthest from the Sun, called its **aphelion.**

After much trial and error, Kepler discovered a way to describe how fast a planet moves anywhere along its orbit. This discovery, also published in *New Astronomy,* is illustrated in Figure 2-9. Suppose that it takes 30 days for a planet to go from point A to point B. During that time, the line joining the Sun and the planet sweeps out a nearly triangular area

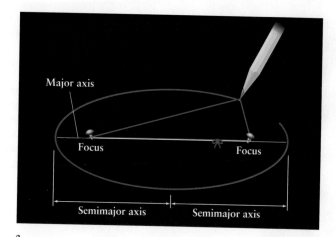

Figure 2-8 **Ellipses** (a) The construction of an ellipse: At all places along an ellipse, the sum of the distances to the two foci is a constant. An ellipse can be drawn with a pencil, a loop of string, and two thumbtacks, as shown. If the string is kept taut, the pencil traces out an ellipse. The two thumbtacks are located at the two foci of the ellipse. (b) A series of ellipses with different eccentricities, e. Eccentricities range between 0 (circle) to just under 1.0 (almost a straight line). Note that all eight planets have eccentricities of less than 0.21.

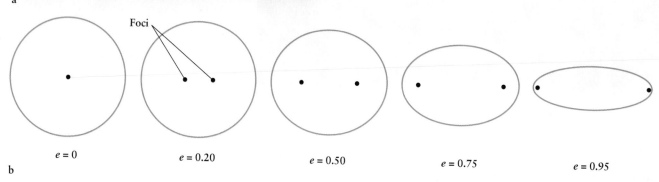

(shaded in Figure 2-9). Kepler discovered that the line joining the Sun and the planet sweeps out the same area during any other 30-day interval. In other words, if the planet also takes 30 days to go from point C to point D, then the two shaded segments in Figure 2-9 are equal in area. Kepler's second law, also called the **law of equal areas,** can be stated thus:

Kepler's second law: *A line joining a planet and the Sun sweeps out equal areas in equal intervals of time.*

A consequence of Kepler's second law is that each planet's speed decreases as it moves from perihelion to aphelion. The speed then increases as the planet moves from aphelion toward perihelion.

Kepler was also able to relate a planet's year to its distance from the Sun. This discovery, published in 1619, is Kepler's third law. It predicts the planet's sidereal period if we know the length of the semimajor axis of the planet's orbit:

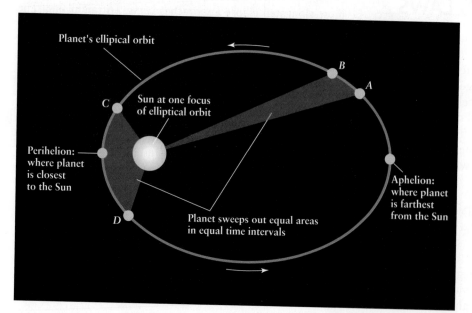

Figure 2-9 **Kepler's First and Second Laws** According to Kepler's first law, every planet travels around the Sun along an elliptical orbit, with the Sun at one focus. According to his second law, the line joining the planet and the Sun sweeps out equal areas (the burgundy-colored regions) in equal intervals of time (time from A to B equals time from C to D). Note: This drawing shows a highly elliptical orbit, with e = 0.74. Even though this is a much greater eccentricity than that of any planet in the solar system, the concept still applies to all planets and other orbiting bodies.

Gravitation and the Motion of the Planets **37**

TABLE 2-3 A Demonstration of Kepler's Third Law				
	Sidereal period P (year)	Semimajor axis a (AU)	P²	= a³
Mercury	0.24	0.39	0.06	0.06
Venus	0.61	0.72	0.37	0.37
Earth	1.00	1.00	1.00	1.00
Mars	1.88	1.52	3.53	3.51
Jupiter	11.86	5.20	140.7	140.6
Saturn	29.46	9.54	867.9	868.3
Uranus	84.01	19.19	7,058	7,067
Neptune	164.79	30.06	27,160	27,160

Kepler's third law: *The square of a planet's sidereal period around the Sun is directly proportional to the cube of the length of its orbit's semimajor axis.*

The relationship is easiest to use if we let P represent the sidereal period in Earth years and a represent the length of the semimajor axis measured in **astronomical units (AU)**. One astronomical unit is the average distance from Earth to the Sun. It is commonly used in discussing distances to various objects in the solar system, because no powers of 10 need be used with it, as they would if distances between planets were referred to in kilometers or miles. (See Appendix H-1 for more details of astronomical distance units.) Now we can write Kepler's third law as

$$P^2 = a^3$$

4 This equation says that a planet closer to the Sun has a shorter year than does a planet farther from the Sun. Using this equation with Kepler's second law reveals that planets closer to the Sun move more rapidly than those farther away. Using data from Tables 2-1 and 2-2, we can demonstrate Kepler's third law as shown in Table 2-3.

When Newton derived Kepler's third law using the law of gravitation, discussed later in this chapter, he discovered that the mass of the planet affects the period of its orbit around the Sun. However, this effect of the planet's mass is vanishingly small for all the planets in the solar system, which is why the equation for Kepler's third law, as shown in Table 2-3, gives such good results for the planets' orbits even though it does not take the masses into account. When calculating the motion of pairs of stars orbiting each other, the effects of the masses must be taken into account.

Kepler's three laws apply not only to the planets orbiting the Sun, but also whenever any object orbits another under the influence of their mutual gravitational attraction. Thus, Kepler's laws apply to moons orbiting planets, artificial satellites orbit-

ing Earth, and (with the above caveat) even two stars revolving around each other.

We saw in Chapter 1 that the Moon's orbit around Earth is not circular. Where in its orbit is the Moon moving fastest, and where is it moving slowest?

2-6 Galileo's discoveries strongly supported a heliocentric cosmology

 While Kepler was in central Europe working on the laws of planetary orbits, an Italian physicist was making dramatic observations in southern Europe. Galileo Galilei did not invent the telescope, but he was one of the first people to point the new device toward the sky and publish his observations. He saw things that no one had ever imagined—mountains on the Moon and spots on the Sun. He also discovered that the apparent size of Venus as seen through his telescope was related to the planet's phase (Figure 2-10). Venus appears smallest at gibbous phase and largest at crescent phase. These observations were a big chink in the geocentric cosmology's armor, as it could not explain why Venus has phases or changes size, but a heliocentric cosmology explains both. Galileo's observations, therefore, supported the conclusion that Venus orbits the Sun, not Earth.

In 1610, Galileo (see Guided Discovery: Astronomy's Foundation Builders) also discovered four moons near Jupiter. Today, in honor of their discoverer, these are called the **Galilean moons** (or **satellites**, another term for moon). Galileo concluded that the moons orbit Jupiter because they move across from one side of the planet to the other. Confirming observations were made in 1620 (Figure 2-11). These observations all provided further proof that Earth is

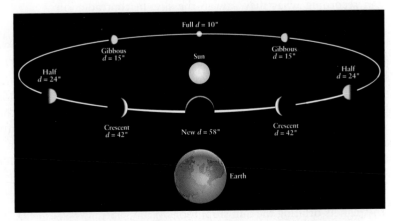

Figure 2-10 The Changing Appearance of Venus This figure shows how the appearance (phase) of Venus changes as it moves along its orbit. The number below each view is the angular diameter (*d*) in arcseconds of the planet as seen from Earth. Note that the phases correlate with the planet's angular size and its angular distance from the Sun, both as seen from Earth. These observations clearly support the idea that Venus orbits the Sun.

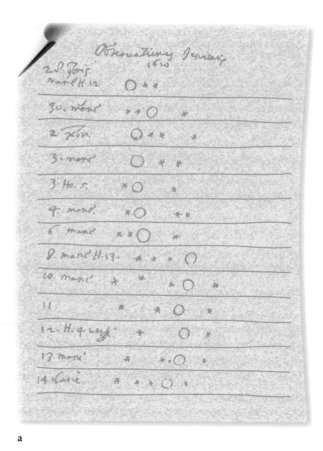

a

b

$\boxed{\text{R}\ \text{I}\ \text{V}\ \text{U}\ \text{X}\ \text{G}}$

Figure 2-11 Jupiter and Its Largest Moons In 1610, Galileo discovered four "stars" that move back and forth across Jupiter. He concluded that they are four moons that orbit Jupiter just as our Moon orbits Earth. (a) Observations made by Jesuits in 1620 of Jupiter and its four visible moons. (b) This photograph, taken by amateur astronomer C. Holmes, shows the four Galilean satellites alongside an overexposed image of Jupiter. Each satellite would be bright enough to be seen with the unaided eye were it not overwhelmed by the glare of Jupiter. *(b: Courtesy of C. Holmes)*

not at the center of the universe. Like Earth in orbit around the Sun, Jupiter's four moons obey Kepler's third law: The square of a moon's orbital period around Jupiter is directly proportional to the cube of its average distance from the planet.

Galileo's telescopic observations constituted the first fundamentally new astronomical data since humans began recording what they saw in the sky. In contradiction to then-prevailing opinions, these discoveries strongly supported a heliocentric view of the universe. Because Galileo's ideas could not be reconciled with certain passages in the Bible or with the writings of Aristotle and Plato, the Roman Catholic Church condemned him, and he was forced to spend his later years under house arrest "for vehement suspicion of heresy." In 1992, Pope John Paul II stated that the church erred in condemning him.

A major stumbling block prevented seventeenth-century thinkers from accepting Kepler's laws and Galileo's conclusions about the heliocentric cosmology. Once anything on Earth is put in motion, it quickly comes to rest. Why don't the planets orbiting the Sun stop, too?

What is the shape of the orbit around Earth of the International Space Station?

The scientific method clarified most of the issues surrounding planetary orbits, leading to the equations and laws developed by the brilliant and eccentric (for example, he believed in alchemy) scientist Isaac Newton, who was born on Christmas Day in 1642, less than a year after Galileo died. In the decades that followed, Newton revolutionized science more profoundly than any person before him, and in doing so, he found physical and mathematical proofs of the heliocentric cosmology.

INSIGHT INTO SCIENCE

Theories and Explanations Scientific theories (or laws) based on observations can be useful for making predictions even if the reasons that these theories work are unknown. The explanation for Kepler's laws came decades after Kepler deduced them, when Newton derived them in 1665 with his mathematical expression for gravitation, the force that holds the planets in their orbits.

2-7   Newton formulated three laws that describe fundamental properties of physical reality

Until the mid-seventeenth century, virtually all mathematical astronomy was done empirically. That is, astronomers from Ptolemy to Kepler created equations directly from data and observations.

WEB LINK 2.6 Isaac Newton (see Guided Discovery: Astronomy's Foundation Builders) introduced a new approach. He began with three physical assumptions, now called **Newton's laws of motion,** which led to equations that have since been tested and shown to be correct in many everyday situations. He also found a formula for the force of **gravity** (or **gravitation**), the attraction between all objects due to their masses. Putting the assumptions into mathematical form and combining them with the equation for gravity, Newton was able to derive Kepler's three laws and use them to predict the orbits of bodies such as comets and other objects in the solar system. Newton was also able to use these same equations to predict the motions of bodies on and near Earth, such as the path of a projectile or the speed of a falling object.

Newton's first law—the law of inertia: *Inertia is the property of matter that keeps an object at rest or moving in a straight line at a constant speed unless acted upon by a net external force.*

5 If all of the external forces acting on an object do not cancel each other out, then there is a *net* external force acting on the object. This is the same as saying that there is an unbalanced external force. Here is an example: If you put a soccer ball between your hands and press on it so that it doesn't move, your hands represent a balanced pair of forces acting on the ball. In that case, you are exerting no net external force on it. Conversely, when your foot hits a soccer ball and the ball sails away, your foot *has* exerted a net external force on the ball.

At first, this law might seem to conflict with your everyday experience. For example, if you shove a chair, it does not move at a constant speed forever but comes to rest after sliding only a short distance. From Newton's viewpoint, however, a net external force does indeed act on the moving chair—namely, friction between the chair's legs and the floor. Without friction, the chair would continue in a straight path at a constant speed. A net external force changes the motion of an object.

Newton's first law tells us that there must be an outside force acting on the planets to continually change their directions and keep them in orbit. If there were no force acting on them, they would move away from the Sun along straight-line paths at constant speeds. Because this does not happen, Newton concluded that some force confines the planets to their elliptical orbits. As we shall see, that force is gravity.

Newton's second law describes quantitatively how a force changes the motion of an object. To better appreciate the concepts of force and motion, we must first understand two related quantities: velocity and acceleration.

Imagine an object motionless in space. Push on it and it begins to move. At any moment, you can describe the object's motion by specifying both its speed and direction. Speed and direction of motion together constitute an object's **velocity.** If you continue to push on the object, its speed will increase—it will accelerate.

Acceleration is the rate at which velocity changes with time. Because velocity involves both speed and direction, a slowing down, a speeding up, or a change in direction are all forms of acceleration.

Suppose an object revolved around the Sun in a perfectly circular orbit. As this object moved along its orbit, its speed would remain constant, but its direction of motion would be continuously changing. This body would have acceleration that involved only a change of direction.

Newton's second law—the force law: *The acceleration of an object is directly proportional to the net force acting on it and is inversely proportional to its mass.*

In other words, the harder you push on something that can move, the faster it will accelerate. Also, an object of greater mass accelerates more slowly when acted on by a force than does an object of lesser mass acted on by the same force. That is why you can accelerate a child's wagon faster than you can accelerate a car by pushing on them equally hard.

Newton's second law can be succinctly stated as an equation. If a **force** acts on an object, the object will experience an acceleration such that

$$\text{Force} = \text{mass} \times \text{acceleration}$$

The **mass** of an object is a measure of the total number 6 of particles that it contains and is expressed in units of kilograms. For example, the mass of the Sun is 2×10^{30} kg, the mass of a hydrogen atom is 1.7×10^{-27} kg, and the mass of the author of this book is 83 kg. At rest, the Sun, a hydrogen atom, and I have these same masses regardless of where we happen to be in the universe. It is important not to confuse the concept of mass with the concept of weight. Your **weight** is the force with which you push down on a scale due to the gravitational attraction of the world on which you stand.

Sitting in a moving car, how can you experimentally verify that your body has inertia?

Force is usually expressed in pounds or newtons. For example, the force with which I am pressing down on the ground is 183 lb. But I weigh 183 lb only on Earth. I would weigh 30.5 lb on the Moon, which has less mass and so pulls me down with less gravitational force. Orbiting in the Space Shuttle, my apparent weight (measured by standing on a scale in the shuttle), would be 0, but my mass would be the same as when I am on Earth. Because I still have inertia in the shuttle, an astronaut there would have to push me with a force to get me to float across the cabin. Whenever we describe the properties of planets, stars, or galaxies, we speak of their masses, never of their weights.

Newton's final assumption, called Newton's third law, is the law of action and reaction.

Newton's third law—the law of action and reaction: *Whenever one object exerts a force on a second object, the second object exerts an equal and opposite force on the first object.*

For example, I weigh 183 lb on Earth, and so I press down on the floor with a force of 183 lb. Newton's third law says that the floor is also pushing up against me with an equal force of 183 lb. (If it were less, I would fall through the floor, and if it were more, I would be lifted upward.) In the same way, Newton realized that because the Sun is exerting a force on each planet to keep it in orbit, each planet must also be exerting an equal and opposite force on the Sun. As each planet accelerates toward the Sun, the Sun in turn accelerates toward each planet.

Because the Sun is pulling on the planets, why don't they fall onto it? Conservation of angular momentum provides the answer. **Angular momentum** is a measure of how much energy is stored in an object due to its rotation and revolution. The details are presented in Appendix H-2. As the orbiting planets fall toward the Sun, their angular momentum provides them with motion perpendicular to that infall, meaning that the planets continually fall toward the Sun, but they continually miss it. Because their angular momentum is conserved, planets neither spiral into the Sun nor fly away from it. Angular momentum remains constant unless acted on by an external torque (also defined in Appendix H-2).

If you are on a freely spinning merry-go-round, what will happen to it as you move toward the center?

---

7   It seems plausible that astronauts in the International Space Station do not feel any force of gravity from Earth, but they do. Orbiting 330 km (approximately 200 mi) above Earth's surface, they feel 90% as much gravitational force from the planet as we do standing on it. They are weightless, however, because as they fall earthward, their angular momentum carries them around the planet at just the right rate to continually miss it.

Angular momentum depends on three things: how fast an object rotates or revolves, how much mass it has, and how spread out that mass is. The greater an object's angular motion or mass, or the more the mass is spread out, the greater its angular momentum. Consider, for example, a twirling ice skater. She rotates with a constant mass, practically free of outside forces. When she wishes to rotate more rapidly, she decreases the spread of her mass distribution by pulling her arms and outstretched leg in closer to her body (Figure 2-12). According to conservation of angular momentum, as the spread of mass decreases, the rotation rate must increase. In astronomy, we encounter many instances of the same law, as giant objects, such as stars, contract.

a   b   R I V U X G

Figure 2-12 **Conservation of Angular Momentum** As this skater brings her arms and outstretched leg in, she must spin faster to conserve her angular momentum. *(Getty Images)*

We have now reconstructed the central relationships between matter and motion. Scientific explanation of the heliocentric cosmology still requires a force to hold the planets in orbit around the Sun and the moons in orbit around the planets. Newton identified that, too.

2-8   Newton's description of gravity accounts for Kepler's laws

Isaac Newton did not invent the idea of gravity. An observant seventeenth-century person would understand that some force pulls things down to the ground. It was Newton, however, who gave us a quantitative description of the action of gravity. Using his first two laws, Newton proved mathematically that the force acting on each of the planets is directed toward the Sun. He expanded this result to the idea that the nature of the force pulling a falling apple straight down to the ground is the same as the nature of the force on the planets from the Sun. More generally, the gravitational force from every object acts to pull every other object directly toward it.

Newton succeeded in formulating a mathematical model that describes the behavior of the gravitational force that keeps the planets in their orbits.

Newton's law of universal gravitation: *Two objects attract each other with a force that is directly proportional to the product of their masses and inversely proportional to the square of the distance between them.*

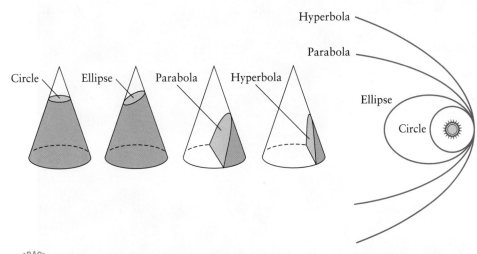

Figure 2-13 **Conic Sections** A conic section is any one of a family of curves obtained by slicing a cone with a plane, as shown. The orbit of one body around another can be an ellipse, a parabola, or a hyperbola. Circular orbits are possible because a circle is just an ellipse for which both foci are at the same point.

In other words, gravitational force decreases with distance: Move twice as far away from an object and you feel only one-quarter of the force from it that you felt before. Despite its weakening, the force of gravity from each object extends throughout the universe. Also, an object that has twice the mass of another object exerts twice the gravitational force as the less massive object.

Using his law of gravity along with his three laws stated earlier, Newton found that he could mathematically explain Kepler's three laws. For example, whereas Kepler discovered by trial and error that the period of orbit, P, and average distance between the Sun and planet, a, are related by $P^2 = a^3$, Newton mathematically derived this equation (corrected by including a tiny contribution due to the mass of the planet, as mentioned earlier). Bodies in elliptical orbits are bound by the force of gravity to remain in orbit.

Newton also discovered that objects can have nonelliptical, unbound orbits. His equations led him to conclude that orbits can also be **parabolas** or **hyperbolas** (Figure 2-13). In both cases, such bodies would make only one pass close to the Sun and then travel out of the solar system, never to return. To date, all of the objects observed in the solar system have begun their existence with elliptical orbit, but some comets (small bodies of rock and ice) have received enough energy from the pull of planets or from jets of gas shooting out of them to develop parabolic or hyperbolic orbits.

Using the equations Newton derived, the orbits of the planets and their satellites could be calculated with unprecedented precision. In the spirit of the scientific method, Newton's laws and mathematical techniques were used to predict new phenomena. For example, Edmond Halley was intrigued by historical records of a comet that was sighted about every 76 years. Using his friend Newton's methods, Halley worked out the details of the comet's orbit and predicted its return in 1758. It was first sighted on Christmas night of that year, and to this day the comet bears Halley's name (Figure 2-14).

Perhaps the most dramatic early use of the scientific method with Newton's ideas was their role in the discovery of the eighth planet in our solar system. The seventh planet, Uranus, had been discovered by William Herschel in 1781 during a systematic telescopic survey of the sky. Fifty years later, however, it was clear that Uranus was not following the orbit predicted by Newton's laws. Two mathematicians, John Couch Adams in England and Urbain-Jean-Joseph Leverrier in France, independently calculated that the deviations of Uranus from its predicted orbit could be explained by the gravitational pull of a then-unknown, more distant planet. Each man predicted that the planet would be found at a certain location in the constellation of Aquarius in September 1846. A telescopic search on September 23, 1846, revealed Neptune less than 1° from its calculated position. Although sighted with a telescope, Neptune was really discovered with pencil and paper.

RIVUXG

Figure 2-14 **Halley's Comet** Halley's Comet orbits the Sun with an average period of about 76 years. During the twentieth century, the comet passed near the Sun twice—once in 1910 and again, as shown here, in 1986. The comet will pass close to the Sun again in 2061. Although dim in 1986, it nevertheless spread more than 5° across the sky, or 10 times the diameter of the Moon. (*Harvard College Observatory/Photo Researchers, Inc.*)

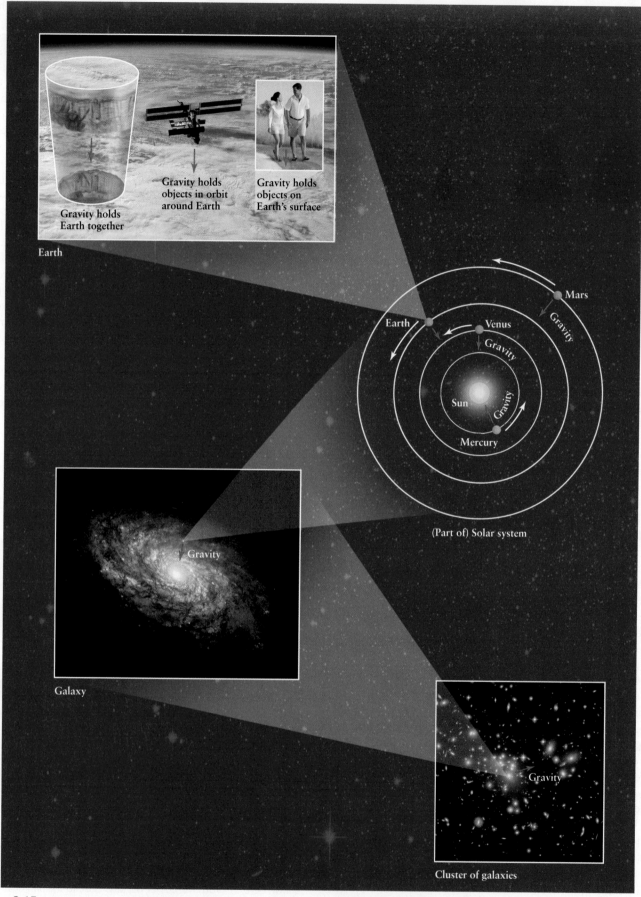

Figure 2-15 **Gravity Works at All Scales** This figure shows a few of the effects of gravity here on Earth, in the solar system, in our Milky Way Galaxy, and beyond. *top: Space station (NASA); couple* *holding hands (Paul Burns/Digital Vision/Getty Images); center: black hole (NASA); bottom: galaxy cluster (ESA, NASA, J.-P. Kneib [Caltech/ Observatoire Midi-Pyrénées] and R. Ellis [Caltech])*

INSIGHT INTO SCIENCE

Quantify Predictions Mathematics provides a language that enables science to make quantitative predictions that can be checked by anyone. For example, in this chapter, we have seen how Kepler's third law and Newton's universal law of gravitation correctly predict the motion of objects under the influence of the Sun's gravitational attraction.

It is a testament to Newton's genius that his three laws were precisely the basic ideas needed to understand so much about the natural world. Newton's process of deriving Kepler's laws and the universal law of gravitation helped secure the scientific method as an invaluable tool in our process of understanding the universe. Figure 2-15 shows some of the effects of gravity at the scales of planets, stars, and galaxies.

THE SOLAR SYSTEM CONTAINS HEAVY ELEMENTS, FORMED FROM AN EARLIER GENERATION OF STARS

How did the solar system form? How were its varied building blocks of rock, metal, ice, and gas created? How has the solar system changed since its formation, and what does its history tell us about the planets we see today? Within the past few decades, telescopes and space probes, along with the theories of modern science, have finally provided some of the answers to these age-old questions. The driving force for the formation was gravity.

2-9   Stars transform matter from lighter elements into heavier ones

Our study of the formation of the solar system begins with an inconsistency, namely that Earth, the Moon, and many other objects that orbit the Sun are composed primarily of heavy elements, such as oxygen, silicon, aluminum, iron, carbon, and calcium. Furthermore, most of these bodies contain extremely little hydrogen and helium. However, observations of the spectra of the Sun, other stars, and interstellar clouds reveal that hydrogen and helium are by far the most abundant elements in the universe. These two elements account for about 99.9% of all observed atoms (or, equivalently, 98% of observed mass). All of the other elements combined account for less than 1% of the observed atoms (and hence 2% of the observed mass of the universe). How is it that Earth, the Moon, Mars, Venus, Mercury, and many smaller bodies in the solar system

typically contain less than 0.15% hydrogen and helium? Somehow, these familiar objects formed from matter that had been enriched with the heavier elements and depleted in hydrogen and helium.

There is a good reason for the overwhelming abundance of hydrogen and helium throughout the universe. Astronomers believe that the universe formed about 13.7 billion years ago with a violent event called the *Big Bang*. Only the lightest elements—hydrogen, helium, and a tiny amount of lithium—were created in the process. The first stars, composed only of these three elements, condensed out of this primeval matter, probably within a few hundred million years after the Big Bang.

The difference in chemical composition between the early universe and the solar system shows that the solar system did not form as a *direct* result of the Big Bang. Let us briefly explore this chemical transformation, which began deep inside the first stars. Once formed, gravity compresses the matter in the central *cores* of stars so much that the hydrogen there is transformed into helium in a process called *fusion*. Hydrogen fusion has the interesting property that some of the hydrogen's mass is converted into electromagnetic energy (photons), a portion of which eventually leaks out through the star's surface, enabling it to shine. The core's helium subsequently fuses to create carbon, also converting some mass into energy. If the star has enough pressure in its core, the carbon transforms into even heavier elements, such as nitrogen, oxygen, neon, silicon, and iron. We explore details of the creation of this heavier matter in Chapters 9 and 10.

Stars also shed matter into space (Figure 2-16). Some of the heavy elements formed inside them are eventually ejected, along with most of the hydrogen and helium remaining in their outer layers. The outer layers of stars are expelled at different rates, ranging from continuous outflows called *stellar winds* (Figure 2-16a), to more energetic bursts called *planetary nebulae*—so named not because they have any direct connections to real planets, but rather because they looked like planets in early, low-resolution telescopes (Figure 2-16b)—to spectacular detonations called *supernovae* (Figure 2-16c). Supernovae are such powerful events that fusion occurs during the explosion, creating even denser elements that are also expelled from the star. Planetary nebulae and supernovae leave only tiny stellar cores, remnants of once bright and hot stars.

Of the matter ejected from stars, some of it leaves as small particles of dust, such as the soot formed in a fire or in diesel exhaust. Over time, enough gas and dust were emitted by enough stars to form interstellar clouds rich in hydrogen and helium, but also containing metals. (In a curious alternative use of a common word, astronomers define **metals** to be all of the elements in the universe other than hydrogen and helium.) It is to these clouds that we turn next in understanding the formation of the solar system and the dense planets like Earth.

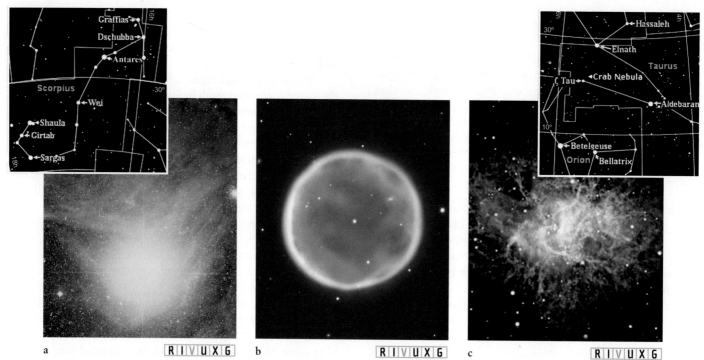

a · RIVUXG   b · RIVUXG   c · RIVUXG

Figure 2-16 **How Stars Lose Mass** (a) The brightest star in Scorpius, Antares, is nearing the end of its existence. Strong winds from its surface are expelling large quantities of gas and dust, creating this nebula reminiscent of an Impressionist painting. The scattering of starlight off this material makes it appear especially bright, even at a distance of 604 light-years. (b) The planetary nebula Abell 39 is 7000 light-years from Earth. In this relatively gentle emission of matter, the central star shed its outer layers of gas and dust in an expanding spherical shell now about 6 light-years across. (c) A supernova is the most powerful known mechanism for a star to shed mass. The Crab Nebula, even though it is about 6000 light-years from Earth, was visible during the day for 3 weeks during 1054. *(a: David Malin/Anglo-Australian Observatory; b: George Jacoby [WIYN Obs.] et al., WIYN, AURA, NOAO, NSF; c: Malin/ Pasachoff/Caltech)*

2-10 Gravity, rotation, and heat shaped the young solar system

The solar system formed from a fragment of a vast cloud of interstellar gas and dust. The region of the cloud from which the solar system developed is called the **solar nebula.** It has a diameter of at least 100 AU and a total mass about 2 to 3 times the mass of the Sun. Matter collapses to form star systems, such as our solar system, for at least four reasons. First, winds from some stars compress nearby gas and dust. Second, the explosive force of a nearby supernova compresses regions of preexisting interstellar clouds. Third, pairs of clouds sometimes collide and compress each other. Fourth, regions of gas and dust radiate away enough heat to cool and condense without outside help. In all of these situations, when a cloud fragment gets sufficiently dense, the mutual gravitational attraction of gas and dust particles in it forces it to collapse, thereby creating new stars and planetary systems (Figure 2-17).

Stellar evolution and star formation are ongoing processes. You can see a cloud in the process of forming stars today with your naked eye, namely the middle "star" in Orion's sword (see Figure 1-4), which looks like a fuzzy blob in the night sky. Called the Orion Nebula or Great Nebula of Orion, it is actually part of a gigantic interstellar cloud region in which new stars and planets are being created.

Which elements on Earth may have been unchanged since the universe began?

Initially, the solar nebula was a very cold collection of gas and dust—well below the freezing point of water. Although most of it was hydrogen and helium, ice and ice-coated dust grains composed of heavy elements were scattered abundantly across this vast volume. Deep inside the nebula, gravitational attraction caused the gas and dust there to fall rapidly toward its center. (In terms of the physics presented earlier, the cloud's gravitational potential energy was being converted into kinetic energy—energy of motion.) As a result, the density, pressure, and temperature at the center of the nebula began to increase, producing a concentration of matter called the **protosun.**

As the protosun continued to increase in mass and to contract, atoms in it collided with one another with increasing speed and frequency. Such collisions created heat, causing the protosun's temperature to soar. Therefore, the first heat and light emitted by our Sun came from colliding gas, not from nuclear fusion, as it does today.

Rotation played a key role in the evolution of the solar system. When it first formed, the solar nebula was a very slowly rotating ensemble of particles. This rotation occurred because the formation of interstellar clouds is tur-

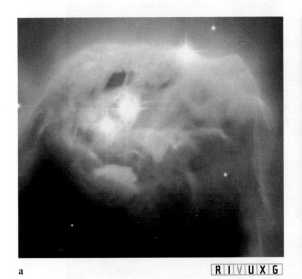

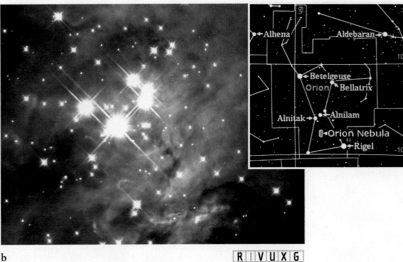

a   R I V U X G   b   R I V U X G

Figure 2-17 **Dusty Regions of Star Formation** (a) These three bright young stars in the constellation Monoceros are still surrounded by much of the gas and dust from which they formed. This is a tiny part of a much larger cloud, known as the Cone Nebula. Astronomers hypothesize that the solar system formed from a similar fragment of an interstellar gas and dust cloud. (b) Newly formed stars in the Orion Nebula. Although visible light from many of the stars is blocked by the nebula, their infrared emission travels through the gas and dust to us. *(a: ACS Science & Engineering Team, NASA; b: NASA; K. L. Luhman/Harvard-Smithsonian Center for Astrophysics and G. Schneider, E. Young, G. Rieke, A. Cotera, H. Chen, M. Rieke, R. Thomson/Steward Observatory, University of Arizona, Tucson)*

bulent, meaning that these clouds are created with many slowly swirling regions, like so much smoke rising from a fire. As a result of this rotation, the nebula had angular momentum (see Appendix H-2), which prevented its outer regions from collapsing into the protosun. A nonswirling collapse would have created a star without any planets or other orbiting matter. Conservation of angular momentum (see Section 2-7) tells us that rotating matter falling inward in the solar nebula revolved faster and faster, like the skater in Figure 2-12 when she pulls her arms in. Even the debris that formed the protosun was orbiting as it fell inward, like water spiraling down a drain. Therefore, the protosun was spinning and the Sun that formed from it should be spinning. It is, as we will see in Chapter 7.

The outer parts of the solar nebula collapsed inward more slowly than the central matter that formed the protosun. This outer region was probably very ragged, but mathematical studies show that the combined effects of gravity, collisions, and rotation transform even an irregular cloud fragment into a rotating disk with a warm center and cold edges (Figure 2-18). You can see an analogous effect of rotation in the way an ice skater's dress is forced into a plane as she spins.

Although we cannot see our solar system as it was before planets formed, astronomers have found what we think are similar disks of gas and dust surrounding other young stars, including those shown in Figure 2-19. Called **protoplanetary disks** or **proplyds,** these systems are undergoing the same initial stage of evolution that we have described here for our solar system.

The protosun's temperature increased as it became denser, and more and more atoms in it collided more often. Radiating more and more heat, the temperature around it began to increase. The increasing temperature vaporized all common icy substances in the inner region of the solar nebula and pushed light gases, like hydrogen and helium, outward. In this inner region, where Earth orbits today, mostly heavier elements remained. These heavy elements eventually came together and formed the four inner planets: Mercury, Venus, Earth, and Mars. The removal of the lighter elements by the protosun from their immediate neighborhood explains why the inner planets are composed mostly of heavy elements. *In other words, we are literally made of stardust!*

2-11 Collisions in the early solar system led to the formation of planets

Although we have used the word *planet* in its everyday usage so far in this book, it is now appropriate to introduce its formal definition. A **planet** is an object that directly orbits a star (rather than orbits another body, like the Moon orbits Earth). A planet also must have enough mass to pull itself into an essentially spherical shape and enough gravitational force so that it has cleared its orbital neighborhood of most of the smaller debris there. If the planet is rotating, its spherical shape will be modified so that it is wider at its equator than at its poles. Bodies that orbit larger objects, which in turn orbit stars, are called **moons** or **natural satellites.**

The formation of the inner planets was the result of innumerable impacts of small, rocky particles in the solar system's protoplanetary disk. Initially, neighboring dust grains and pebbles in the solar nebula collided and stuck together. Then, over a period of a few million years, these accumulations of dust and pebbles coalesced into larger objects, called **planetesimals,** with diameters of a few kilometers or miles (see Figure 2-18).

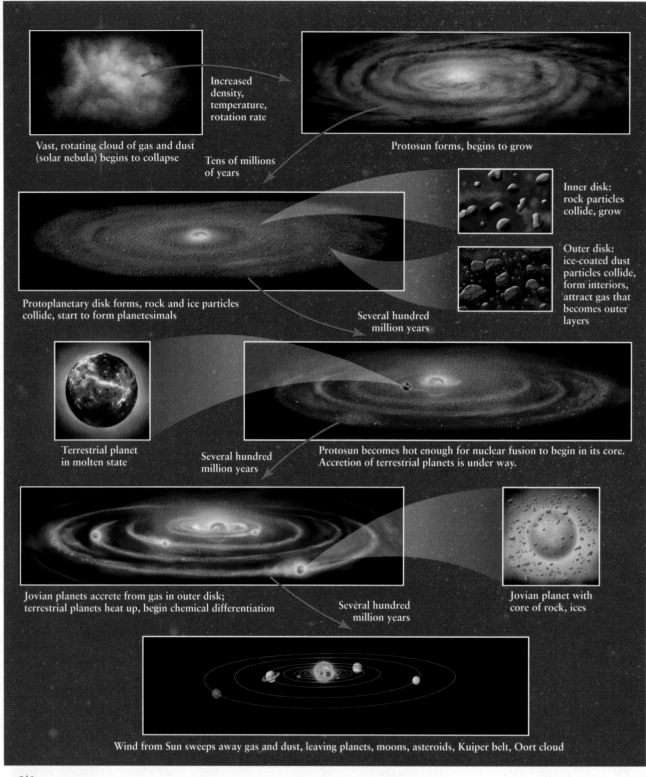

Vast, rotating cloud of gas and dust (solar nebula) begins to collapse

Increased density, temperature, rotation rate

Tens of millions of years

Protosun forms, begins to grow

Inner disk: rock particles collide, grow

Outer disk: ice-coated dust particles collide, form interiors, attract gas that becomes outer layers

Protoplanetary disk forms, rock and ice particles collide, start to form planetesimals

Several hundred million years

Terrestrial planet in molten state

Several hundred million years

Protosun becomes hot enough for nuclear fusion to begin in its core. Accretion of terrestrial planets is under way.

Jovian planets accrete from gas in outer disk; terrestrial planets heat up, begin chemical differentiation

Several hundred million years

Jovian planet with core of rock, ices

Wind from Sun sweeps away gas and dust, leaving planets, moons, asteroids, Kuiper belt, Oort cloud

Figure 2-18 **The Formation of the Solar System** This sequence of drawings shows stages in the formation of the solar system.

What force ensured that once the solar nebula began collapsing, it continued to do so?

The planetesimals collided with anything in their paths. Many were pulverized, many fell into the Sun, but eventu- ally, their mutual gravitational attraction brought a few of them together to form larger objects, called **protoplanets.** Computer simulations of the inner solar system, based on Newton's laws (see Section 2-7), suggest that **accretion**—the coming together of smaller pieces of matter to form larger ones—continued for over 100 million years and should lead

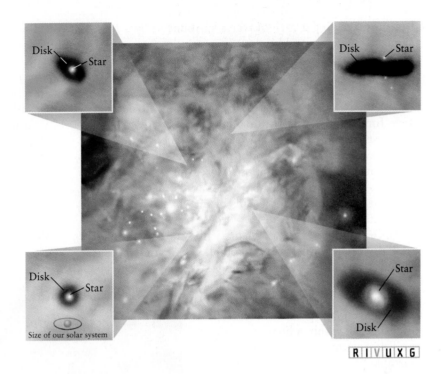

Figure 2-19 **Young Circumstellar Disks of Matter** The heart of the Orion Nebula as seen through the Hubble Space Telescope. The four insets are false-color images of protoplanetary disks within the nebula. A recently formed star is at the center of each disk. The disk in the upper right is seen nearly edge-on. Our solar system is drawn to scale in the lower left image. *(C. R. O'Dell and S. K. Wong, Rice University; NASA)*

to the formation of fewer than half a dozen rocky planets (Figure 2-20). The agreement between predictions of the number of planets and their actual number in our inner solar system (four) is encouraging. Observations made in 2004 of star systems that underwent all stages of planet formation indicate that this entire process, often delayed by collisions between large protoplanets, typically takes several hundred million years to complete.

INSIGHT INTO SCIENCE

Computer-Aided Analysis Many equations are so complex that they require computer analysis to enable us to understand their implications. An example is the physics of the formation of the planets and the Sun. Observations then lead to refined computer models.

The traditional theory of giant planet formation, called the **core-accretion model,** posits that while the inner regions of the solar system were heating up, temperatures in the outer regions of the solar nebula remained quite cool. Ice and ice-coated dust, along with hydrogen and helium gas, were able to survive in these cooler regions. The outer planets—Jupiter, Saturn, Uranus, and Neptune—began to form from the accretion of planetesimals. For each, a rocky core even bigger than Earth apparently served as a "seed," and for about a million years the core became coated with additional rock, gas, and ice. When the gas-rich envelope became as massive as the rocky core, the gas and ice began to accumulate at a runaway pace as the growing planet's gravitational attraction pulled more of it onto the surface. Thereafter, the envelope pulled in most of the gas and ice in

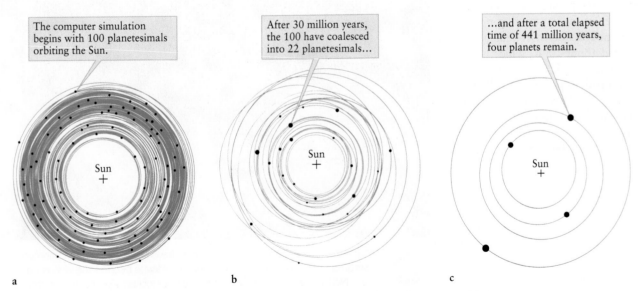

The computer simulation begins with 100 planetesimals orbiting the Sun.

After 30 million years, the 100 have coalesced into 22 planetesimals...

...and after a total elapsed time of 441 million years, four planets remain.

a

b

c

Figure 2-20 **Accretion of the Inner Planets** This computer simulation shows the formation of the inner planets over time.

its vicinity, creating a shell of water surrounded by a huge volume of hydrogen and helium.

The outer planets now contain different amounts of hydrogen, helium, methane, ammonia, and water. Many of their moons are also partially composed of these low-density substances. Discovered in 1930 by American astronomer Clyde Tombaugh, Pluto was originally listed as a planet—a title it lost in 2006. It is now called a *dwarf planet*. The difference between dwarf planets and planets is that dwarf planets do not have enough gravitational force (in other words, mass) to clear their neighborhoods of other orbiting debris. See Appendix F for details of the change in Pluto's classification.

Since the 1980s, a second theory, called the **gravitational instability model** and based on complex computer simulations, has presented a completely different view of the formation of the giant planets. These simulations indicate that the giant planets condensed into existence within a hundred thousand years without massive terrestrial (Earthlike) cores acting as seeds of their formation.

In systems such as our own solar system, it is likely that the core-accretion model is closer to being correct, but as we will see shortly, many stars have giant planets much closer to them than Mercury is to the Sun. For these systems, the core-accretion model works too slowly, whereas the gravitational instability model may allow such planets to form before the gas near the star has evaporated.

During the millions of years that the inner planets were forming, the Sun was also evolving. Throughout this time, the temperature and pressure at the center of the contracting proto-sun continued to increase. Finally, its core reached about 10^7 K, which is hot enough to start hydrogen fusion. This process stopped the protosun's collapse, and the Sun was born. Fusion continues today and provides Earth with the energy necessary to sustain life.

Sunlike stars take approximately 100 million years to form from a nebula and settle down, which means that the Sun probably became a full-fledged star before the accretion of the inner planets was completed. Radiation from the Sun heated and disbursed the remaining gases in the solar system, thereby limiting the sizes of the planets and preventing new planetesimal formation.

Because the planets formed in a disk, they should all orbit in more or less the same plane as Earth (that is, nearly in the plane of the ecliptic). Observations bear out this conclusion. The angles of the orbital planes of the other planets, with respect to the ecliptic, are called their **orbital inclinations,** all of which are 7° or less (Figure 2-21).

Furthermore, according to this model, all of the planets should orbit the Sun in the same "sense" as does Earth. Traveling way above Earth's North Pole and looking down at Earth's orbit, we see that our planet is going counterclockwise around the Sun. All the other planets also orbit counterclockwise as observed from the same vantage point.

To find out more about the solar system's origins, astronomers study the leftover interplanetary debris: asteroids, meteoroids, and comets. Many of these bodies are

9

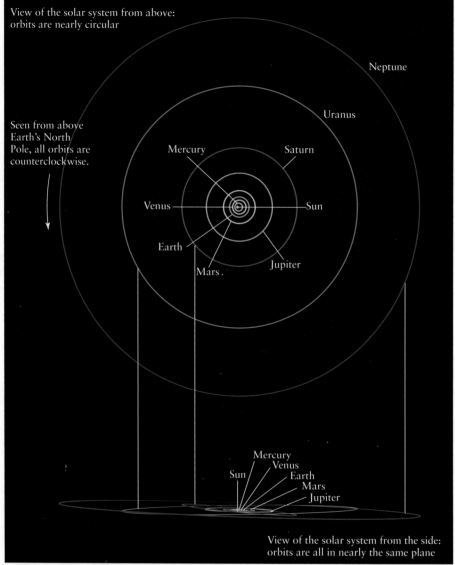

View of the solar system from above: orbits are nearly circular

Seen from above Earth's North Pole, all orbits are counterclockwise.

Neptune

Uranus

Mercury   Saturn

Venus   Sun

Earth

Mars

Jupiter

Mercury
Venus
Sun   Earth
Mars
Jupiter

View of the solar system from the side: orbits are all in nearly the same plane

Figure 2-21 **The Solar System** This scale drawing shows the distribution of planetary orbits around the Sun. All orbits are counterclockwise because the view is from above Earth's North Pole. The four terrestrial planets are located close to the Sun; the four giant planets orbit at much greater distances. Seen from above the disk of the solar system, most of the orbits appear nearly circular. Mercury has the most elliptical orbit of any planet.

R I V U X G

Figure 2-22 **Our Moon** This photograph, taken by astronauts in 1972, shows thousands of craters produced by impacts of leftover rocky debris from the formation of the solar system. Age-dating of lunar rocks brought back by the astronauts indicates that the Moon is about 4.5 billion years old. Most of the lunar craters were formed during the Moon's first 700 million years of existence, when the rate of bombardment was much greater than it is now. *(NASA)*

thought to be virtually unchanged remnants of the solar system's formation. According to radioactive dating of the oldest debris ever discovered from space (Appendix H-3), the solar system was roughly in the form we know today just under 4.6 billion years ago.

Most of the moons in the solar system are planetesimals captured by the gravitational attractions of the planets. The larger ones that orbit Jupiter and Saturn probably formed from disks of gas and dust that orbited those worlds, analogous to the disk that orbited the Sun in which the planets coalesced. One moon, ours, was created as the result of a collision between Earth and a Mars-sized body in a highly elliptical orbit (see Chapter 4).

Impacts dominated the early surface development of the planets and moons for hundreds of millions of years. Young Earth, along with every other body that orbited the Sun, was pounded by countless strikes, creating scars, called **craters.** Although most of the craters on our planet's surface have been erased, as we will study in the next chapter, we can still see them on our Moon (Figure 2-22). Indeed, the Moon's virtually airless environment has preserved important information about the early history of the solar system. (Note that we capitalize the word *moon* only when we are referring to Earth's Moon.)

Radioactive dating of Moon rocks brought back by the Apollo astronauts indicates that the rate of impacts declined dramatically about 3.8 billion years ago. Since that time, impact cratering has proceeded at a very low level. Thus, most of the craters on the Moon and planets were formed

during the first 800 million years of the solar system's history, as the young planets pulled down rocky debris left over from the formative years of the solar nebula.

Conditions developed on the young Earth, especially the presence of liquid water, so that biological evolution could begin to occur there. Strong evidence indicates that liquid water once existed on Mars and may still exist inside it and inside Jupiter's moons Europa and Ganymede, raising the hotly debated question of whether simple life evolved on those worlds, too.

What two elements comprise most of Saturn's outer layer?

2-12 Minor debris from the formation of the solar system still exists

In addition to the eight planets and their moons, many other objects that still orbit the Sun were created when the solar system formed. Between the orbits of Mars and Jupiter are believed to be millions of **asteroids,** bodies composed mostly of rock and metal and that are typically less than a kilometer across. This region is called the **asteroid belt.** The largest asteroid, Ceres, has a diameter of about 900 km. The next largest, Pallas and Vesta, are each about 500 km in diameter. Still smaller ones are increasingly numerous. More than 100,000 kilometer-sized asteroids have been observed. A close-up picture of the asteroid Gaspra is shown in Figure 2-23.

Other asteroids have highly elliptical orbits that take them across the paths of some of the planets. Some asteroids

R I V U X G

Figure 2-23 **An Asteroid** This picture of the asteroid Gaspra was taken in 1991 by the *Galileo* spacecraft on its way to Jupiter. The asteroid measures 12 km × 20 km × 11 km. Millions of similar chunks of rock orbit the Sun between the orbits of Mars and Jupiter. *(NASA)*

have their own moons, such as the asteroid Ida, which is orbited by the heavily cratered Dactyl.

Pieces of rocky and metallic debris even smaller than asteroids are called **meteoroids.** Typical meteoroids are boulder-sized or smaller and apparently exist throughout the disk of the solar system, although most are in the asteroid belt. Chemical studies show that many meteoroids are small fragments of asteroids that broke off when larger bodies collided.

Beyond the orbit of Neptune lie hundreds of billions of chunks of rock and ice, called **comets,** or more accurately, *comet nuclei.* Many of these comet nuclei orbit in a doughnut-shaped region beyond Neptune's orbit, called the *Kuiper belt,* which is centered on the ecliptic. They are called *Kuiper belt objects,* or *KBOs.* Others are believed to be distributed even farther out from the Sun in a spherical distribution, called the *Oort comet cloud.* Many comet nuclei have highly elongated orbits that occasionally bring them close to the Sun. When this happens, the Sun's radiation vaporizes some of their ices, producing the long, flowing tails that we associate with comets (Figure 2-24).

As you are likely to have heard in the media, objects similar to Pluto have been discovered in the Kuiper belt. Like Pluto, they have more elliptical orbits than the planets. It is likely that the KBOs are composed of rock and ice. Collisions probably break up many objects in the Kuiper belt and Oort cloud, thereby creating the comet nuclei and sending some of them into the inner solar system to become comets. Examples of especially large KBOs that have been observed are Pluto, Eris, Quaoar (pronounced *kwa-o-ar*), and Sedna.

R I V U X G

Figure 2-24 **A Comet** The comet nucleus, the solid part of a comet, is an irregularly shaped chunk of ice and rocky debris typically 10 km in diameter. When a comet passes near the Sun, solar radiation vaporizes some of the comet's ices, and the resulting gases and dust form one or two tails millions of kilometers long. *(Dennis diCicco/CORBIS)*

COMPARATIVE PLANETOLOGY

There are two ways of studying objects in the solar system. One is to compare and contrast one feature of all similar objects, such as all of the atmospheres or all of the surfaces of planets, and then move on to the next feature. This has the advantage of showing the "big picture" for each feature as it arises. The other approach is to take an object and explore all of its properties before moving on to the next object. This has the advantage of connecting all of the properties of each object directly to each other, so that you do not, say, connect the atmosphere of Venus with the surface of Mercury. Because both approaches are valuable, we do both.

2-13 Comparisons among the eight planets show distinct similarities and significant differences

The planets that emerged from the accretion of planetesimals in the young solar system were Mercury, Venus, Earth, Mars, Jupiter, Saturn, Uranus, and Neptune. Before exploring the properties of the individual planets in the upcoming chapters, it is instructive to compare a variety of their orbital and physical properties.

Orbits Table 2-4 lists some orbital characteristics of the eight planets. As shown in Figure 2-21, the four inner planets—Mercury, Venus, Earth, and Mars—are crowded close to the Sun. In contrast, the orbits of the four large outer planets—Jupiter, Saturn, Uranus, and Neptune—are widely spaced at greater distances from the Sun. Kepler's laws showed us that all of the planets have elliptical orbits. Even so, most of their orbits are nearly circular. The exception is Mercury, whose orbit is noticeably oval.

How are asteroids related to many of the meteoroids in the solar system?

TABLE 2-4 Orbital Characteristics of the Planets		
	Average distance from Sun		Orbital period
	(AU)	(10⁶ km)	(year)
Mercury	0.39	58	0.24
Venus	0.72	108	0.62
Earth	1.00	150	1.00
Mars	1.52	228	1.88
Jupiter	5.20	778	11.86
Saturn	9.54	1,427	29.46
Uranus	19.19	2,871	84.01
Neptune	30.06	4,497	164.79

The four inner planets are small, have high densities, and are made of rocky materials.

The four outer planets are large, have low densities, and are made primarily of light elements.

R I V U X G

Figure 2-25 **The Sun and the Planets** This figure shows the eight planets drawn to size scale in order of their distances from the Sun (distances not to scale). The four planets that orbit nearest the Sun (Mercury, Venus, Earth, and Mars) are small and made of rock and metal. The next two planets (Jupiter and Saturn) are large and composed primarily of hydrogen and helium. Uranus and Neptune, also mostly hydrogen and helium, are intermediate in size and contain much water. *(Calvin J. Hamilton and NASA/JPL)*

Size The planets fall into three size groups (Figure 2-25). The four inner planets form one group; Jupiter and Saturn form the second group; and Uranus and Neptune are the third, intermediate-sized group. Jupiter is the largest planet, with a diameter about 11 times bigger than that of Earth, whereas Mercury, with about a third of Earth's diameter, is the smallest planet. Indeed, two moons, Ganymede and Titan, are larger than Mercury. Neptune and Uranus are each about 4 times larger in diameter than Earth, the largest of the four inner planets. The diameters of the planets are given in Table 2-5.

Mass Mass, a measure of the total amount of matter an object contains, is another characteristic that distinguishes the inner planets from the outer planets. The four inner planets have small masses compared to the giant outer ones. Again, first place goes to Jupiter, whose mass is 318 times greater than Earth's (see Table 2-5).

Density Size and mass can be combined in a useful way to provide information about the chemical composition of a planet (or any other object). Matter composed of heavy elements, such as iron or lead, has more atoms packed into the same volume than does matter composed of light elements, such as hydrogen, helium, or carbon. Therefore, objects made primarily of heavier elements have a greater average density than objects composed primarily of lighter elements, where average density is given by the equation

$$\text{Average density} = \frac{\text{Total mass}}{\text{Total volume}}$$

The chemical composition (kinds of elements) of any object, therefore, determines its **average density**—how much mass the object has in a unit of volume. For example, a kilogram of copper takes up less space (is denser) than a kilogram of water, even though both have the same mass (Figure 2-26). Average density is expressed in kilograms per cubic meter. To help us grasp this concept, we often compare the average densities of the planets to the density of something familiar, namely liquid water, which has a density of 1000 kg/m³.

TABLE 2-5 Physical Characteristics of the Planets					
	Diameter		Mass		Average density
	(km)	(Earth = 1)	(kg)	(Earth = 1)	(kg/m³)
Mercury	4,878	0.38	3.3×10^{23}	0.06	5,430
Venus	12,100	0.95	4.9×10^{24}	0.81	5,250
Earth	12,756	1.00	6.0×10^{24}	1.00	5,520
Mars	6,786	0.53	6.4×10^{23}	0.11	3,950
Jupiter	142,984	11.21	1.9×10^{27}	317.94	1,330
Saturn	120,536	9.45	5.7×10^{26}	95.18	690
Uranus	51,118	4.01	8.7×10^{25}	14.53	1,290
Neptune	49,528	3.88	1.0×10^{26}	17.14	1,640

R I V U X G

Figure 2-26 **The Volumes of Objects with Different Densities**
All of the objects in this image have the same mass (total number
of particles). However, the chemicals from which they form have
different densities (number of particles per volume), so they all take
up different amounts of space (volume). *(© 2002 Richard Megna/
Fundamentals Photographs, NYC)*

The four inner planets have high average densities compared to water (see Table 2-5). In particular, the average density of Earth is 5520 kg/m³. Because the density of typical
surface rock is only about 3000 kg/m³, its high density implies that Earth contains a large amount of material inside
it, namely iron and nickel, that is much denser than surface
rock. The fact that the inner four planets have densities similar to Earth's (in Latin, *terra*) means that they are composed
of similar chemicals. Consequently, these are called **terrestrial
planets.** We will explore them in Chapter 5.

In sharp contrast, Jupiter, Saturn, Uranus, and Neptune
have relatively low average densities (see Table 2-5). Indeed,
Saturn's average density is less than that of water. Their low
densities support the belief that the giant outer planets contain significant amounts of the lightest elements, hydrogen
and helium, as discussed earlier in this chapter. We will see
that Jupiter and Saturn share similar compositions, primarily hydrogen and helium, while Uranus and Neptune both
contain vast quantities of water as well as much hydrogen
and helium. These four are sometimes called *Jovian planets*
(the Roman god Jupiter was also called Jove), because they
all have relatively low densities and high masses compared
to Earth's. However, the differences in the chemistries of
Jupiter and Saturn compared to those of Uranus and Neptune make the name *giant planets* (giant compared to Earth)
more appropriate in describing these four. We will explore
them as two groups in Chapter 5.

Spectra Further information about the chemical composition of bodies in the solar system is obtained from their
spectra. For solar system objects, this radiation is primarily sunlight scattered off the surface or clouds that surround
each object. The spectra provide us with details of an object's
surface or atmospheric chemical composition. From spectra,
we can confirm that the outer layers of the giant planets are
primarily hydrogen and helium, and that the surface of Mars
is rich in iron oxides, among many other things.

INSIGHT INTO SCIENCE

Astronomical Measurement Astronomers use the laws
of physics to infer things we cannot measure directly.
For example, we can get an overall idea of the chemical
compositions of planets from their masses and volumes.
Kepler's laws enable us to determine the mass of each
planet from the periods of their moons' orbits. The measured diameters of the planets (determined by their distances from Earth and their angular sizes in the sky) yield
their volumes. Dividing total mass by total volume yields
the average density. Comparing this density to the densities of known substances gives us information about the
planets' interior chemistries.

Albedo The surfaces or the upper cloud layers of the various planets scatter (send in all directions) different amounts
of light. The fraction of incoming light returning directly
into space is called a body's **albedo.** An object that scatters no light has an albedo of 0.0; for example, powdered
charcoal has an albedo of nearly 0.0. An object that scatters
all of the light that strikes it (a high-quality mirror comes
close) has an albedo of 1.0. The albedo multiplied by 100
gives the percentage of light directly scattered off that body.
Three planets (Mercury, Earth, and Mars) have albedos of
0.37 or less. Such albedos result from dark, dry surfaces, or
from a mixture of light (water and clouds) and dark (continents) surfaces exposed to space. In contrast, Venus, Jupiter,
Saturn, Uranus, and Neptune have albedos of 0.47 or more.
High albedos like these imply bright materials exposed to
space. The albedos of Venus, Jupiter, Saturn, Uranus, and
Neptune are high because they are all completely enshrouded by clouds.

Why is Earth's albedo continually changing?

Moons Every planet, except Mercury and Venus, has moons. There are at least 166 planetary
moons known to exist in the solar system (up from
99 known in 2001), and more are still being discovered.
Unlike our Moon, most moons are irregularly shaped, more
like potatoes than spheres. We will discover that there is as
much variety among moons as there is among planets.

 10

PLANETS OUTSIDE OUR SOLAR SYSTEM

Understanding the process by which the solar system was formed is being aided by observations of different stages of planet formation occurring around other stars. Although astronomers have been searching for planets outside the solar system for centuries, it was only in the 1980s that observations leading to their discovery were first made.

2-14   Planets that orbit other stars have been discovered

As indicated earlier in the chapter, astronomers have discovered stars in early stages of formation that are surrounded by proplyds: disks of gas and dust. These disks are similar to what the solar nebula was believed to have been like over four and a half billion years ago. Indirect observational evidence for the existence of planets outside the solar system, called *extrasolar planets* or *exoplanets*, first came from the distortion of proplyd disks.

If one or more planets orbit in a proplyd disk around a young star, then the planet's gravitational pull will affect the disk of gas and dust around it, causing the disk to warp or become off-center from its star. An example of serious warping can be seen in the edge-on disks surrounding the star Beta Pictoris (Figure 2-27). This star and the material that orbits it formed only 20 million years ago. In 2006, astronomers resolved (saw details of) the second disk, labeled

in Figure 2-27b. Computer models show that a Jupiterlike planet that orbits out of the plane of the big disk would attract debris from that disk, thereby forming the smaller disk in the plane of its orbit. Beta Pictoris also has millions of comet nuclei, which were discovered in 2001. By studying such systems as Beta Pictoris, we can begin addressing such questions as whether Earth acquired its water from comets in the young solar system.

The interaction between planets and disks is seen in other systems. The nearby star Fomalhaut has an off-centered disk of gas and dust. This star and its entourage are only some 200 million years old. Observations suggest that this distortion (Figure 2-28) is caused by a planet tugging on the gas as it orbits. The pull of the planet is so great that the center of the disk is displaced 15 AU from the star.

Along with disks of gas and dust and large numbers of comets, asteroids have also been discovered orbiting other stars. The first system discovered with asteroids is Zeta Leporis, a young star just 70 light-years from Earth. That asteroid belt, located about the same distance from its star as is our asteroid belt, is estimated to contain about 200 times as much mass as ours. The pieces of asteroid debris were not seen directly but were inferred by the existence of hot (room-temperature) dust surrounding Zeta Leporis that is being generated by the collisions of asteroids.

Observing extrasolar planets directly as they orbit other stars is extremely challenging because even high-albedo planets like Jupiter are typically very dim compared to the bodies that they orbit. Although no planets have yet been so observed, one object labeled 2M1207b *almost* qualified.

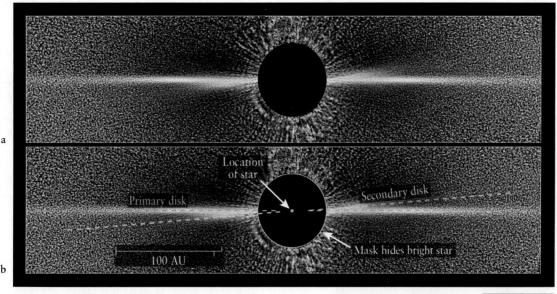

R I V U X G

Figure 2-27 **A Circumstellar Disk of Matter** Hubble visible-light images of two edge-on disks of material orbit the star Beta Pictoris (blocked out in these images) 50 light-years from Earth. The longer one is 225 billion km (140 billion mi) across. These disks are believed to be composed primarily of iceberglike bodies that orbit the star. The smaller, secondary disk is believed to have been formed by the gravitational pull of a roughly Jupiter-mass planet in that orbit. Because the secondary disk is so dim, the labeling for this image is added in (b). *(NASA, ESA, and D. Golimowski)*

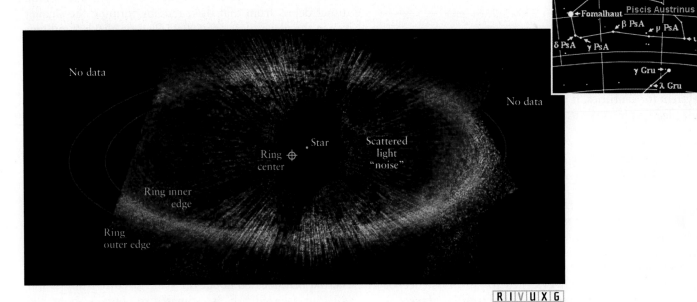

R I V U X G

Figure 2-28 Off-Centered Disk The star Fomalhaut, blocked out so that its light does not obscure the disk, is surrounded by gas and dust in a ring whose center is separated from the star by 15 AU, nearly as far as Uranus is from the Sun. This offset is believed to be due to the gravitational effects of a giant planet orbiting Fomalhaut. This system is 25 light-years from Earth. The debris in that system, and in space between us and it, scatters light that is considered "noise" in such images. *(NASA, ESA, P. Kalas and J. Graham [University of California, Berkeley], and M. Clampin)*

This body (Figure 2-29), first observed in 2004, has a mass between 5 and 8 times the mass of Jupiter. It is orbiting a slightly bigger object, which itself is too small to sustain fusion in its core. Hence, neither object shines very brightly. Observations indicate that both objects were formed together, rather than 2M1207b forming from a disk of gas and dust orbiting the bigger one. Bodies that form simultaneously, even if they have masses too small to be stars, are not considered planets!

Since 1995, most extrasolar planets have been discovered by their effects on the stars they orbit (Figure 2-30a). Some planets are discovered by measuring changes in the velocities of their stars toward or away from us. Radial motion causes the wavelengths of all the light emitted by an object to shift, an effect called the **Doppler shift,** which is calculated from the stars' spectra (Figure 2-30b). This motion toward or away from Earth is created by the gravitational pull of the planet. The Doppler shift changes cyclically as the star and planet circle their common center of mass, and the length of time of one cycle is the period of the planet's orbit. Modern spectroscopic techniques are so good that we can detect stars approaching or receding at speeds as low as 4 km/h (2.5 mph).

Other planets are discovered from variations in a star's proper motion, or motion among the background stars. A planet's attraction causes its star to deviate from motion in a straight line. This *astrometric* method of discovery looks for just such a wobble (Figure 2-30c).

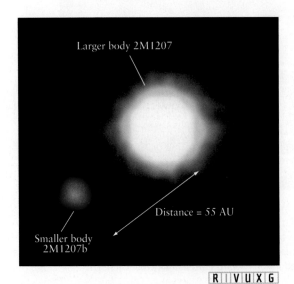

R I V U X G

Figure 2-29 Image of an Almost-Extrasolar Planet This infrared image, taken at the European Southern Observatory, shows the two bodies 2M1207 and 2M1207b. Neither is quite large nor massive enough to be a star, and because evidence suggests that 2M1207b did not form from a disk of gas and dust surrounding the larger body, it is not a planet. This system is about 170 light-years from the solar system in the constellation Hydra. *(ESO/VLT/NACO)*

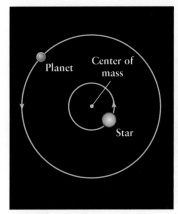
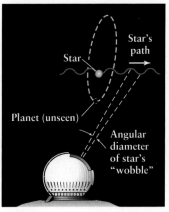
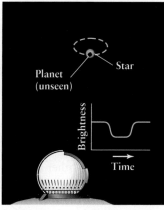

a. A star and its planet **b.** The radial velocity method **c.** The astrometric method **d.** The transient method

 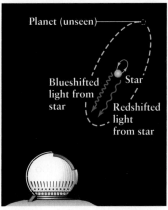

Figure 2-30 Detecting Planets That Orbit Other Stars
(a) A planet and its star both orbit around their common center of mass, always staying on opposite sides of that point. The planet's motion around the center of mass often provides astronomers with the information that a planet is present. (b) As a planet moves toward or away from us, its star moves in the opposite direction. Using spectroscopy, we can measure the Doppler shift of the star's spectrum, which reveals the effects of the unseen planet

or planets. (c) If a star and its planet are moving across the sky, the motion of the planet causes the star to orbit its center of mass. This motion appears as a wobbling of the star across the celestial sphere. (d) If a planet happens to move in a plane that takes it across its star (that is, the planet transits the star), as seen from Earth, then the planet will hide some of the starlight, causing the star to dim. This change in brightness will occur periodically and can reveal the presence of a planet.

Yet other planets are located by observing changes in the various brightnesses of their stars. As a planet passes or transits between us and its star, it partially eclipses the star (Figure 2-30d). In 2003, astronomers used this *transit method* to detect a planet only one-sixteenth as far from its star as Mercury is from the Sun. That planet orbits once every 28 h, 33 min.

Using this method, astronomers can also study the spectrum of a planet's atmosphere. As a gas giant passed in front of the star HD 209458 (Henry Draper catalog star number 209458), hydrogen, sodium, oxygen, and carbon in the planet's atmosphere absorbed certain wavelengths of starlight. The Hubble Space Telescope further revealed that the outer layers of this planet are being heated so much that at least 10,000

tons of hydrogen are evaporating into space from it every second. At this lower limit of mass loss, the planet has lost about 0.1% of its mass over its lifetime of 5 billion years.

The transit method has also revealed that some extrasolar planets have much more high-density material in their cores than do our giant planets. The changing intensity of light from the star HD 149026 showed that the planet passing in front of it, which has a mass of 115 Earth masses, is much smaller than other planets of that mass. Calculations reveal that about 70 Earth masses (roughly two-thirds) of that planet are rocky material in its core. For comparison, Jupiter's terrestrial core is about 4% of its total mass.

As you can see, many giant planets orbit surprisingly close to their stars (Figure 2-31), much closer than Earth

Figure 2-31 Planets and Their Stars This figure shows the separation between extrasolar planets and their stars. The corresponding star names are given on the left of each line. Note that many systems have giant planets that orbit much closer than 1 AU from their stars. (M_J is shorthand for the mass of Jupiter.) For comparison, the solar system is shown at top. *(California and Carnegie Planet Search)*

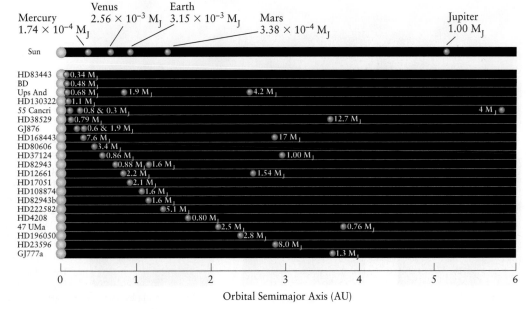

Mercury 1.74×10^{-4} M$_J$ · Venus 2.56×10^{-3} M$_J$ · Earth 3.15×10^{-3} M$_J$ · Mars 3.38×10^{-4} M$_J$ · Jupiter 1.00 M$_J$

Sun

HD83443 0.34 M$_J$
BD 0.48 M$_J$
Ups And 0.68 M$_J$ 1.9 M$_J$ 4.2 M$_J$
HD130322 1.1 M$_J$
55 Cancri 0.8 & 0.3 M$_J$ 4 M$_J$
HD38529 0.79 M$_J$ 12.7 M$_J$
GJ876 0.6 & 1.9 M$_J$
HD168443 7.6 M$_J$ 17 M$_J$
HD80606 3.4 M$_J$
HD37124 0.86 M$_J$ 1.00 M$_J$
HD82943 0.88 M$_J$ 1.6 M$_J$
HD12661 2.2 M$_J$ 1.54 M$_J$
HD17051 2.1 M$_J$
HD108874 1.6 M$_J$
HD82943b 1.6 M$_J$
HD222582 5.1 M$_J$
HD4208 0.80 M$_J$
47 UMa 2.5 M$_J$ 0.76 M$_J$
HD196050 2.8 M$_J$
HD23596 8.0 M$_J$
GJ777a 1.3 M$_J$

0   1   2   3   4   5   6
Orbital Semimajor Axis (AU)

to the Sun. If the core accretion model of how giant planets form is correct, then they are spiraling inward after having formed farther away from their stars than they are now.

In 2004, astronomers began finding extrasolar planets by using a property of space that Albert Einstein discovered. His theory of general relativity, explored in detail in Chapter 10, shows that matter warps the surrounding space, causing, among other things, passing light to change direction. This is quite analogous to how light passing through a lens changes direction and is focused—light can also be focused by gravity, an effect called **microlensing** (or *gravitational microlensing*). Figure 2-32 shows how a star with a planet, passing between Earth and a distant star, can focus the light from the distant star, causing it to appear to change brightness. This occurs twice, once as the distant star's light is focused toward us by the closer star and again when the light is focused toward us by the planet.

The vast majority of planets that have been discovered orbiting stars that shine like the Sun have masses ranging between a half and a few times the mass of Jupiter (see, for example, Figure 2-31). The smallest-mass planet known with confidence has about twice the mass of our Moon. However, it, along with at least two other planets, orbits the remnant of a star that exploded, destroying any life that may have existed on its planets. The two other planets in that system each have about four times Earth's mass.

11 In 2005, microlensing was used to detect a terrestrial planet orbiting a shining star. This planet has five-and-a-half Earth's masses. A similar-mass planet, Gliese 581c, was discovered in 2007. While these two worlds are likely to be composed of rock and metal, with water on their surfaces, neither of the stars they orbit are Sunlike. Both stars are so

cool compared to the Sun that to be inhabitable, the planets would have to be closer to their stars than Mercury is to the Sun. In those situations, the planets would be in synchronous rotation, like the Moon is around Earth. One side of each planet would be permanently frozen, while the water on the other side would be permanently vaporized.

In 1999, astronomers began observing some stars wobbling in ways too complicated to be caused by a single planet, implying the presence of several planets. In a multiple-planet system, each planet contributes a tug on the star. By combining the effects of two or more planets, the observed pattern of the star's motion (Doppler or astrometric) can be reproduced. The first star discovered with multiple planets was Upsilon Andromedae (Figure 2-33). One of its planets, labeled *Upsilon Andromedae B,* is 10 times closer to its star than Mercury is to the Sun. Using the infrared-sensitive Spitzer Space Telescope, astronomers have determined that the difference in temperature between the daytime and nighttime sides of Upsilon Andromedae B is 1400 K. The star 55 Cancri A, in the constellation Cancer, has at least four planets. So far, 20 stars have been observed with two or more planets that orbit them; several are depicted in Figure 2-31.

A dramatic consequence of the inward spiraling of planets was discovered in 2001, when the remnants of at least one planet were discovered in the atmosphere of a star that still has at least two other planets in orbit around it. This star's atmosphere contained a rare form of lithium that is found in planets, but which is destroyed in stars within 30 million years after they form. The presence of this isotope, ^{6}Li, means that at least one planet spiraled so close to the star that it was vaporized.

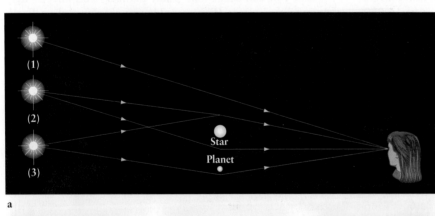

a

(1) No microlensing

(2) Microlensing by star

(3) Microlensing by star and planet

b

Figure 2-32 Microlensing Reveals an Extrasolar Planet (a) Gravitational fields cause light to change direction. As a star with a planet passes between Earth and a more distant star, the light from the distant star is focused toward us, making the distant star appear brighter. (b) The focusing of the distant star's light occurs twice, once by the closer star and once by its planet, making the distant star change brightness. For these observations, the closer star and planet are 17,000 light-years away, while the distant star is 24,000 light-years away.

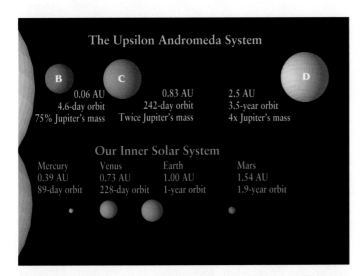

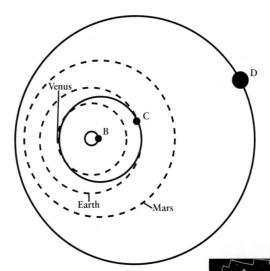

Figure 2-33 **A Star with Three Planets** (a) The star Upsilon Andromedae has at least three planets, discovered by measuring the complex Doppler shift of the star. This star system is located 44 light-years from Earth, and the planets all have masses similar to Jupiter's. (b) The orbital paths of the planets, labeled B, C, and D, along with the orbits of Venus, Earth, and Mars, are drawn for comparison. *(Adam Contos, Harvard-Smithsonian Center for Astrophysics)*

2-15 Dust and extrasolar planets orbit a breathtaking variety of stars

While some planets are orbiting stars similar to the Sun, many planets orbit burned-out remnants of stars and even orbit pairs or trios of stars. Most of the planets we have found are around stars in the disk of the Milky Way Galaxy (which is where our solar system resides), but a few have been found in large clumps of stars outside our Galaxy's disk. These clumps, called *globular clusters,* contain very old stars that formed shortly after the universe began. Because it is in a globular cluster, the oldest known planet dates back to the cluster's formation, 13 billion years ago.

The vast majority of the more than 310 extrasolar planets discovered so far have highly elliptical orbits, with eccentricities up to at least to $e = 0.71$ (see Section 2-5). Star systems with massive planets on highly eccentric orbits are unlikely to have life-sustaining Earthlike planets, because the changing location of the massive planet is likely to prevent smaller planets from staying in stable orbits.

Given the growing number of extrasolar planets that are being discovered orbiting a wide range of stars, astronomers are beginning to make estimates for the number of Jupiterlike planets and the number of them orbiting Sunlike stars in our Milky Way Galaxy. These numbers range between 1 billion and 30 billion planets, of which up to 2 billion are believed to be in orbit around Sunlike stars. All of these numbers will be refined as better estimates are determined for the percentage of star systems with planets.

Although astronomers had expected to find disks of gas and dust orbiting young stars, consistent with the theory explaining the formation of the solar system, they have also begun finding disks of dusty debris orbiting stars as old as the Sun, over 4 billion years. These stars are all known to have planets, and it is most likely that the rubble in them was formed by the collisions of asteroids and comets with each other and perhaps with the planets in these systems.

We should start finding Earthlike planets that orbit Sunlike stars in the coming years. The problem, until recently, has been the sensitivity of our technologies. The difficulty in finding such planets is primarily that they are so dim, small, and low in mass compared to their stars. Their dimness makes it exceedingly hard for current telescopes to see the starlight they scatter compared to the monumental brightness of the stars themselves. When these planets eclipse their stars, the small sizes of the orbiting bodies make the change in the brightness of their star also very hard to detect. Likewise, the small masses of the planets compared to their stars mean that they cause very small Doppler shifts of just a few meters per second, which we are just now able to detect.

When we do start detecting terrestrial planets orbiting other stars, which will be the best candidates for supporting life? We are most likely to discover extraterrestrial life on extrasolar planets orbiting at about 1 AU from stars similar in mass (within about a factor of two) to the Sun. At this distance, such stars provide enough heat to keep water liquid, and these stars shine long enough for life to evolve. We discuss the issues related to other life in the universe in Chapter 13.

2-16 Frontiers yet to be discovered

The science related to forces and orbits described in this chapter was well established by the beginning of the nineteenth century. However, questions remain. Why are the inertial mass (defined in Newton's second law) and the mass used in Newton's law of gravitation identical?

The formations of the solar system and other planetary systems are so complex that we are continually making refinements to our models of how they started and evolved. Our understanding of the development of the solar system is likely to be greatly advanced by further study of proplyds and planets that orbit other stars. For example, observations of these systems should help us pin down the time scales involved in forming planetary systems. One of the global issues in our solar system that remains to be confirmed is how the different chemical elements became distributed at the distances they are found—why are the inner planets in our solar system terrestrial in composition, while planets even closer to other stars are Jupiterlike gas giants? Is the core-accretion model correct and does it universally explain planet formation, or do some systems form along the lines of the gravitational instability model?

Another major challenge in solar system astronomy is identifying and determining the properties of the myriad Kuiper belt and Oort comet cloud objects.

The extrasolar planets and their environs present many mysteries, such as how Jupiter-mass planets came to be closer to their stars than any such bodies in our solar system. How common are Earthlike planets, and around what kinds of stars do they typically form?

We begin a detailed exploration of the solar system in Chapter 4 by examining the two bodies we know best—Earth and its Moon. By understanding them, we will be better able to make some sense of the remarkably alien neighboring worlds we encounter in the following chapters.

SUMMARY OF KEY IDEAS

Science: Key to Comprehending the Cosmos
• The ancient Greeks laid the groundwork for progress in science by stating that the universe is comprehensible.

• The scientific method is a procedure for formulating theories that correctly predict how the universe behaves.

• A scientific theory must be testable, that is, capable of being disproved.

• Theories are tested and verified by observation or experimentation and result in a process that often leads to their refinement or replacement and to the progress of science.

• Observations of the cosmos have led astronomers to discover some fundamental physical laws of the universe.

Origins of a Sun-Centered Universe
• Early Greek astronomers devised a geocentric cosmology, which placed Earth at the center of the universe.

• Copernicus's heliocentric (Sun-centered) theory simplified the general explanation of planetary motions compared to the geocentric theory.

• The heliocentric cosmology refers to motion of planets and smaller debris orbiting the Sun. Other stars do not orbit the Sun.

• The sidereal orbital period of a planet is measured with respect to the stars. It determines the length of the planet's year. Its synodic period is measured with respect to the Sun as seen from the moving Earth (for example, from one opposition to the next).

Kepler's and Newton's Laws
• Ellipses describe the paths of the planets around the Sun much more accurately than do the circles used in previous theories. Kepler's three laws give important details about elliptical orbits.

• The invention of the telescope led Galileo to new discoveries, such as the phases of Venus and the moons of Jupiter, that supported a heliocentric view of the universe.

• Newton based his explanation of the universe on three assumptions, now called Newton's laws of motion. These laws and his law of universal gravitation can be used to deduce Kepler's laws and to describe most planetary motions with extreme accuracy.

• The mass of an object is a measure of the amount of matter in it; weight is a measure of the force with which the gravity of a world pulls on an object's mass when the two objects are at rest with respect to each other (or, equivalently, how much the object pushes down on a scale).

• The path of one astronomical object around another, such as that of a comet around the Sun, is an ellipse, a parabola, or a hyperbola. Ellipses are bound orbits, while objects with parabolic and hyperbolic orbits fly away, never to return.

Formation of the Solar System
• Hydrogen, helium, and traces of lithium, the three lightest elements, were formed shortly after the creation of the universe. The heavier elements were produced much later by stars and are cast into space when stars die. By mass, 98% of the observed matter in the universe is hydrogen and helium.

• The solar system formed 4.6 billion years ago from a swirling, disk-shaped cloud of gas, ice, and dust, called the solar nebula.

• The four inner planets formed through the accretion of dust particles into planetesimals and then into larger protoplanets. The four outer planets probably formed through the runaway accretion of gas and ice onto rocky protoplanetary cores over millions of years, but possibly by gravitational collapse in under 100,000 years.

• The Sun formed at the center of the solar nebula. After about 100 million years, the temperature at the protosun's center was large enough to ignite thermonuclear fusion reactions.

• For 800 million years after the Sun formed, impacts of asteroidlike objects on the young planets dominated the history of the solar system.

Comparative Planetology

• The four inner planets of the solar system share many characteristics and are distinctly different from the four giant outer planets.

• The four inner, terrestrial planets are relatively small, have high average densities, and are composed primarily of rock and metal.

• Jupiter and Saturn have large diameters and low densities and are composed primarily of hydrogen and helium. Uranus and Neptune have large quantities of water as well as much hydrogen and helium.

• Pluto, once considered the smallest planet, has a size, density, and composition consistent with other Kuiper belt objects.

• Asteroids are rocky and metallic debris in the solar system, are larger than about a kilometer in diameter, and are found primarily between the orbits of Mars and Jupiter. Meteoroids are smaller pieces of such debris. Comets are debris that contains both ice and rock.

Planets Outside Our Solar System

• Astronomers have observed disks of gas and dust orbiting young stars.

• At least 310 extrasolar planets have been discovered orbiting other stars.

• Most of the extrasolar planets that have been discovered have masses roughly the mass of Jupiter.

• Extrasolar planets are discovered indirectly as a result of their effects on the stars they orbit.

WHAT DID YOU THINK?

1 *What makes a theory scientific?* A theory is an idea or set of ideas proposed to explain something about the natural world. A theory is scientific if it makes predictions that can be objectively tested and potentially disproved.

2 *What is the shape of Earth's orbit around the Sun?* All planets have elliptical orbits around the Sun.

3 *Do the planets orbit the Sun at constant speeds?* No. The closer a planet is to the Sun in its elliptical orbit, the faster it is moving. The planet moves fastest at perihelion and slowest at aphelion.

4 *Do all of the planets orbit the Sun at the same speed?* No. A planet's speed depends on its average distance from the Sun. The closest planet moves fastest, the most distant planet moves slowest.

5 *How much force does it take to keep an object moving in a straight line at a constant speed?* Unless an object is subject to an outside force, such as friction, it takes no force at all to keep it moving in a straight line at a constant speed.

6 *How does an object's mass differ when measured on Earth and on the Moon?* Assuming the object doesn't shed or collect pieces, its mass remains constant whether on Earth or on the Moon. Its weight, however, is less on the Moon.

7 *Do astronauts orbiting Earth feel the force of gravity from our planet?* Yes. They are continually pulled earthward by gravity, but they continually miss it because of their motion around it. Because they are continually in free-fall, they feel weightless.

8 *Were the Sun and planets among the first generation of objects created in the universe?* No. All matter and energy were created by the Big Bang. However, much of the material that exists in our solar system was processed inside stars that evolved before the solar system existed. The solar system formed billions of years after the Big Bang occurred.

9 *How long has Earth existed, and how do we know this?* Earth formed along with the rest of the solar system about 4.6 billion years ago. The age is determined from the amount of radioactive decay that has occurred on Earth.

10 *What typical shape(s) do moons have, and why?* Although some moons are spherical, most look roughly like potatoes. Those that are spherical are held together by the force of gravity, pulling down high regions. Those that are potato-shaped are held together by the electromagnetic interaction between atoms, just like rocks. These latter are too small to be reshaped by gravity.

11 *Have any Earthlike planets been discovered orbiting Sunlike stars?* Not really. Most extrasolar planets are Jupiterlike gas giants. The planets similar in mass and size to Earth are either orbiting remnants of stars that exploded or, in the case of Gliese 581, a star much less massive and much cooler than the Sun.

Review Questions

1. Who wrote down the equation for the law of gravitation? **a.** Copernicus, **b.** Brahe, **c.** Newton, **d.** Galileo, **e.** Kepler

2. Which of the following most accurately describes the shape of Earth's orbit around the Sun? **a.** circle, **b.** ellipse, **c.** parabola, **d.** hyperbola, **e.** square

3. Of the following planets, which takes the longest time to orbit the Sun? **a.** Earth, **b.** Uranus, **c.** Mercury, **d.** Jupiter, **e.** Venus

4. What is a Sun-centered model of the solar system called?

5. How long does it take Earth to complete a sidereal orbit of the Sun?

6. How did Copernicus explain the retrograde motions of the planets?

7. Which planets can never be seen at opposition? Which planets never pass through inferior conjunction?

8. At what configuration (superior conjunction, greatest eastern elongation, etc.) would it be best to observe Mercury or Venus with an Earth-based telescope? At what configuration would it be best to observe Mars, Jupiter, or Saturn? Explain your answers.

9. What are the synodic and sidereal periods of a planet?

10. What are Kepler's three laws? Why are they important?

11. In what ways did the astronomical observations of Galileo support a heliocentric cosmology?

12. How did Newton's approach to understanding planetary motions differ from that of his predecessors?

13. What is the difference between mass and weight?

14. Why was the discovery of Neptune a major confirmation of Newton's universal law of gravitation?

15. Why does an astronaut have to exert a force on a weightless object to move it?

16. Which of the following times is closest to the age of the universe when our solar system formed? **a.** 0 years (they formed together), **b.** a million years, **c.** 10 million years, **d.** a billion years, **e.** 10 billion years

17. Pluto is most similar in composition to which of the following objects? **a.** Eris, **b.** Jupiter, **c.** our Moon, **d.** Earth, **e.** the Sun

18. Describe four methods for discovering extrasolar planets.

Observing Projects

19. Use your *Starry Night Enthusiast*™ software to observe the phases of Venus. Select **Favourites > Guides > Atlas** from the menu. Open the **Find** pane and double-click on the entry for *Venus*.

Close the **Find** pane and use the **Zoom** buttons in the toolbar to adjust the field of view to about 7 arcmin (7') wide. **a.** Draw the current shape (phase) of Venus. Set the **Time Flow Rate** to *30 days* and single step forward in time ▶, drawing Venus to scale at each step. Try to be accurate in both size and shape. Make a total of 20 timesteps and drawings. **b.** From your drawings, determine when the planet is nearer or farther from Earth than is the Sun. **c.** Deduce from your drawings when Venus is coming toward us or is moving away from us. **d.** Explain why Venus goes through this particular cycle of phases.

20. Using the newly invented telescope in 1610, Galileo observed four moons orbiting Jupiter and this convinced him that not all objects in the universe followed orbits around Earth. Newton's theory of gravity explained this observation by assuming correctly that all objects exert forces on other objects. In Jupiter's case, this meant that its gravity causes the moons to move in orbits around it. You can observe these moons and discover some of the properties of their motions around Jupiter. **a.** Open *World Wide Telescope*, click on the **View** tab and then the **Pause** button in the **Observing Time** pane to stop time. In the **Observing Location** pane, remove the checkmark from the checkbox adjacent to **View from this location**. This will prevent the horizon from interfering with your observations. Open **Explore**, click on the **Solar System** thumbnail to display the options and click on **Jupiter** to center and lock the view on this planet. Open the **Settings** menu and, in the **Solar System Options** pane, click the **Show Small Moons as Points** so that this option is **On**. This will ensure that Jupiter's moons can be seen easily. **Zoom** out from this magnified view until Jupiter's diameter is about 1/10 of the vertical extent of the screen. You should now see several of Jupiter's moons against the star background. In the **View** tab, set the **Time Rate** to **×10000** and watch these moons orbit Jupiter as the planet moves across your sky against the star background. You can stop **Time** and identify any one of the moons with a circle on the screen by moving the cursor over the appropriate thumbnail of the selected moon. Which of these four Galilean moons appears to move furthest from Jupiter in its orbit? Which moon remains closest to the planet? **b.** You can watch the motions of Jupiter's moons without the confusion produced by the background stars by replacing them by a black background. Open the drop-down list under **Imagery** and click on **Black Sky Background**. To watch several specific events, double-click on the Jupiter thumbnail to center and lock it in the center of the view. Click the **View** tab and then the **Pause** button to stop time flow. In the **Observing Time** pane, click the UTC checkbox to turn this option **on**, adjust the **Date** to 2008/06/22 and the **Time** to 23:50:00 and then click the **OK** button. Zoom out so that all four of the Galilean moons are visible in the view. Hover the mouse cursor over the thumbnails of the moons at the bottom of the screen to verify that Io and Callisto appear to the right (west) of Jupiter and Ganymede and Europa appear to the east of the planet. Zoom in to a field of view of 2 arcminutes (the display above the constellation overview thumbnail at the bottom right of the screen should read roughly 00:02:00). Run time forward at **×1000** and watch Io, Europa and Ganymede as they approach Jupiter. Why do you think that Io disappears before it reaches the limb of the planet? What do you think are the moving dark circles on Jupiter that precede Ganymede and Europa as these moons cross in front of the planet? From which direction is the Sun shining upon Jupiter at this time?

RIVUXG

Three of the Four Buildings Containing the Very Large Telescope at the Paranal Observatory, Atacama, Chile *(ESO)*

WHAT DO YOU THINK?

1 What is light?

2 Which type of electromagnetic radiation is most dangerous to life?

3 What is the main purpose of a telescope?

4 Why do all research telescopes use mirrors, rather than lenses, to collect light?

5 Why do stars twinkle?

6 Which is hotter, a "red-hot" or a "blue-hot" object?

7 What color does the Sun emit most brightly?

Answers to these questions appear in the text beside the corresponding question numbers in the margins and at the end of the chapter.

With our eyes alone we can see visible light from several thousand stars. Until the seventeenth century, few people even dreamed that there were more of them. It was then that telescopes revolutionized human understanding of the universe, showing for the first time how

little of the cosmos we normally see. The process of discovery continues, as telescopes reveal new things in space nearly every day. We can think of the radiation emitted by objects out there as the medium of natural cosmic communication, and telescopes as the means by which we gather and read those cosmic messages.

We have also discovered that visible light is only a tiny fraction of the energy emitted by objects in space. Indeed, such things as interstellar clouds of gas and dust, the bodies lying behind those clouds, newly forming stars, intergalactic gas clouds, and a variety of exotic objects (such as black holes and neutron stars) are nearly invisible to even our best optical telescopes. However, these things strongly emit a variety of nonvisible radiations (namely radio waves, microwaves, infrared and ultraviolet radiations, X rays, and gamma rays) that we now have the technology to detect. Today, we use telescopes on the ground, floating in the air, orbiting Earth, or traveling elsewhere in the solar system to see these myriad "stealth" objects in space.

In this chapter you will discover

• the connection between visible light, radio waves, and other types of electromagnetic radiation

• the debate over what light is and how Einstein resolved it

• how telescopes collect and focus light

• why different types of telescopes are used for different types of research

• the limitations of telescopes, especially those that use lenses to collect light

• what the new generations of land-based and space-based high-technology telescopes being developed can do

• how astronomers use the entire spectrum of electromagnetic radiation to observe the stars and other astronomical objects and events

• the origins of electromagnetic radiation

• the structure of atoms

• that stars with different surface temperatures emit different intensities of electromagnetic radiation

• that astronomers can determine the chemical compositions of stars and interstellar clouds by studying the wavelengths of electromagnetic radiation that they absorb or emit

• how to tell whether an object in space is moving toward or away from Earth

THE NATURE OF LIGHT

So far in this text we have used the word *light* in its everyday sense—the stuff to which our eyes are sensitive. This is more properly called *visible light,* and it is a form of **electromagnetic radiation,** which has properties of both particles, called **photons,** and waves. Detecting electromagnetic radiation using telescopes is the essence of observational astronomy. Although human perception of objects here on Earth and in space comes primarily from the visible light that our eyes detect, this is only a tiny fraction of all the electromagnetic radiation emitted by objects in the universe. The rest of this radiation is invisible to our eyes, so it has to be observed through specially designed telescopes. In this chapter, we examine telescopes for all types of electromagnetic radiation. To understand how telescopes work, we begin by exploring the properties of the electromagnetic radiation that they collect.

3-1   Newton discovered that white light is not a fundamental color and proposed that light is composed of particles

From the time of Aristotle until the late seventeenth century, most people believed that white is the fundamental color of light. The colors of the rainbow (or, equivalently, the colors created by light passing through a prism) were believed to be added or created somehow as white light went from one medium through another. Isaac Newton performed experiments during the late 1600s that disproved these beliefs. He started by passing a beam of sunlight through a glass prism, which spread the light out into the colors of the rainbow (Figure 3-1a). The change in direction as light travels from one medium into another is called **refraction,** and the resulting spread of colors (complete or with colors missing) is called a **spectrum** (plural, **spectra**).

Newton then selected a single color and sent it through a second prism (Figure 3-1b). The light that emerged from the first prism was refracted by the second prism, but *it remained the same color.* The fact that individual colors of refracted light were unchanged by the second prism led Newton to conclude that the colors of a full spectrum (shades of red, orange, yellow, green, blue, and violet) were in fact properties of the light itself, and that white light is a mixture of colors. To prove this last point, he recombined all the spectrum colors, thereby recreating white light. Thus, different colors of the spectrum are different entities. But, Newton wondered, what was the nature of light that it could break apart into distinct colors and then be completely reconstituted?

Back in the mid-1600s, the Dutch scientist Christiaan Huygens proposed that light travels in the form of waves. Newton, on the other hand, performed many experiments in optics that convinced him that light is composed of tiny particles of energy. It turns out that both ideas were right.

In 1801, the English physicist Thomas Young demonstrated that light is indeed composed of waves. Young sent light of a single color through two parallel slits (Figure 3-2a). He reasoned that if the light were waves, then these waves would behave the way waves on the surface of water behave, flowing through similar gaps (Figure 3-2b). In particular, the light waves from each slit would interact with light waves from the other. For example, when two light waves meet, where one wave is going up and the other is going down, they would interfere with each other and partially or totally cancel each other out, leaving dark regions on the screen. When two waves that are both moving up or both moving down meet, they would reinforce each other and create bright regions. As you can see in Figure 3-2, this is precisely what happened. (If light were just random particles going through the slits, they would not interfere with each other in this way, and the pattern on the screen would just yield two bright regions, one behind each slit.) The analogy with water waves ends here. For example, light waves do not have to travel through a medium, unlike water waves, which travel on water. Further insight into the wave character of light came from calculations by the Scottish physicist James Clerk Maxwell in the 1860s. Maxwell unified the descriptions of the basic properties of electricity and magnetism into four equations. By combining these equations, he demonstrated that electric and magnetic effects should travel through space together in the form of coupled waves (Figure 3-3) that have equal amplitudes. Maxwell's suggestion that some of these waves, now called electromagnetic radiation, are observed as light was soon confirmed by a variety of experiments. Despite the name electromagnetic radiation, visible light is not electrically charged.

Newton showed that sunlight is composed of all the colors of the rainbow. Young, Maxwell, and others showed that light travels as waves. What makes the colors of the rainbow distinct from each other? The answer is surprisingly simple: Different colors are waves with different wavelengths. A **wavelength,** usually designated by λ, the lowercase Greek letter lambda, is the distance between two successive wave crests.

The wavelengths of all colors are extremely small, less than a thousandth of a millimeter. To express these tiny distances conveniently, scientists use a unit of length called the nanometer (nm), where $1 \text{ nm} = 10^{-9}$ m. Another unit you might encounter when talking to astronomers is the angstrom, Å, where $1 \text{ Å} = 0.1 \text{ nm} = 10^{-10}$ m. Experiments demonstrate that visible light has wavelengths ranging from about 400 nm for the shortest wavelength of violet light to about 700 nm for the longest wavelength of red light. Intermediate colors of the rainbow fall between these wavelengths (see Figure 3-1a). The complete spectrum of colors from the longest wavelength to the shortest is red, orange, yellow, green, blue, and violet. Referring back to Figure 3-1, you can also see that the amount of refraction that different colors undergo depends on their wavelengths: The shorter the wavelength, the more the light is refracted.

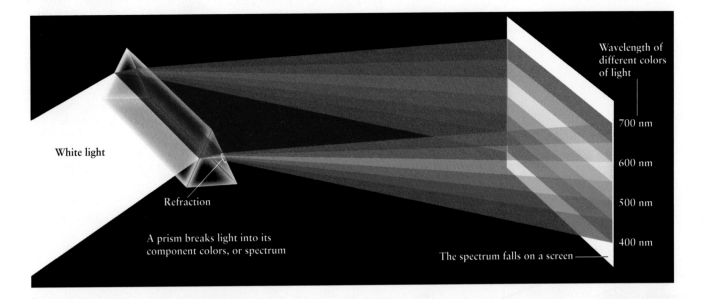

a

b

Figure 3-1 **Prisms and a Spectrum** (a) When a beam of white light passes through a glass prism, the light is separated or refracted into a rainbow-colored band called a spectrum. The numbers on the right side of the spectrum indicate wavelengths in nanometers (1 nm = 10^{-9} m). (b) Drawing of Newton's experiment showing that glass does not add to the color of light, but only changes its direction. Because color is not added, this experiment shows that color is an intrinsic property of light.

3-2 Light travels at a finite, but incredibly fast, speed

The fact that we see lightning before we hear the accompanying thunderclap tells us that light travels faster than sound. But does that mean that light travels instantaneously from one place to another, or does it move with a measurable speed?

The first evidence for the finite speed of light came in 1675, when Ole Rømer, a Danish astronomer, carefully timed eclipses of Jupiter's moons (Figure 3-4). Rømer discovered that the moment at which a moon enters Jupiter's shadow depends on the distance between Earth and Jupiter. When Jupiter is in opposition—that is, when Jupiter and Earth are on the same side of the Sun—the Earth–Jupiter distance is relatively short compared to when Jupiter is near conjunction. At opposition, Rømer found that eclipses occur slightly earlier than predicted by Kepler's laws, and they occur slightly later than predicted when Jupiter is near conjunction.

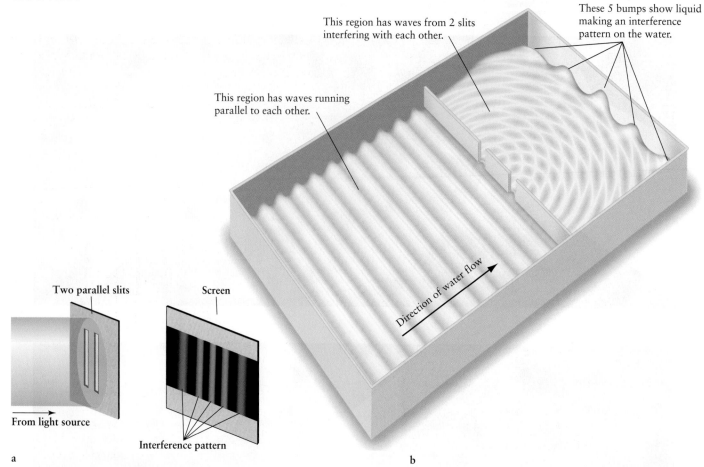

This region has waves from 2 slits interfering with each other.

These 5 bumps show liquid making an interference pattern on the water.

This region has waves running parallel to each other.

Direction of water flow

Two parallel slits

Screen

From light source

Interference pattern

a

b

Figure 3-2 Wave Travel (a) Electromagnetic radiation travels as waves. Thomas Young's interference experiment shows that light of a single color passing through a barrier with two slits behaves as waves that create alternating light and dark patterns on a screen. (b) Water waves passing through two slits in a ripple tank create interference patterns. As with light, the water waves interfere with each other, creating constructive interference (crests) and destructive interference (troughs) throughout the right side of the tank and on the far right wall. *(University of Colorado, Center for Integrated Plasma Studies, Boulder, CO)*

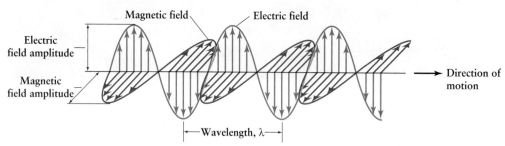

Magnetic field

Electric field

Electric field amplitude

Magnetic field amplitude

Direction of motion

Wavelength, λ

Figure 3-3 Electromagnetic Radiation All forms of electromagnetic radiation (radio waves, microwaves, infrared radiation, visible light, ultraviolet radiation, X rays, and gamma rays) consist of electric and magnetic fields oscillating perpendicular to each other and to the direction they move. In empty space, this radiation travels at a speed of 3×10^5 km/s. These fields are the mathematical description of the electric and magnetic effects. The distance between two successive crests, denoted by λ, is called the wavelength of the light.

Rømer correctly concluded that light travels at a finite speed, and so it takes more time to travel longer distances across space. The greater the distance to Jupiter, the longer the image of an eclipse takes to reach our eyes. From his timing measurements, Rømer concluded that it takes 162 min for visible light to traverse the diameter of Earth's or-

bit (2 AU). Incidentally, Rømer's interpretation of the data requires a heliocentric cosmology—that both Earth and Jupiter orbit the Sun.

What color is refracted least?

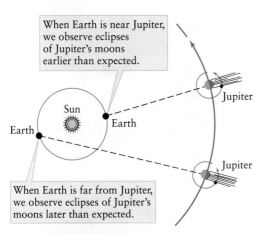

When Earth is near Jupiter, we observe eclipses of Jupiter's moons earlier than expected.

When Earth is far from Jupiter, we observe eclipses of Jupiter's moons later than expected.

Figure 3-4 **Evidence That Light Travels at a Finite Speed** The times of the eclipses of Jupiter's moons as seen from Earth depend on the relative positions of Jupiter, Earth, and the Sun. Rømer correctly attributed the variations in these times to the variations in the time that it takes light from these events to reach Earth.

Rømer's subsequent calculation of the speed of light was off by 25%, because the value for the astronomical unit (the average distance from Earth to the Sun) that existed at that time was highly inaccurate. Nevertheless, he proved his main point—light travels at a finite speed. The first accurate laboratory measurements of the speed of visible light were performed in the mid-1800s.

Maxwell's equations also reveal that light of all wavelengths travels at the same speed in a vacuum (a region that contains no matter), and, despite a few atoms per cubic meter, the space between planets and stars is a very good vacuum. The constant speed of light in a vacuum, usually designated by the letter c, has been measured to be 299,792.458 km/s, which we generally round to

$$c = 3.0 \times 10^5 \text{ km/s} = 1.86 \times 10^5 \text{ mi/s}$$

(Standard abbreviations for units of speed, such as *km/s* for kilometers per second and *mi/s* for miles per second, will be used throughout the rest of this book.) Light that travels through air, water, glass, or any other substance always moves more slowly than it does in a vacuum.

The value c is a fundamental property of the universe. The speed of light appears in equations that describe, among other things, atoms, gravity, electricity, magnetism, distance, and time. It has extraordinary universal properties. For example, if you were traveling in space at 99% of the speed of light, $0.99c$, you would still measure the speed of any light beam moving toward you as c, which is also the speed you would measure for any light beam moving away from you!

In 1842, Christian Doppler, a professor of mathematics in Prague, deduced that wavelength, hence color, is affected by motion. As objects move toward or away from you, they change color, analogous to how the pitch of a siren changes as it passes you. As shown for the observer on the left in Figure 3-5, the wavelengths of electromagnetic radiation from an approaching source are compressed. The circles represent consecutive wave peaks (S1 through S4) emitted in all directions as the source moves along. Because each successive wave is sent out from a position slightly closer to her, she sees a shorter wavelength than she would if the source were stationary. All of the colors in the spectrum of an approaching source, such as a star, are therefore shifted toward the short-wavelength (blue) end of the spectrum, regardless of its distance. This phenomenon is called a **blueshift.**

Conversely, electromagnetic waves from a receding source are stretched out. The observer on the right in Figure 3-5 sees a longer wavelength than he would if the source were stationary. All of the colors in the spectrum of a receding source, regardless of its distance, are shifted toward the longer-wavelength (red) end of the spectrum, producing a **redshift.** A blueshift or a redshift is also called a **Doppler shift.** The amount of Doppler shift varies directly with approaching or receding speed: When the speeds are small compared to the speed of light, an object that approaches twice as fast as another has all of its colors (or wavelengths) blueshifted twice as much as does the slower-moving object. An object moving away twice as fast as another object has its colors redshifted twice as much as the slower-moving object.

The speed of sound is about 0.34 km/s (0.21 mi/s). How can this value and the information in this section be used to determine a person's distance from a lightning strike?

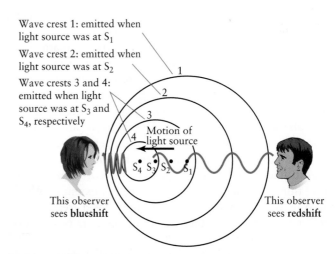

Wave crest 1: emitted when light source was at S_1

Wave crest 2: emitted when light source was at S_2

Wave crests 3 and 4: emitted when light source was at S_3 and S_4, respectively

Motion of light source

S_4 S_3 S_2 S_1

This observer sees **blueshift**

This observer sees **redshift**

  Figure 3-5 **The Doppler Shift** Wavelength is affected by motion between the light source and the observer. A source of light is moving toward the left. The four circles (numbered 1 through 4) indicate the location of light waves that were emitted by the moving source when it was at points S_1 through S_4, respectively. Note that the waves are compressed in front of the source but stretched out behind it. Consequently, wavelengths appear shortened (blueshifted) if the source is moving toward the observer and lengthened (redshifted) if the source is moving away from the observer. Motion perpendicular to the observer's line of sight does not affect wavelength.

3-3   Einstein showed that light sometimes behaves as particles that carry energy

By 1905, scientists were comfortable with the wave nature of light. However, in that year, Albert Einstein proposed that light sometimes acts as particles, creating what is now called the *wave–particle duality*. He used this idea to explain the *photoelectric effect*. Physicists knew that electrons are bound onto a metal's surface by electric forces and that it takes energy to overcome those forces. Shorter wavelengths of light can knock some electrons off the surfaces of metals, while longer wavelengths of light cannot, no matter how intense the beam of long-wavelength light. Because some colors (or, equivalently, wavelengths) can remove the electrons and others cannot, the electrons must receive different amounts of energy from different colors of light. But how? Einstein proposed that light travels as waves enclosed in discrete packets, now called **photons,** and that photons with different wavelengths have different amounts of energy. Specifically, *the shorter the wavelength, the higher a photon's energy.*

$$\text{Photon energy} = \frac{\text{Planck's constant} \times \text{The Speed of light}}{\text{Wavelength}}$$

where Planck's constant (named for the German physicist Max Planck) has the value 6.67×10^{-34} J s, where J is the unit of energy called a joule, and wavelength, the distance between wave crests or troughs, is shown in Figure 3-3. Einstein's concept of light, confirmed in numerous experiments, means that light can act both as waves (as when going through slits) and as particles (as when striking matter).

The waves shown in Figure 3-3 are moving to the right. If you count the number of wave crests that pass a given point per second, you have found the **frequency** of the photon. The unit of frequency is *hertz,* named in honor of German physicist Heinrich Hertz. One hertz means that one cycle per second—or that one wave crest per second—passes any point. A thousand hertz means a thousand cycles per second, and so on. The frequency is used, among many other things, to identify radio stations. For example, WCPE radio in Wake Forest, North Carolina, has a frequency of 89.7 megahertz (a megahertz is a million hertz or a million cycles per second).

All photons with the same wavelength are identical to each other, and, therefore, every photon of a given wavelength carries the same amount of energy as every other photon with that wavelength. The energy delivered by a photon is either enough to eject an electron from the surface of the metal or it is not; there is no middle ground. Extensive testing in the twentieth century confirmed both the wave and particle properties of light.

Although the energy of a single photon is fixed by its wavelength, the total number of photons passing per second from that source with a given energy determines the *intensi-*

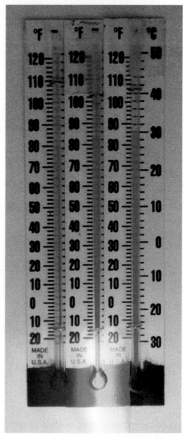

RIVUXG

Figure 3-6 Experimental Evidence for Infrared Radiation This photograph shows the visible colors separated by a prism. The two thermometers in the regions illuminated by visible light have temperatures less than the thermometer to the right of red. Therefore, there must be more radiation energizing (heating) the warmest thermometer. This energy is what we call infrared radiation—invisible to the human eye but detectable as heat. *(NASA/JPL-Caltech)*

ty of the electromagnetic radiation at that wavelength (that is, how bright the object appears to be). The more photons detected, the higher the intensity, and vice versa. However, the intensity of light does not change the energy *per* photon. If one photon is unable to eject an electron from a metal, then billions of photons with that energy will still be unable to remove that electron.

How are the frequencies of ocean waves determined?

3-4   Visible light is only one type of electromagnetic radiation

We have said that visible light has a narrow range of wavelengths, from about 400 to 700 nm. However, Maxwell's equations place no length restrictions on the wavelengths of electromagnetic radiation. What lies on either side of this interval? Around 1800, the British astronomer William Herschel discovered **infrared radiation** in an experiment with a prism. When he held a thermometer just beyond the red end of the visible spectrum, the thermometer registered a temperature increase, indicating that it was being heated by an invisible form of energy (Figure 3-6). Infrared radiation, discovered before Maxwell's equations were formulated, was later identified as electromagnetic radiation with wavelengths slightly longer than red light. Our bodies detect infrared radiation as heat.

During experiments with electric sparks in 1888, Hertz succeeded in producing electromagnetic radiation a few centimeters in wavelength, now known as **radio waves.** At wavelengths shorter than those of visible light, **ultraviolet (UV) radiation** extends from about 400 nm to 10 nm. In 1895, Wilhelm Roentgen invented a machine that produces electromagnetic radiation with wavelengths shorter than 10 nm, now called **X rays.** Modern versions of Roentgen's machine are found in medical and dental offices and airport security checkpoints. X rays have wavelengths between about 10 and 0.01 nm. Shorter still are **gamma rays.** These boundaries are all arbitrary and are primarily used as convenient divisions in the electromagnetic spectrum, which is actually continuous. We now know that visible light occupies only a tiny fraction of the full range of possible wavelengths, collectively called the **electromagnetic spectrum.** As shown in Figure 3-7, the electromagnetic spectrum stretches from the longest-wavelength radio waves, through infrared radiation, visible light, ultraviolet radiation, and X rays, to the shortest-wavelength photons, gamma rays. On the long-wavelength side of the visible spectrum, infrared radiation covers the range from about 700 nm to 1 mm. Astronomers interested in infrared radiation often express wavelength in *micrometers* or *microns* (abbreviated μm), where 1 μm = 1000 nm = 10^{-6} m. From roughly 1 mm to 10 cm is the range of microwaves. These are sometimes considered as a separate class of photons and sometimes categorized as infrared radiation or radio waves. Formally, radio waves are all electromagnetic waves longer than 10 cm.

The various types of electromagnetic radiation share many basic properties. For example, they are all photons, they all travel at the same speed, and they all sometimes behave as particles and sometimes as waves. But, because of their different wavelengths (and therefore different energies), they interact very differently with matter. For example, X rays penetrate deeply into your body tissues, while

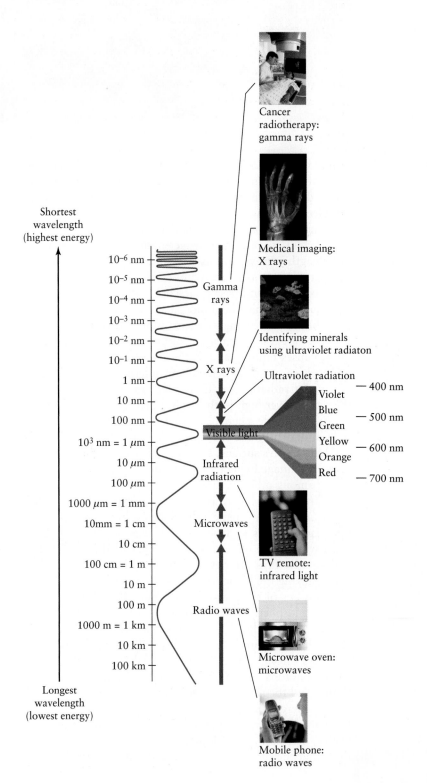

Figure 3-7 The Electromagnetic Spectrum
The full array of all types of electromagnetic radiation is called the electromagnetic spectrum. It extends from the longest-wavelength radio waves to the shortest-wavelength gamma rays. Visible light forms only a tiny portion of the full electromagnetic spectrum. Note that 1 μm (micrometer) is 10^{-6} m, and 1 nm (nanometer) is 10^{-9} m. The **insets** show how we are now using all parts of the electromagnetic spectrum here on Earth. *(Will and Deni McIntyre/Science Photo Library; Edward Kinsman/Photo Researchers, Inc.; Chris Martin-Bahr/Science Photo Library; Bill Lush/Taxi/Getty; Michael Porsche/Corbis; Ian Britton, Royalty-Free/Corbis)*

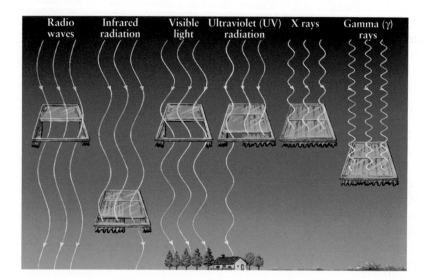

Figure 3-8 "Windows" through the Atmosphere Earth's atmosphere allows different types of electromagnetic radiation to penetrate into it in varying amounts. Visible light, radio waves, short-wavelength infrared, and long-wavelength ultraviolet reach all the way to Earth's surface. The other types of radiation are absorbed or scattered by the gases in the air at different characteristic altitudes (indicated by heights of windows). Although the atmosphere does not have actual "windows," astronomers use the term to characterize the passage of radiation through it.

visible light is mostly stopped and scattered by the surface layer of skin; your eyes respond to visible light but not to infrared radiation; and your radio detects radio waves but not ultraviolet radiation.

Earth's atmosphere is relatively transparent to visible light, radio waves, short-wavelength infrared, and long-wavelength ultraviolet. As a result, these radiations pass through the atmosphere without much loss and can be detected by ground-based telescopes sensitive to them. Astronomers say that the atmosphere has *windows* for these parts of the electromagnetic spectrum (Figure 3-8).

The longest-wavelength ultraviolet radiation, called UVA, causes tanning and sunburns. Ozone (O_3) in Earth's atmosphere normally screens out intermediate-wavelength ultraviolet radiation, or UVB. Until recently, the ozone in the *ozone layer* high in the atmosphere was being depleted by human-made chemicals, such as chlorofluorocarbons (CFCs) and bromine-rich gases. As a result, more UVB is reaching Earth's surface, and these highly energetic photons severely damage living tissue, causing skin cancer and glaucoma, among other diseases.

Earth's atmosphere is completely opaque to the other types of electromagnetic radiation, meaning that they do not reach Earth's surface. (This is a good thing, because short-wavelength ultraviolet radiation, called UVC, X rays, and gamma rays are devastating to living tissue. Gamma rays, packing the highest energies, are the deadliest.) Direct observations of these wavelengths must be performed high in the atmosphere or, ideally, from space.

As noted earlier, photons with different energies interact with matter in different ways. Higher-energy photons will pass through or rip apart material from which lower-energy photons will bounce off. Telescope designs for collecting and focusing radiation, therefore, differ, depending on the energy or, equivalently, the wavelength of interest. Knowing the energies of photons and their effects on matter enables astronomers to design both the telescopes to collect them and the devices used to record their presence. In the

next section we will consider the lengths to which astronomers have gone to capture visible and invisible (or nonoptical) electromagnetic radiation.

OPTICS AND TELESCOPES

Since the time of Galileo, astronomers have been designing instruments to collect more light than the human eye can gather on its own. Collecting more light enables us to see things more brightly, in more detail, and at a greater distance. There are two basic types of telescopes—those that collect light through lenses, or **refracting telescopes,** and those that collect it from mirrors, or **reflecting telescopes.** The earliest telescopes, such as Galileo's, used lenses, which have a variety of shortcomings as light-gathering devices. Consequently, all modern research telescopes use mirrors to collect light. Lenses are still used in the eyepieces of home telescopes to straighten the gathered light, so when we look through the telescopes directly, our brains can accurately interpret what we see. Lenses are also used to collect light in binoculars and cameras. Reflecting and refracting telescopes have the same main purpose—to collect as many photons as possible to better observe the sky (see Section 3-6). We begin exploring telescopes by discussing how reflecting telescopes work. Then we consider how lenses collect light and, finally, how astronomers have developed telescopes to see nonvisible electromagnetic radiation.

3-5 Reflecting telescopes use mirrors to concentrate incoming starlight

The first reflecting telescope was built in the seventeenth century by Isaac Newton (Figure 3-9). To understand how these telescopes work, consider a flow of photons, more commonly called a *light ray,* moving toward a flat mirror. In Figure 3-10a, the light ray strikes the mirror, and

Figure 3-9 Replica of Newton's Reflecting Telescope Built in 1672, this reflecting telescope has a spherical primary mirror 3 cm (1.3 in) in diameter. Its magnification was 40×. *(Royal Greenwich Observatory/Science Photo Library)*

we imagine a perpendicular line coming out of the mirror at that point. According to the principle of **reflection,** the angle between the incoming light ray and the perpendicular (dashed line) is always equal to the angle between the outgoing, reflected light ray and the perpendicular. This principle is often stated as "The angle of incidence equals the angle of reflection." This rule also applies if the mirror is curved (Figure 3-10b).

Which has more energy, an infrared photon or an ultraviolet photon?

Using this principle, Newton determined that a concave (hollowed-out) mirror ground, in the shape of a parabola, causes all parallel incoming light rays that strike the mirror to converge to a **focal point** (see Figure 3-10b). The distance between this **primary mirror** and the focal point, where the image of the distant object is formed, is called the **focal length** of the mirror. Focal points exist for light from sources that are extremely far away, like the stars. (Figure 3-11 shows why stars can be considered "far away.") If the object is larger than a point, such as the Moon or a planet, then the light will converge to a plane, called the **focal plane,** located at the distance of the focal length.

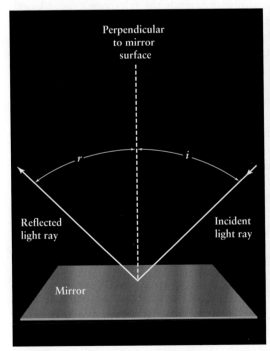

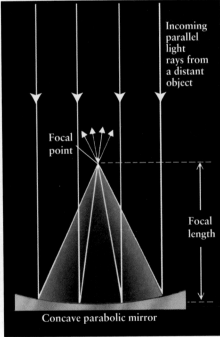

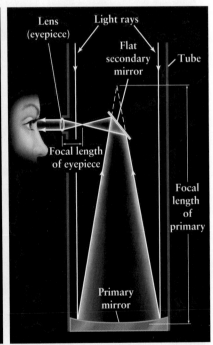

a   b   c

Figure 3-10 Reflection (a) The angle at which a beam of light strikes a mirror (the angle of incidence, *i*) is always equal to the angle at which the beam is reflected from the mirror (the angle of reflection, *r*). (b) A concave parabolic mirror causes parallel light rays to converge and meet at the focal point. The distance between the mirror and focal point is the focal length. (c) A Newtonian telescope uses a flat mirror, called the secondary mirror, to send light toward the side of the telescope. The light rays are made parallel again by passing through a lens, called the eyepiece. The dashed line shows where the focal point of this primary mirror would be if the secondary mirror were not in the way.

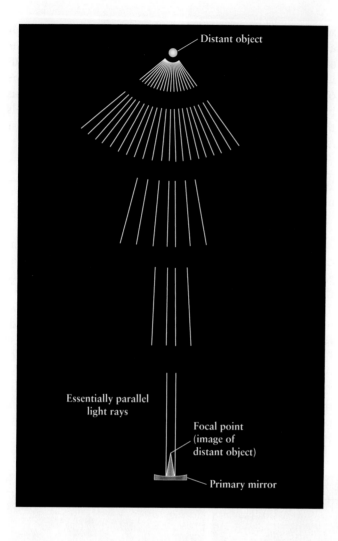

Distant object

Essentially parallel
light rays

Focal point
(image of
distant object)

Primary mirror

Figure 3-11 Parallel Light Rays from Distant Objects As light travels away from any object, the light rays, all moving in straight lines, separate. By the time light has traveled trillions of kilometers or miles, only the light rays moving in virtually parallel tracks are still near each other.

To view the image, Newton placed a small, flat mirror at a 45° angle between the primary mirror and the focal point, as sketched in Figure 3-10c and Figure 3-12a. This **secondary mirror** reflects the light rays to one side of the telescope, and the astronomer views the image through an **eyepiece lens.** We will discuss how this lens works in Section 3-8. A telescope with this optical design is still called a **Newtonian reflector.**

Newtonian telescopes are popular with amateur astronomers because they are convenient to use while the observer is standing up. However, they are not used in research observatories because they are lopsided. If astronomers attach their often heavy and bulky research equipment onto the side of a Newtonian telescope, the telescope twists and distorts the image in unpredictable ways.

Three basic designs exist for the reflecting telescopes used in research. In the first, a hole is drilled directly through the center of the primary mirror. A convex (outwardly curved), secondary mirror placed between the primary mirror and its focal point reflects the light rays back through the hole (Figure 3-12b). This design is called the **Cassegrain focus.** This secondary mirror extends the telescope's focal length. Compact, relatively low-weight equipment is bolted to the bottom of the telescope, and the light is brought into focus down there. This design has an advantage over Newtonian telescopes in that the attached equipment is balanced and does not distort the telescope frame and, hence, the image.

The second design that astronomers use has two variations. First, heavier or bulkier optical equipment that requires firmer mounting on the telescope is placed at the *Nasmyth focus* (after Scottish engineer James Nasmyth, who developed it). Second, for observations that benefit from an extremely long focal length, the equipment is placed at the **coudé focus** (named after a French word meaning "bent like an elbow"). In both designs (Figure 3-12c), a curved tertiary (third) mirror channels the light rays away from the telescope to a remote focal point. Equipment that profits

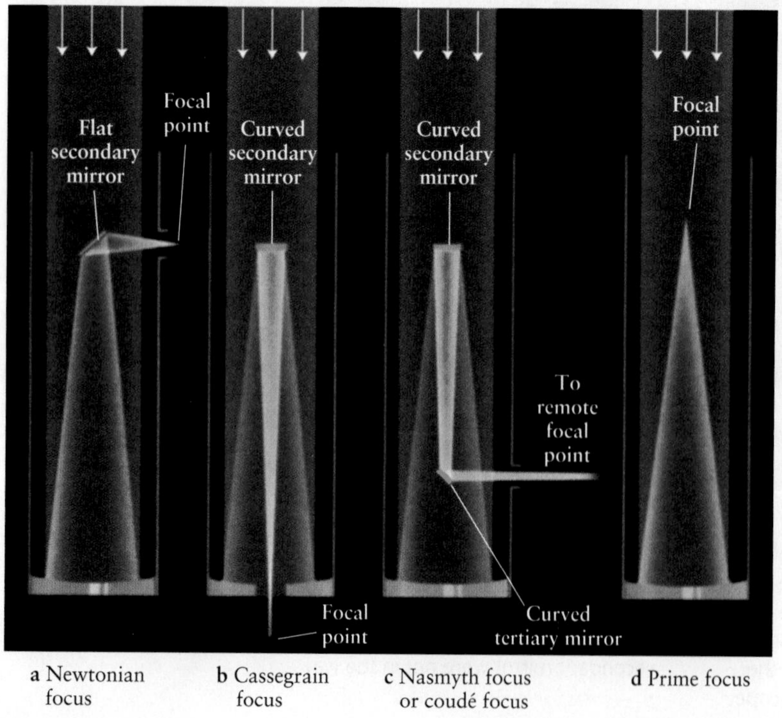

Flat
secondary
mirror

Focal
point

Curved
secondary
mirror

Curved
secondary
mirror

Focal
point

To
remote
focal
point

Focal
point

Curved
tertiary mirror

a Newtonian
focus

b Cassegrain
focus

c Nasmyth focus
or coudé focus

d Prime focus

Figure 3-12 Reflecting Telescopes Four of the most common optical designs for reflecting telescopes: (a) Newtonian focus (popular among amateur astronomers) and the three major designs used by researchers—(b) Cassegrain focus, (c) Nasmyth focus or coudé focus, and (d) prime focus.

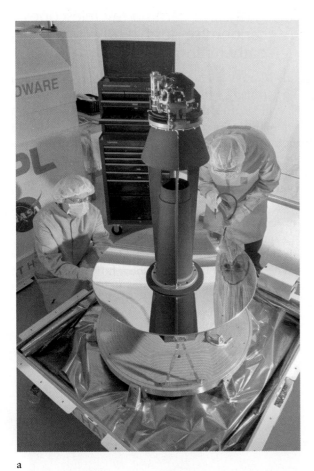

a

b

Figure 3-34 **Spitzer Space Telescope** (a) The mirror assembly for the Spitzer Space Telescope, showing the 85-cm objective mirror. (b) Launched in 2003, this Great Observatory is taking images and spectra of planets, comets, gas, and dust around other stars and in interstellar space, galaxies, and the large-scale distribution of matter in the universe. **Inset:** An infrared image of a region of star formation invisible to optical telescopes. *(a: Balz/SIRTF Science Center; b: NASA/JPL-Caltech; inset: NASA/JPL)*

a RIVUXG

b RIVUXG

Figure 3-35 **Views of the Milky Way's Central Regions** (a) An optical image in the direction of Sagittarius, toward the Milky Way's center. The dark regions are interstellar gas and dust clouds that prevent visible light from beyond them from reaching us. (b) An infrared image of the same area of the sky, showing many more distant stars whose infrared radiation passes through the clouds and is collected by our telescopes. *(a & b: Howard McCallon and Gene Kopan/2 MASS Project)*

ultraviolet radiation that astronomers can use to study the chemistries of these cosmic bodies. Therefore, in 1999, astronomers launched the Far Ultraviolet Spectroscopic Explorer (FUSE). It is providing us with information about such things as how much deuterium (hydrogen nuclei with one neutron) was created when the universe formed, the location of particularly hot interstellar gas and dust, and the chemical evolution of galaxies.

Telescopes dedicated to studying the Sun have observed it from the ground since 1941, and solar telescopes have studied it continuously from space since 1984. Of the latter, SOHO (the Solar and Heliospheric Observatory) and Hinode have ultraviolet detectors and are still in operation. They observe the Sun's outer layers rising and sinking, along with a wide range of energetic activity emanating from the Sun, including particles and radiation racing outward from it, among other things. In Chapter 7 we will discuss these observations and their physical origins.

3-16 X-ray and gamma-ray telescopes cannot use normal reflectors to gather information

X rays and gamma rays from space interact strongly with the particles in Earth's atmosphere, preventing these dangerous radiations from reaching our planet's surface. Therefore, direct observations of astronomical sources that emit these extremely short wavelengths must be made from space. Astronomers got their first look at X-ray sources during brief rocket flights in the late 1940s. Several small satellites, launched during the early 1970s, viewed the sources of both X rays and gamma rays in space, revealing hundreds of previously unknown objects, including several black holes (see Chapter 10). X-ray telescopes have also been carried on Space Shuttle missions. The insets in Figure 3-36 show how different our Sun appears when seen through X-ray and visible-light "eyes."

R I V U X G

R I V U X G

a   R I V U X G

b   R I V U X G

Figure 3-36 Nonvisible and Visible Radiation (a) This X-ray telescope was carried aloft in 1994 by the Space Shuttle. The X-ray telescope was prepared by the Lockheed-Martin Solar and Astrophysics Laboratory, the National Astronomical Observatory of Japan, and the University of Tokyo, with the support of NASA and ISAS. The **inset** shows an X-ray image of the Sun, from the Yohkoh mission of ISAS, Japan. (b) The McMath-Pierce Solar Telescope at

Kitt Peak Observatory near Tucson, Arizona (the inverted V-shaped structure), takes visible-light photographs of the Sun, such as the one shown in the **inset**. Comparing the images in the two insets reveals how important observing nonvisible radiation from astronomical phenomena is to furthering our understanding of how the universe operates. (Insets: L. Golub, Naval Observatory, IBM Research, NASA; a: NASA; b: NOAO)

X-ray photons are tricky to collect. Because of their high energies, X rays penetrate nearly head on even highly polished surfaces that they meet. Therefore, normal reflecting mirrors cannot be used to focus them. Instead, X-ray telescopes are designed to deflect photons at a fairly shallow angle, because X rays that are barely skimming or grazing a surface can be reflected and thereby focused. This is analogous to skipping a flat rock on water; throw it at a steep angle and it immediately sinks; throw it at a shallow angle and it will skip off the surface. Figure 3-37 shows the design of such "grazing incidence" X-ray telescopes.

After being focused, X rays are detected in several ways. As at infrared, visible, and UV wavelengths, CCDs can detect these photons. Other devices, called *scintillators,* detect the visible light created as X rays pass through them. Still other instruments, called *calorimeters,* detect the heat generated by X rays passing through the detector.

At least seven X-ray telescopes are orbiting Earth today. Among them, the Chandra X-ray Observatory, a NASA Great Observatory named after the Nobel laureate astronomer Subrahmanyan Chandrasekhar, provides images with better than 1″ resolution, and the European Space Agency's XMM-Newton has 6″ resolution. Other X-ray telescopes are carried high into the atmosphere by balloons. They float at altitudes up to 40 km (25 mi) above Earth's surface and often stay up for several weeks. Whereas low-energy X rays do not penetrate that far into the atmosphere, higher-energy X rays do, giving these balloon-borne telescopes a wide range of objects to study.

More than 10,000 X-ray sources have been discovered all across the sky. Among these are planetary atmospheres, stars (see the Sun in Figure 3-36), stellar remnants, vast clouds of intergalactic gas, jets of gas emitted by galaxies, black holes, quasars, clusters of galaxies, and a diffuse X-ray glow that fills the universe.

The electromagnetic radiation with the shortest wavelengths and the most energy is the gamma ray. In 1991, the Compton Gamma Ray Observatory, also a NASA Great Observatory, was carried aloft by the Space Shuttle. Named in honor of Arthur Holly Compton, an American physicist who made important discoveries about gamma rays, this orbiting observatory carried four instruments that performed a variety of observations, giving us tantalizing views of the gamma-ray sky until the telescope failed in 2000. At least 10 gamma-ray telescopes are presently in orbit, and several have been flown on balloon missions.

Several thousand gamma-ray sources have been discovered. However, gamma rays are too powerful even for grazing incidence telescopes. Therefore, astronomers have devised other methods of detecting them and determining their origin. These include absorbing them in crystals, allowing them to pass through tiny holes called collimators whose directions are well determined, and using chambers in which the gamma rays transform into electrons and positrons (positively charged electrons), leaving a track whose direction can be determined. These techniques are not nearly as precise as those used in other parts of the spectrum,

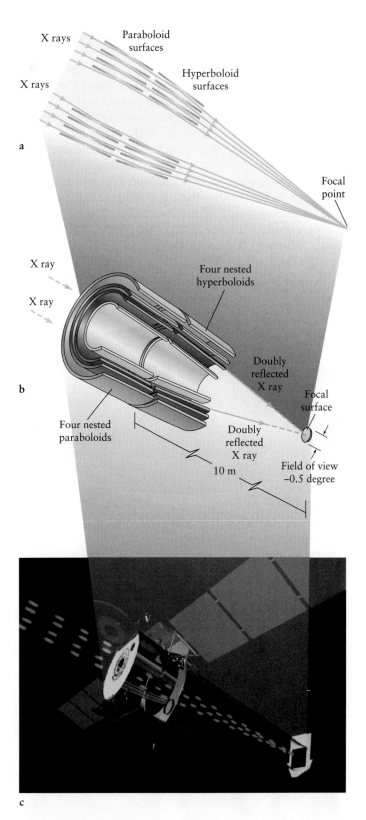

Figure 3-37 **Grazing Incidence X-ray Telescopes** X rays penetrate objects they strike head on. To focus them, X rays have to be gently nudged by skimming off cylindrical "mirrors." The shapes of the mirrors optimize the focus. The bottom diagram shows how X rays are focused in the Chandra X-ray Telescope. *(a & b: NASA/JPL-Caltech; c: NASA/Chandra X-ray Observatory Center/Smithsonian Astrophysical Observatory)*

and the best resolution gamma-ray instruments are only accurate to about 5 arcmin.

Astronomers are also taking advantage of the interaction between very high-energy gamma rays and Earth's atmosphere to build telescopes on the ground that indirectly detect the gamma rays. When these photons strike the atmosphere, they often cause particles in the air to move faster than the speed of light in air (but still slower than the speed of light in empty space, the ultimate speed limit). These particles quickly slow down, emitting short bursts of blue light that travel in the same direction as the incoming gamma rays; by detecting this light, it is possible to determine the direction from which the gamma rays came.

 We now have telescopes with which we can see the energy in the universe from virtually all parts of the electromagnetic spectrum (Figure 3-38).

Telescopes provide us with more than just stunning images of objects in space. They also provide information about the chemistry of stars and interstellar gas and dust, the motion of objects toward or away from us, whether stars are rotating, and whether they are alone in space or orbiting a companion, among other things. We will explore how we get this information when we study the various objects in space.

Until recently, photons were the only sources of detailed astronomical information that we had. However, in the past four decades, physicists have built detectors for waves and particles whose nature is not electromagnetic. Chapters 7 and 10 discuss two of these devices: neutrino detectors and gravity wave detectors.

So far in this chapter, we have explored the basic properties of electromagnetic radiation and the techniques

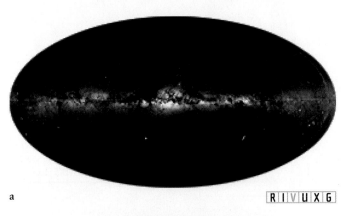

a   R I V U X G

Figure 3-38 Survey of the Universe in Various Parts of the Electromagnetic Spectrum By mapping the celestial sphere onto a flat surface (like making a map of Earth), astronomers can see the overall distribution of strong or nearby energy sources in space. The center of our Galaxy's disk cuts these images horizontally in half. Because most of the emissions shown in these diagrams fall in this region, we know that most of the strong sources of various electromagnetic radiation as seen from Earth (except X rays) are in our Galaxy: (a) visible light, (b) radio waves, (c) infrared radiation, (d) X rays, and (e) gamma rays. *(GFSC/NASA)*

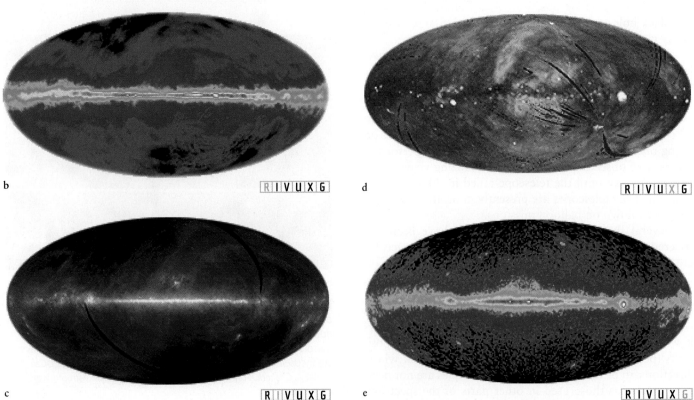

b   R I V U X G

d   R I V U X G

c   R I V U X G

e   R I V U X G

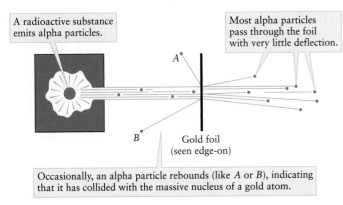

A radioactive substance emits alpha particles.

Most alpha particles pass through the foil with very little deflection.

A

B   Gold foil (seen edge-on)

Occasionally, an alpha particle rebounds (like *A* or *B*), indicating that it has collided with the massive nucleus of a gold atom.

Figure 3-48 **Rutherford Scattering Experiment** Most helium nuclei (originally called alpha particles) entering a thin foil scatter slightly as they pass through the medium, but some scatter backward, indicating that they have encountered very dense compact objects. Such experiments were the first evidence that most matter is concentrated in what are now called atomic nuclei.

a **radioactive** element naturally and spontaneously transforms into another element by emitting particles. Certain radioactive elements, such as uranium and radium, were known to emit such particles with considerable speed. It seemed plausible that a beam of these high-speed particles would penetrate a thin sheet of gold. Rutherford and his associates found that almost all of the particles did pass through the gold sheet with little or no deflection. To their surprise, however, an occasional particle bounced right back (Figure 3-48). It must have struck something very dense, indeed.

Within a few decades, the structure of atoms became evident. That dense "something" is now called the **nucleus** of the atom, and it consists of particles called **protons** and **neutrons**. Surrounding the nucleus, one or more **electrons** normally orbit. Newton's second law (see Section 2-7) explains that electrons move much more than nuclei because the electrons have only 1/2000 the mass of a proton or neutron.

Unlike the planets, whose gravitational interaction keeps them orbiting the Sun, electrons orbit nuclei because electrons and protons have a property called *electric charge*. All protons have the exact same positive charge, while all electrons have a negative charge equal in strength to the

proton's charge. The terms *positive* and *negative* are arbitrary and just indicate that they are opposite to each other. Particles with opposite charges attract each other. Therefore, protons attract electrons, and it is this attraction that keeps electrons in orbit around nuclei. The interaction between charged particles is said to be a result of the **electromagnetic force,** the second of four fundamental forces in nature. (Gravitation is the other fundamental force we have discussed.)

Particles with the same type of charge, such as a pair of protons or a pair of electrons, repel each other. For atoms that have more than one proton in their nuclei (and many do), the protons are pushing away from each other. For nuclei to have more than one proton, there must be an attractive force stronger than the repulsion of protons to keep them glued together. The electrically neutral neutrons in the nucleus help provide that attractive force, called the **strong nuclear force.** This is the third of the four fundamental forces. Many scientists find it fascinating that all of the interactions between matter and energy in nature occur as a result of just four forces, these three plus the **weak nuclear force.** The weak nuclear force is involved in some radioactive decays, such as when a neutron transforms into a proton. In Chapter 12, we will further explore the weak nuclear force in our study of the evolution of the whole universe. The properties of all four fundamental forces are summarized in Table 3-2.

The number of protons in an atom's nucleus, called the **atomic number,** determines the element of that atom. All hydrogen nuclei have one proton; all helium nuclei have two, and so forth. There are 92 different types of elements that form naturally. Uranium is the most massive, with 92 protons (see Appendix E).

In contrast, the nuclei of most elements can have different numbers of neutrons. For example, hydrogen always has 1 proton, but it can have 0, 1, or 2 neutrons; oxygen, with an atomic number of 8, always has 8 protons, but it can have 8, 9, or 10 neutrons. Each different combination of protons and neutrons is called an **isotope.** There are three isotopes of hydrogen and three isotopes of oxygen. Hydrogen with no neutrons is the most common hydrogen isotope, while oxygen with 8 neutrons is by far the most abundant isotope of oxygen.

TABLE 3-2 The Four Fundamental Forces of Nature		
Name	Strength (compared to strong force)	Range of effect (from each object)
Strong force	1	Inside atomic nuclei
Electromagnetic force	1/137	Throughout the universe
Weak force	10^{-5}	Inside atomic nuclei
Gravitational force	6×10^{-39}	Throughout the universe

Some isotopes are stable, meaning that the numbers of protons and neutrons in their nuclei do not change. However, many elements have isotopes that are unstable and come apart spontaneously. These isotopes are called radioactive. For example, carbon with 6 neutrons, ^{12}C ("carbon twelve"), is stable, while carbon with 8 neutrons, ^{14}C ("carbon fourteen"), is unstable. ^{14}C decays into nitrogen with 7 neutrons, ^{14}N. To learn how radioactive decay is used to determine the ages of different objects, see Appendix H-3. Using radioactive age-dating techniques on the oldest rocks brought back from the Moon, astronomers are able to determine roughly when it formed (4.5 billion years ago). Applying the same technique to space debris found on Earth (meteorites), we can determine that the solar system, consisting of the Sun and everything orbiting it, formed some 4.62 billion years ago.

Normally, the number of electrons that orbit an atom is equal to the number of protons in its nucleus, thus making the atom electrically neutral. Astronomers denote neutral atoms by writing the atomic symbol followed by the Roman numeral I. For example, neutral hydrogen is written as H I and neutral iron is Fe I.

When an atom contains a different number of electrons than protons, the atom is called an **ion.** The process of creating an ion is called **ionization.** Ions are denoted by the atomic symbol followed by a Roman numeral that is one greater than the number of missing electrons. Positively ionized hydrogen (missing its one electron) is denoted H II, while positively ionized iron with seven electrons missing is denoted Fe VIII. It is worth noting that negative ions also exist, in which nuclei have more electrons orbiting than they have protons.

Atoms can share electrons and, by doing so, become bound together. Such groups of atoms are called **molecules.** They are the essential building blocks of all complex structures, including life.

3-22   Spectra occur because electrons absorb and emit photons with only certain wavelengths

Because protons and neutrons have masses about 2000 times greater than the mass of an electron, over 99.95% of the mass of any atom is concentrated in its nucleus. The electron orbits are far from the nucleus, typically 10,000 times farther away than the radius of the nucleus. This is why you may have heard the statement that matter is mostly empty space.

Be careful not to imagine electrons as miniature planets orbiting the nucleus as a "miniature Sun." Protons, neutrons, and electrons are not tiny solid bodies. Rather, like photons, they all have both wave and particle properties. The science that accurately describes their complex behavior is called **quantum mechanics.**

Quantum mechanics explains that electrons in atoms can exist in only certain *allowed orbits* around their nuclei,

except when they are making a **transition** from one allowed orbit to another. These orbital conditions are completely unlike planets, which can exist at any distance (in any orbit) around the Sun. Each allowed electron orbit has a well-defined energy associated with it, and every different type of atom and molecule has a unique set of allowed orbits. These orbits and the transitions between them are the key to understanding the spectra that we have been discussing.

Consider an atom of the simplest hydrogen isotope, which contains just a single proton in its nucleus orbited by one electron. (This discussion generalizes directly to all of the other atoms and isotopes, made more complex only because they have more than one electron in orbit.) Figure 3-49 shows this hydrogen atom's lowest energy levels.

Normally, the electron is in the lowest-energy allowed orbit or energy level, commonly called the **ground state;** this is labeled $n = 1$ in Figure 3-49. Each allowed orbit with successively higher energy is labeled $n = 2, 3, 4$, and so on. When an electron is in an orbit with more energy than the lowest energy state available to it, it is said to be in an **excited state.**

Electrons change orbits by absorbing or emitting photons. However, electrons cannot absorb just any photon that they encounter. Electrons can absorb only those photons with energies exactly enough to boost them up to a higher-energy allowed orbit. That is, the photons that are absorbed are those with energies equal to the difference

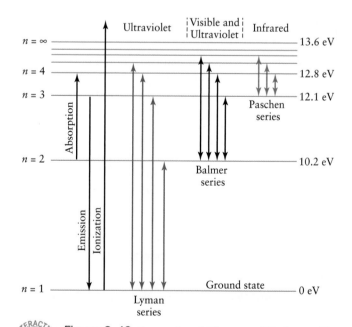

Figure 3-49 **Energy Level Diagram of Hydrogen** The activity of a hydrogen atom's electron is conveniently displayed in a diagram showing some of the energy levels at which it can exist. A variety of electron jumps, or transitions, are also shown, including those that produce the most prominent lines in the hydrogen spectrum. For another perspective on the same information, you can try the linked Interactive Exercise.

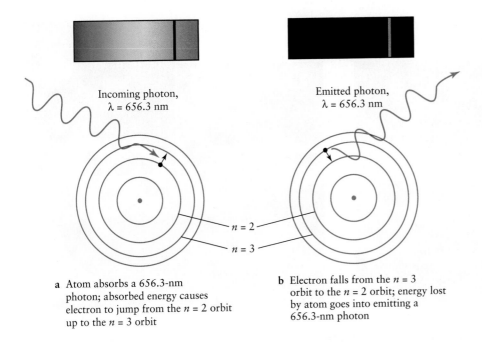

a Atom absorbs a 656.3-nm photon; absorbed energy causes electron to jump from the $n = 2$ orbit up to the $n = 3$ orbit

b Electron falls from the $n = 3$ orbit to the $n = 2$ orbit; energy lost by atom goes into emitting a 656.3-nm photon

Figure 3-50 **The Absorption and Emission of an H_α Photon** This schematic diagram of hydrogen's four lowest allowed orbits shows what happens when a hydrogen atom absorbs or emits an H_α photon, which is red and has a wavelength of 656.28 nm. The spectral lines by such events are also shown. (a) A photon is absorbed by the electron, causing the electron to transition from orbit $n = 2$ up to orbit $n = 3$. (b) A photon is emitted as the electron makes a transition from orbit $n = 3$ down to orbit $n = 2$.

between the energies of two allowed orbits. All photons that do not satisfy this condition pass straight through the atom. If the electron starts in its ground state, then it must get exactly the energy necessary to move it to an excited state. If it is already in an excited state, then by absorbing a photon, it must transition to a higher-energy excited state.

For example, referring to Figure 3-49 and Figure 3-50, transitions between the $n = 2$ and the $n = 3$ energy levels require the electron to absorb a photon with energy equal to 12.1 eV − 10.2 eV = 1.9 eV (electron volt, a measure of energy). Recall that a photon's energy corresponds to a certain wavelength. In this case, the photon absorbed by the electron has a wavelength of 656.3 nm (Figure 3-50a); it is a red photon (recall Figure 3-7).

Although some absorption occurs at optical wavelengths, it also happens in many other parts of the electromagnetic spectrum. Indeed, most of the transitions in hydrogen, shown in Figure 3-49 as the Lyman series (up from and down to $n = 1$) and the Paschen series (up from and down to $n = 3$), are nonvisible photons. Only a few of the Balmer series (up from and down to $n = 2$) transitions are visible, with the rest being ultraviolet. We will refer to Balmer transitions when we study stars. They are named after Johann Balmer, who first calculated their wavelengths. The longest-wavelength Balmer line is called H_α

(H alpha), the second H_β (H beta), the third H_γ (H gamma), and so forth, ending with the shortest-wavelength Balmer line, H_∞ (H infinity). (The first dozen lines of the series have Greek-letter subscripts; the remainder are identified by numerical subscripts.) Part of the spectrum of a star having Balmer absorption lines H_α through H_θ (H theta) is shown in Figure 3-51.

Absorption lines are therefore caused by photons being taken out of the stream of light by electrons, which thereby move into higher-energy allowed orbits. An observer on the right in Figure 3-47 looking at the hot blackbody through the cloud of cooler gas sees dark absorption lines where photons have been absorbed by the cooler gas.

Electrons in excited states are unstable. They lose energy by bumping into other particles and suddenly having either too much or too little energy than is necessary to be in that state. When this happens, the electron is forced to descend to a lower-energy allowed state, either the ground state or some allowed state between where it starts and the ground state. In either case, to make this transition down, the electron must *emit* a photon with energy once again equal to the difference between the energies of the starting and final allowed orbits (see Figure 3-50b). The emitted photons have the same set of wavelengths

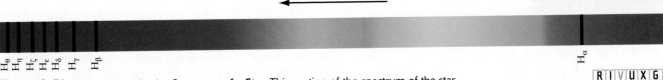

Shorter wavelength

H_θ H_η H_ζ H_ϵ H_δ H_γ H_β H_α

R I V U X G

Figure 3-51 **Balmer Lines in the Spectrum of a Star** This portion of the spectrum of the star Vega shows eight Balmer lines, from H_α at 656.3 nm through H_θ at 388.9 nm. *(NOAO)*

Figure 3-52 Emission Spectra from Interstellar Gas Clouds
(a) Stars in this interstellar gas cloud (NGC 2363 in the constellation Camelopardus, the Giraffe) emit blackbody spectra. Electrons in the cloud's hydrogen gas absorb and reemit the red light from these stars. NGC 2363 is located some 10 million light-years away. (b) Part of the Rosette Nebula (NGC 2237), an interstellar gas cloud in the constellation Monoceros (the Unicorn). The green glow is generated by doubly ionized oxygen atoms (O III; oxygen atoms missing two electrons) in the cloud that emits 501-nm photons. The Rosette is 3000 light-years away. *(a: L. Drissen, J.-R. Roy, and C. Robert/Département de Physique and Observatoire du Mont Mégantic, Université Laval; and NASA; b: T. A. Rector, B. Wolpa, M. Hanna, AURA/NOAO/NSF)*

as the absorbed photons. Furthermore, the emitted photons are sent out in all directions, which create the glowing, emission-line spectra of hot objects. The color of the gas depends on the atoms and molecules that it contains. Hydrogen-rich gas clouds glow red because of the H_α emission (Figure 3-51 and Figure 3-52a). Oxygen emits many green photons, so oxygen-rich gas clouds glow green (Figure 3-52b).

Which Balmer line in Figure 3-49 is H_α?

If an electron orbiting in a hydrogen atom encounters a photon with more than 13.6 eV, that photon is absorbed and knocks the electron completely out of orbit and away from the atom. This process is called *photoionization*. Each type of atom has different photoionization energies, above which all electrons are kicked out of the atoms. Photoionization occurs in stars and interstellar nebulae (Figure 3-52b). Indeed, many spectral lines from stars and nebulae are those of ionized atoms that retain at least one electron.

3-23 Spectral lines shift due to the relative motion between the source and the observer

The speed of an object toward or away from us is called its **radial velocity** (Figure 3-53) because the motion is along our line of sight or, put another way, along the "radius"

drawn from Earth to the object seen on the celestial sphere. Of course, the star or other object may also have a velocity perpendicular or transverse to our line of sight, such as a car passing a pedestrian. Both the radial velocity and the **transverse velocity** are measured in kilometers per second.

Astronomers cannot measure the transverse velocity directly. Instead, they measure the angle that the object moves among the stars on the celestial sphere as seen from Earth. This is called its **proper motion** and is often measured in arcseconds per year or arcseconds per century. The star with the greatest proper motion as seen from Earth is Barnard's star (Figure 3-54). Proper motion does not affect the perceived wavelength and cannot be determined by Doppler shift. It must be measured by comparing the locations of objects with higher proper motion to the positions of objects with lower proper motions. Planets moving among the relatively fixed stars on the celestial sphere are examples of bodies that have large proper motions.

Stellar proper motions are so small that they can be measured for only relatively nearby stars in our Galaxy. However, the radial velocity of virtually every object in space can be determined. Indeed, even the slow motions of rising and sinking gases on the Sun's surface can be measured by their spectral Doppler shifts. Doppler shift measurements of stars in double-star systems give crucial data about the speeds of the stars orbiting each other. Doppler measurements of planets, the Sun, and other stars reveal that many of these

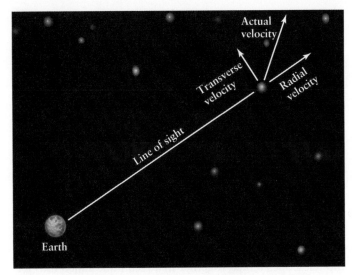

Figure 3-53 Radial and Transverse Velocities of a Star The actual velocity of a star (or other object) can be separated into radial and transverse motions. The speed of the star toward or away from Earth in kilometers per hour is its radial velocity. This motion creates a Doppler shift in the star's spectrum. The speed in kilometers per hour perpendicular to the radial velocity is called its transverse velocity, which does not affect the star's spectrum. The angular motion of the star across the celestial sphere (that is, in the direction of its transverse velocity) is called proper motion.

bodies are rotating. This is determined when the Doppler shift shows that half of a surface is heading toward us, while the other half is simultaneously moving away from us. Furthermore, Doppler measurements reveal that all of the very distant galaxies are moving away from us, meaning that the universe is expanding. The spectra of distant galaxies enable us to determine the rate of that expansion. In later chapters, we will refer to the Doppler shift whenever we need to convert an observed wavelength shift into a speed toward or away from us.

Does a police siren approaching you sound higher or lower in pitch than the siren at rest relative to you?

INSIGHT INTO SCIENCE

Remote Science Astronomical objects are so remote and their activities are often so complex that astrophysicists must use a tremendous amount of physics to interpret observations. For example, a single spectrum can contain information about stars, extrasolar planets, interstellar gas, Earth's motion and atmosphere, and the performance of the observing telescope and the equipment attached to it. All of these factors must be understood theoretically and accounted for.

With the knowledge of spectroscopy, atomic and nuclear physics, and the Doppler shift, we can now list many of the vital properties of matter in space that are available

to us. These include the chemical compositions of stars, interstellar gas clouds, and other objects; the temperatures of these objects; the rotation of objects in space; the motion of objects toward or away from Earth; the presence of planets and dim companion stars; and the rate at which the universe is expanding. Knowing these properties, astrophysicists have developed models that provide numerous insights into the ages, masses, distances, internal activity, rotation rates, and companion objects of stars, among other things.

Modern physics was born when Newton set out to understand the motions of the planets. Two-and-a-half centuries later, scientists—including Maxwell, Planck, Einstein, Rutherford, and Doppler, among many others—discovered the basic properties of electromagnetic radiation and the structures of atoms. As we will see in the following chapters, the fruits of their labors have important implications for astronomy even today.

3-24 Frontiers yet to be discovered

Before the twentieth century, astronomers were like the blind men in the fable who are trying to describe an elephant. Their perceptions were piecemeal; we could describe parts of the universe but not the whole. Our ancestors did not have the technology that could enable them to see the big picture.

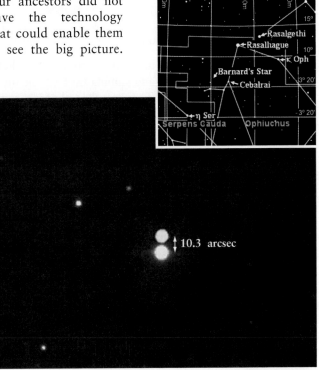

R I V U X G

Figure 3-54 Proper Motion of Barnard's Star Two images of Barnard's star, taken a year apart in 2000 and 2001, show the proper motion of the star during that time. In addition to having the largest known proper motion (10.3″ per year), Barnard's star is one of the closest stars to Earth. *(John Sanford/Science Photo Library/Photo Researchers)*

However, we are beginning to see it, and as you will learn in the chapters that follow, our understanding of the cosmos is therefore increasing dramatically. A vast amount of observational information remains to be gathered. Indeed, literally every planet, moon, piece of interplanetary debris, star, stellar remnant, gas cloud, galaxy, quasar, cluster of galaxies, and supercluster of galaxies has stories to tell. Observational astronomy is so new an activity that we are making new and often unexpected discoveries almost daily.

Spectra of objects in space continue to yield new insights into the chemical composition of the planets, moons, local space debris, stars, interstellar gas and dust, and galaxies. Furthermore, as we obtain more accurate Doppler measurements of objects in our Galaxy, in other nearby galaxies, and of entire galaxies, we are better able to determine the rate at which the objects in the universe are moving relative to one another. The more we understand of the chemistries and motions of objects in space, the more accurately we are able to explain the evolution of the solar system, stars, galaxies, and the universe. There is still a lot to discover about the elephant.

SUMMARY OF KEY IDEAS

The Nature of Light
• Photons—units of vibrating electric and magnetic fields—all carry energy through space at the same speed, the speed of light (300,000 km/s in a vacuum, slower in any medium).

• Radio waves, microwaves, infrared radiation, visible light, ultraviolet radiation, X rays, and gamma rays are the forms of electromagnetic radiation. They travel as photons, sometimes behaving as particles, sometimes as waves.

• Visible light occupies only a small portion of the electromagnetic spectrum.

• The wavelength of a visible-light photon is associated with its color. Wavelengths of visible light range from about 400 nm for violet light to 700 nm for red light.

• Infrared radiation, microwaves, and radio waves have wavelengths longer than those of visible light. Ultraviolet radiation, X rays, and gamma rays have wavelengths that are shorter.

Optics and Telescopes
• A telescope's most important function is to gather as much light as possible. When possible, it resolves (reveals details) and magnifies objects.

• Reflecting telescopes, or reflectors, produce images by reflecting light rays from concave mirrors to a focal point or focal plane.

• Refracting telescopes, or refractors, produce images by bending light rays as they pass through glass lenses. Glass impurity, opacity to certain wavelengths, and structural difficulties make it inadvisable to build extremely large refractors. Reflectors are not subject to the problems that limit the usefulness of refractors.

• Earth-based telescopes are being built with active and adaptive optics. These advanced technologies yield resolving power comparable to the Hubble Space Telescope.

Nonoptical Astronomy
• Radio telescopes have large, reflecting antennas (dishes) that are used to focus radio waves.

• Very sharp radio images are produced with arrays of radio telescopes linked together in a technique called interferometry.

• Earth's atmosphere is fairly transparent to most visible light and radio waves, as well as to some infrared and ultraviolet radiation arriving from space, but it absorbs much of the electromagnetic radiation at other wavelengths.

• For observations at other wavelengths, astronomers mostly depend upon telescopes carried above the atmosphere by rockets. Satellite-based observatories are giving us a wealth of new information about the universe and permitting coordinated observation of the sky at all wavelengths.

• Charge-coupled devices (CCDs) record images on many telescopes used between infrared and X-ray wavelengths.

• By studying the wavelengths of electromagnetic radiation emitted and absorbed by an astronomical object, astronomers can learn about the object's temperature, chemical composition, rotation rate, companion objects, and movement through space.

Blackbody Radiation
• A blackbody is a hypothetical object that perfectly absorbs electromagnetic radiation at all wavelengths. The relative intensities of radiation that it emits at different wavelengths depend only on its temperature. Stars closely approximate blackbodies.

• Wien's law states that the peak wavelength of radiation emitted by a blackbody is inversely proportional to its temperature—the higher its temperature, the shorter the peak wavelength. The intensities of radiation emitted at various wavelengths by a blackbody at a given temperature are shown as a blackbody curve.

• The Stefan-Boltzmann law shows that a hotter blackbody emits more radiation at every wavelength than does a cooler blackbody.

• The motion of an object toward or away from an observer causes the observer to see all of the colors from the object blueshifted or redshifted, respectively. This effect is generically called a Doppler shift.

Discovering Spectra

• Spectroscopy—the study of electromagnetic spectra—provides important information about the chemical composition of astronomical objects.

• Kirchhoff's three laws of spectral analysis describe the conditions under which absorption lines, emission lines, and a continuous spectrum can be observed.

• Spectral lines serve as distinctive "fingerprints" that identify the chemical elements and compounds comprising a light source.

Atoms and Spectra

• An atom consists of a small, dense nucleus (composed of protons and neutrons) surrounded by electrons. Atoms of different elements have different numbers of protons, while different isotopes have different numbers of neutrons.

• Quantum mechanics describes the behavior of particles and shows that electrons can only be in certain allowed orbits around the nucleus.

• The nuclei of some atoms are stable, while others (radioactive ones) spontaneously split into pieces.

• The spectral lines of atoms of a particular element correspond to the various electron transitions between allowed orbits with different energy levels of those atoms. When an electron shifts from one energy level to another, a photon of the appropriate energy (and hence a specific wavelength) is absorbed or emitted by the atom.

• The spectrum of hydrogen at visible wavelengths consists of part of the Balmer series, which arises from electron transitions between the second energy level of the hydrogen atom and higher levels.

• Every different atom, isotope, and molecule has a different set of spectral lines.

• When a neutral atom loses or gains one or more electrons, it is said to be charged. The atom loses an electron when the electron absorbs a sufficiently energetic photon, which rips it away from the nucleus.

• The equation that describes the Doppler effect states that the size of a wavelength shift is proportional to the radial velocity between the light source and the observer.

WHAT DID YOU THINK?

1 *What is light?* Light—more properly, "visible light," is one form of electromagnetic radiation. All electromagnetic radiation (radio waves, microwaves, infrared radiation, visible light, ultraviolet radiation, X rays, and gamma rays) has both wave and particle properties.

2 *Which type of electromagnetic radiation is most dangerous to life?* Gamma rays have the highest energies of all photons, so they are the most dangerous to life. However, ultraviolet radiation from the Sun is the most common everyday form of dangerous electromagnetic radiation that we encounter.

3 *What is the main purpose of a telescope?* A telescope is designed primarily to collect as much light as possible.

4 *Why do all research telescopes use mirrors, rather than lenses, to collect light?* Telescopes that use lenses have more problems, such as chromatic aberration, internal defects, complex shapes, and distortion from sagging, than do telescopes that use mirrors.

5 *Why do stars twinkle?* Rapid changes in the density of Earth's atmosphere cause passing starlight to change direction, making stars appear to twinkle.

6 *Which is hotter, a "red-hot" or a "blue-hot" object?* Of all objects that glow visibly from heat generated or energy stored inside them, those that glow red are the coolest.

7 *What color does the Sun emit most brightly?* The Sun emits all wavelengths of electromagnetic radiation. The colors it emits most intensely are in the blue-green part of the spectrum. Because the human eye is less sensitive to blue-green than to yellow, and because Earth's atmosphere scatters blue-green wavelengths more readily than longer wavelengths, we normally see the Sun as yellow.

Review Questions

The answers to all computational problems, which are preceded by an asterisk (*), appear at the end of the book.

1. Describe reflection and refraction. How do these processes enable astronomers to build telescopes?

2. Give everyday examples of refraction and reflection.

*3. How much more light does a 3-m-diameter telescope collect than a 1-m-diameter telescope?

4. Explain some of the advantages of reflecting telescopes over refracting telescopes.

5. What are the three major functions of a telescope?

6. What is meant by the angular resolution of a telescope?

7. What limits the ability of the 5-m telescope at Palomar Observatory to collect starlight? There are several correct answers to this question.

8. Why will many of the very large telescopes of the future make use of multiple mirrors?

9. What is meant by adaptive optics? What problem does adaptive optics overcome?

10. Compare an optical reflecting telescope to a radio telescope. What do they have in common? How are they different?

11. Why can radio astronomers observe at any time of the day or night whereas optical astronomers are mostly limited to observing at night?

12. Why must astronomers use satellites and Earth-orbiting observatories to study the heavens at X-ray wavelengths?

13. What are NASA's four Great Observatories, and in what parts of the electromagnetic spectrum do (or did) they observe?

14. Why did Rømer's observations of the eclipses of Jupiter's moons support the heliocentric, but not the geocentric, cosmology?

15. A blackbody glowing with which of the following colors is hottest? **a.** yellow, **b.** red, **c.** orange, **d.** violet, **e.** blue.

16. Of the following photons, which has the lowest energy? **a.** infrared, **b.** gamma rays, **c.** visible light, **d.** ultraviolet, **e.** X ray.

17. The spectrum of which of the following objects will show a blueshift? **a.** an object moving just eastward on the celestial sphere, **b.** an object moving just northward on the celestial sphere, **c.** an object moving directly toward Earth, **d.** an object moving directly away from Earth, **e.** an object that is not moving relative to Earth.

18. What is a blackbody? What does it mean to say that a star appears almost like a blackbody? If stars appear to be like blackbodies, why are they not black?

19. What is Wien's law? How could you use it to determine the temperature of a star's surface?

20. What is the Stefan-Boltzmann law? How do astronomers use it?

21. Using Wien's law and the Stefan-Boltzmann law, state the changes in color and intensity that are observed as the temperature of a hot, glowing object increases.

22. What color will an interstellar gas cloud composed of hydrogen glow, and why?

23. What is an element? List the names of five different elements, and briefly explain what makes them different from each other.

24. How are the three isotopes of hydrogen different from each other?

25. Explain how the spectrum of hydrogen is related to the structure of the hydrogen atom.

26. Why do different elements have different patterns of lines in their spectra?

27. Explain why the Doppler shift tells us only about the motion directly along the line of sight between a light source and an observer, but not about motion across the celestial sphere.

28. What is the Doppler shift, and why is it important to astronomers?

Observing Projects

29. Determine what fraction of stars visible to the naked eye you can see from your location. This exercise is best done outdoors with a laptop computer or through a window at night in a very dark room. On a dark, clear night, run *Starry Night Enthusiast™*. In the Menu, select **View > Hide Daylight**. Open the **Options** side pane, select **Local View**, and click on **Local Light Pollution**. Place the cursor over the words **Local Light Pollution** and click on the **Local Light Pollution Options** button that appears. Slide the pollution level bar until the view matches your night sky. The goal is to estimate what fraction of the visible stars you are seeing compared to the total number of stars you could be seeing under ideal conditions. Carefully do the following on the screen: Using a roughly 8 cm (3 in) square section of the sky on the screen (a square about half the length of a typical pen on each side), count and record the number of stars with your present setting. Now set the **Local Light Pollution** to **less** (which essentially gives you ideal conditions) and repeat the count of visible stars. Divide the first number you count by the second. What fraction of the stars were you seeing? To see how much light pollution occurs in large cities, adjust the slide bar in the **Local Light Pollution Options** to the far right for maximum light pollution and repeat the star-counting within the same limited sky region. What fraction of the stars are viewers in large cities seeing?

30. You can use *WorldWide Telescope* (WWT) to explore **M31**, the **Andromeda Galaxy**, under conditions that simulate the use of binoculars or a telescope. You can use your observations to evaluate the effects of the greater light-gathering power and improved resolution that are provided by optical aids.

Launch WWT and click the **Search** tab. Type "M31" in the edit box and click the thumbnail for this galaxy. **a.** Zoom out to a field of view of about 60°. You can open the **View** menu and click **Constellation Figures** and **Boundaries on** to locate the position of this galaxy in your sky. This distant galaxy is visible to the unaided eye from a dark location. Click the constellation guidelines **off** to see M31 and its surroundings without these imaginary lines. If the crosshairs (+ symbol) are on and covering M31, grab the sky and move the galaxy slightly off of them. In the center of the view you can see M31 as it might appear to the naked eye from a very dark-sky site. Make a sketch of this galaxy as it appears in the view, along with a background of several bright stars, adding notes about its appearance. **b.** Zoom in to a field of view of about 15° to simulate the appearance of this galaxy as it would appear through a pair of powerful binoculars. Again, sketch and describe the appearance of M31. **c.** Press the "+" key on the keyboard twice to zoom in to a field of view of about 2°. This simulates the appearance of M31 as it would appear through a small telescope. Again, sketch and describe the appearance of the galaxy, noting particularly any details now apparent at this higher magnification. Locate and identify the two galaxies M32 and M110 in this image. Is the resolution of this telescope image better, the same, or worse than that seen through binoculars or the unaided eye?

 31. Use WWT to observe the center of the **Milky Way Galaxy** in different wavelengths of electromagnetic radiation. Click the **Guided Tours** tab, followed by the thumbnail of the **Galaxies** folder. Then click the thumbnail labeled **Center of the Milky Way** and watch the tour. When it is finished, you may elect to click the **Watch Again** button to see the tour once more. To pause the tour, move the cursor over the menu tabs at the top of the screen and click the **Pause** button that appears to the left of the tour summary thumbnails. Click the **Play** button to resume the tour. As you watch the tour, answer the following questions: **a.** In which constellation is the center of the Milky Way, as seen from Earth? **b.** Describe the appearance of the center of the Milky Way as seen in visible light. What interferes with the viewing of the actual center of the Galaxy in visible light? **c.** Is the view of the center of the Milky Way clearer when seen at near infrared wavelengths than at visible-light wavelengths or less clear? **d.** At which wavelengths are the clearest and sharpest details of the center of the Milky Way visible? Can you explain why this is so?

Stars atop a Silent Volcano

THE LARGEST ASTRONOMICAL OBSERVATORY IN THE WORLD SITS ON MAUNA KEA IN HAWAII. THE VIEWS, UP OR DOWN, ARE SPECTACULAR   BY MARGUERITE HOLLOWAY

Marguerite Holloway, "Stars atop a Silent Volcano," *Scientific American,* July 2004, 112–113.

CANADA-FRANCE-HAWAII TELESCOPE was built in 1979. Although the catwalk is not open to visitors, a similar view can be had from the publicly accessible catwalk at the University of Hawaii's 2.2-meter telescope. Both telescopes have been instrumental in recent discoveries of Jupiter's many satellites.

It is with some trepidation that I turn onto Saddle Road after a half-hour drive north from Kona on Hawaii's Big Island. Only one rental car company—Harper's Car and Truck—will let renters use their vehicles on this road without limiting coverage, so I am expecting a pockmarked dirt nightmare. But Saddle Road, which crosses the top part of the island, is smooth and paved. Its 50 miles are merely narrow, empty of gas stations and stores, and unlit at night. A short distance from two major cities and several smaller ones, Saddle Road seems isolated, cut off from the world.

Turning off at mile marker 28 only intensifies that feeling, as it becomes possible to peer beyond this world. It is here that more than 106,000 people come every year to visit Mauna Kea, a long-quiet volcano, sacred to Hawaiians, atop which astronomers are using the world's largest collection of telescopes to better understand the origin and evolution of the universe and the composition of everything in it.

Mauna Kea, which means "white mountain," is an ideal spot for studying the cosmos. It is unusually clear at night because clouds are drawn away by a downdraft. At 4,200 meters, it is above 40 percent of the earth's atmosphere and also above a layer of clouds that separates it from the lower atmosphere's wet ocean air, so the summit air is free of dust, smog and moisture. Because of these attributes and the fact that the volcano has not been active for 4,500 years or so (the mantle hot spot is now enlivening Mauna Loa 30 miles away as well as a submarine

RICHARD J. WAINSCOAT

KECK I AND II are among the best-known telescopes on Mauna Kea (*left*). Some of the observatory's findings have been made with HIRES (*white box on platform in image at right*). Using this powerful instrument, astronomers have discovered over half of the more than 120 known extrasolar planets.

site even farther to the south), astronomers had their eyes on the mountain for many years. Development started in the late 1960s, and today 13 telescopes of various kinds aid research, including the famous Keck I and Keck II.

Before guests can drive to the summit and see all this powerful equipment, they must spend some time at the visitor center—named after *Challenger* crew member and Hawaiian Ellison S. Onizuka—to acclimate to the altitude. (Official summit tours are offered only on weekends at 1 P.M., but stargazing programs are available every night at 6; see www.ifa.hawaii. edu/mko) This particular Sunday, after watching a film about the history of the site and the ongoing research, a tour group of about 25 people climbs into four-wheel drives for the long, winding, unpaved ride up the volcano. The altitude makes engines inefficient, so filling the tank just before entering Saddle Road is essential; the high elevation also prevents brakes from working well, so the road-gripping ability of a four-wheel drive is mandatory.

The first stop on the barren, brown, wind-scoured mountain is Keck I, one of the twin optical and infrared telescopes that make up the W. M. Keck Observatory. They are the largest such telescopes: each weighs 300 tons and stands eight stories tall. Astronomers have used the instruments to study early stars and how they formed in ancient clouds of gas just after the Big Bang, some 13.7 billion years ago. They have discovered planets outside our solar system orbiting their own stars. And recently researchers using the Keck II telescope were able to discern methane in the atmosphere of Mars, which could have been introduced by a comet, released by volcanic activity on the planet itself or, some speculate, generated by the metabolism of bacteria.

The summit is chilly, but inside the telescope it seems even colder. Our guide explains that it costs $35,000 a month to keep the Keck domes near freezing—important for eliminating any chance of heat distorting the 10-meter mirror or the metal equipment. (For those who want to visit on weekdays, Keck I's viewing area and gallery are open from 10 A.M. until 4 P.M. Monday through Friday; see www2. keck.hawaii.edu)

We next visit the University of Hawaii's 2.2-meter telescope, and from the catwalk around it we look down on the white bubble domes of the Kecks and a sea of clouds below and behind them. This is the only other telescope that is open to visitors on the summit. The 2.2-meter was instrumental, along with several others at Mauna Kea, in the discoveries of dozens of additional satellites around Jupiter. The planet's total now stands at 63. The telescope was also used in the discovery this February of a nearby—that is, only 33 light-years distant—planet-forming red dwarf star. By examining this star, dubbed AU Mic, astronomers hope to better understand how planets are formed.

Although visitors can't go inside many of the telescopes and can't see the work of astronomers up close, it is still a remarkable feeling to be tiny against these great white structures, to feel a mote against the backdrop of such a monumental search. Many of the tour takers are amateur astronomers who spend every vacation stargazing and visiting observatories. "Mauna Kea Observatory is clearly the best," sighs one as we head back to the cars.

It is late afternoon, the tour is over, and the air is becoming even colder. The four-wheel drives fall into an informal convoy and creep down the mountain. I am heading back along Saddle Road as the sun sets. The light catches small flowers, which blaze like small lanterns along the road. Tufts of white grass glow against the black lava fields; the land looks like a bleached blonde showing her roots. Everywhere light leaps out against the dark.

SA

Earth and Moon

R I V U X G

Earth and the Moon As Seen from Space *(NASA)*

WHAT DO YOU THINK?

1. Can Earth's ozone layer, which is now being depleted, be replenished?

2. Who was the first person to walk on the Moon, and when did this event occur?

3. Do we see all parts of the Moon's surface at some time throughout the lunar cycle?

4. Does the Moon rotate and, if so, how fast?

5. What causes the ocean tides?

6. When does the spring tide occur?

Answers to these questions appear in the text beside the corresponding numbers in the margins and at the end of the chapter.

In this chapter, we will explore Earth and the Moon, noting what makes Earth suitable for life. Our detailed and rapidly developing knowledge of Earth allows us to present material on topics that range from the magnetic fields surrounding it, to its atmosphere and climate, to the depth and composition of its crust and the properties extending to its metal-rich core. As you will see, our knowledge about the Moon is much more limited. In the next chapter, we explore other worlds with many different properties from Earth, leading to vastly different environments from our own.

In this chapter you will discover

- why Earth is such an ideal environment for life

- that Earth is constantly in motion inside and out

- how Earth's magnetic field helps protect us

- what made the craters on the Moon

- how the Sun and the Moon cause Earth's tides

- that both Earth and the Moon have two (different) major types of surface features

- that water ice may have been found at the Moon's poles

EARTH: A DYNAMIC, VITAL WORLD

Looking at Earth from outer space (Figure 4-1), one first notices that its clouds, water, and land combine to scatter, on average, 37% of the sunlight it receives right back into space. That amount of scattered light (an albedo of 0.37) and its variation throughout the year are unique among the planets in our solar system. In contrast, cloud-covered worlds, such as Venus, scatter much more light, while bone-dry worlds, such as Mars and our Moon, scatter much less. Furthermore, Earth's albedo varies, consistent with changing cloud and ice layers, as well as with different amounts of land and water facing the Sun. Water, the nearly universal solvent in which life on Earth first formed, covers about 71% of our planet's surface. (A solvent is a liquid in which various substances can dissolve.)

Earth is a geologically active, ever-changing world. Earthquakes shake many regions of it. Volcanoes pour huge quantities of molten rock from just under the crust (the outer layer) onto the surface, and gas from within it vents into the oceans and atmosphere. Some mountains are still rising, while others are wearing away. Flowing water erodes topsoil and carves river valleys. Rain and snow help rid the atmosphere of dust particles—and life teems virtually everywhere, making Earth unique in the solar system.

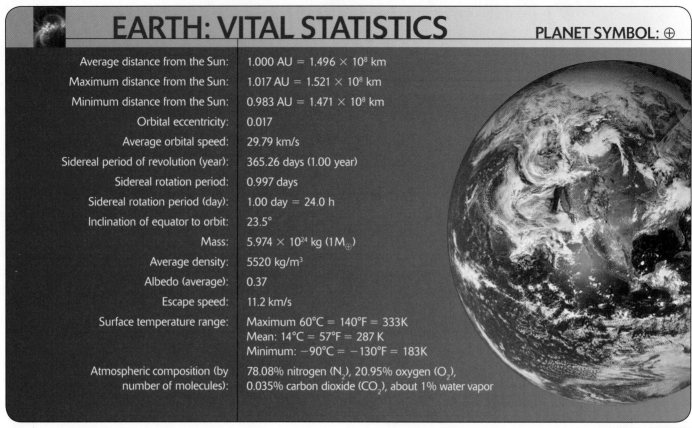

EARTH: VITAL STATISTICS

PLANET SYMBOL: ⊕

Average distance from the Sun:	1.000 AU = 1.496×10^8 km
Maximum distance from the Sun:	1.017 AU = 1.521×10^8 km
Minimum distance from the Sun:	0.983 AU = 1.471×10^8 km
Orbital eccentricity:	0.017
Average orbital speed:	29.79 km/s
Sidereal period of revolution (year):	365.26 days (1.00 year)
Sidereal rotation period:	0.997 days
Sidereal rotation period (day):	1.00 day = 24.0 h
Inclination of equator to orbit:	23.5°
Mass:	5.974×10^{24} kg ($1 M_\oplus$)
Average density:	5520 kg/m³
Albedo (average):	0.37
Escape speed:	11.2 km/s
Surface temperature range:	Maximum 60°C = 140°F = 333K Mean: 14°C = 57°F = 287 K Minimum: −90°C = −130°F = 183K
Atmospheric composition (by number of molecules):	78.08% nitrogen (N_2), 20.95% oxygen (O_2), 0.035% carbon dioxide (CO_2), about 1% water vapor

R I V U X G

FIGURE 4-1 **A View of Earth's Surface** An oasis in the forbidding void of space, Earth is a world of unsurpassed beauty and variety. Ever-changing cloud patterns drift through its skies. More than two-thirds of its surface is covered with oceans. This liquid water, in combination with a huge variety of chemicals from its lands, led to the formation and evolution of life over most of the planet's surface. The symbol ⊕ is astronomers' shorthand for Earth. *(NASA)*

4-1 Earth's atmosphere has evolved over billions of years

Earth's atmosphere is a key part of our planet's dynamic activity and its ability to sustain life. Essential to the existence of complex life is air that contains oxygen molecules. The air we breathe is predominantly a 4-to-1 mixture of nitrogen to oxygen, two gases that are found only in small amounts in the atmospheres of other planets. The theory of planet formation and the models that attempt to reproduce the early history of Earth indicate that this is the third atmosphere to envelop our planet.

Earth's earlier atmospheres contained no free oxygen. Composed of trace remnants of hydrogen and helium left over from the formation of the solar system, the first atmosphere did not last very long. The atoms and molecules in these gases were too light to stay near Earth. Heated by sunlight (meaning that they moved faster and faster), these particles gained enough kinetic energy to overcome Earth's gravitational pull and fly away into space.

The gases of the second atmosphere came from inside Earth, vented through volcanoes and cracks in Earth's surface. That atmosphere was composed primarily of carbon dioxide and water vapor, along with some nitrogen. There was roughly 100 times as much gas in that second atmosphere as there is in the air today. Carbon dioxide and water store a great deal of heat, helping create what is called the *greenhouse effect,* about which we will say more shortly. The high concentration of these gases early in Earth's life stored more heat than the atmosphere does today. Earth back then would have been a serious hothouse had the Sun reached its full intensity at that time. Evidence indicates it did not, and so the high concentrations of carbon dioxide and water in the air prevented the young Earth from freezing.

Oceans began forming when water precipitated out of this atmosphere, as well as from impacts of water-rich space debris, such as the comet nuclei mentioned in Chapter 2. These events occurred within 300 million years of Earth's formation. The rain absorbed and carried down lots of carbon dioxide, and the oceans eventually absorbed about half the carbon dioxide in that second atmosphere. (Water holds a good deal of carbon dioxide, as you can see by the amount of carbon dioxide fizz in a can of soda.) As life evolved in the oceans, much of that dissolved carbon dioxide was transformed into the shells of many organisms, and then, as the organisms died, their carbon-rich shells sank to the ocean bottoms. Shells piled up and compressed the shells underneath them into rock, such as limestone.

Early plant life in the oceans and then on the shores removed most of the remaining carbon dioxide in the air by converting it into oxygen plus the nutrients that the plants needed to survive. The most efficient of such conversion mechanisms is photosynthesis, which today helps maintain the balance between carbon dioxide and oxygen in the air. What remained in the air were mostly nitrogen and water vapors.

Oxygen that entered the air early in the life of Earth came from oxygen-rich molecules, such as carbon dioxide, that were broken apart by the Sun's ultraviolet radiation. Then came oxygen, released as a by-product of ocean plant-life activity. In both cases, this oxygen did not stay in the air for long. Oxygen is highly reactive, and early atmospheric oxygen combined quickly with many elements on Earth's surface, notably iron, to form new compounds, such as iron oxides (otherwise known as rust).

About 2 billion years ago, after all of the surface minerals that could combine with oxygen had done so, the atmosphere began to fill with this gas. With the carbon dioxide gone, the air became the nitrogen–oxygen mixture that we breathe today. Although nitrogen was only a tiny fraction of that earlier atmosphere, it now dominates the thinner air of today.

Because air has mass, it is pulled down toward Earth by gravity, thereby creating a pressure on the surface. We define pressure as a force acting over some area:

$$\text{Pressure} = \frac{\text{Force}}{\text{Area}}$$

The weight of the air pushing down creates a pressure at sea level of 14.7 pounds per square inch, which is commonly denoted as 1 atmosphere (atm). The atmospheric pressure decreases with increasing altitude, falling by roughly half with every gain of 5.5 km of altitude. Seventy-five percent of the mass of the atmosphere lies within 11 km (approximately 7 mi, or 36,000 ft) of Earth's surface.

Lower Layers of the Atmosphere All of Earth's weather—clouds, rain, sleet, and snow—occurs in the lowest layer of the atmosphere, called the troposphere. Commercial jets generally fly at the top of this layer to avoid having to plow through the denser air below. If Earth were the size of a typical classroom globe, the troposphere would be only as thick as a sheet of paper wrapped around its surface.

Starting at Earth's surface, atmospheric temperature initially decreases as you go upward, reaching a minimum of 218 K (−67°F) at the top of the troposphere. Above this level is the region called the stratosphere, which extends from 11 to 50 km (approximately 7 to 31 mi) above Earth's surface. This is the realm of the ozone layer.

INSIGHT INTO SCIENCE

What's in a word? The word "layer" in the term *ozone layer* suggests a narrow region, perhaps a few meters thick. Not so. The ozone layer extends in altitude over some 40 km of the atmosphere!

The Ozone Layer The ozone layer has been much in the news in recent years. What is it, and how threatened is it? Ozone molecules (three oxygen atoms, O_3, bound together) are created in the stratosphere when short-wavelength ultraviolet radiation from the Sun occasionally splits normal oxygen molecules, O_2, into two separate oxygen atoms. Each resulting oxygen atom then combines with different O_2 molecules to create ozone, O_3. Once created, ozone efficiently absorbs intermediate-wavelength solar ultraviolet rays, thereby heating the air in this layer while preventing much of this lethal radiation from reaching Earth's surface.

The troposphere (lowest part of the atmosphere) is heated primarily by Earth, so the temperature in this part of the atmosphere decreases with altitude—the farther away from the heat source of Earth, the cooler the air. However, because the ozone in the stratosphere stores significant heat from the Sun above it, the temperature in the stratosphere actually rises with altitude to about 285 K (50°F) at its top (Figure 4-2).

Not much ozone actually exists in the ozone layer. If all of it were compressed to the density of the air we breathe, it would be a layer only a few millimeters (about ⅛ in.) thick. It is enough, however, to protect us from most of the Sun's lethal ultraviolet radiation. Ultraviolet radiation has

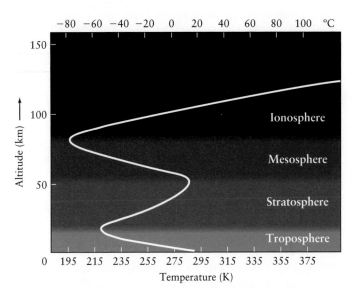

FIGURE 4-2 **Temperature Profile of Earth's Atmosphere** The atmospheric temperature changes with altitude because of the way sunlight and heat from the Earth's surface interact with various gases at different heights.

been a double-edged sword in the history of Earth. Because Earth's early atmosphere lacked ozone, ultraviolet radiation penetrated to the planet's surface, where it provided much of the energy necessary for life to evolve in the oceans. After life developed, however, the formation of the ozone layer by the same ultraviolet radiation was essential in lessening the amount of ultraviolet radiation that reached the surface, thereby allowing life-forms to leave the oceans and survive on land.

Today, the ozone layer is still vital, because the energy it absorbs would otherwise cause skin cancer and various eye diseases. In sufficiently high doses, ultraviolet radiation even damages the human immune system by destroying some of the molecules that normally repair cells in our bodies. However, over the past few decades, the ozone layer has been depleted by human-made chemicals, such as chlorofluorocarbons (CFCs). Until the past few years, ozone levels in the mid-latitudes of the northern hemisphere have been decreasing by about 4% per decade. There are some hints that the ozone is decreasing more slowly, but this is still unclear. Much more rapid and dramatic drops in the ozone layer over Earth's poles, especially over the South Pole, have been observed for more than two decades. These seasonal *ozone holes* typically extend over 24 million sq km, or 4.6% of Earth's surface area. There has been some improvement in the past few years, but the holes persist.

Fortunately, as explained earlier, sunlight creates ozone. *If* current international efforts aimed at banning ozone-depleting compounds continue, the ozone layer will naturally replenish itself in a century or so, and the ozone holes will probably stop occurring.

Upper Layers of the Atmosphere Note in Figure 4-2 that above the stratosphere lies the mesosphere. Atmospheric temperature again declines as you move up through the mesosphere, reaching a minimum of about 200 K (−103°F) at an altitude of about 80 km (50 mi). This minimum marks the bottom of the ionosphere, also called the thermosphere, above which the Sun's ultraviolet light heats and ionizes atoms, producing charged particles that reflect radio waves. In this era of burgeoning satellite usage and human space travel, the effects of the Sun's electromagnetic radiation, along with the flow of particles from it and from other sources, on Earth's atmosphere and on people in space are increasingly important. As a result, *space weather* satellites are now continually monitoring our neighborhood in space and warning about solar events that could imperil humans and machines out there.

Carbon Dioxide and the Greenhouse Effect Today Perhaps you have had the experience of parking your car in the sunshine on a warm summer day. You roll up the windows, lock the doors, and go on an errand. After a few hours, you return to discover that the interior of your automobile has become considerably hotter than the outside air temperature.

What happened to make your car so warm? First, sunlight entered it through the windows. This radiation was absorbed by the dashboard and the upholstery, increasing their temperatures. Because they become warm, your dashboard and upholstery emit infrared radiation, which you detect as heat and which your car windows do not permit to escape. This energy is therefore trapped inside your car and absorbed by the air and interior surfaces. As more sunlight comes through the windows and is trapped, the temperature continues to increase. The same trapping of sunlight warms a greenhouse and Earth. The greenhouse effect created by the carbon dioxide (CO_2) in the second atmosphere strongly heated the young Earth. When the carbon dioxide was dramatically reduced, that effect also decreased, but it never went away completely.

 The greenhouse effect in our air works like this: All wavelengths of electromagnetic radiation from the Sun approach Earth's atmosphere. Some of this energy is scattered from clouds, land, and water right back into space (Figure 4-3a), while some is absorbed in the atmosphere, thereby heating it. Some visible light, along with some ultraviolet radiation and radio waves, penetrate to Earth's surface and is absorbed by it. This energy helps heat the planet (along with heat from inside) until the surface material can store no further energy. Behaving nearly like a blackbody, Earth radiates the excess energy, primarily as infrared radiation (where its blackbody spectrum peaks).

Unlike most visible light, which passes relatively easily through the gases in our atmosphere, infrared radiation is absorbed by certain gases, especially water, carbon dioxide, and ozone in the air. The energy from these photons further heats the atmosphere. Carbon dioxide levels are increasing in Earth's atmosphere, primarily as a result of human activity, such as the burning of fossil fuels. Increased atmospheric carbon dioxide stores more heat in the air, and the air becomes warmer. Normally, carbon dioxide is removed from the air (and oxygen added) by the process of photosynthesis in green leaves. With increased deforestation and the expansion of cities, fewer plants and trees are left, so the carbon dioxide in the air remains there longer, overheating the air and creating *global warming*. Figure 4-3b shows how the CO_2 concentration has changed over the past millennium.

The consequences of recent global warming are quite complex and serious. They include the melting of polar ice caps, rising sea levels, changes in ocean heating and circulation patterns, changes in weather patterns worldwide, and increases in undesirable weather events, such as hurricanes, tornadoes, droughts, floods, heat waves, and shoreline erosion. With scientific data on the human effects that add to global warming in hand, many nations of the world are working together to decrease its causes.

An increase in the vegetation on Earth will have what effect on the carbon dioxide level in the atmosphere?

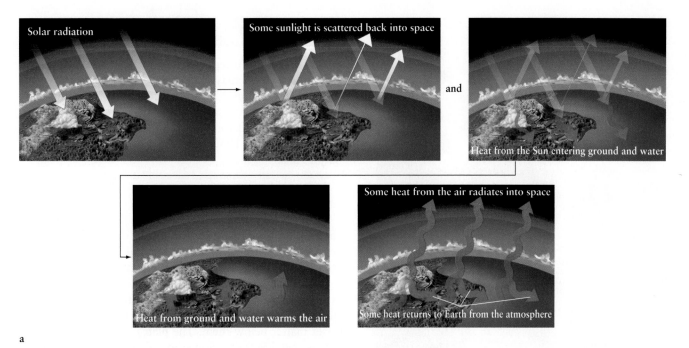

Solar radiation

Some sunlight is scattered back into space

and

Heat from the Sun entering ground and water

Heat from ground and water warms the air

Some heat from the air radiates into space

Some heat returns to Earth from the atmosphere

a

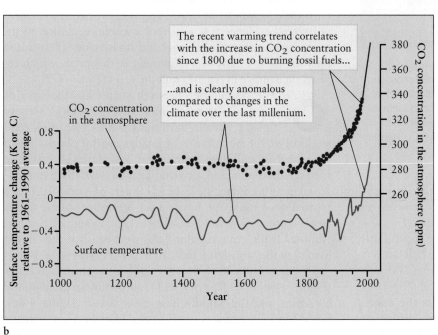

The recent warming trend correlates with the increase in CO₂ concentration since 1800 due to burning fossil fuels...

...and is clearly anomalous compared to changes in the climate over the last millenium.

CO₂ concentration in the atmosphere

Surface temperature

b

FIGURE 4-3 The Greenhouse Effect
(a) Sunlight and heat from Earth's interior warm Earth's surface, which in turn radiates energy, mostly as infrared radiation. Much of this radiation is absorbed by atmospheric carbon dioxide and water, heating the air, which in turn increases Earth's temperature even further. In equilibrium, Earth radiates as much energy as it receives. (b) The amount of carbon dioxide in our atmosphere since 1000 A.D. has been determined. The increase in carbon dioxide since 1800 due to burning fossil fuels and decreases in forestation has caused a dramatic temperature increase.

4-2 Plate tectonics produce major changes on Earth's surface

Large land masses called continents protrude through Earth's oceans. These land masses are regions of Earth's outermost layer, or **crust,** and are composed of relatively low-density rock that floats on denser material below it. By studying the various kinds of rocks at a particular location, geologists can deduce the history of that site. For example, whether an area was once covered by an ancient sea or flooded by lava from volcanoes is readily apparent from the kinds of rocks found there at present.

The crust has distinct boundaries (for example, as shown in Figure 4-4), many occurring beneath Earth's oceans. These divisions suggest that the continents are separate bodies. Earthquakes and volcanoes often occur near the boundaries, hinting at activity taking place below Earth's crust.

Anyone who carefully examines the shapes of Earth's continents might conclude that landmasses move. Eastern South America, for example, looks like it would fit snugly against western Africa. Between 1912 and 1915, German scientist Alfred Wegener published his observations on this remarkable fit between landmasses on either side of the Atlantic Ocean, an idea proposed by Isaac Newton two

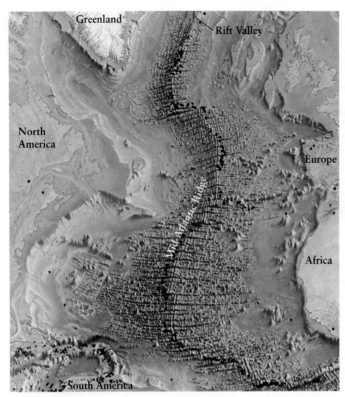

FIGURE 4-4 **The Mid-Atlantic Ridge** This artist's rendition of the bottom of the North Atlantic Ocean shows an unusual mountain range in the middle of the ocean floor. Called the Mid-Atlantic Ridge, these mountains are created by lava seeping up from Earth's interior along a rift that extends from Iceland to Antarctica. The black dots indicate locations of earthquakes. *(Courtesy of M. Tharp and B. C. Heezen)*

centuries earlier. Wegener was inspired to advocate the theory of **continental drift**—the idea that the continents on either side of the Atlantic Ocean have moved apart.

Wegener argued that a single, gigantic supercontinent called *Pangaea* (a Greek word for "all lands") began to break up and drift apart some 200 million years ago. Initially, most geologists ridiculed Wegener's ideas. Although it was generally accepted that the continents do float on denser rock beneath them, few geologists could accept that entire continents move across Earth's face at speeds as great as several centimeters per year. Wegener and other "continental drifters" could not explain what forces would shove the massive continents around or how less dense continental rock had gotten through the dense rock of the ocean floor. Then, in the mid-1950s, geologists discovered long, underwater mountain ranges. The Mid-Atlantic Ridge, for example, stretches from Iceland to Antarctica (see Figure 4-4). Careful examination revealed that molten rock from Earth's interior is being forced upward there, causing the ocean floor to slide apart on either side of the ridge. This **seafloor spreading** is pushing the Americas away from Europe and Africa at a speed of roughly 3 cm per year. The relative motion of the plates worldwide ranges from about 1.5 to 7 cm per year.

Seafloor spreading provided just the physical mechanism that had been missing from the theory of continental drift. It established Wegener's theory of crustal motion, which was developed into the modern theory of **plate tectonics.**

Earth's surface is made up of about a dozen major thick rafts of rock called *tectonic plates* that move relative to one another. These plates are floating on the slowly moving, denser matter below them. The plates are not always separate, as they are today. Sometimes they merge into a single mass called a *supercontinent*. In recent years, geologists have uncovered evidence that points to a whole succession of supercontinents that broke apart and reassembled. Pangaea is only the most recent supercontinent in this cycle, which repeats on average every 500 million years (Figure 4-5).

The boundaries between plates are the sites of some of the most impressive geological activity on our planet. Earthquakes and volcanoes tend to occur at the boundaries of Earth's crustal plates, where the plates are colliding (convergent boundaries), separating (divergent boundaries), or grinding against each other (slip-strike fault boundaries). If the epicenters of earthquakes are plotted on a map, the boundaries between tectonic plates (see Figure 4-4) stand out clearly. For example, the San Andreas Fault, running along the west coast of the United States, is the slip-strike fault that occurs where the North American and Pacific plates are rubbing against each other (Figure 4-6b). The Red Sea and Gulf of Suez are expanding as Africa and the Middle East move apart (Figure 4-6c). Great mountain ranges, such as the Himalayas, are thrust up by ongoing collisions between the Indian–Australian and Eurasian plates (Figure 4-6d). During such collisions, one plate moves downward (subducts) into Earth under the other. Where old crust subducts, we find deep trenches, such as those off the coasts of Japan and Chile.

Tectonic plate motion helps explain why Earth's oceans have so few craters today: Countless impacts occurred on the oceans, which cover most of our planet. Some of the impacting bodies were slowed or obliterated before cratering the ocean floor, but others were big enough or fast enough to hit the floors. The resulting craters in the oceans have been drawn underground by tectonic activity and thereby erased. (Most craters exposed to the atmosphere were eroded away by wind and water, with the larger ones being drawn underground by plate motion.)

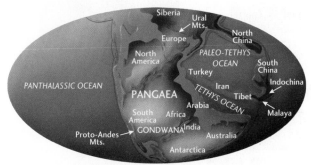

a   237 million years ago: the supercontinent Pangaea

b   152 million years ago: the breakup of Pangaea

c   The continents today

FIGURE 4-5 **The Supercontinent Pangaea** (a) The present continents are pieces of what was once a bigger, united body called Pangaea. (b) Geologists believe that Pangaea must have first split into two smaller supercontinents, which they call Laurasia and Gondwanaland. (c) These bodies later separated into the continents of today. Gondwanaland split into Africa, South America, Australia, and Antarctica, while Laurasia divided to become North America and Eurasia.

What powers all of this activity? Some enormous energy source has caused shifting among land masses for billions of years. The answer to that question leads us to a scientific model of Earth's interior.

4-3   Earth's interior consists of a rocky mantle and an iron-rich core

Geologists have calculated that Earth was entirely liquid soon after its formation, about 4.6 billion years ago. The violent impact of space debris–along with energy released by the breakup of radioactive elements–heated, melted, and kept the young Earth molten. The process of radioactive atoms breaking apart, called *nuclear fission,* also creates the heat used to generate electricity in nuclear power plants. This differs from the energy generated by *nuclear fusion* in the Sun, wherein lighter atoms are fused together, as mentioned in Chapter 2 and discussed in detail in Chapter 7.

Most of the iron and other dense elements sank toward the center of the young, liquid Earth, just as a rock sinks in a pond. At the same time, most of the less dense materials were forced upward toward the surface (Figure 4-7a). This process, called **planetary differentiation,** produced a layered structure within Earth: a very dense central **core** surrounded by a **mantle** of less dense minerals, which, in turn, was surrounded by a thin crust of relatively light minerals (Figure 4-7b).

Differentiation explains why most of the rocks you find on the ground are composed of lower-density elements, such as silicon and aluminum. Much of the iron, gold, lead, and other denser elements found on Earth today actually had to return to the surface, emerging through volcanoes and other lava flows. Despite this return of heavier elements to the surface, the average density of crustal rocks is 3000 kg/m^3, considerably less than the average density of Earth as a whole (5520 kg/m^3).

To understand Earth's interior, geologists have created a mathematical model that includes the effects of differentiation, gravity, radioactivity, chemistry, heat transfer, rotation, and other factors. The model predicts that the temperature in Earth increases steadily from about 290 K on the surface to nearly 5000 K (9000°F) at the center. Radioactivity in Earth's interior helps replenish heat lost through the planet's surface. In addition to temperature, pressure also increases with increasing depth below Earth's surface. The deeper you go, the more mass presses down on you. These factors shape Earth's inner layers.

What type of boundary exists between Africa and the Middle East?

You might think that a temperature of 5000 K would melt virtually any substance. The mantle, in particular, is composed largely of minerals rich in iron and magnesium, both of which have melting points of only slightly more than 1250 K at Earth's surface. However, the melting point of anything also depends on the pressure to which it is subjected: the higher the pressure, the higher a material's melting point. The high pressures within Earth (1.4 million atm at the bottom of the mantle) keep the mantle solid to a depth of about 2900 km and also allow for a solid iron core in the center of the planet. In other words, the inner core's pressure is so high that it forces the very hot atoms there into a solid.

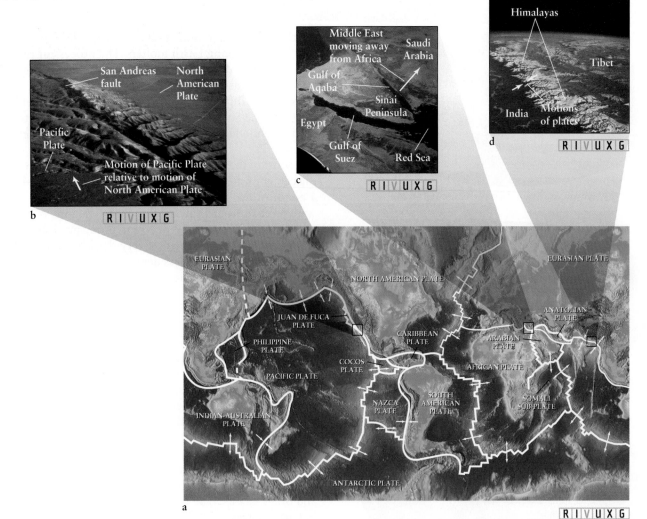

FIGURE 4-6 **Earth's Major Tectonic Plates**
(a) Earth's surface is divided into a dozen or so rigid plates that move relative to one another. The boundaries of the plates are the scenes of violent seismic and geologic activity, such as earthquakes, volcanoes, rising mountain ranges, and sinking seafloors. The arrows indicate whether plates are moving apart (←—→), together (→←), or sliding past one another (↑↓). (b) Rubbing of two plates. The San Andreas Fault, running up the west coast of North America, formed because the Pacific Plate is moving northwest along the North American Plate.

(c) The separation of two plates. The plates that carry Egypt and Saudi Arabia are moving apart, leaving the trench that contains the Red Sea. This view, taken by astronauts in 1966, shows the northern Red Sea on the right. (d) The Collision of Two Plates. The plates that carry India and China are colliding. As a result, the Himalayas are being thrust upward. In this photograph, taken by astronauts in 1968, Mount Everest is one of the snow-covered peaks near the center. *(a: Digital image by Peter W. Sloss, NOAA-NESDIS-NGDC; b: Craig Aurness/ Corbis; c: Gemini 12, NASA; d: Apollo 7, NASA)*

Geologists have tested their models of Earth's interior by studying the response of the planet to earthquakes. These events produce a variety of **seismic waves,** vibrations that travel through Earth either as ripples, such as ocean waves, or by compressing matter, such as sound waves. Geologists use sensitive **seismographs** to detect and record these vibratory motions. The varying density and composition of Earth's interior affects the direction of seismic waves traveling through Earth. By studying the deflection of these waves, geologists have been able to determine the structure of the planet's interior.

Analysis of seismic recordings led to the discovery that Earth's solid core has a radius of about 1250 km (775 mi),

surrounded by a molten iron core with a thickness of about 2250 km (1375 mi). For comparison, the overall radius of our planet is 6378 km (3963 mi). The cores are composed of roughly 80% iron and 20% lighter elements, such as nickel, oxygen, and sulfur. The interior of our planet therefore has a curious structure: a liquid region sandwiched between a solid inner core and a solid mantle (see Figure 4-7b).

A sophisticated computer model of Earth's interior predicts that the solid inner core is rotating faster than the rest of the planet. Seismic evidence supports the model that the core goes around about 2° more per year than the surface. Refined models and observations like these are revealing more of the secrets of Earth's interior.

FIGURE 4-7 **Cutaway Model of Earth** Early Earth was initially a homogeneous mixture of elements with no continents or oceans. (a) Molten iron sank to the center and light material floated upward to form a crust. (b) As a result, Earth has a dense iron core and a crust of light rock, with a mantle of intermediate density between them.

The tremendous heat and molten rock inside Earth provide the energy that drives the motion of the tectonic plates in Earth's crust. That heat, from the pressure of the overlying rock and from the decay of radioactive elements, affects Earth's surface today. This heat transport is accomplished by **convection**. During this process a liquid or gas is heated from below, and as a result, it expands, becomes buoyant, and therefore rises, carrying the heat it received with it. Eventually, it releases the heat, cools, condenses, and sinks back down. This fluid motion causes

As a result of differentiation, Earth has the layered structure that we see today

a circular current, called a *convection current*, to arise. You witness convection every time you see liquid simmering in a pot. Heated from below, blobs of hot fluid move upward. The rising liquid releases its heat at its surface (Figure 4-8a).

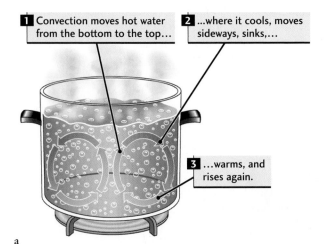

1 Convection moves hot water from the bottom to the top…
2 …where it cools, moves sideways, sinks,…
3 …warms, and rises again.

FIGURE 4-8 **Convection and the Mechanism of Plate Tectonics** (a) Heat supplied by the heating coil warms the water at the bottom of the pot. The heated water consequently expands and thereby decreases its density. This lower-density water rises (like bubbles in soda) and transfers its heat to the cooler surroundings. When the hot rising water gets to the top of the pot, it loses a lot of heat into the room, becomes denser, and sinks back to the bottom of the pot to repeat the process. (b) Convection currents in Earth's interior are responsible for pushing around rigid plates on its crust. New crust forms in oceanic rifts, where magma oozes upward between separating plates. Mountain ranges and deep oceanic trenches are formed where plates collide and crust sometimes sinks back into the interior. Note that not all tectonic plates move together or apart—some scrape against each other.

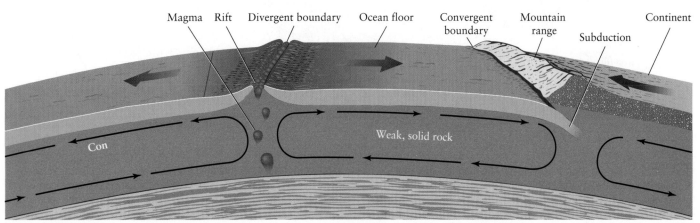

b

Heated from below, the rocks of Earth's upper mantle, although solid, are weak and hot enough to have an oozing, plastic flow (like stretching Silly Putty). As sketched in Figure 4-8b, molten rock, or *magma,* seeps upward along cracks or rifts in the ocean floor, where convection currents of mantle rock are carrying plates apart. Crustal rock sinks back down into Earth where plates collide. The continents ride on top of the plates as they are pushed around by the convection currents circulating beneath them.

Supercontinents like Pangaea apparently sow the seeds of their own destruction by blocking the flow of heat from Earth's interior. As soon as a supercontinent forms, temperatures beneath it increase. As heat accumulates, the supercontinent domes upward and cracks. Overheated molten rock wells up to fill the resulting rifts, which continue to widen as pieces of the fragmenting supercontinent are pushed apart.

Earth's molten, iron-rich interior affects Earth's exterior in another important way, one that has fewer obvious signs than tectonic plate motion. This is the creation of a magnetic field that extends through Earth's surface and out into space. As we will see next, that magnetic field results from the convection of molten metal inside Earth combined with Earth's rotation.

4-4   Earth's magnetic field shields us from the solar wind

If you have ever used a compass, you have seen the effect of Earth's magnetic field. This planetary field is quite similar to the field that surrounds a bar magnet (Figure 4-9a). Magnetic fields are created by electric charges in motion. Thus, the motion of your compass is caused by electric charges moving inside Earth. The electric wires in your house also carry moving charges called *electric currents* that, in turn, create magnetic fields of their own.

Many geologists believe that convection of molten iron in Earth's outer core, combined with our planet's rotation, creates the electric currents, which, in turn, create Earth's magnetic field. The details of this so-called **dynamo theory** of Earth's magnetic field are still being developed.

Heated from below, gases also rise. What does this suggest about Earth's atmosphere?

Magnetic fields nearly always form complete loops. (The unconnected lines in Figure 4-9b actually are completed outside of the boundaries of the figure.) The magnetic fields near Earth's south rotation pole loop out tens of thousands

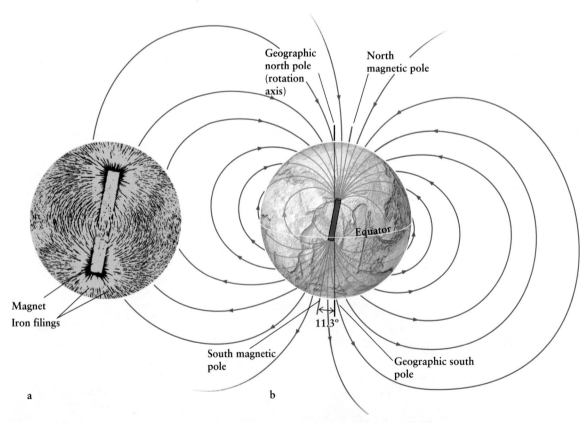

Geographic north pole (rotation axis)

North magnetic pole

Equator

Magnet
Iron filings

11.3°

South magnetic pole

Geographic south pole

a

b

FIGURE 4-9 **Earth's Magnetic Field** (a) The magnetic field of a bar magnet is revealed by the alignment of iron filings on paper. (b) Generated in Earth's molten, metallic core, Earth's magnetic field extends far into space. While Earth does not contain a bar magnet, its rotation, combined with moving electric charges in its core, create an equivalent field. Note that the field is not aligned with Earth's rotation axis. By convention, the magnetic pole near Earth's north rotation axis is called the magnetic north pole even though it is actually the south pole on a magnet! We will see similar misalignments and flipped magnetic fields when we study other planets. *(a: Jules Bucher/Photo Researchers)*

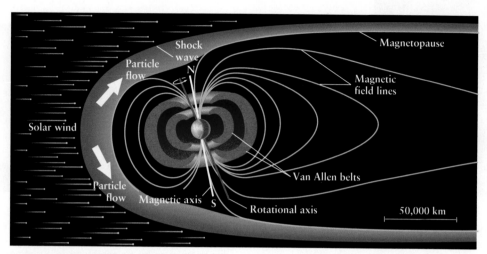

FIGURE 4-10 **Earth's Magnetosphere** A slice through Earth's magnetic field, which surrounds the entire planet, carves out a cavity in space that excludes charged particles ejected from the Sun, called the solar wind. Most of the particles of the solar wind are deflected around Earth by the fields in a turbulent region colored blue in this drawing. Because of the strength of Earth's magnetic field, our planet traps some charged particles in two huge, doughnut-shaped rings called the Van Allen belts (in red).

of kilometers into space and then return near Earth's north rotation pole. The places where the magnetic fields pierce Earth's surface most intensely are called the north and south magnetic poles. These poles move daily. Evidence in solidified lava also reveals that Earth's magnetic field completely reverses or flips (the north magnetic pole becomes the south magnetic pole, and vice versa) on an irregular schedule, ranging upward from a few tens of thousands of years. Each flip takes a few thousand years to complete. The last reversal was about 600,000 years ago. Today, the magnetic and rotation poles of Earth are about 11.3° apart; some planets have much larger angles between their magnetic and rotation poles. Figure 4-10 is a scale drawing of the magnetic fields around Earth, which comprise Earth's magnetosphere.

Our planet's magnetic field protects Earth's surface from bombardment by particles flowing from the Sun (see Chapter 7) called the **solar wind,** an erratic flow of electrically charged particles ejected from the Sun's upper atmosphere. Near Earth, the particles in the solar wind move at speeds of approximately 400 km/s, or about a million miles per hour.

Magnetic fields can change the direction of moving charged particles. Thus, the solar wind particles heading toward Earth are deflected by our planet's magnetic field. Many particles, therefore, pass around Earth. However, the field also traps some of these charged particles in two huge, doughnut-shaped rings, called the **Van Allen radiation belts,** which surround Earth (see Figure 4-10). The region between the two belts fills up with charged particles, only to be drained of them as a result of lightning on Earth, making that region unstable. These belts, discovered in 1958 during the flight of the first successful U.S. Earth-orbiting satellite, were named after the physicist James Van Allen, who insisted that the satellite carry a Geiger counter to detect charged particles.

INSIGHT INTO SCIENCE

Pace of Discovery Discoveries, such as the Van Allen belts, remind us of how much remains to be learned, even about our own planet. Scientists continue to make such fundamental discoveries for many reasons. For example, the technology necessary to make certain observations may just have been developed, or our equations that describe various phenomena may become more complete. Sometimes these equations are so complex that their implications are not immediately clear. Progress in science requires unquenchable curiosity and unfailing open-mindedness.

VIDEO 4.2 Sometimes the Van Allen belts overload with particles. The particles then leak through the magnetic fields near the poles and cascade down into Earth's upper atmosphere, usually in a ring-shaped pattern. As these high-speed, charged particles collide with gases in the upper atmosphere, the gases fluoresce (give off light) like the gases in a neon sign. The result is a beautiful, shimmering display, called the **northern lights (aurora borealis)** or the **southern lights (aurora australis),** depending on the hemisphere in which the phenomenon is observed (Figure 4-11).

Occasionally, a violent event on the Sun's surface, called a **coronal mass ejection,** sends a burst of protons and electrons straight through the Van Allen belts and into the atmosphere. These events deplete the ozone layer and cause spectacular auroral displays that can often be seen over a wide range of latitudes and longitudes. Coronal mass ejections also disturb radio transmissions and can damage communications satellites and electric transmission lines.

What happens to the Van Allen belts when Earth's magnetic field is flipping?

120

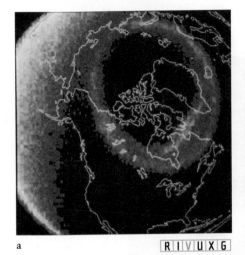

a RIVUXG

b RIVUXG

c RIVUXG

FIGURE 4-11 **The Northern Lights (Aurora Borealis)** A deluge of charged particles from the Sun can overload the Van Allen belts and cascade toward Earth, producing auroras that can be seen over a wide range of latitudes. (a) View of an aurora from the deep space satellite *Dynamics Explorer 1*. The green lines show the locations of land masses under the aurora. Auroras typically occur 100 to 400 km above Earth's surface. (b) View of aurora from space. Part of a Space Shuttle is visible at the bottom left of the picture. (c) Aurora borealis in Alaska. The gorgeous aurora seen here is mostly glowing green due to emission by oxygen atoms in our atmosphere. Some auroras remain stationary for hours, while others shimmer, like curtains blowing in the wind. (*a: L. A. Frank and J. D. Crave, University of Iowa; b: MTU/Geological & Mining Engineering & Sciences; c: J. Finch/Photo Researchers, Inc.*)

THE MOON AND TIDES

Earth's only natural satellite provides one of the most dramatic sights in the nighttime sky. The Moon is so large and so close to us that some of its surface features are readily visible to the naked eye. Even without a telescope, you can easily see dark gray and light gray areas that cover vast expanses of the Moon (Figure 4-12). People have dreamed of visiting our nearest neighbor in space for millennia. One of the first movies ever made (1902) was *A Trip to the Moon*,

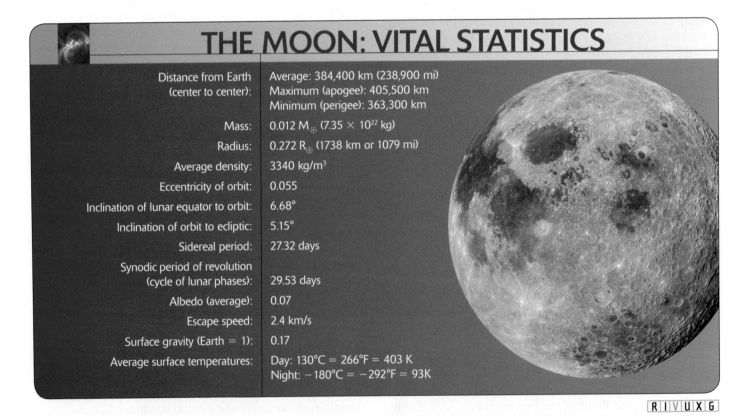

THE MOON: VITAL STATISTICS

Distance from Earth (center to center):	Average: 384,400 km (238,900 mi) Maximum (apogee): 405,500 km Minimum (perigee): 363,300 km
Mass:	0.012 $M_\oplus$ (7.35×10^{22} kg)
Radius:	0.272 $R_\oplus$ (1738 km or 1079 mi)
Average density:	3340 kg/m^3
Eccentricity of orbit:	0.055
Inclination of lunar equator to orbit:	6.68°
Inclination of orbit to ecliptic:	5.15°
Sidereal period:	27.32 days
Synodic period of revolution (cycle of lunar phases):	29.53 days
Albedo (average):	0.07
Escape speed:	2.4 km/s
Surface gravity (Earth = 1):	0.17
Average surface temperatures:	Day: 130°C = 266°F = 403 K Night: −180°C = −292°F = 93K

RIVUXG

FIGURE 4-12 **The Moon** Our Moon is one of seven large satellites in the solar system. The Moon's diameter of 3476 km (2160 mi) is slightly less than the distance from New York to San Francisco. This photograph is a composite of first-quarter and last-quarter views, in which long shadows enhance the surface features. (*NASA*)

based on Jules Verne's book *From the Earth to the Moon* (1866). The wonder of actually sending people there for the first time was one of the few things that has ever caused humanity to stop its quarreling long enough to participate, even vicariously, in a great adventure. To date, 12 people have walked on the Moon. Our exploration of it begins with a study of its surface.

4-5 The Moon's surface is covered with craters, plains, and mountains

Perhaps the most familiar and characteristic features on the Moon are its craters (Figure 4-13a). As discussed in Chapter 3 on the formation of the solar system, all the lunar craters that we have examined are the result of bombardment by meteoritic material (rather than volcanic activity). Nearly all of these craters are circular, indicating that they were not merely gouged out by moderate-speed rocks, which would typically create oval craters. Instead, the high-speed (upward of 180,000 km/h or 112,000 mi/h) collisions with the Moon's surface vaporized the rapidly moving debris, often with the power of a large nuclear bomb. The resulting explosions of hot gas and rubble produced the round-rimmed craters that we observe today (Figures 4-13 b, c, and d).

Meteoritic impact causes material from the crater site to be ejected onto the surrounding surface. This pulverized rock is called an **ejecta blanket.** You can see some of the more recent ejecta blankets, which are lighter-colored than the older ones (see Figure 4-13 and Figure 4-14). The ejecta blankets darken with time as their surfaces roughen from impacts by the solar wind and other particles from space. Therefore, the lightness of its ejecta is one clue astronomers use to determine how long ago a crater formed.

Craters larger than about 20 km often form *central peaks* (see Figure 4-13). These occur because the impact compresses the crater floor so much that, afterward, the crater rebounds and pushes the peak upward. As the peak goes up, the crater walls collapse and form *terraces*.

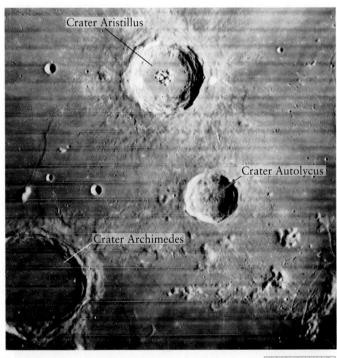

Crater Aristillus

Crater Autolycus

Crater Archimedes

a   R I V U X G

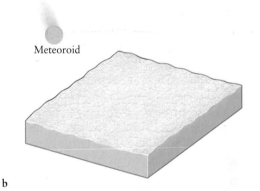

Meteoroid

b

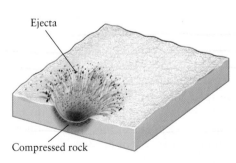

Ejecta

Compressed rock

c

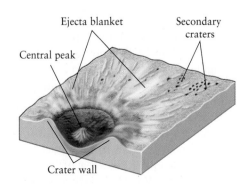

Ejecta blanket   Secondary craters

Central peak

Crater wall

d

FIGURE 4-13 **Lunar Craters** (a) This photograph, taken from lunar orbit by astronauts, includes the crater Aristillus. Note the crater's central peaks; collapsed, terraced crater wall; and ejecta blanket. Numerous smaller craters resulting from the impact pockmark the surrounding lunar surface. The following three drawings show the crater formation process. (b) An incoming meteoroid, (c) upon impact, is pulverized and the surface explodes outward and downward. (d) After the impact, the ground rebounds, creating the central peak and causing the crater walls to collapse. The lighter region is the ejecta blanket. *(a: 2004 Lunar and Planetary Institute/ Universities Space Research Association)*

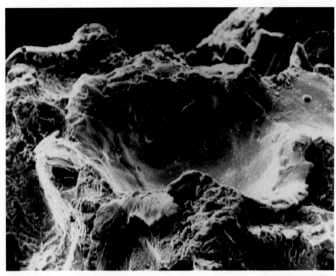

R I V U X G

FIGURE 4-14 **A Microscopic Lunar Crater** This photograph, made with a microscope, shows tiny microcraters less than 1 mm across on a piece of moon rock. *(NASA)*

Through an Earth-based telescope, some 30,000 craters, with diameters ranging from 1 km to more than 100 km, are visible. Following a tradition established in the seventeenth century, the most prominent craters are named after astronomers, physicists, mathematicians, and philosophers, such as Kepler, Copernicus, Pythagoras, Plato, and Aristotle. Close-up photographs from lunar orbit reveal millions of craters too small to be seen with Earth-based telescopes. Indeed, extreme close-up photos of the Moon's surface reveal countless microscopic craters (see Figure 4-14).

Earth, by comparison, has only about 215 known craters caused by impacts. Many craters on Earth have been pulled inside our planet by plate tectonic motion and obliterated. Many other craters of a few kilometers in diameter and smaller have been worn away (eroded) by the weathering effects of wind and water. Earth's atmosphere vaporized countless space rocks that would otherwise have created small craters here. Moreover, those that do hit are slowed down by the atmosphere so that they produce smaller craters than they otherwise would have. We will discuss Earth's craters further in Chapter 6.

Besides its craters, the most obvious characteristic of the Moon visible from Earth is that its surface is various shades of gray (see Figure 4-12). Most prominent are the large, dark gray plains called **maria** (pronounced *mar-ee-uh*). The singular form of this term, **mare** (pronounced *mar-ay*), means "sea" in Latin and was introduced in the seventeenth century when observers who used early telescopes thought that these features were large bodies of water. We know now that no water exists in liquid form on our satellite. Nevertheless, we retain these poetic, fanciful names, including Mare Tranquillitatis (Sea of Tranquility), Mare Nubium (Sea of Clouds), Mare Nectaris (Sea of Nectar),

and Mare Serenitatis (Sea of Serenity). As we will discuss shortly, maria are basins on the Moon that were filled in with lava after most cratering ended.

The largest of the maria is Mare Imbrium (Sea of Showers). It is roughly circular and measures 1100 km (700 mi) in diameter (Figure 4-15). Although the maria seem quite smooth in telescopic views from Earth, close-up photographs from lunar orbit and from the surface reveal that they contain small craters and occasional cracks, called **rilles** (Figure 4-16). Rilles are the collapsed roofs of underground lava rivers or the beds of long-gone surface lava rivers. When flowing underground lava rivers solidified, the rock they formed shrank and the roofs fell in on the spaces created.

As we will discuss further in Section 4-8, the same side of the Moon always faces Earth. One of the surprises stemming from the earliest lunar exploration is that the far side is much more heavily cratered than the near side. A few tiny maria, including Mare Moscoviense, Mare Ingenii, and Mare Australe, plus one moderate-sized mare, Orientale, grace the Moon's far side, the side we never see from Earth (Figure 4-17). Otherwise, craters and mountains cover the entire far side of the Moon.

Detailed measurements by astronauts in lunar orbit demonstrated that the maria on the Moon's Earth-facing side are 2 to 5 km below the average lunar elevation on that side, whereas the ones on the far side are 4 to 5 km above the

Mare: fewer craters, hence relatively young

500 km

Surrounding light-colored terrain: more craters, so relatively old

R I V U X G

FIGURE 4-15 **Mare Imbrium and the Surrounding Highlands** Mare Imbrium, the largest of the 14 dark plains that dominate the Earth-facing side of the Moon, is ringed by lighter-colored highlands strewn with craters and towering mountains. The highlands were created by asteroid impacts pushing land together. *(NASA)*

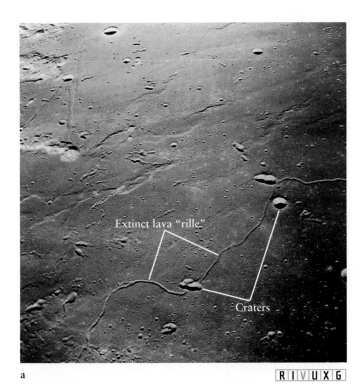

a   R I V U X G

b   R I V U X G

FIGURE 4-16 Details of a Lunar Mare (a) Close-up views of the lunar surface reveal rilles and numerous small craters on the maria. Astronauts in lunar orbit took this photograph of Mare Tranquillitatis (Sea of Tranquility) in 1969 while searching for potential landing sites

for the first human landing. At 1100 km (700 mi) across, this mare is the same size as the distance from London to Rome or from Chicago to Philadelphia. (b) Astronaut David Scott on Hadley's rille during the *Apollo 15* mission to the Moon. *(a: NASA; b: Kennedy Space Center/NASA)*

average lunar elevation. The flat, low-lying, dark gray maria cover only 17% of the entire lunar surface. The remaining 83% is composed of light gray, heavily cratered mountainous regions, called **highlands** (see Figure 4-12). The difference in heights of the maria on the two sides of the Moon apparently stems from the way they were formed. Those on the near side were created by flows of magma that filled low-lying basins (enormous craters) surrounding mountains (highlands) pushed up by large impacts. The small maria on the far side were more likely lava flows that pushed up and out relatively high in the mountains.

Which of the three named craters in Figure 4-13a is the youngest?

Lunar mountain ranges that rim the vast maria basins suggest ancient impacts far more violent than those that produced typical craters. One such mountain range can be seen around the edge of Mare Imbrium (see Figure 4-15). Although they take their names from famous terrestrial ranges, such as the Alps, the Apennines, and the Carpathians, the lunar mountains were not formed through plate tectonics, as mountains were on Earth. Analyses of Moon rocks and observations from lunar orbit indicate that violent impacts of huge asteroids thrust up the surrounding land to create the lunar highlands.

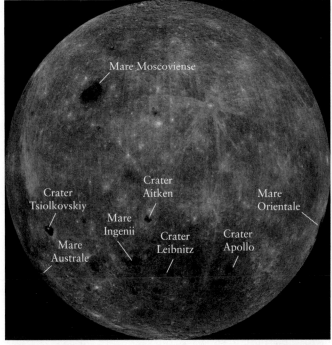

R I V U X G

FIGURE 4-17 The Far Side of the Moon A composite image from the *Galileo* spacecraft. Mare Orientale lies on the boundary between the near and far sides. *(NASA)*

4-6   Visits to the Moon yielded invaluable information about its history

Reaching the Moon became national obsessions for the United States and the former Soviet Union in the early 1960s. A series of successful space missions by the former Soviet Union spurred President John F. Kennedy's goal of landing a person on the Moon by the end of that decade.

The first of six manned lunar landings, *Apollo 11*, set down on Mare Tranquillitatis on July 20, 1969. Astronaut Neil Armstrong was the first human to set foot on the Moon. In the same year, *Apollo 12* also landed on a mare. Human voyages into space encountered a setback when *Apollo 13* experienced a nearly fatal explosion en route to the Moon. Fortunately, a heroic effort by the astronauts and ground personnel brought it safely home. *Apollo 14* through *Apollo 17* took on progressively more challenging lunar terrain. The period of human visitation to our Moon lasted less than 4 years, culminating in 1972, when *Apollo 17* landed amid mountains just east of Mare Serenitatis (Figure 4-18).

Astronauts discovered that the lunar surface is covered with fine powder and rock fragments. This layer, ranging in thickness from 1 to 20 m, is called the regolith (from the Greek, meaning "blanket of stone"; Figure 4-19). It formed during 4.5 billion years of relentless micrometeorite bombardment that pulverized the surface rock. We do not refer to this layer as soil, because the term *soil* suggests the pres-

R I V U X G

FIGURE 4-19 The Regolith The Moon's surface is covered by a layer of powdered rock that was created over billions of years as a result of bombardment by space debris. Called *regolith*, this surface material sticks together like wet sand, as illustrated by this *Apollo 11* astronaut's bootprint. *(NASA)*

ence of decayed biological matter, which is not found on the Moon's surface. The rough, powdered regolith absorbs most of the light incident on it. Therefore, the Moon's albedo is only 0.07, meaning that it scatters only 7% of the light striking it.

A major goal of the Apollo program was to study the geology of the Moon and help determine its history. The astronauts brought back a total of 382 kg (842 lb) of lunar rocks, which have provided important information

R I V U X G

FIGURE 4-18 An Apollo Astronaut on the Moon *Apollo 17* astronaut Harrison Schmitt enters the Taurus-Littow Valley on the Moon. The enormous boulder seen here slid down a mountain to the right of this image, fracturing on the way. This final Apollo mission landed in the most rugged terrain of any Apollo flight. *(NASA)*

R I V U X G

FIGURE 4-20 Mare Basalt This 1.53-kg (3.38-lb) specimen of mare basalt was brought back by *Apollo 15* astronauts in 1971. Small holes that cover about a third of its surface suggest that gas was dissolved in the lava from which this rock solidified. When the lava reached the airless lunar surface, bubbles formed as the pressure dropped and the gas expanded. Some of the bubbles were frozen in place as the rock cooled. *(NASA)*

about the early history of the Moon. By carefully measuring trace amounts of radioactive elements, geologists determined that lunar rock from the maria is solidified lava that dates back only 3.1 to 3.8 billion years. This indicates that the maria developed about a billion years after the Moon formed and that the maria were formed from molten rock. Indeed, these Moon rocks are composed mostly of the same minerals that are found in volcanic rocks in Hawaii and Iceland. The rock of these low-lying lunar plains is called **mare basalt** (Figure 4-20). About 17% of the Moon's surface rock is basalt.

In contrast to the dark mare basalt, the lunar highlands are covered with a light-colored rock called **anorthosite** (Figure 4-21). On Earth, anorthositic rock is found only in very old mountain ranges, such as the Adirondacks in the eastern United States. Compared to the mare basalts, which have more of the heavier elements, such as iron, manganese, and titanium, anorthosite is rich in calcium and aluminum. Anorthosite is, therefore, less dense than basalt. Many highland rocks brought back to Earth are **impact breccias** (Figure 4-22), which are composites of different rock fused together as a result of meteorite impacts. Typical anorthositic specimens from the highlands are between 4.0 and 4.3 billion years old. These ancient specimens of highland rock probably represent samples of the Moon's original crust (see Figure 4-17).

Four of the five Apollo missions that landed on the Moon set up scientific packages containing seismic detectors (Figure 4-23). These devices are sensitive to moonquakes, motion of the Moon's surface analogous to earthquakes. Between 1969 and 1977, they detected more than 12,500 events, including a few quakes strong enough to knock books off shelves (if there had been shelves or books there

R I V U X G

FIGURE 4-22 **Impact Breccias** These rocks are created from shattered debris fused together under high temperature and pressure. Such conditions prevail immediately following impacts of space debris on the Moon's surface. *(NASA)*

at the time). Moonquakes occur for any of four reasons: impacts, motion of the lunar surface due to changing tides created by Earth's gravity, expansion and contraction due to heating and cooling of the Moon's surface, and other activity inside the Moon.

R I V U X G

FIGURE 4-21 **Anorthosite** The lunar highlands are covered with this ancient type of rock, which is believed to be the material of the original lunar crust. This sample's dimensions are 18 cm × 16 cm × 7 cm. Although this sample is medium gray, other anorthosites retrieved from the Moon have been white, while others are darker gray than this one. This one was brought back by *Apollo 16* astronauts. *(NASA)*

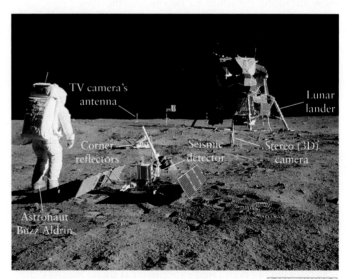

R I V U X G

FIGURE 4-23 *Apollo 11* **Landing Site** On the Moon's Sea of Tranquility, Astronaut Buzz Aldrin stands next to the package of equipment containing the seismic detector. The corner reflectors are used, even today, to determine the distance from Earth to the Moon. The stereo camera took pairs of images of the Moon's surface. Seeing them through special glasses gives a 3-D close-up view of the Moon's surface. The bottom half of the lander is still on the Moon's surface. The top half brought astronauts Neil Armstrong and Buzz Aldrin back into lunar orbit, where they transferred to the command module to fly home. *(NASA)*

With several seismic detectors working simultaneously, geologists were able to deduce information about the Moon's interior from the timing and strength of the vibrations that were detected. The data suggested that the Moon has a small iron core, possibly a layer of hot rock above it, and a thick layer of solid rock above that, surrounded by the crust. The presence of the iron core, amounting to only 3% or 4% of the Moon's total mass, was confirmed by data from the *Lunar Prospector* satellite in 1998. This compares to Earth's core, which has about 30% of our planet's total mass.

Since spacecraft began orbiting the Moon, scientists have noted that their orbits do not follow the expected elliptical paths. Every once in a while, usually over a circular mare, the spacecraft momentarily dip Moonward, as they are pulled by a higher than normal concentration of mass. These *mass concentrations,* or **mascons,** are local regions of relatively dense rock and metal near the Moon's surface. The mascons found in maria are believed to be magma that filled the bottom of the mare after the basins were formed by impacts of very large bodies. The magma solidified after emerging from inside the Moon. New mascons continue to be discovered, including seven found by the *Lunar Prospector* spacecraft in 1998.

In 1994, the lunar-orbiting *Clementine* spacecraft sent radio signals to the Moon's surface. These signals scattered off the materials there, and then they were received and analyzed on Earth. The results were intriguing, suggesting that frozen water exists in craters near the Moon's south pole. If so, this ice is not in danger of being evaporated by sunlight, because, just as is the case with Earth's poles, the Sun deposits very little energy at the lunar poles (Figure 4-24). The *Lunar Prospector,* using different technology, found evidence for ice at the Moon's north pole; however, when the same spacecraft was intentionally crashed into a crater at the south pole, no water vapor was released. This isolated experiment is by no means definitive. However, in 2003, radio experiments from Earth similar to those run on *Clementine* also failed to see indications of water. Clearly, the issue of whether there are large quantities of water on the Moon is unresolved. If it exists there, the ice was probably deposited on the Moon during comet and asteroid impacts; estimates based on the first observations are that more than 3 billion tons of water may be encased there, either mixed in with the regolith or in layers below it.

The presence of water ice would greatly simplify human settlement of the Moon, because ice can provide us

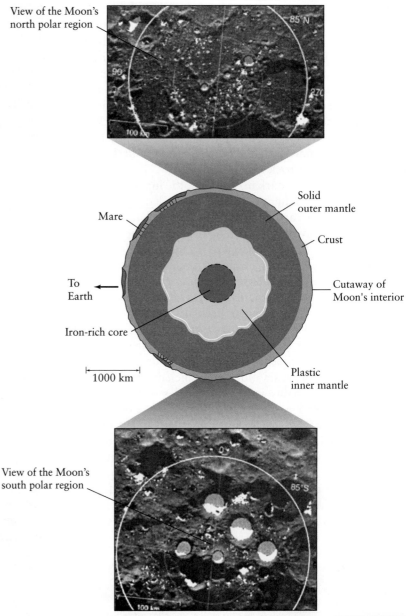

R I V U X G

FIGURE 4-24 The Moon's Interior Based on seismic experiments left on the Moon by Apollo astronauts and the studies done by *Lunar Prospector,* we know that the Moon has a crust, mantle, and core. The lunar crust has an average thickness of about 60 km on the Earth-facing side but about 100 km on the far side. The crust and solid upper mantle extend to about 800 km, where the nonrigid inner mantle begins. The Moon's core has a radius of between 300 and 425 km. Although the main features of the Moon's interior are analogous to those of Earth, the proportions and details are quite different. The *Clementine* spacecraft revealed that the south pole has a significant basin. The crust was apparently stripped away by an impact. Top inset: A radar image of the Moon's north polar region. The areas computer-colored in white and light gray are regions where the Sun never shines and that may contain water. Bottom inset: A radar image of the Moon's south polar region, also showing regions of permanent or near-permanent darkness. *(Insets: NASA/Galaxy)*

not only with liquid water, but also with oxygen and rocket fuel. For example, the Space Shuttle's primary rocket engine is fueled by liquid oxygen and liquid hydrogen. Separating the Moon's water into liquid oxygen and liquid hydrogen would create the same fuel. An existing water reservoir would make establishing a lunar colony much less expensive and more feasible than lugging all that water from Earth or dragging a water-rich comet to the Moon.

Astronauts found no traces of life on the Moon. Life as we know it requires liquid water, of which the Moon has none. The Moon has an atmosphere of sodium gas, so thin as to be considered an excellent vacuum compared to the density of the air we breathe. An atmosphere is held around a planet or moon by that body's gravitational force. If the atmospheric gases are too hot (moving too fast) for the gravitational force to hold them, they leak into space. Our Moon has so little gravitational attraction (about one-sixth what we feel at Earth's surface) that it is unable to retain its tenuous atmosphere. The atmosphere is renewed by sunlight striking the Moon, releasing more gas from its rocks. Earth, in contrast, has enough mass to retain such gases as nitrogen, oxygen, carbon dioxide, water, and argon.

The Moon's slow rotation rate and small iron core suggest that unlike Earth, the Moon should not have a strong magnetic field. Spacecraft orbiting the Moon have confirmed this. However, the *Lunar Prospector* has discovered local areas where the magnetic fields are strong enough to keep the solar wind away from the Moon's surface. These anomalous regions have yet to be explained.

What type of rock did astronauts from *Apollo 11,* the first flight to land on the Moon, bring back?

4-7   The Moon probably formed from debris cast into space when a huge asteroid struck the young Earth

Before the Apollo program, scientists debated three different theories about the origin of the Moon. The **fission theory** holds that the Moon was pulled out from a rapidly rotating proto-Earth. This theory does not explain why the Moon's overall chemistry is like that of Earth's outer layers and not its core, and that the Moon has virtually no water incorporated in its rocks, as revealed by the bone-dry samples brought back by Apollo astronauts. The **capture theory** posits that the Moon was formed elsewhere in the solar system and then drawn into orbit around Earth by gravitational forces. However, it is physically very difficult for a planet to capture such a large moon. Furthermore, because bodies formed in different places have different overall chemical compositions, this theory does not explain the similar geochemistries of the Moon and Earth's surface. The **co-creation theory** proposes that Earth and the Moon were formed near each other at the same time. As with the fission theory, this theory fails to explain why, compared

to Earth, the Moon has less of the denser elements, such as iron, and why certain types of rocks found on the Moon are not found on Earth.

A fourth theory, now held to be correct by most astronomers, was first proposed in the 1980s. It applies the established fact of age-old collisions in the solar system to the formation of the Moon. Called the **collision-ejection theory,** or *large impact theory,* it proposes that the newly formed Earth was struck at an angle by a Mars-sized asteroid that literally splashed some of Earth's surface layers into orbit around the young planet. Evidence from Moon rocks places this event at around 4.5 billion years ago, within the first 100 million years of Earth's existence. A computer simulation of this cataclysm is shown in Figure 4-25. This orbiting matter became a short-lived ring that eventually condensed into the Moon, just as planetesimals condensed to form Earth.

The collision-ejection theory is consistent with the known facts about the Moon. For example, rock vaporized by the impact would have been depleted of *volatile* (easily evaporated) elements and water, leaving the Moon rocks parched, as we now know they are. Also, the Moon has little heavy iron-rich matter because that material had sunk deep into Earth before the asteroid struck. The material from Earth that splashed into orbit and formed the Moon was mostly the lighter rock floating on Earth's surface. Furthermore, most of the debris from the collision would orbit near the plane of the ecliptic as long as the orbit of the impacting asteroid had been near that plane. Also, the impact of an object large enough to create the Moon could have tipped Earth's axis of rotation and so inaugurated the seasons (Figure 4-26). The fact that this theory is both consistent with observations and explains two seemingly separate things (the existence of the Moon and the tilt of Earth) makes it especially appealing to scientists.

The surface of the newborn Moon probably remained molten for millennia, due to heat released during the impact of rock fragments falling onto the young satellite and the decay of short-lived radioactive isotopes inside it. As the Moon gradually cooled, low-density lava floating on its surface began to solidify into the anorthositic crust that exists today in the highlands.

By correlating the ages of Moon rocks with the density of craters at the sites where they were collected, geologists have found that the rate of impacts on the Moon changed over the ages. The ancient, heavily cratered lunar highlands are evidence of intense bombardment of asteroids that dominated the Moon's early history. This barrage extended from 4.5 billion years ago, when the Moon's surface first solidified, until about 3.8 billion years ago. This period of impacts was punctuated by a severe pounding 4 billion years ago, part of a solar-system–wide reign of terror.

The frequency of impacts gradually tapered off as meteoroids and planetesimals were swept up by the newly formed planets, as discussed in Chapter 2. Recorded among the final scars at the end of this crater-making era are the impacts of more than a dozen objects, each measur-

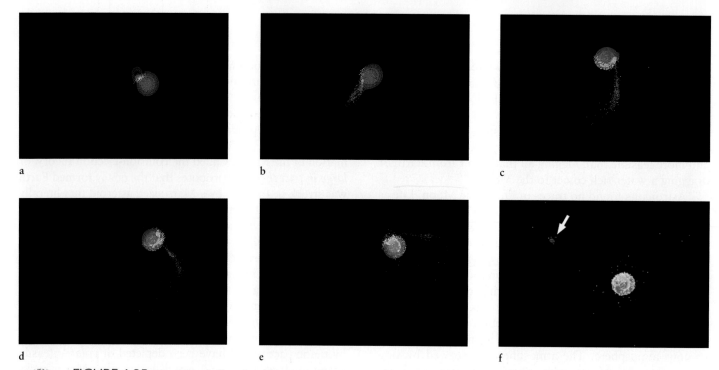

a b c

d e f

FIGURE 4-25 **The Moon's Creation** This computer simulation models the creation of the Moon from material ejected by the impact of a large planetlike body with the young Earth. To follow the ejected material as it moves away from Earth, successive views (a–f) pan out to show increasingly larger volumes of space. Blue and green indicate iron from the cores of Earth and the asteroid; red, orange, and yellow indicate rocky mantle material. In this simulation, the impact ejects both mantle and core material, but most of the dense iron falls back onto Earth. The surviving ejected rocky matter coalesces to form the Moon (indicated by arrow). *(W. Benz)*

ing at least 100 km across. As these huge rocks slammed into the young Moon, they blasted gigantic craters in regions that would later become the maria. Meanwhile, heat from the decay of long-lived radioactive elements, such as uranium and thorium, again began to melt the inside of the Moon. Then, from 3.8 to 3.1 billion years ago, great floods of molten rock gushed up from the lunar interior through the weak spots in the crust created by these large

impacts. Lava filled the impact basins and created the maria we see today. Because most cratering was over by then, maria have few craters.

The reason that the far side of the Moon has only one significant mare apparently stems from the Moon's crust on the side facing Earth being thinner than the crust on the far side. This difference was created when the Moon was young and molten. At that time, the Moon was much closer to Earth (we will explore this shortly). Its rapid orbit and the tides created on it by Earth caused much of the low-density, crustal lava to flow away from Earth to the lunar far side, where it remained when the crust solidified. It was therefore easier for the molten rock inside the Moon to escape through the large craters on the thin crust of our side than through those on the thicker-crusted far side.

Relatively little large-scale activity has happened on the Moon since those ancient times. Although only a few fresh large craters have been formed, the small-sized debris still plentiful in the solar system continues to create small

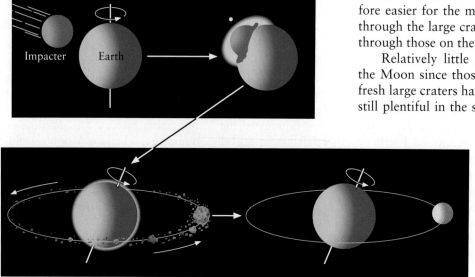

FIGURE 4-26 **Tilting Earth's Axis** The collision that created the Moon could have also knocked Earth's rotation axis over so that today it has a 23½° tilt, thereby creating the seasons.

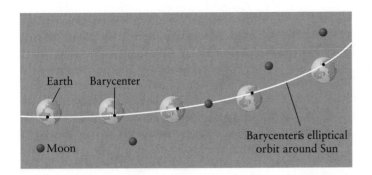

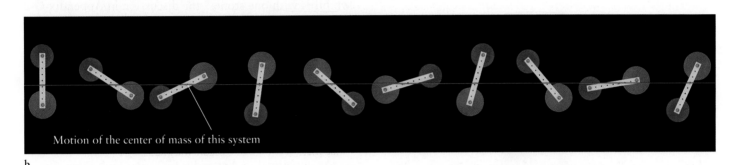

FIGURE 4-27 **Motion of the Earth–Moon System** (a) The paths of Earth and the Moon as their barycenter follows an elliptical orbit around the Sun. (b) Analogously, this time-lapse image shows a pair of disks connected by a piece of wood sliding across a table. Note that its center of mass moves in a straight line, while its other parts follow curved paths. (Tom Patagnes)

ones. Back in 1953, an impact was observed that has been connected with the formation of a 1.5-km (0.93-mi) diameter crater. Smaller impacts are observed nearly every year when large groups of space debris are known to be crossing the Moon's path. By and large, however, the astronauts visited a world that has remained largely unchanged for more than 3 billion years. Unlike Earth, which is still quite geologically active, any molten rock inside the Moon has apparently stopped directly affecting that world's surface. Astronomers hope to learn more about the Moon's interior when modern seismographs are placed on its surface in the not-too-distant future.

Why are we fortunate on Earth that the Moon is rarely being struck today?

4-8 Tides have played several important roles in the history of Earth and the Moon

Gravitational attraction keeps Earth and the Moon together, while the angular momentum (see Appendix H-2) given to the Moon when it was formed keeps the Moon from being pulled back onto Earth. Contrary to appearances, the Moon does not orbit Earth. Rather, both bodies are in motion around their center of mass, called their *barycenter* (Figure 4-27a), just as a pair of objects tied together and sent spinning across an air table move around a similar common point (Figure 4-27b).

The same surface features on the Moon always face Earth, regardless of what phase the Moon is in (see Figure

1-21). This occurs because the Moon is in **synchronous rotation** around Earth: The Moon rotates on its axis at the same rate that it and Earth orbit their barycenter (Figure 4-28). To see why the Moon must be rotating if it keeps the same

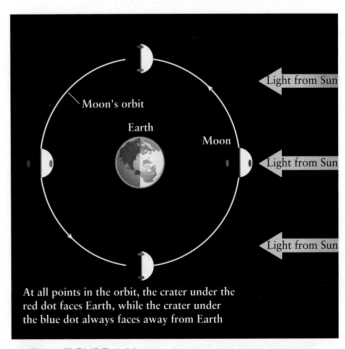

At all points in the orbit, the crater under the red dot faces Earth, while the crater under the blue dot always faces away from Earth

FIGURE 4-28 **Synchronous Rotation of the Moon** The motion of the Moon around Earth as seen from above Earth's north polar region (ignoring Earth's orbit around the Earth–Moon barycenter). For the Moon to keep the same side facing Earth as it orbits our planet, the Moon must rotate on its axis at precisely the same rate that it revolves around Earth.

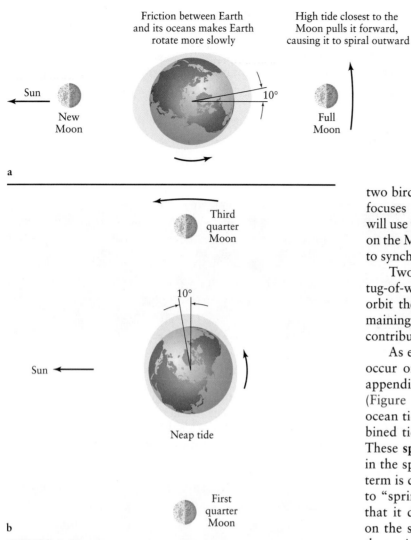

Friction between Earth and its oceans makes Earth rotate more slowly

High tide closest to the Moon pulls it forward, causing it to spiral outward

Sun

New Moon

10°

Full Moon

a

Third quarter Moon

10°

Sun

Neap tide

First quarter Moon

b

FIGURE 4-29 **Tides on Earth** The gravitational forces of the Moon and the Sun deform the oceans. Due to Earth's rapid rotation, the high tide closest to the Moon leads it around Earth by about 10°. (a) The greatest deformation (spring tides) occurs when the Sun, Earth, and Moon are aligned with the Sun and Moon either on the same or opposite sides of Earth. (b) The least deformation (neap tides) occurs when the Sun, Earth, and Moon form a right angle.

side facing us, imagine standing up, holding your arm out in front of you with your palm facing you, and twirling yourself around. You are Earth, while your palm represents the side of the Moon facing Earth. The only way your palm can always face you (be in synchronous rotation) is if your hand (the Moon) rotates at exactly the same rate that it swings (revolves). So the Moon must be rotating.

We have seen the Moon's "far side" only from spacecraft (see Figure 4-17). The far side of the Moon is often, and incorrectly, called its "dark side." The dark side of the Moon is exactly like the dark side of Earth—it is the side where the Sun is below the horizon. Whenever we view less

than a full Moon, we see part of its dark side and some sunlight is then striking part of its far side.

The origin of the Moon's synchronous rotation is in the tidal force that acts between Earth and the Moon. This tidal force also creates ocean tides on Earth. To understand how the Moon came to have synchronous rotation, we investigate the forces acting between Earth and the Moon. To "kill two birds with one stone," the discussion in Appendix G-2 focuses on how the Moon creates the tides on Earth. We will use this information to see how Earth once created tides on the Moon that changed the Moon's original rotation rate to synchronous rotation.

Two-thirds of our tides result from the gravitational **5** tug-of-war between Earth and the Moon as the two bodies orbit their barycenter. The Sun generates most of the remaining third of the tides, with Jupiter and the other planets contributing a tiny fraction.

As explained in Appendix G-2, two high tides always **6** occur on opposite sides of Earth (see the figure in that appendix). When the Sun, Moon, and Earth are aligned (Figure 4-29a), the Sun and Moon create pairs of high ocean tides in the same directions, and the resulting combined tides are the highest high tides of the lunar cycle. These **spring tides** (so called not because they only occur in the spring—they occur in all seasons—but because the term is derived from the German word *springen,* meaning to "spring up") occur at every new and full Moon. Note that it does not matter whether the Sun and Moon are on the same or opposite sides of Earth when generating the spring tides, because the Sun and Moon both create high tides on opposite sides of Earth. At first quarter and third quarter, the Sun and Moon form a right angle as seen from Earth (Figure 4-29b). The gravitational forces they exert compete with each other, so the tidal distortion is the least pronounced. The especially small tidal shifts on these days are called **neap tides.**

Finally, we can explain the Moon's synchronous rotation: When the Moon was young, it was molten and Earth's gravity created huge tides of molten rock up to 18 m (60 ft) high there. Friction generated by these tides caused the Moon to speed up or slow down (we do not know which) until one high tide there was fixed directly between the Moon's center and the center of Earth. As a result, the Moon's rotation rate became the same as its revolution rate. When it solidified, the Moon was locked into this synchronous rotation.

Why haven't the ocean tides on Earth put our planet into synchronous rotation with respect to the Moon?

4-9   The Moon is moving away from Earth

A century ago, Sir George Darwin (son of the evolutionist Charles Darwin) proposed that the Moon is moving away from Earth. To understand why, Figure 4-29 shows that the high tide nearest the Moon is actually 10° ahead of the Moon in its orbit around Earth. This offset occurs because friction between the ocean and the ocean floor causes the rapidly spinning Earth to drag the tidal bulge ahead of the line between the center of Earth and the center of the Moon. The gravitational force from the water in the high ocean tide nearest the Moon acts back on the Moon, pulling it ahead in its orbit, giving it energy and thereby forcing it to spiral outward. This is similar to the effect you would get if you tied a ball on a string, let it hang down, and then started spinning yourself around. Your hand would act through the string, giving the ball energy and starting it to spin and move farther and farther away from your body.

To test the concept that the Moon is moving away from Earth, *Apollo 11, 14,* and *15* astronauts placed on the Moon sets of reflectors (see Figure 4-23), similar to the orange and red ones found on cars. These reflectors are specially designed so that light striking them from any direction is reflected back toward its source. Pulses of laser light were fired at the Moon from Earth (Figure 4-30), and the time it took the light to reach the Moon, bounce off a reflector array, and return to Earth was carefully recorded. Knowing the speed of light and the time it takes for the light pulse to make the round trip, astronomers can calculate the distance to the Moon with an error of only a few millimeters. From such measurements over time, astronomers have established that the Moon is currently spiraling away from Earth at a rate of 3.8 cm per year.

This means, of course, that the Moon used to be closer to Earth. Although the initial distance of the Moon when it coalesced is not known, its present rate of recession tells us that it was at least half its present distance; more accurate calculations suggest one-tenth. At that distance, the tides on Earth were a thousand times higher than they are today.

Where does the Moon get the energy necessary to spiral away from Earth? The answer is from Earth itself. As the tides move westward to try to stay under the Moon, they encounter the eastward-moving landmasses, which disrupt the water's motion. Because the oceans move westward, while the planet rotates eastward, the oceans actually push on the continents opposite to Earth's direction of rotation. The result is that the tides are slowing Earth's rotation. The energy lost by Earth as it spins down is gained by the Moon. At present, the day is slowing down by about one-thousandth of a second each century. Based on fossil and other evidence, geologists calculate that when Earth first formed, the day was only between 5 h and 6 h long.

Will the Moon ever leave Earth completely as a result of these dynamics? In fact, it will not. Rather, Earth's rotation will slow down until it is in synchronous rotation with respect to the Moon. Thereafter, the Moon will remain at a fixed distance over one side of Earth. However, billions of years before that happens, the Sun is going to self-destruct!

What effects would the original tides on Earth have had on the geology of our young planet?

4-10   Frontiers yet to be discovered

Earth and the Moon still hold many mysteries. We have yet to understand the different rotation rate of Earth's core from its surface, as well as the details of the dynamo effect that generates Earth's magnetic field. We do not know precisely how fast Earth was rotating initially. We also need to better understand the impact of humans on Earth and its atmosphere.

The Moon still inspires many questions and hides many secrets. The six American-manned missions and three Soviet soft, robotic lunar landings have barely scratched the Moon's surface, bringing back samples from only nine locations. We still know very little about the Moon's far side. Also, is water really buried at the poles? Is the Moon's interior molten? How massive is the Moon's iron-rich core? How old are the youngest lunar rocks? Did lava flows occur over western Oceanus Procellarum only 2 billion years ago, as crater densities there suggest? Is the Moon really geologically dead, or does it just appear dead because our examination of the lunar surface has been so cursory? Such questions can be answered only by sending new and improved science experiments to the Moon.

RIVUXG

FIGURE 4-30 Lunar Ranging Beams of laser light are fired through three telescopes at the Observatoire de la Côte d'Azur, France. The light is then reflected back by the corner reflectors placed on the Moon by Apollo astronauts. From the time it takes the light to reach the Moon and return to Earth, astronomers can determine the distance to the Moon to within a few millimeters. *(Observatoire de la Côte d'Azur)*

SUMMARY OF KEY IDEAS

Earth: A Dynamic, Vital World

• Earth's atmosphere is about four-fifths nitrogen and one-fifth oxygen. This abundance of oxygen is due to the biological processes of life-forms on the planet.

• Earth's atmosphere is divided into layers named the troposphere, stratosphere, mesosphere, and ionosphere. Ozone molecules in the stratosphere absorb ultraviolet light rays.

• The outermost layer, or crust, of Earth offers clues to the history of our planet.

• Earth's surface is divided into huge plates that move over the upper mantle. Movement of these plates, a process called plate tectonics, is caused by convection in the mantle. Also, upwelling of molten material along cracks in the ocean floor produces seafloor spreading. Plate tectonics is responsible for most of the major features of Earth's surface, including mountain ranges, volcanoes, and the shapes of the continents and oceans.

• Study of seismic waves (vibrations produced by earthquakes) shows that Earth has a small, solid inner core surrounded by a liquid outer core. The outer core is surrounded by the dense mantle, which in turn is surrounded by the thin, low-density crust on which we live. Earth's inner and outer cores are composed primarily of iron. The mantle is composed of iron-rich minerals.

• Earth's magnetic field produces a magnetosphere that surrounds the planet and blocks the solar wind.

• Some charged particles from the solar wind are trapped in two huge, doughnut-shaped rings called the Van Allen radiation belts. A deluge of particles from a coronal mass ejection by the Sun can initiate an auroral display.

The Moon and Tides

• The Moon has light-colored, heavily cratered highlands and dark-colored, smooth-surfaced maria.

• Many lunar rock samples are solidified lava formed largely of minerals also found in Earth rocks.

• Anorthositic rock in the lunar highlands was formed between 4.0 and 4.3 billion years ago, whereas the mare basalts solidified between 3.1 and 3.8 billion years ago. The Moon's surface has undergone very little geologic change over the past 3 billion years.

• Impacts have been the only significant "weathering" agent on the Moon; the Moon's regolith (pulverized rock layer) was formed by meteoritic action. Lunar rocks brought back to Earth contain no water and are depleted of volatile elements.

• Frozen water may have been discovered at the Moon's poles.

• The collision-ejection theory of the Moon's origin, accepted by most astronomers, holds that the young Earth was struck by a huge asteroid, and debris from this collision coalesced to form the Moon.

• The Moon was molten in its early stages, and the anorthositic crust solidified from low-density magma that floated to the lunar surface. The mare basins were created later by the impact of planetesimals and were then filled with lava from the lunar interior.

• Gravitational interactions between Earth and the Moon produce tides in the oceans of Earth and set the Moon into synchronous rotation. The Moon is moving away from Earth, and, consequently, Earth's rotation rate is decreasing.

WHAT DID YOU THINK?

1 *Can Earth's ozone layer, which is now being depleted, be naturally replenished?* Yes. Ozone is created continuously from normal oxygen molecules by their interaction with the Sun's ultraviolet radiation.

2 *Who was the first person to walk on the Moon, and when did this event occur?* Neil Armstrong was the first person to set foot on the Moon. He and Buzz Aldrin flew on the *Apollo 11* spacecraft piloted by Michael Collins. Armstrong and Aldrin set down the Eagle Lander on the Moon on July 20, 1969.

3 *Do we see all parts of the Moon's surface at some time throughout the lunar cycle of phases?* No. Because the Moon's rotation around Earth is synchronous, we always see the same side. The far side of the Moon has been seen only from spacecraft that pass or orbit it.

4 *Does the Moon rotate and, if so, how fast?* The Moon rotates at the same rate that it revolves around Earth. If the Moon did not rotate, then, as it revolved, we would see its entire surface from Earth, which we do not.

5 *What causes the ocean tides?* The tides are created by orbital and gravitational forces, primarily from the Moon and, to a lesser extent, from the Sun.

6 *When does the spring tide occur?* Spring tides occur twice monthly, during each full and new Moon.

Review Questions

1. The Moon's surface is best described as: **a.** fine-grained powder, **b.** solid rock, **c.** rocky rubble, **d.** liquid water oceans and dry land, **e.** molten rock.

2. What type of chemical or molecule is most common in Earth's atmosphere? **a.** carbon dioxide, **b.** oxygen, **c.** water, **d.** nitrogen, **e.** hydrogen

3. Why does Earth's albedo change daily? Seasonally?

4. Why is Earth's surface not riddled with craters as is that of the Moon?

5. List the layers of Earth's atmosphere.

6. Describe the process of plate tectonics. Give specific examples of geographic features created by plate tectonics.

7. How do we know about Earth's interior, given that the deepest wells and mines extend only a few kilometers into its crust?

8. Describe the interior structure of Earth.

9. Why is the center of Earth not molten?

10. Describe Earth's magnetosphere.

11. What are the Van Allen belts?

12. What kind of features can you see on the Moon with a small telescope?

13. On the basis of lunar rocks brought back by the astronauts, explain why the maria are dark-colored, but the lunar highlands are light-colored.

14. Why are there so few craters on the maria?

15. Briefly describe the main differences and similarities between Moon rocks and Earth rocks.

16. How do we know that the maria were formed after the lunar highlands?

17. What is a tidal force? How do tidal forces produce tides in Earth's oceans?

18. What is the difference between spring tides and neap tides? During which phase(s) do each occur?

19. Why do most scientists support the collision-ejection theory for the Moon's formation?

20. Why are virtually all of the craters on the Moon circular, even though many impacts there were not head-on?

Observing Projects

21. Use the *Starry Night Enthusiast*™ program to view Earth as it appears from the Moon. Select **Favourites > Solar System > Moon > Eagle has Landed** to view Earth from the *Apollo 11* landing site. Stop time flow and press the "k" key on the keyboard to remove the constellation display from the view. Click the **Now** button in the toolbar. Use the Hand Tool to locate Earth in the lunar sky. If Earth is near the "new" phase, it will be hard to see, but the label will help. With the cursor over the image of Earth in the view, click the right mouse button (**Ctrl–Click** on a Macintosh) to open the object contextual menu for Earth and select **Magnify**. Change the **Time Flow Rate** to **10 minutes** and run time forward for a while. **a.** Observe the surface features of Earth that you see. Do they change? If so, explain why, considering that there is no significant change in the face of the Moon as seen from Earth. **b.** Run time for a much longer period and describe the phases of Earth, ignoring the changes in features.

22. You can use *WorldWide Telescope* (WWT) to observe the changing appearance of the Moon under different illumination by the Sun, and examine some of its surface features in great detail. **a.** Open WWT, click

on **Explore**, click on the **Solar System** thumbnail, and click on the **Moon** thumbnail. Open **View** and ensure that you are at your home location in the **Observing Location** pane. To prevent the horizon from obstructing the view, click **off** the **View from this location** option in the **Observing Location** pane. Check that **UTC** is **off** and click **Now** to display the Moon at its current phase, against the background of stars. (You might try comparing this view with the real Moon if this is visible from your observing site.) Run **Time** forward and adjust the time rate to watch the motion of the Moon against the background stars and the change of its phases as the boundary between dark and light, the *terminator,* moves across the Moon's surface (start the time rate at ×1000 and increase as necessary). Do you see all of the Moon at some time during the cycle of phases, or do you always see the same side? What conclusion can you draw about the rotation of the Moon as it revolves in its orbit around Earth? From these observations, is there any meaning to the phrase "dark side of the Moon"? **b.** Run **Time** forward until the Moon is very close to quarter phase and note the **Date**. To measure the approximate length of the lunar cycle, run time forward at ×100,000 until you reach the same phase again, with the light and dark quadrants in the same alignment. Note the **Date** and calculate the difference between the two dates, which is the length of the lunar cycle. **c.** You can examine the detailed surface of the Moon with WWT. Open the **Look At** drop-down list and select **Planet**. Open the **Imagery** drop-down list and click on **Moon**. Zoom in until the Moon almost fills the view. You can rotate this image in any direction by holding the left mouse button while moving the mouse. Rotate the Moon around its equator and examine the distribution of structure on the Moon's surface. Is the structure uniformly distributed or are there two distinct hemispheres? If the latter is true, then describe these differences and determine which of these hemispheres is facing Earth, on the basis of your examination of the Moon in earlier sections or of photographs of the Moon taken from Earth in this textbook. Use the Longitude (LNG) and Latitude (Lat) in the lower right side of the window to give the sides facing toward and away from Earth, if such is the case. **d.** The Moon image in WWT can be viewed at high resolution. You can therefore examine features of the surface in great detail. Zoom in on the smooth regions of the Moon, the so-called mare, and examine **craters**, both large and small. Describe their characteristics in terms of crater wall structure, central features, shape (for example, are they mostly circular, or are there oval craters?) and compare and contrast these characteristics for large and small craters. Look for valleys, known as rilles, threading their way between mountain ranges. One example is at latitude 0°, longitude +24°. (These coordinates are displayed at the bottom right of the screen). Are these rilles mostly straight, or do they wind around and show twists and turns? Look for evidence of flowing lava at some time in the history of the Moon, for example in the long but narrow ranges of low "hills" that appear across the maria, as if lava flowed for some distance and then set in place, or in the incomplete craters, which lava might have filled the maria at some time after their formation.

The Other Planets and Moons

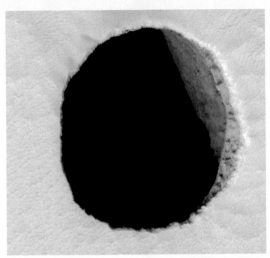

R I V U X G

A Shaft, Photographed by the *Mars Reconnaissance Orbiter,* That Descends At Least 78 m (255 ft) into the Side of the Volcano Arsia Mons *(NASA/JPL/University of Arizona)*

WHAT DO YOU THINK?

1 Which terrestrial planet—Mercury, Venus, Earth, or Mars—has the coolest surface temperature?

2 Which planet is most similar in size to Earth?

3 Which terrestrial planet—Mercury, Venus, Earth, or Mars—has the hottest surface temperature?

4 What is the composition of the clouds that surround Venus?

5 Does Mars have liquid water on its surface today? Did it have liquid surface water in the past?

6 Is life known to exist on Mars today?

7 Is Jupiter a "failed star"? Why or why not?

8 What is Jupiter's Great Red Spot?

9 Does Jupiter have continents and oceans?

10 Is Saturn the only planet with rings?

11 Are the rings of Saturn solid ribbons?

Answers to these questions appear in the text beside the corresponding numbers in the margins and at the end of the chapter.

The planets in our solar system have similarities and differences that astronomers are learning to understand. For example, there are two basic groups of planets: terrestrial worlds similar in size and chemistry to Earth (Mercury, Venus, and Mars), and the much larger giant or Jovian worlds (Jupiter, Saturn, Uranus, and Neptune). In this chapter, we explore the properties of these planets individually, as well as compare the planets to each other.

In this chapter you will discover

• Mercury, a Sun-scorched planet with dormant volcanoes, a heavily cratered surface, and a substantial iron core

• Venus, perpetually shrouded in thick, poisonous clouds and mostly covered by gently rolling hills

• Mars, a red, dusty planet that once had running water on its surface and may still have liquid water underground

• Jupiter, an active, vibrant, multicolored world more massive than all of the other planets combined

• Jupiter's diverse system of moons

• Saturn, with its spectacular system of thin, flat rings and numerous moons, including bizarre Enceladus and Titan

• what Uranus and Neptune have in common and how they differ from Jupiter and Saturn

MERCURY

The closest planet to the Sun, Mercury, is a truly inhospitable world of temperature extremes (Figure 5-1). Its incredibly thin atmosphere and weak magnetic field leave it virtually unprotected from countless impacts and a continuous bath of deadly solar radiation. Despite having a surface nearly as dark as coal, Mercury sometimes appears as one of the brightest objects in our sky, due to its proximity to the Sun. We can see it from Earth only for a few hours before sunrise or a few hours after sunset because its angle from the Sun (its elongation) is always less than 28°.

5-1 Photographs from *Mariner 10* and the *Messenger* spacecraft reveal Mercury's lunarlike surface

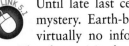

Until late last century, Mercury was a complete mystery. Earth-based telescopic studies provided virtually no information about the planet's surface. The three visits by *Mariner 10* photographed only

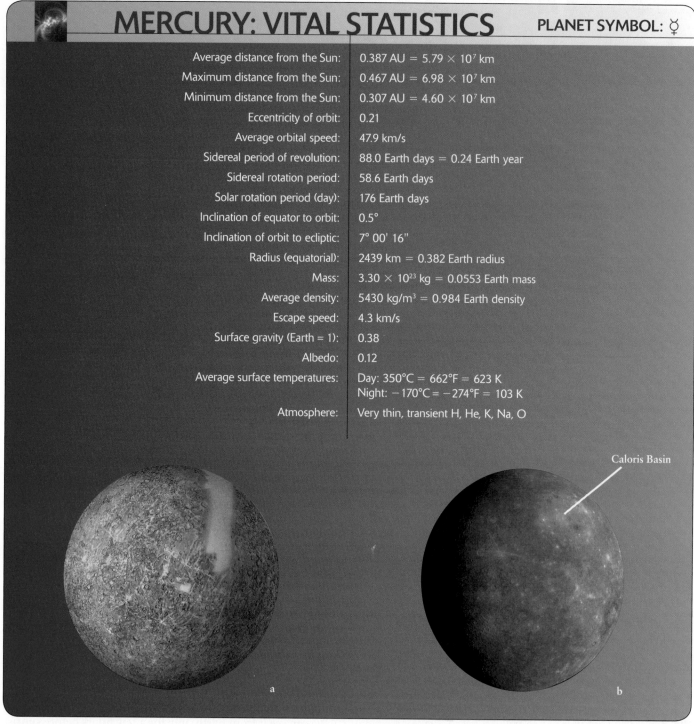

MERCURY: VITAL STATISTICS

PLANET SYMBOL: ☿

Average distance from the Sun:	0.387 AU = 5.79 × 10⁷ km
Maximum distance from the Sun:	0.467 AU = 6.98 × 10⁷ km
Minimum distance from the Sun:	0.307 AU = 4.60 × 10⁷ km
Eccentricity of orbit:	0.21
Average orbital speed:	47.9 km/s
Sidereal period of revolution:	88.0 Earth days = 0.24 Earth year
Sidereal rotation period:	58.6 Earth days
Solar rotation period (day):	176 Earth days
Inclination of equator to orbit:	0.5°
Inclination of orbit to ecliptic:	7° 00' 16"
Radius (equatorial):	2439 km = 0.382 Earth radius
Mass:	3.30 × 10²³ kg = 0.0553 Earth mass
Average density:	5430 kg/m³ = 0.984 Earth density
Escape speed:	4.3 km/s
Surface gravity (Earth = 1):	0.38
Albedo:	0.12
Average surface temperatures:	Day: 350°C = 662°F = 623 K
	Night: −170°C = −274°F = 103 K
Atmosphere:	Very thin, transient H, He, K, Na, O

Caloris Basin

a

b

R I V U X G

FIGURE 5-1 **Mercury: Vital Statistics** Heavily cratered Mercury was visited three times by the *Mariner 10* spacecraft in 1974 and 1975 and twice by *Mercury Messenger* in 2008. (a) This image, taken by *Mariner 10*, shows one side of Mercury, while (b) this image, taken by *Messenger*, shows most of the other side. *Messenger* will settle into orbit around Mercury on March 18, 2011. *(NASA/John Hopkins University Applied Physics Laboratory/Carnegie Institution of Washington)*

WEB LINK 5.2

79 percent of Mercury's surface. It was only in 2008 that the *Messenger* spacecraft imaged the rest of it. *Messenger* is short for MErcury Surface, Space ENvironment, GEochemistry, and Ranging.

Unlike our Moon, Mercury has a variety of volcanic craters out of which poured large amounts of lava that filled smaller impact craters and created smooth regions between larger ones. Like the Moon, Mercury has an exceptionally

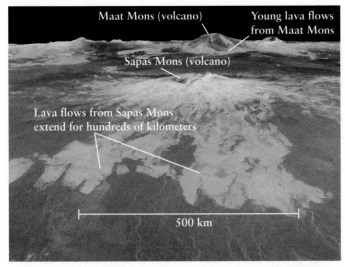

Maat Mons (volcano)   Young lava flows from Maat Mons

Sapas Mons (volcano)

Lava flows from Sapas Mons extend for hundreds of kilometers

500 km

R I V U X G

VIDEO 5.2

FIGURE 5-14 A Venusian Landscape A computer combined radio images to yield this perspective view of Venus as you would see it from an altitude of 4 km (2.5 mi). The color results from light being filtered through Venus's thick clouds. The brighter color of the extensive lava flows indicates that the flows reflect radio waves more strongly. The vertical scale has been exaggerated 10 times to show the gentle slopes of Sapas Mons and Maat Mons, volcanoes named for ancient Phoenician and Egyptian goddesses, respectively. *(NASA, JPL Multimission Image Processing Laboratory)*

radar signals through the clouds surrounding Venus. Like radar used by police, some of the radar signals bounced off Venus and returned to the spacecraft. By measuring the time delay of the radar echo, scientists determined the heights and depths of Venus's hills and valleys. As a result, astronomers have been able to construct a three-dimensional map of the planet. The detail visible on this map (also called *resolution*) is about 75 m. You could see a football stadium on Venus if there were any (none were detected).

Venus is remarkably flat compared to Earth. More than 80% of Venus's surface is covered with volcanic plains and gently rolling hills created by numerous lava flows. Figure 5-14 is an image using the *Magellan* radar data to create a view of a Venusian landscape.

Global radar images revealed just two large highlands, or "continents," rising well above the generally level surface of the planet (Figure 5-15). The continent at high latitudes in the northern hemisphere is called *Ishtar Terra* (after the Babylonian goddess of love) and is approximately the same size as Australia. Ishtar Terra is dominated by a high plateau ringed by towering mountains. The highest mountain is Maxwell Montes, whose summit rises to an altitude of 11 km above the average surface. For comparison, Mount Everest on Earth rises 9 km above sea level.

The larger Venusian continent, *Aphrodite Terra* (named after the Greek goddess Venus by the Romans), is a vast belt of highlands that straddles the equator. Aphrodite is 16,000 km (10,000 mi) in length and 2000 km (1200 mi) wide, giving it an area about one-half that of Africa. The global view of the surface of Venus in Figure 5-16 shows

indicate that Venusian rock is quite similar to lava rocks, called *basalt*, which are common on Earth and the Moon.

WEB LINK 5.4
The entire Venusian surface was mapped by the *Magellan* spacecraft, which arrived at Venus in 1990. In orbit about the planet, *Magellan* sent

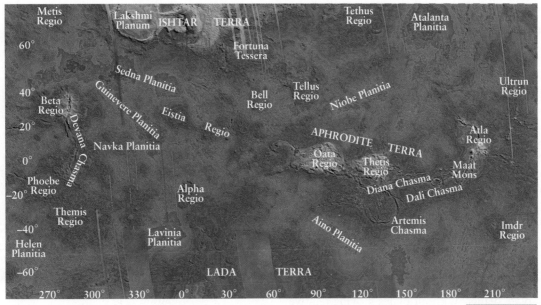

R I V U X G

WEB LINK 5.5
FIGURE 5-15 A Map of Venus This false-color radar map of Venus, analogous to a topographic map of Earth, shows the large-scale surface features of the planet. The equator extends horizontally across the middle of the map. Color indicates elevation—red for highest, followed by orange, yellow, green, and blue for lowest. The planet's highest mountain is Maxwell Montes on Ishtar Terra. Scorpion-shaped Aphrodite Terra, a continentlike highland, contains several spectacular volcanoes. Do not confuse the blue and green for oceans and land. *(Peter Ford/MIT, NASA/JPL)*

R I V U X G

FIGURE 5-16 A "Global" View of Venus A computer using numerous *Magellan* images creates a simulated globe. Color is used to enhance small-scale structures. Extensive lava flows and lava plains cover about 80% of Venus's relatively flat surface. The bright band running almost east–west is the continentlike highland region Aphrodite Terra. *(NASA)*

that most of Aphrodite Terra is covered by vast networks of faults and fractures.

Venus has more than 1600 major volcanoes and volcanic features and fewer than a thousand impact craters (Figure 5-17), as compared to the hundreds of thousands seen on the Moon and Mercury. Today, Venus's thick atmosphere heats, and thereby vaporizes, much of the infalling debris that would otherwise create craters. Although pieces larger than about a kilometer in diameter would blast through even this thick atmosphere, most of the rubble entering Venus's atmosphere is turned into dust by friction before it reaches the ground. However, there should have been a period shortly after Venus formed and before its atmosphere developed (from gases escaping the planet's interior) when impacts and cratering were common.

Accepting that the young Venus was once cratered like every other solid object in the solar system, the small number of craters on it needs to be explained. The problem is that, unlike Earth, Venus lacks the large-scale tectonic plates that divide up and refresh Earth's surface. Put another way, Venus is a one-plate planet. Without plates moving relative to each other, how can the craters be erased? Several mechanisms to explain this are being explored. Most theories posit that the lack of surface motion has forced Venus's crust to become thicker than Earth's crust. In one theory, heat from radioactive elements inside the planet causes mantle convection without tectonic plate motion. The temperatures inside

the planet eventually become high enough in the regions of rising mantle to create massive volcanoes out of which pour enormous amounts of magma that cover large areas of the surface. A variation of this theory proposes that the thickening crust is suddenly thinned from below by the horizontal convective motion of the mantle, like stripping layers off of plywood. When so thinned, heat is able to rise through it and cause many volcanoes to form.

Another theory proposes that, episodically, the one-plate surface of Venus is heated below enough to crack and begin Earthlike tectonic plate motion until it cools down and again becomes a single plate. During the period of plate motion, the surface is erased and replaced.

In yet another theory, the mantle becomes so hot that large sections of the crust melt more or less simultaneously, destroying the old crust and its craters. Indeed, that theory has been expanded to periodic global meltdowns occurring every 700 million years or so. For all of these models, when enough heat is released from the planet's interior, its surface once again solidifies, and, with the possible exception of local volcanic activity, the crust rethickens.

The sulfur content of the air and the traces of active volcanoes on the surface are compelling indicators that, like Earth, Venus has a molten interior. Because the average density of Venus is similar to that of Earth, its core is predominantly iron. Currents in the molten iron should generate a magnetic field; however, none of the spacecraft sent there has detected one. In general, magnetic fields are created by rotation of currents (see Section 4-4). Therefore, Venus's apparent lack of a magnetic field is plausible only if it rotates exceptionally slowly. In fact, the planet takes 116.8 Earth

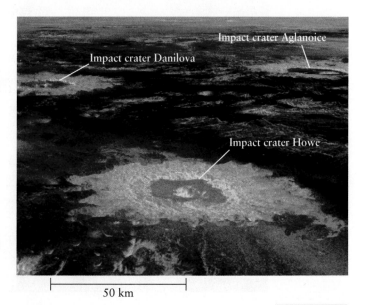

R I V U X G

FIGURE 5-17 **Craters on Venus** These three impact craters, with extensive ejecta surrounding each, are located on Venus's southern hemisphere. They were imaged using radar by the *Magellan* spacecraft. The colors are based on the *Venera* images (see Figure 5-15). *(NASA/Magellan Images, JPL)*

days to get from one sunrise to the next, if you could see the Sun from Venus's surface.

Unlike Earth and Mercury, Venus exhibits **retrograde rotation.** This means that the direction of Venus's orbit around the Sun (counterclockwise as seen from space far above Earth's North Pole) is *opposite* the direction of its rotation (clockwise as seen from the same vantage point). In other words, sunrise on Venus occurs in the west. Venus's rotation axis is tilted more than 177°, compared to Earth's 23½° tilt. Because Venus's axis is within 3° of being perpendicular to the plane of its orbit around the Sun, the planet has no seasons. Although we do not know the cause of Venus's retrograde rotation, one likely explanation is that a monumental impact flipped the rotation axis early in the planet's existence.

Why does little cratering occur on Venus today, even compared to the present low rate of cratering on our Moon?

MARS

Mars's distinctive rust-colored hue makes it stand out in the night sky (Figure 5-18). For centuries, Mars has generated more excitement as a possible home for alien life than any other world in our solar system. The idea began following the first well-documented telescopic observations of Mars in 1659 by the Dutch physicist Christiaan Huygens. Huygens identified a prominent, dark surface feature that reemerged about every 24 h, suggesting a rate of rotation similar to that of Earth.

Belief that advanced life exists on Mars skyrocketed at the end of the nineteenth century after Giovanni Virginio Schiaparelli, an Italian astronomer, reported in 1877 seeing 40 lines crisscrossing the Martian surface (Figure 5-19). He called these dark features *canali,* an Italian term meaning "channels." It was soon mistranslated into English as *canals,* implying the existence on Mars of intelligent creatures capable of engineering feats. This speculation led

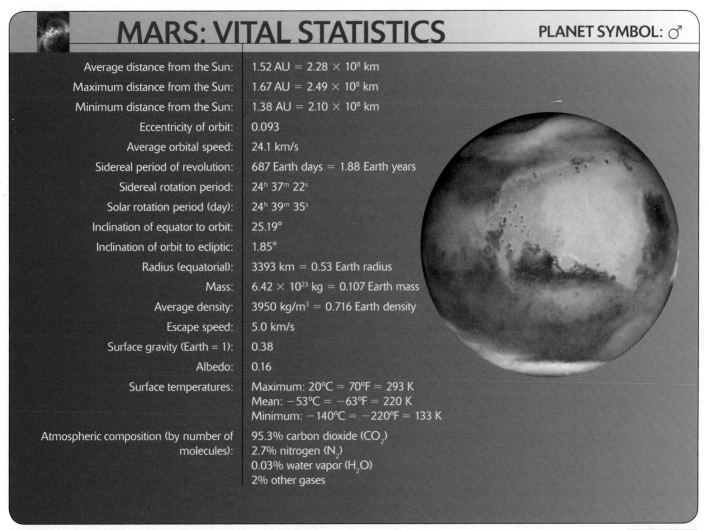

MARS: VITAL STATISTICS   PLANET SYMBOL: ♂

Average distance from the Sun:	1.52 AU = 2.28×10^8 km
Maximum distance from the Sun:	1.67 AU = 2.49×10^8 km
Minimum distance from the Sun:	1.38 AU = 2.10×10^8 km
Eccentricity of orbit:	0.093
Average orbital speed:	24.1 km/s
Sidereal period of revolution:	687 Earth days = 1.88 Earth years
Sidereal rotation period:	$24^h\ 37^m\ 22^s$
Solar rotation period (day):	$24^h\ 39^m\ 35^s$
Inclination of equator to orbit:	25.19°
Inclination of orbit to ecliptic:	1.85°
Radius (equatorial):	3393 km = 0.53 Earth radius
Mass:	6.42×10^{23} kg = 0.107 Earth mass
Average density:	3950 kg/m^3 = 0.716 Earth density
Escape speed:	5.0 km/s
Surface gravity (Earth = 1):	0.38
Albedo:	0.16
Surface temperatures:	Maximum: 20°C = 70°F = 293 K Mean: −53°C = −63°F = 220 K Minimum: −140°C = −220°F = 133 K
Atmospheric composition (by number of molecules):	95.3% carbon dioxide (CO_2) 2.7% nitrogen (N_2) 0.03% water vapor (H_2O) 2% other gases

FIGURE 5-18 **Mars: Vital Statistics** This photograph, taken from space, shows the Arabia Terra (in light orange) and carbon dioxide snow at the planet's poles. *(NASA, The Hubble Heritage Team)*

R I V U X G

Percival Lowell, who came from a wealthy Boston family, to finance a major new observatory near Flagstaff, Arizona. By the end of the nineteenth century, Lowell had allegedly observed 160 Martian "canals."

By the beginning of the twentieth century, it was fashionable to speculate that the Martian canals formed an enormous, planetwide irrigation network to transport water from melting polar ice caps to vegetation near the equator. In view of the planet's reddish, desertlike appearance, Mars was thought to be a dying planet whose inhabitants must go to great lengths to irrigate their farmlands. No doubt the Martians would readily abandon their arid ancestral homeland and invade Earth for its abundant resources.

INSIGHT INTO SCIENCE

Perception and Reality Our tendency to see patterns, even where they do not exist, makes it easy for us to leap to unjustified conclusions. Scientific images and photographs require objective analysis. This is why scientific discoveries are not considered to be authentic until they have been reproduced and verified by independent researchers.

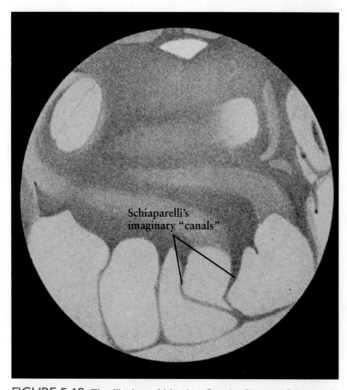

FIGURE 5-19 The Illusion of Martian Canals Giovanni Schiaparelli examined Mars through a 20-cm (8-in) diameter telescope, the same size used by many amateur astronomers today. His drawings of the red planet showed features perceived by Percival Lowell and others as irrigation canals. Higher resolution images from Earth and spacecraft visiting Mars failed to show the same features. *(Michael Hoskin, ed., The Cambridge Illustrated History of Astronomy, Cambridge University Press, 1997, p. 286. Illustration by G. V. Schiaparelli. Courtesy of Institute of Astronomy, University of Cambridge, UK)*

5-8 Mars's global features include plains, canyons, craters, and volcanoes

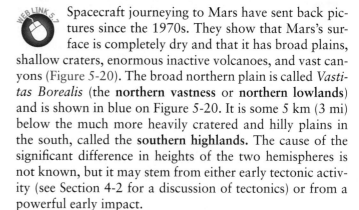

Spacecraft journeying to Mars have sent back pictures since the 1970s. They show that Mars's surface is completely dry and that it has broad plains, shallow craters, enormous inactive volcanoes, and vast canyons (Figure 5-20). The broad northern plain is called *Vastitas Borealis* (the **northern vastness** or **northern lowlands**) and is shown in blue on Figure 5-20. It is some 5 km (3 mi) below the much more heavily cratered and hilly plains in the south, called the **southern highlands.** The cause of the significant difference in heights of the two hemispheres is not known, but it may stem from either early tectonic activity (see Section 4-2 for a discussion of tectonics) or from a powerful early impact.

Between the two hemispheres, *Mariner 9* discovered a vast canyon now called *Valles Marineris*, which runs roughly parallel to the Martian equator (Figure 5-21). Valles Marineris stretches over 4000 km, about one-fifth the circumference of Mars. If this canyon were located on Earth, it would stretch from New York to Los Angeles. The individual canyons that make up Valles Marineris are up to 6 km (4 mi) deep and 190 km (120 mi) wide. Valles Marineris begins with heavily fractured terrain in the west and ends with ancient-cratered terrain in the east. Geologists believe that Valles Marineris is a large crack that formed as the planet cooled. It was enhanced by nearby rising crust to its west and widened further by erosion. Some of the eastern parts of this system appear to have been formed almost entirely by water flow, similar to Earth's Grand Canyon.

Most impact craters are located on Mars's southern hemisphere (Figure 5-22). This suggests that the northern vastness has been resurfaced by some process that eradicated ancient lowland craters, consistent with either of the earlier explanations of the differences in hemispheres.

Most of the volcanoes on Mars are located in the northern hemisphere. The largest volcano, Olympus Mons (Figure 5-23a), covers an area as big as the state of Missouri and rises 26 km (16 mi) above the surrounding plains—nearly 3 times the height of Mount Everest. The highest volcano on Earth, Mauna Loa in the Hawaiian Islands, has a summit only 17 km above the ocean floor.

Around Olympus Mons and other volcanoes, orbiting spacecraft have observed numerous cratered cone-shaped features that are typically the size of a football field (Figure 5-23b). These cones are believed to have been created by ice below the Martian surface that was liquefied and vaporized by lava from the nearby volcanoes. The expanding water pushed its way up to the surface, deforming the land before exploding out the centers of these "ice" volcanoes. Similar clusters of cones on Earth, formed by the same mechanism, are found in Iceland.

Planetwide, high-resolution photographs of Mars over the past 40 years have failed to show one canal of a size consistent with those allegedly seen from Earth (see Figure 5-19). We now know for certain that Schiaparelli's *canali*

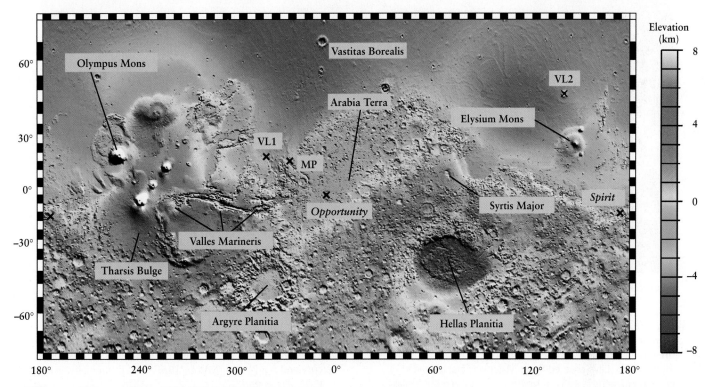

ANIMATION 5.5   WEB LINK 5.8   VIDEO 5.5

FIGURE 5-20 The Topography of Mars The color coding on this map of Mars shows elevations above (positive numbers) or below (negative numbers) the planet's average radius. To produce this map, an instrument on board *Mars Global Surveyor* fired pulses of laser light at the planet's surface, then measured how long it took each reflected pulse to return to the spacecraft. The *Viking 1 Lander* (VL1), *Viking 2 Lander* (VL2), *Mars Pathfinder* (MP), *Opportunity,* and *Spirit* landing sites are each marked with an **X**. The volcanoes depicted in white are higher than the scale on the right can show. *(MOLA Science Team, NASA/GSFC)*

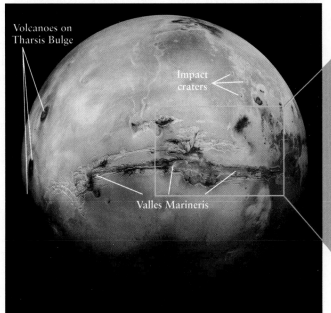

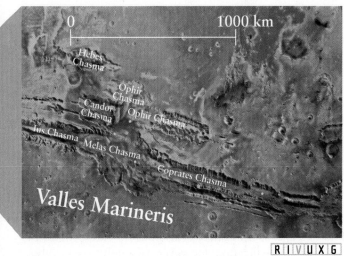

R I V U X G

R I V U X G

VIDEO 5.6

FIGURE 5-21 Martian Terrain This high-altitude photograph shows a variety of the features on Mars, including broad, towering volcanoes (left) on the highland, called Tharsis Bulge; impact craters (upper right); and vast, windswept plains. The enormous Valles Marineris canyon system crosses horizontally just below the center of the image. **Inset:** Details of the Valles Marineris, which is about 100 km (60 mi) wide. The canyon floor has two major levels. The northern (upper) canyon floor is 8 km (5 mi) beneath the surrounding plateau, whereas the southern canyon floor is only 5 km (3 mi) below the plateau. *(USGS/NASA; inset: NASA/GSFP/LTP)*

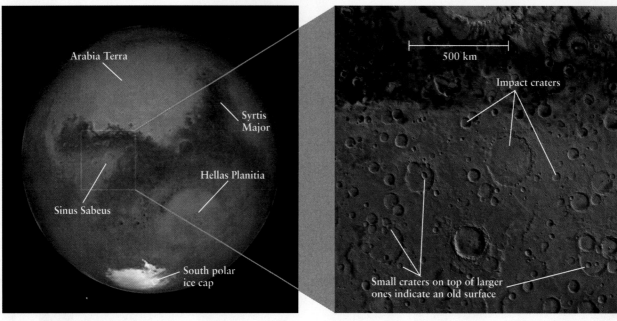

Mars from the Hubble Space Telescope   Closeup of Sinus Sabeus region

R I V U X G

FIGURE 5-22 **Craters on Mars** This image was made during the opposition of 2003. Arabia Terra looks like the Arabian peninsula on Earth. It is an old, highly eroded region dotted with numerous flat-bottom craters. Lava-covered Syrtis Major was first identified by Christiaan Huygens in 1659. A single impact carved out Hellas Planitia, which is five times the size of Texas. *Inset:* This mosaic of images from the *Viking 1* and *2* orbiter spacecraft shows an extensively cratered region located south of the Martian equator. Note how worn down these craters are compared to those on Mercury and our Moon. (*NASA; J. Bell, Cornell University; and M. Wolff, SSI; inset: USGS*)

were optical illusions, and the science fiction writers who wrote about advanced cultures there were completely off the mark.

INSIGHT INTO SCIENCE

Science Fact and Fiction Science fiction may foreshadow real scientific discoveries. (Can you think of some examples?) But science also reins in the fantasies of the science fiction writer. A century ago, millions of people, including top scientists, believed that technologically advanced life existed on Mars. However, scientific evidence says no.

5-9 Although no canals exist on Mars, it does have some curious natural features

Mars not only lacks canals, it also has no cities, roads, or other signs of civilization. However, astronomers have photographed several surface features that, at first glance, could have been crafted by intelligent life-forms. In 1976, a hundred years after Schiaparelli's alleged discovery of canali, the *Viking 1* orbiter spacecraft photographed a feature that appeared to be a humanlike face (Figure 5-24a). However, when the *Mars Orbiter* photographed the surface in greater detail and from a different angle in 1998 (Figure 5-24b), it found no facial features. If anything, the same spot looked more like a giant heel print.

Other photographs show a collection of pyramids, a skull (Figure 5-24c), and a "happy face" (Figure 5-24d). Were any of these things relics of an advanced civilization? Scientists universally believe that they are all naturally formed features. The first face is a region of hard rock around which softer debris is being worn away. Similarly, the pyramids and skull are consistent with winds eroding softer rock around harder rock pushed upward from inside Mars billions of years ago. The same thing happens on Earth. The happy face is made from a large impact crater and several small natural features inside it.

What two major activities created Valles Marineris?

Have you ever seen something that you interpreted as one thing, but it turned out to be something else?

INSIGHT INTO SCIENCE

Keep It Simple II Was the Martian "face" made by intelligent life or by wind-sculpted hills? When two or more alternative interpretations present themselves, scientists apply the principle of Occam's razor (page 27) and always choose the explanation that requires the fewest unproven assumptions.

a

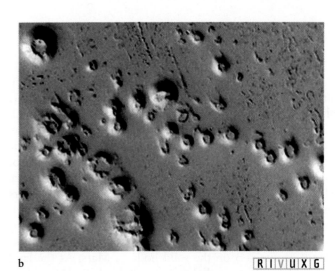

b   R I V U X G

FIGURE 5-23 **The Olympus Caldera** (a) This view of the summit of Olympus Mons is based on a mosaic of six pictures taken by one of the *Viking* orbiters. The caldera consists of overlapping volcanic craters and measures about 70 km across. The volcano is wreathed in mid-morning clouds brought upslope by cool air currents. The

cloud tops are about 8 km below the volcano's peak. (b) These cones on Mars may have been created when lava from Olympus Mons heated underground ice, causing the resulting water and vapor to expand, raise the planet's surface, and burst out. *(Peter Lanagan, JPL, U. Arizona, et al., MOC, MGS, NASA)*

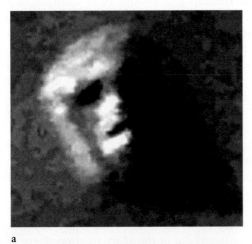

a

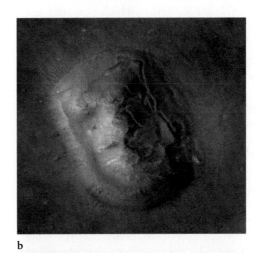

b

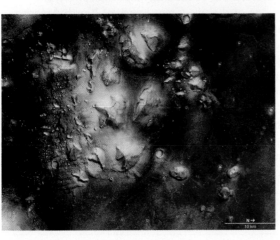

c

d   R I V U X G

FIGURE 5-24 **In the Eye of the Beholder** Images (a) and (b) are of the same site on Mars, taken 22 years apart. They show how, at high resolution, the apparent face in (a) changed to a more "natural-looking" feature in (b). (c) In the same

region of Mars, other erosion features also appear to be pyramids and skulls. (d) The Galle crater and its interior features combine to give the impression of a "happy face." *(a, b: NSSDC/NASA; NASA/JPL/Malin Space Science Systems; c, d: ©ESA/DLR/FU Berlin [G. Neu Kum])*

WEB LINK 5.9

5-10 Mars's interior is less molten than the inside of Earth

Part of the Valles Marineris was apparently formed when large regions of Mars's surface moved apart due to tectonic plate motion. This activity is generated by the flow of rock under the planet's surface (see Section 4-2 for more on plate tectonics). However, there is no evidence that plate tectonic activity occurs on Mars today. This indicates that more of Mars's interior is cooled, solid rock than exists inside Earth. This conclusion is supported by the observation that Mars lacks a global magnetic field (see Section 4-4 for discussion of the field-generating dynamo effect). However, observations made in 2003 revealed that the Sun creates tiny tides (less than a centimeter) on the solid body of Mars. As the land rises and sinks ever so slightly through the day on Mars, friction created by this motion in the planet's interior provides enough heat to keep some of the interior molten. (Rub your hands vigorously together to see how friction creates heat.)

Although a global magnetic field on Mars is absent today, there are remnants of such a field. In 1997, the *Mars Global Surveyor* (which operated until 2006) discovered patterns of local surface magnetic fields in nine places on Mars, similar to places on Earth where tectonic plates separate and lock Earth's magnetic field in the rock (see Section 4-4). Geologists propose that when Mars was young, its internal "dynamo" created a strong magnetic field and that tectonic plates helped craft its surface, while locking traces of its changing magnetic field in the then-molten rock. However, the planet quickly cooled, its global magnetic field vanished, and tectonic-plate activity apparently ceased nearly 4 billion years ago.

A more molten interior early in its existence is also consistent with the enormous volcanoes that exist on Mars. Volcanoes are places where molten rock oozes or bursts from a body's interior. The volcanoes on Mars are the type where molten rock oozed out, like the volcanoes that created the Hawaiian Islands. Such volcanoes are very different from Mount Saint Helens in Washington State, or Mount Etna in Sicily, where the material from inside Earth is ejected violently. On Earth, the Hawaiian Islands are only the most recent additions to a long chain of volcanoes. They resulted from **hot-spot volcanism,** a process by which molten rock rises to the surface from a fixed hot region far below. The Pacific tectonic plate is slowly moving northwest at a rate of several centimeters per year. As a result, new volcanoes are created above the hot spot, while older ones move off and become extinct, eventually disappearing beneath the ocean.

Because Mars's surface is apparently frozen in place due to the lack of tectonic plate motion, one hot spot can keep pumping lava upward through the same vent for millions of years. One result is Olympus Mons—a single giant volcano, rather than a long chain of smaller ones (see Figure 5-23a). This volcano's summit has collapsed to form a volcanic crater, called a **caldera,** large enough to contain the state of Rhode Island. Because Mars's solid crust and mantle (the region directly below the surface) are now so

thick, they insulate the surface, preventing any more molten rock from escaping. Therefore, we do not expect to see any more large-scale volcanic activity there.

By observing the orbit of *Mars Global Surveyor,* astronomers have concluded that the Martian crust is about 40 km thick under the northern lowlands and about 70 km thick under the southern highlands. However, they find that the boundary between the thin and thick crusts does not line up with the boundary between high and low terrain. Current models of Mars suggest that it has a core about 3400 km in diameter and that at least some of that core is molten. A detailed understanding of Mars's interior waits for the placing of seismic detectors on the planet's surface.

5-11 Martian air is thin and often filled with dust

To understand more about the activity on Mars's surface, we must examine its atmosphere, which has only 0.6% the pressure of Earth's atmosphere. As on Venus, some 95% of Mars's thin atmosphere is composed of carbon dioxide. The remaining 5% consists of nitrogen, argon, and some traces of oxygen. Unlike the gravitational attraction of tiny Mercury, which is too weak to hold any gases as a permanent atmosphere, Mars's gravitational force is just strong enough to prevent carbon dioxide, nitrogen, argon, and oxygen from escaping into space, but not strong enough to hold down water vapor.

A growing body of evidence indicates that there was once a vast water-filled ocean in the northern hemisphere of Mars. The observations include layers of rock deposited by water and channels in which water from the south flowed northward. However, much of that water has evaporated and drifted into space, never to return. Today, the concentration of water vapor in Mars's atmosphere is 30 times lower than it is in Earth's air. If all of the water vapor could somehow be squeezed out of the Martian atmosphere, it would not fill even one of the five Great Lakes of North America. The remainder of the water that was on Mars's surface has turned to ice, much of which remains there today. Since 2002, satellites orbiting Mars have discovered enough water ice under Mars's south pole to cover the entire planet to a depth of 11 m (36 ft). In 2008, the *Phoenix Mars Lander* dug into the Martian regolith and observed water ice just under the surface. Chemical tests made by it confirmed this discovery. Water ice is being discovered by satellites at many latitudes from the polar regions to the equator, and it is certain that much more water remains to be found inside Mars.

If the Hawaiian Islands were fixed over their hot spot, rather than moving, what would happen to them?

Despite the low density of Mars's air, the red planet experiences Earthlike seasons because of a striking coincidence, first noted in the late 1700s by the German-born English astronomer William Herschel. Just as Earth's equatorial plane is tilted 23½° from the plane of its

FIGURE 5-25 **Changing Seasons on Mars** During the Martian winter, the temperature decreases so much that carbon dioxide freezes out of the Martian atmosphere. A thin coating of carbon dioxide frost covers a broad region around Mars's north pole. During the summer in the northern hemisphere, the range of this north polar carbon dioxide cap decreases dramatically. During the summer, a ring of dark sand dunes is exposed around Mars's north pole. *(S. Lee/J. Bell/M. Wolff/ Space Science Institute/NASA)*

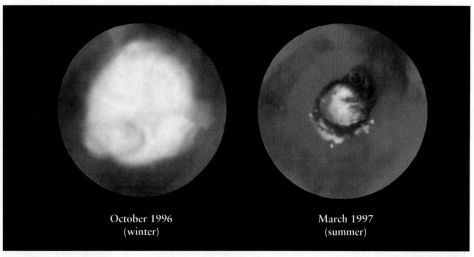

October 1996
(winter)

March 1997
(summer)

R I V U X G

orbit, Mars's equator makes an angle of about 25° from the plane of its orbit. If, as is now suspected, the northern hemisphere of Mars is lower than the southern hemisphere because the north was struck by a large body, that impact could have also tilted the planet and initiated the seasons there. The Martian seasons last nearly twice as long as Earth's, because Mars takes nearly 2 Earth years to orbit the Sun. When Mars is near opposition, even current telescopes for home use reveal its daily changes. Dark seasonal markings on the Martian surface can be seen to vary, and prominent polar caps shrink noticeably during the spring and summer months (Figure 5-25).

Although sometimes blue (Figure 5-26a), Mars's atmosphere is often pastel red, sometimes turning shades of pink and russet (Figure 5-26b). All of these latter colors are due to the fine dust blown from the planet's desertlike surface during windstorms. The dust is iron oxide, familiar here on Earth as rust. Mars's sky also changes color because the amount of dust in the air varies with the season. During the winter, carbon dioxide ice adheres to the dust particles and drags them to the ground. This helps clear and lighten the air. In the summer months, the carbon dioxide is not frozen, and the dust blown by surface winds remains aloft longer.

Hill, 1km (0.6 mi) away

Mars Sojourner rover

a   R I V U X G

b   R I V U X G

FIGURE 5-26 **The Atmosphere of Mars** (a) When Mars's sky is relatively free of dust, it appears similar in color to our sky, as shown in this sunset photo taken by the rover *Opportunity*. The darker, brown color in which the Sun is immersed is due to lingering dust in the sky. When less dust is present, the Sun looks almost white during Martian sunsets. (b) Most of the images we have of Mars's sky show colors like that seen here. Taken by the *Mars Pathfinder,* this photograph shows the *Sojourner* rover snuggled against a rock named Moe on the Ares Vallis to run tests on the rock. At the top of the image, the pink color of the Martian sky is evident. *(NASA/Lunar and Planetary Institute)*

a

FIGURE 5-27 **Martian Dust Devil** (a) This *Spirit* image is one in a movie sequence of a dust devil moving left to right across the surface of Mars. The rovers have filmed several such events. (b) These dark streaks are the paths of dust devils on the Argyre Planitia of Mars. The tracks cross hills, sand dunes, and boulder fields, among other features on the planet's surface. *(a: NASA/JPL; b: NASA/JPL/Malin Space Science Systems)*

The sky color also changes over periods of many years. In 1995, for example, the amount of dust was observed to have dropped dramatically compared to that observed in the 1970s. The reason for such long-term changes is still under investigation.

As on Earth, the temperature changes with the seasons on Mars. For example, one day at the landing site of the rover *Spirit,* near the equator, the temperature ranged from 188 K to 243 K, whereas 100 days later it ranged from 200 K to 263 K (see Figure 5-20 for the site location). During the afternoons, heat from the planet's surface warms the air and sometimes creates whirlwinds, called **dust devils** (Figure 5-27). A similar phenomenon occurs in dry or desert terrain on Earth. Martian dust devils reach altitudes of 6 km (20,000 ft). Spacecraft on Mars detect drops in air pressure as dust devils sweep past, just as on Earth. Some dust devils are large enough to be seen by orbiting spacecraft.

The winds on Mars help explain why its surface is eroding. Over the past two centuries, astronomers have seen faint surface markings disappear under a reddish-orange haze as thin Martian winds have stirred up finely powdered dust from the Martian regolith. Some storms obscure the

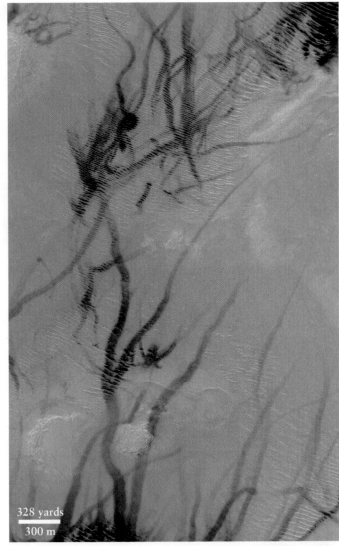

328 yards
300 m

b   R I V U X G

entire planet, as happened in 2001. Over the ages, winds have worn down crater walls, and deposits of dust have filled in the crater bottoms (Figure 5-28). But the Martian atmosphere is so thin that 3 billion years of sporadic storms have not carried enough of Mars's extremely fine-grained powder to eradicate the craters completely.

R I V U X G

FIGURE 5-28 **Crater Endurance** Photographed by the rover *Opportunity,* this crater on Mars is about 130 m (430 ft) across. Rocks are visible in the crater and in vertical cliffs along its walls. By studying such rocks, astronomers hope to understand more of the history of water on Mars. *(JPL/NASA)*

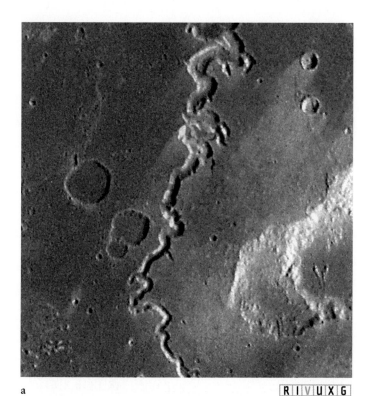

a   R I V U X G

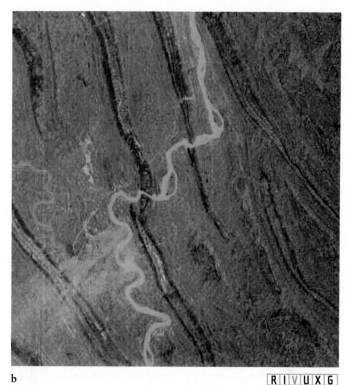

b   R I V U X G

FIGURE 5-29 Rivers on Mars and Earth (a) Winding canyons on Mars, such as the one in this *Viking 1* orbiter image, appear to be due to sustained water flow. This belief is supported by the terraces seen on the canyon walls in high-resolution *Mars Orbiter* images. Long periods of water flow require that the planet's atmosphere was once thicker and its climate more Earthlike. (b) Yangtze River near Chongqing, China. Typical of rivers on Earth, it shows the same snakelike curve as the river channels on Mars. *(a: NASA; b: TMSC/NASA)*

Why do you think it is easier to stand up in a dust devil on Mars than in the same equivalent-speed event here on Earth?

5-12   Surface features indicate that water once flowed on Mars

Despite disproving the theory that Mars has broad canals or any other liquid surface water, Mars-orbiting spacecraft did reveal many dried-up riverbeds (Figure 5-29a); lakes (Figure 5-30a); river deltas (Figures 5-30a, b), where water and debris emptied into lakes and oceans; sedimentation laid down by water flow (Figure 5-31); and other water-related features. Some of the riverbeds include intricate branched patterns and delicate channels meandering among flat-bottomed craters. Rivers on Earth invariably follow similarly winding courses (Figure 5-29b).

Surface rovers *Spirit* and *Opportunity* found strong evidence that water did indeed create many of these features (Figure 5-32), including the presence of water-formed hematite—iron-rich rocks that look like blueberries—and rocks with extremely high levels of salts that had previously been dissolved in liquid water. Such salts take millions of years to seep into rocks in the quantities they have been found, implying that liquid water was on the red planet's surface for at least that length of time. Although water existed on Mars's surface for long periods, there are indications in several places that large bodies of water have been gone for billions of years. The evidence for this is the discovery by the rovers of lava rocks that contain olivine and pyroxene, rocks that dissolve quickly in water. The origin of at least some of the water that was on Mars is believed to have been ice-rich bodies that struck the surface, releasing their water into the atmosphere, and later raining down on the surface.

In 2001, astronomers reported evidence that small-scale volcanic activity may still be causing episodic flows of both water and lava on Mars's surface. *Mars Global Surveyor* images show channels apparently carved by water flowing down the walls of pits or craters, including one event that occurred sometime between 1999 and 2005 (Figure 5-33a, b). What makes these observations especially intriguing is that they appear to be geologically young, indicating that liquid water may still exist under the Martian surface. In 2008, the *Mars Reconnaisance Orbiter* photographed several avalanches in progress. What caused them is still under investigation.

Although the amount of water that created the features shown in Figure 5-33b is relatively modest, large quantities of water—lakes or even oceans—are thought to be still trapped under the Martian surface. If so, volcanic heating or the heat generated by impacts from space debris could open fissures, enabling water to melt, escape, and even carve new riverbeds today.

Outlet   Shoreline

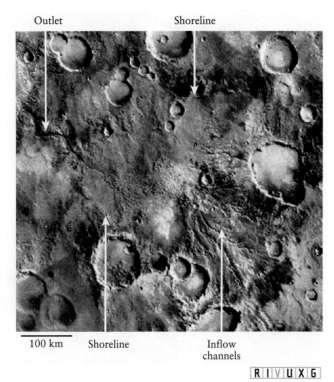

100 km   Shoreline   Inflow
                              channels

a   R I V U X G

Ancient
dried
riverbeds

50 km

b   R I V U X G

FIGURE 5-30 **Evidence of Water on Mars** (a) This dry Martian lake, photographed by the *Mars Global Surveyor* with a resolution of 1.5 m, is an excellent example of how geology and astronomy overlap. The features of this dry lake are consistent with those found on lakebeds on Earth. (b) This network of dry riverbeds is located on Mars's cratered southern hemisphere. *(a: NASA/JPL; b: NASA)*

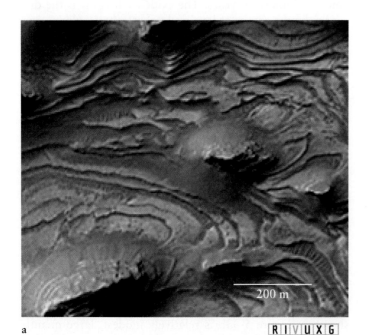

200 m

a   R I V U X G

b   R I V U X G

FIGURE 5-31 **Ancient Oceans and Lakes on Mars** (a) This *Mars Global Surveyor* image of a portion of Valles Marineris reveals terrain with "stair-step layers." Such terrain is likely to have been created by sedimentation at the bottom of an ancient body of water. (b) This image of Burns Cliff, photographed by the rover *Opportunity,* shows a close-up of layers of rock laid down on a body of water on Mars that went through wet and dry periods. *(a: Malin Space Science Systems/JPL/ NASA; b: NASA/JPL/Cornell)*

a   R I V U X G

b   R I V U X G

FIGURE 5-32 **Layers of Rock Laid Down by Water** (a) This close-up image taken by the rover *Opportunity* shows a small section of rock layers in a location called the *Dells*. The angled and curved layering seen here is created only on Earth by water flow, strongly suggesting that this sediment was also deposited by water. The nearly spherical rocks, called blueberries because they are dark, have been chemically identified as hematite, an iron-rich mineral that is usually formed in water. The rovers have found blueberries strewn in a wide variety of locations. (b) This iron meteorite, dubbed the *Heat Shield Rock* because it was discovered behind the Mars rover *Sojourner's* heat shield, is surrounded by hematite blueberries. *(a: NASA/JPL/USGS; b: NASA/JPL/Cornell)*

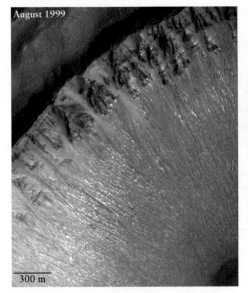

a

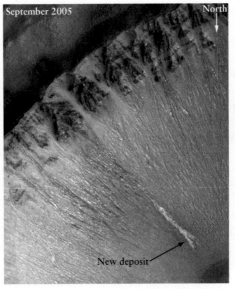

b

c   R I V U X G

FIGURE 5-33 **Martian Gullies** (a, b) Two images of the same southern hemisphere crater on Mars taken 6 years apart. Although not incontrovertible, they provide strong evidence for recent water flow from inside the red planet. The reason the deposit occurred now is still under investigation. (c) Polygonal cracks are visible in this *Opportunity* image of Escher Rock. They were believed to have formed when this area was flooded. Water seeped into the rock, which cracked as the water froze and expanded, then the water evaporated away. *(a, b: NASA/JPL/Malin Space Science Systems; c: NASA/JPL/Cornell)*

a   R I V U X G

b   R I V U X G

FIGURE 5-34 A Piece of Mars on Earth Mars's thin atmosphere does little to protect it from impacts. Some of the debris ejected from impact craters there apparently traveled to Earth. (a) SNC meteorite recovered in Antarctica. The meteorite shows strong evidence of having been exposed to liquid water on Mars, perhaps for hundreds of years. (b) Possible fossil remains of primitive bacterial life on Mars, although theories of nonbiological origins have also been presented. *(a: NASA; b: NASA/JPL)*

In 2004, the rover *Opportunity* discovered Martian rock (Figure 5-33c) with polygon-shaped cracks similar to dried mud flats found on Earth. The best explanation for these features on Mars is that long after the lakes and oceans dried there, an impact released water that again briefly flooded the surface rock. The water seeped into the rock and, upon drying, it expanded, froze, and cracked the rock. This scenario would require vast amounts of water inside Mars for several reasons. First, heated by the impact, most of the liberated water would have left as vapor. Second, liquid water on the surface would not stay there very long because of the low air pressure. When air pressure is very low, molecules easily escape from a liquid's surface, causing the water to vaporize. Because the pressure of Mars's atmosphere is only 0.6% that of Earth's atmosphere, any liquid water on Mars today would quickly transform into ice or furiously boil and evaporate into the thin Martian air, and then be lost into space. Thus, if liquid water did cause the rock formations seen by *Opportunity*, there must have been a lot of it to start with for enough to cover large areas of the Martian surface that have such cracks.

The total amount of water that existed on Mars's surface is not known. However, by studying flood channels that exist there, geologists estimate that there was once enough water to cover the planet to a depth of 500 m (1500 ft). For comparison, Earth has enough water to cover our planet to a depth of 2700 m (8900 ft), assuming that Earth's surface is everywhere a uniform height.

Further evidence that water once flowed on Mars comes from so-called *SNC meteorites* found on Earth (Figure 5-34a). These space rocks are believed to have once been pieces of Mars because their chemistries are consistent with those of rocks studied on Mars's surface and because they contain trace gases in relative amounts found only in the current Martian atmosphere. The meteorites were ejected into space during especially powerful impacts on that planet's surface. They also contain water-soaked clay, which is not expected to be found on any objects in the solar system besides Mars, Earth, and Jupiter's moons Europa, Ganymede, and Callisto. At least 57 meteorites from Mars have been identified.

Some of the water that was on Mars is still near the surface today, frozen in several places on the red planet from its poles to its equator. Recent satellite measurements indicate that as much as 90% of the ice at the poles is water ice, the remainder being dry ice (carbon dioxide ice). If thawed, this polar ice would create a global ocean 11 m (33 ft) deep. The temperature at Mars's poles, typically 160 K (–170°F), keeps the water permanently frozen, like the permafrost found in northern Asia and Antarctica. Seasonal variations at Mars's poles (see Figure 5-25) are created by the freezing and evaporating of dry ice, which reaches thickness of 2 m. Indeed, during the winter, about one-third of the carbon dioxide in Mars's atmosphere comes down as dry ice.

The 1997 visit by *Pathfinder* and its little rover (see Figure 5-25b) to the Ares Vallis revealed evidence of several floods at the mouth of a dried flood plain. The evidence includes layers of sediment, clumps of rock and sand stuck together like similar groupings created by water on Earth, rounded rocks worn down as they were dragged by the water, and rocks aligned by water flow. The rover performed some 20 chemical analyses of rocks in the landing area. The soil is a rusty color, laden with sulfur. The source of this sulfur is a mystery, as is the large amount of silicon-rich silica discovered on rocks, such as the one named Barnacle Bill.

There is observational evidence of an ancient shoreline encircling the northern lowlands of Mars. The belief that the northern lowlands were an ocean is supported by the analysis of a 1.2-billion-year-old meteorite from Mars. It has remnants of salt believed to have been deposited there when the Martian ocean dried up.

5-13 Search for microscopic life on Mars continues

The discovery in the 1960s that water once existed on the surface of Mars rekindled speculation about Martian life. Although it was clear that Mars has neither civilizations nor fields of plants, microbial life-forms still seemed possible. Searching for signs of organic matter was one of the main objectives of the ambitious and highly successful *Viking* missions.

WEB LINK 5.12 The two *Viking* spacecraft were launched during the summer of 1975. Each consisted of two modules—an orbiter and a lander. Almost a year later, both *Viking* landers set down on rocky plains north of the Martian equator.

Which ice on Mars changes more actively today, carbon dioxide ice or water ice?

The landers confirmed the long-held suspicion that the red color of the planet is due to large quantities of iron in its soil. Despite the high iron content of its crust, Mars has a smaller average density (3950 kg/m³) than that of other terrestrial planets (more than 5000 kg/m³ for Mercury, Venus,

Sulfur salts   R I V U X G

FIGURE 5-35 **Regolith of Mars** The rover *Spirit's* wheels churned up the regolith of Mars, revealing sulfur-based salts just below the surface. *(NASA/ JPL/Cornell)*

and Earth). Mars must therefore contain overall a smaller percentage of iron than these other planets.

Each *Viking* lander was able to dig into the Martian regolith and retrieve rock samples for analysis, which showed that the rocks at both sites are rich in iron, silicon, and sulfur. The Martian regolith (Figure 5-35) ranges in consistency from clay to sand, although its chemistry is different from clay and sand on Earth. The *Viking* landers (see Figure 13-1) each carried a compact biological laboratory designed to test for microorganisms in the Martian soil. Three biological experiments were conducted, each based on the idea that living things alter their environment: They eat, they breathe, and they release waste products. During each experiment, a sample of the Martian regolith was placed in a closed container, with or without a nutrient substance. The container was then examined for any changes in its contents.

The first data returned by the Viking biological experiments caused great excitement. In almost every case, rapid and extensive changes were detected inside the sealed containers. However, further analysis showed that these changes were due solely to nonbiological chemical processes. Apparently, the Martian regolith is rich in chemicals that effervesce (fizz) when moistened. A large amount of oxygen is tied up in the regolith in the form of unstable chemicals called *peroxides* and *superoxides,* which break down in the presence of water to release oxygen gas.

The chemical reactivity of the Martian regolith comes from the Sun's ultraviolet radiation and from electric activity inside dust devils. Ultraviolet photons easily break apart molecules of carbon dioxide (CO_2) and water vapor (H_2O) by knocking off oxygen atoms, which then become loosely attached to chemicals in the regolith. Ultraviolet photons also produce highly reactive ozone (O_3) and hydrogen peroxide (H_2O_2) near the planet's surface, which become incorporated in the regolith. Dust devils suck up lots of dust from the planet's surface and transform it into a variety of compounds.

That so much ultraviolet radiation from the Sun gets to the surface of Mars indicates what about the Martian atmosphere?

Here on Earth, hydrogen peroxide is commonly used as an antiseptic. When you pour this liquid on a wound, it fizzes and froths as the loosely attached oxygen atoms chemically combine with organic material and destroy germs. The Viking landers may have failed to detect any organic compounds on Mars because the superoxides and peroxides in the Martian regolith act as antiseptics there, making it sterile.

In 2003, astronomers detected methane in Mars's atmosphere. This is an intriguing discovery, since sunlight breaks down methane within a few centuries. Therefore, the gas needs to be continuously replenished. This simple compound, CH_4, is often a by-product of biological activity (cows come to mind in this regard). Methane is also trapped inside planets, like Earth and possibly Mars, during their formation and thereafter leaks out. Although the methane discovered on Mars is suggestive that simple life might be active under the planet's surface today, it is by no means positive proof that such life exists.

Possible signs of ancient life on Mars have been discovered here on Earth. When cut open, several of the Martian meteorites have shown microscopic features that could be fossils of Martian bacterial life and their excretions (see Figure 5-34b). These features, no larger than 500 nm (1/100 the diameter of a human hair), are 30 times smaller than bacteria found on Earth. Detailed analysis of the meteorites in 2001 revealed the presence of several organically created features, including tiny spheres found with some of Earth's bacteria, and magnetite crystals, which are also used by some bacteria on Earth as compasses to find food. The debate as to whether the features in these meteorites are fossil evidence of life on Mars has raged unabated for several years, and it is not yet resolved. (It is worth noting that similar claims of fossils from Mars in the 1960s have been disproved.)

The search for microscopic life on Mars is not over. If there is liquid water under the Martian surface, then it is entirely possible that life has evolved in it and may still exist. This belief is based on the incredibly wide range of places that life has developed in water on Earth, from volcanic vents on the bottoms of oceans to geysers and a lake under the Antarctic polar ice cap.

5-14 Mars's two moons look more like potatoes than spheres

Two tiny moons orbit close to Mars's surface. *Phobos* (Greek, meaning "fear") and *Deimos* ("panic") are so small that they were not discovered until 1877. In mythology, these names are either Mars's sons by Aphrodite or the two horses that pulled Mars's war chariot—take your pick.

Potato-shaped Phobos is the inner and larger of the two moons (Figure 5-36). It is heavily cratered, and its surface has been transformed into dust at least 1 m thick by countless tiny impacts over the eons. Football-shaped Deimos is less cratered than Phobos (Figure 5-36).

Both Phobos and Deimos orbit in the same direction that Mars rotates, just as our Moon orbits Earth. However, Phobos is so close to the planet that this moon goes around three times in one Martian day. This rapid revolution creates the unusual situation wherein Phobos rises in the west, races across the sky in only 5½ hours as seen from Mars's equator, and sets in the east. As seen from Mars, Deimos rises in the east and takes about 3 Earth days to creep from one horizon to the other.

Phobos and Deimos were not formed like our Moon, by splashing off Mars. Rather, they are captured planetesimals. Both moons are in synchronous rotation as they orbit the red planet.

INSIGHT INTO SCIENCE

Imagine the Moon The definitions we use for words based on our everyday experience often fail us in astronomy. For example, the word *moon* usually creates an image of a spherical body, like our Moon. In reality, most of the moons in the solar system are unsymmetrical, like Phobos and Deimos.

5-15 Comparisons of planetary features provide new insights

Now that we have examined the terrestrial planets individually, it can be useful to see how their various features compare to each other and to Earth. Table 5-1 summarizes much of this material.

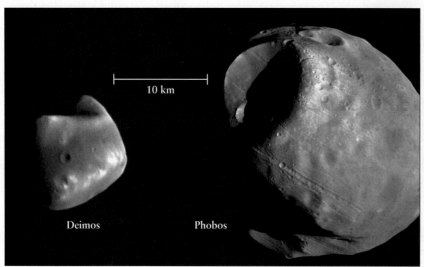

R I V U X G

FIGURE 5-36 **Phobos and Deimos** Phobos, the larger of Mars's two moons, is potato-shaped and measures approximately 28 × 23 × 20 km. It is dominated by crater Stickney, named for discoverer Asaph Hall's wife (Angeline Stickney). Deimos is less cratered than Phobos and measures roughly 16 × 12 × 10 km. *(left: European Space Agency; right: Dr. Edwin V. Bell/ NSSDC/Raytheon ITSS)*

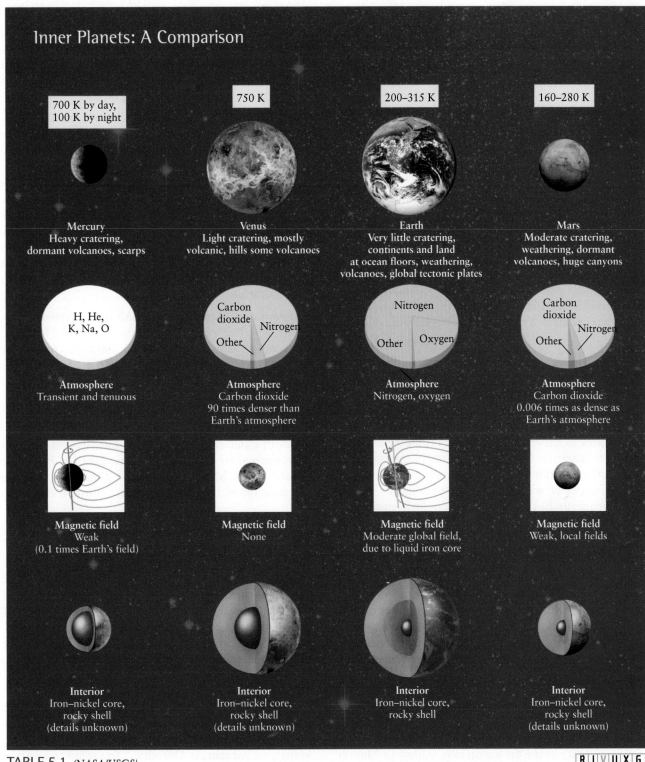

Inner Planets: A Comparison

Mercury
700 K by day, 100 K by night
Heavy cratering, dormant volcanoes, scarps

Venus
750 K
Light cratering, mostly volcanic, hills some volcanoes

Earth
200–315 K
Very little cratering, continents and land at ocean floors, weathering, volcanoes, global tectonic plates

Mars
160–280 K
Moderate cratering, weathering, dormant volcanoes, huge canyons

---

Atmosphere (Mercury)
H, He, K, Na, O
Transient and tenuous

Atmosphere (Venus)
Carbon dioxide, Nitrogen, Other
Carbon dioxide 90 times denser than Earth's atmosphere

Atmosphere (Earth)
Nitrogen, Oxygen, Other
Nitrogen, oxygen

Atmosphere (Mars)
Carbon dioxide, Nitrogen, Other
Carbon dioxide 0.006 times as dense as Earth's atmosphere

Magnetic field (Mercury)
Weak (0.1 times Earth's field)

Magnetic field (Venus)
None

Magnetic field (Earth)
Moderate global field, due to liquid iron core

Magnetic field (Mars)
Weak, local fields

Interior (Mercury)
Iron–nickel core, rocky shell (details unknown)

Interior (Venus)
Iron–nickel core, rocky shell (details unknown)

Interior (Earth)
Iron–nickel core, rocky shell

Interior (Mars)
Iron–nickel core, rocky shell (details unknown)

TABLE 5-1 *(NASA/USGS)*   R I V U X G

Size and Mass Earth is the largest and most massive of all four terrestrial planets. In this regard, Venus is almost the sister planet to Earth, with nearly 95% of Earth's diameter and 82% of Earth's mass. Although Mars is most similar to Earth in other ways, such as its history of surface water and rotation rate, it is only half (53%) of Earth's diameter and 11% of Earth's mass. Mercury, with 38% of Earth's diameter and a scant 5.5% of Earth's mass, is much more similar to Earth's Moon in dimensions. Mercury is only 1.4 times bigger than the Moon.

Surface Features Although all four terrestrial planets have craters, only Mercury has them in large numbers, similar to what we saw on our Moon. Venus has erased most of its craters by melting them or covering them with magma. Mars has removed many of its craters as a result of erosion and weather. Earth has removed most craters by tectonic plate motion and weathering. Like Earth, Venus has continentlike plateaus and ocean bottomlike lowlands. Mars has a lower northern cap than the rest of the planet, but no real continentlike regions. Mercury has relatively uniform height. While all four terrestrial planets have volcanoes, those of Mars and Mercury are extinct. Water erosion occurred just on Mars and Earth.

Would our Moon have to be closer or farther away to orbit in the same direction but rise in the west?

Interior The interior chemistries of the four terrestrial planets are similar, with cores consisting primarily of iron surrounded by rock. Venus, Earth, and Mars are known to have partially molten cores, and it remains to be seen if Mercury's core is also molten. Although Mercury is the smallest terrestrial planet, it has the highest density, meaning that it has the highest percentage of iron of these (and, in fact, of any) planets. This is true probably because Mercury lost more of its outer, rocky layer as a result of impacts than did any other terrestrial planet.

Water Earth contains by far the highest percentage of water of the terrestrial planets. Mars contains water frozen on or near its surface, and we have yet to determine whether it has any liquid water deep inside. Venus contains very little water compared to either Earth or Mars because Venus is so hot that it has evaporated surface water into its atmosphere, and water in its interior has probably been mostly ejected through volcanoes or when the surface periodically melts. Mercury apparently has some water (very little compared to Earth or Mars) frozen at its poles, the result of collisions with water-rich comets. Because its interior is so iron-rich, the water-bearing layers were probably blasted into space by impacts early in Mercury's existence.

Atmosphere Venus has by far the densest atmosphere of the terrestrial planets, with about 90 times as much gas as the air we breathe. Furthermore, Venus's atmosphere is composed primarily of carbon dioxide, with a minor component of nitrogen. The thick atmosphere has protected Venus's surface from all but the most massive infalling space debris. Its thick atmosphere and the possibility that Venus's surface periodically melts explain why Venus has few impact craters. Venus's atmosphere is most similar in composition to the air around Mars, although the density of Mars's air is only 0.6% as dense as the air we breathe. Although Mars's thin atmosphere has enabled many pieces of space debris to strike the planet and form craters, these craters are continually being eroded, primarily by wind.

Earth's atmosphere, once very similar to that of Venus, was transformed by water and life into the nitrogen–oxygen atmosphere we have today. As with Venus, Earth's atmosphere protects the surface from most impacts. Many craters have been removed by our planet's plate tectonic motion, which apparently does not occur today on any other planet. Mercury's gravity is too low to hold any gases as a perma-

nent atmosphere. It is surrounded by a very thin atmosphere of transient gases from the Sun and from the planet's interior. As these drift into space, they are replaced by fresh gas. Because its atmosphere is so thin and its surface unchanged by internal activity for billions of years, Mercury is the most heavily cratered of the terrestrial planets.

Temperature Some of the temperatures on the terrestrial worlds are surprising at first glance. Mercury, which is closest to the Sun, has a hot daytime surface of about 700 K (800°F). Its lack of atmosphere allows a good deal of this heat to escape at night, bringing its nighttime temperature down to a frigid 100 K (−280°F), much colder than on any other terrestrial planet. Venus's thick atmosphere creates a greenhouse effect that keeps that planet at 750 K (890°F), even hotter than Mercury. Earth's surface temperature ranges from about 330 K (140°F) to 180 K (−130°F), and Mars is only slightly colder, with temperatures ranging from 280 K (45°F) down to 160 K (−170°F).

Rotation and Magnetic Fields All of the terrestrial planets rotate, with Earth's solar day being shortest at 24 h. This motion, combined with Earth's liquid iron core, creates a strong magnetic field that surrounds our planet. Mars has virtually the same solar day of 24 h 39 min, but its molten iron core is much smaller than ours and it has no global magnetic field, only local magnetic fields. The solar days of Venus (117 Earth days) and Mercury (176 Earth days) are both extremely long by Earth standards, and, indeed, a solar day on Mercury is 2 Mercurian years long! No magnetic field has been detected around Venus. Mercury has a weak global field that may result from the extremely high amount of iron it contains.

OUTER PLANETS

We can pretty easily imagine a trip to Mars, exploring its vast canyons and icy polar regions. Even Venus and Mercury lend themselves to being compared to Earth and the Moon. However, we find few similarities between Earth and the giant outer planets: Jupiter, Saturn, Uranus, and Neptune. Until recently, Pluto was also considered one of the outer planets. That categorization has changed as described in Appendix F. Now Pluto is categorized as a dwarf planet and we explore Pluto in Chapter 6.

The four giant worlds lack solid surfaces, are so much larger, rotate so much faster, and have such different chemical compositions from our planet that upon seeing them one knows, to paraphrase Dorothy, "You're not on Earth anymore." There is nothing on Earth remotely similar to the swirling red and brown clouds of Jupiter, the ever-changing ring system of Saturn, or the blue-green clouds of Uranus and Neptune. The outer four worlds collectively have at least 163 moons, which have an amazing range of shapes, sizes, surfaces, and properties. Some of these moons have water interiors; some have volcanoes; some have surface ice; one, Titan, has a thick atmosphere; and one, Hyperion, is arguably the most bizarre-looking object in the solar system. We begin our exploration of the outer planets with magnificent Jupiter.

JUPITER

Jupiter's multicolored bands, dotted with ovals of white and brown, give it the appearance of a world unlike any of the terrestrial planets. Viewed even through a small telescope (see Figure 2-11), you can also see up to four of its moons. As the high-resolution images in this chapter reveal, Jupiter is a world of breathtaking beauty.

Jupiter is the largest planet in the solar system: more than 1300 Earths could be packed into its volume (Figure 5-37). Using the orbital periods of its moons in Kepler's laws, astronomers have determined that Jupiter is 318 times more massive than Earth. Indeed, Jupiter has more than 2½ times as much mass as all of the other planets combined. This huge mass has created the urban myth that Jupiter is a failed star, meaning that it has almost enough matter to shine on its own, like the Sun, which we will study in Chapter 7. The fact is that Jupiter would have to be 75 times more massive than it is to generate energy like the Sun, and therefore to be classified as a star. Nevertheless, Jupiter emits approximately twice as much energy as it receives from the Sun. Its extra energy comes from radioactive elements in its core and from an overall contraction amounting to less than 10 cm per century.

5-16 Jupiter's outer layer is a dynamic area of storms and turbulent gases

Jupiter is permanently covered with clouds (see Figure 5-37). Because it rotates about once every 10 h—the fastest of any planet—Jupiter's clouds are in perpetual motion and are confined to narrow bands of latitude. Even through a small telescope, you can see Jupiter's dark, reddish bands called **belts,** alternating with light-colored bands called **zones.** These belts and zones are gases flowing east or west, with very little north–south motion. In contrast, winds on slower-rotating Earth wander over vast ranges of latitude (compare Figure 4-1).

JUPITER: VITAL STATISTICS   PLANET SYMBOL: ♃

Average distance from the Sun:	5.20 AU = 7.78×10^8 km
Maximum distance from the Sun:	5.46 AU = 8.16×10^8 km
Minimum distance from the Sun:	4.95 AU = 7.41×10^8 km
Eccentricity of orbit:	0.048
Average orbital speed:	13.1 km/s
Orbital period:	11.86 years
Rotation period:	9^h 50^m 28^s (equatorial) 9^h 55^m 30^s (internal)
Inclination of equator to orbit:	3.12°
Inclination of orbit to ecliptic:	1.30°
Radius:	71,490 km = 11.21 Earth radii (equatorial) 66,850 km = 10.48 Earth radii (polar)
Mass:	1.90×10^{27} kg = 318 Earth masses
Average density:	1330 kg/m³ = 0.241 Earth density
Escape speed:	60.2 km/s
Surface gravity (Earth = 1):	2.36
Albedo:	0.52
Average temperature at cloud tops:	−108°C = −162°F = 165 K
Atmospheric composition (by number of molecules):	86.2% hydrogen (H_2), 13.6% helium (He), 0.2% methane (CH_4), ammonia (NH_3), water vapor (H_2O), and other gases

Jupiter

Earth

R I V U X G

FIGURE 5-37 This view was sent back from *Voyager 1* in 1979. Features as small as 600 km across can be seen in the turbulent cloud tops of this giant planet. The complex cloud motions that surround the Great Red Spot are clearly visible. Also, clouds at different latitudes have different rotation rates. The inset image of Earth shows its size relative to Jupiter. *(NASA/JPL; inset: NASA)*

a R I V U X G

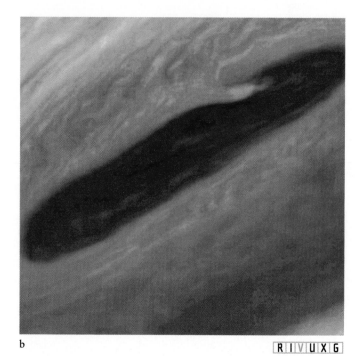

b R I V U X G

FIGURE 5-38 **Close-Ups of Jupiter's Atmosphere** The dynamic winds, rapid rotation, internal heating, and complex chemical composition of Jupiter's atmosphere create its beautiful and complex banded pattern. (a) A *Voyager 2* southern hemisphere image showing a white oval that has existed for over 40 years. (b) A *Voyager 2* northern hemisphere image showing a brown oval. The white feature overlapping the oval is a high cloud. *(NASA)*

Jupiter's belts and zones provide a framework for turbulent swirling cloud patterns, as well as rotating storms similar in structure to hurricanes or cyclones on Earth. These storms are known as *white ovals* and *brown ovals* (Figure 5-38). The white ovals are observed to be cool clouds higher than the average clouds in Jupiter's atmosphere. The brown ovals are warmer and lower clouds, seen through holes in the normal cloud layer. The various oval features last from hours to centuries. Computers show us how the cloud features on Jupiter would look if the planet's atmosphere were unwrapped like a piece of paper (Figure 5-39).

Jupiter's most striking feature is its **Great Red Spot** (Figure 5-40), which is so large that it can be seen through a small telescope. It changes dimensions, and at present is about 25,000 km long by 12,000 km wide—large enough so that two Earths could easily fit side by side inside it. The

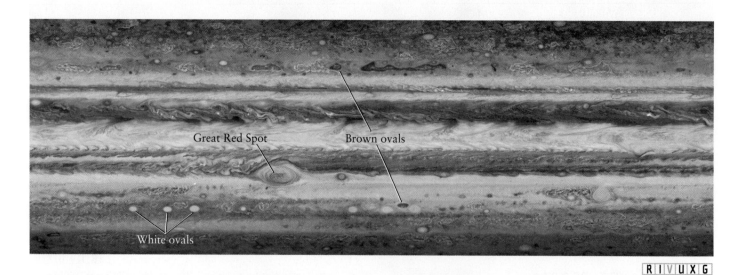

R I V U X G

FIGURE 5-39 **Jupiter Unwrapped** *Cassini* images of Jupiter combined and opened to give a maplike representation of the planet. The banded structure is absent near the poles. *(Courtesy NASA/JPL-Caltech)*

R I V U X G

FIGURE 5-40 **The Great Red Spot** This true-color image of the Great Red Spot, taken by *Galileo* in 1996, shows what this giant storm would look like if you were traveling over it in a spacecraft. The counterclockwise circulation of gas in the Great Red Spot takes about 6 days to make one rotation. The clouds that encounter it are forced to pass around it, and when other oval features are near it, the entire system becomes particularly turbulent, like the batter in a two-bladed blender. *(Courtesy NASA/JPL-Caltech)*

Great Red Spot was first observed around 1656, either by the English scientist Robert Hooke or the Italian astronomer Giovanni Cassini. Because earlier telescopes were unlikely to have been able to see it, the Great Red Spot could well have formed long before that time.

The Great Red Spot is a hurricane or typhoonlike storm of swirling gases. Heat welling upward from inside Jupiter has maintained it for more than 3 centuries. (Consider what life would be like for us if Earth sustained storms for such long periods!) Between 1998 and 2000, three smaller white storms on Jupiter merged (Figure 5-41a-d), creating a larger white storm that became red in 2006 (see Figure 5-41e inset). Named *Red Spot Jr.*, it is similar but somewhat smaller than the Great Red Spot and located at nearly the same latitude (Figure 5-41e). As with the Great Red Spot, the cause of the red color is still being studied.

In 1690, Cassini noticed that the speeds of Jupiter's clouds vary with latitude, an effect called **differential rotation.** Near the poles, the rotation period of Jupiter's atmosphere, 9 h 55 min 30 s, is 5 min longer than at the equator. Furthermore, clouds at different latitudes circulate in opposite directions—some eastward, some westward. At their boundaries, the clouds rub against each other, creating beautiful swirling patterns (see Figure 5-38). The interactions of clouds at different latitudes also help provide stability for storms like the Great Red Spot.

9 Astronomers first determined Jupiter's overall chemical composition from its average density—only 1330 kg/m³. Recall from Section 2-13 that average density is mass divided by volume. We determine Jupiter's mass using Kepler's

third law and the orbital periods of its moons, while the trigonometry of Jupiter's distance from Earth and angular size in our sky reveal its diameter and hence its volume. This low density implies that Jupiter is primarily composed of the lightweight elements hydrogen and helium surrounding a relatively small core of metal and rock. It has no solid continents, islands, or water oceans on its surface.

Spectra from Earth-based telescopes and from the *Galileo* probe sent into Jupiter's upper atmosphere in 1995 give more detail about the chemistry of this giant world's upper level. About 86% of its atoms are hydrogen and 13% are helium. The remainder consists of molecular compounds, such as methane (CH_4), ammonia (NH_3), and water vapor (H_2O). Keeping in mind that different elements have different masses, we can convert these percentages of atoms into the masses of various substances in Jupiter's atmosphere: 75% hydrogen, 24% helium, and 1% other substances. Because the interior contains heavier elements than the surface, the overall mass distribution in Jupiter has been calculated to be 71% hydrogen, 24% helium, and 5% all heavier elements.

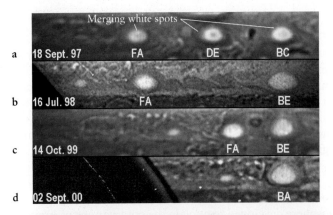

a 18 Sept. 97   FA   DE   BC

b 16 Jul. 98   FA   BE

c 14 Oct. 99   FA   BE

d 02 Sept. 00   BA

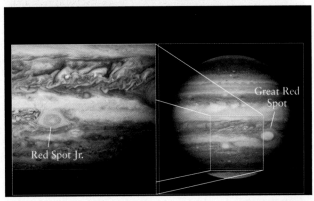

e

R I V U X G

FIGURE 5-41 **Creating Red Spot Jr.** (a-d) For 60 years prior to 1998, the three white ovals labeled FA, DE, and BC traveled together at the same latitude on Jupiter. Between 1998 and 2000, they combined into one white oval, labeled BA, which (e) became a red spot, named Red Spot Jr., in 2006. *(a-d: NASA/JPL/WFPC2; e: NASA, ESA, A. Simon-Miller, NASA/GSFC, and I. de Pater, University of California, Berkeley)*

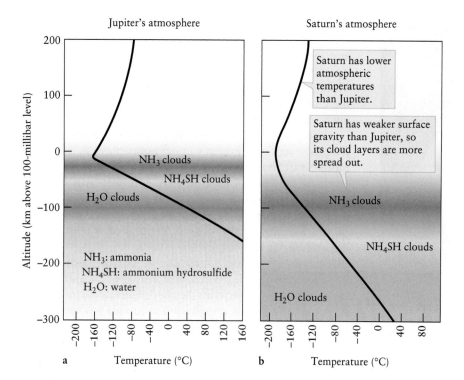

Jupiter's atmosphere

Saturn's atmosphere

Saturn has lower atmospheric temperatures than Jupiter.

Saturn has weaker surface gravity than Jupiter, so its cloud layers are more spread out.

NH₃ clouds

NH₄SH clouds

H₂O clouds

NH₃: ammonia
NH₄SH: ammonium hydrosulfide
H₂O: water

NH₃ clouds

NH₄SH clouds

H₂O clouds

a Temperature (°C)

b Temperature (°C)

FIGURE 5-42 **Jupiter's and Saturn's Upper Layers** These graphs display temperature profiles of (a) Jupiter's and (b) Saturn's upper regions, as deduced from measurements at radio and infrared wavelengths. Three major cloud layers are shown in each, along with the colors that predominate at various depths. Data from the *Galileo* spacecraft indicate that Jupiter's cloud layers are not found at all locations around the planet; there are some relatively clear, cloud-free areas.

gradually gets denser until, 1000 km below the cloud tops, the pressure is high enough for the hydrogen to be what we would consider a liquid.

In introducing the solar system, we noted that the young planets heated up as they coalesced. After they formed, radioactive elements continued to heat their interiors. On Earth, this heat leaks out of the surface through volcanoes and other vents. Jupiter loses heat everywhere on its surface, because, unlike Earth, it has no surface landmasses to block the heat loss.

Heated from deep within Jupiter, blobs of liquid hydrogen and helium move upward inside it. When these blobs reach the cloud tops, they release their heat and descend back into the interior. (The same process, *convection*, drives the motion of Earth's mantle and its tectonic plates, as well as liquid simmering on a stove; see Figure 4-8.) Jupiter's rapid, differential rotation draws the convective gases into bands of winds moving eastward and westward at different speeds around the planet.

The descent of the probe from the *Galileo* spacecraft into Jupiter's atmosphere revealed wind speeds of up to 600 km/h (375 mph), higher-than-expected air density and temperature, and lower-than-expected concentrations of water, helium, neon, carbon, oxygen, and sulfur. This probably occurred because the probe descended into a particularly arid region of the atmosphere called a *hot spot,* akin to the air over a desert on Earth.

Observations from spacecraft visiting Jupiter and its moons, combined with the scientific model of Jupiter's atmosphere developed to explain the observations of Jupiter's clouds and chemistry, indicate that it has three major cloud layers (Figure 5-42a). The uppermost Jovian cloud layer is composed of crystals of frozen ammonia. These crystals and the frozen water in Jupiter's clouds are white, so what chemicals create the subtle tones of brown, red, and orange? The answer is as yet unknown. Some scientists think that sulfur compounds, which can assume many different colors depending on their temperature, play an important role. Others think that phosphorus is involved, especially in the Great Red Spot. The middle cloud layer is primarily ammonium hydrosulfide, and the bottom cloud layer is mostly composed of water vapor.

Below its cloud layer, Jupiter's mantle is entirely liquid. Here on Earth, the distinction between the gaseous air and the liquid oceans is very clear—jump off a diving board and you know when you hit the water. However, the conditions on Jupiter under which hydrogen liquefies are different from anything we normally experience. As a result, on Jupiter there is no definite boundary between the planet's gaseous atmosphere and its liquid mantle. The hydrogen

What are the two most common elements in Jupiter's upper layers?

Astronomers believe that Jupiter's belts and zones result from the combined actions of the planet's convection and rapid differential rotation. Until recently, they believed that the light-colored zones are regions of hotter, rising gas, while the dark belts are regions of cooler, descending gas (Figure 5-43). However, high-resolution observations by the passing *Cassini* spacecraft on its way to Saturn revealed numerous white clouds that are too small to be seen from Earth rising in the dark belts of Jupiter, while the zones are sinking gas—just the opposite of the original model! The correct explanation for the behavior of gases in the belts and zones may be provided by *Cassini* when it studies the analogous belts and zones found on Saturn (see Section 5-24) in greater detail in the next few years.

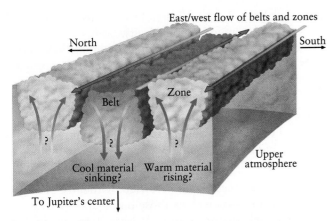

FIGURE 5-43 Original Model of Jupiter's Belts and Zones The light-colored zones and dark-colored belts in Jupiter's atmosphere were believed, until recently, to be regions of rising and descending gases, respectively. In the zones, gases warmed by heat from Jupiter's interior were thought to rise upward and cool, forming high-altitude clouds. In the belts, cooled gases were thought to descend and undergo an increase in temperature; the cloud layers seen there are at lower altitudes than in the zones. Observations by the *Cassini* spacecraft on its way to Saturn suggest that just the opposite may be correct (stay tuned)! In either case, Jupiter's rapid differential rotation shapes the rising and descending gas into bands of winds parallel to the planet's equator. Differential rotation also causes the wind velocities at the boundaries between belts and zones to move predominantly to the east or west.

5-17　Jupiter's interior has four distinct regions

Descending below Jupiter's clouds, we first encounter liquid molecular hydrogen and helium. As in Earth's oceans, the pressure in this fluid increases with depth. The gravitational force created by Jupiter's enormous mass compresses and heats its interior so much that 20,000 km (12,500 mi) below the cloud tops, the pressure is 3 million atm. Below this depth, the pressure is high enough to transform hydrogen into **liquid metallic hydrogen** (Figure 5-44). In this state, hydrogen acts like a metal (similar to the copper wiring in a house) in its ability to conduct electricity and heat. Electric currents that run through this rotating, metallic region of Jupiter generate a powerful planetary magnetic field (Figure 5-45). At the cloud level of Jupiter, this field is 14 times stronger per square meter than Earth's field is at our planet's surface. Jupiter's magnetic field has a region, like Earth's Van Allen belts, where charged particles are stored (depicted in orange in Figure 5-45) and its tail streams outward to Saturn, over 700 million km away.

For Earth-based astronomers, the only evidence of Jupiter's magnetosphere is a hiss of radio static, which varies cyclically over a period of 9 h 55 min 30 s, the planet's internal rotation rate. Beginning with the two *Pioneer* and two *Voyager* spacecraft that journeyed past Jupiter in the 1970s, and continuing with the recent *Galileo* and *Cassini*

spacecraft, the awesome dimensions of Jupiter's magnetosphere have been revealed. It is nearly 30 million km across and envelops the orbits of many of its moons. If Jupiter's magnetosphere were visible from Earth, it would cover an area of the sky 16 times larger than the full Moon.

Although the sequence of events in Jupiter's formation is still being worked out, it appears that a terrestrial (rock and metal) protoplanet formed at Jupiter's orbit and attracted hydrogen and helium to form its outer layers. This terrestrial matter is now Jupiter's core. This core is only 4% of Jupiter's mass, which still amounts to nearly 13 times the mass of the entire Earth. Water, carbon dioxide, methane, and ammonia are likely to have existed as ice in Jupiter's terrestrial protoplanet, and they were also attracted onto the young and growing planet. When astronomers talk in general terms about "ice," they refer to any or all of these four compounds.

The tremendous crushing weight of the bulk of Jupiter above the core—equal to the mass of 305 Earths—compresses

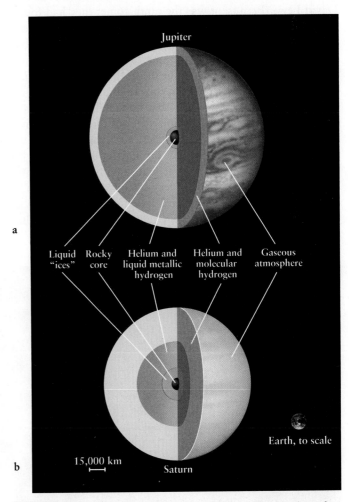

FIGURE 5-44 Cutaways of Jupiter and Saturn The interiors of both Jupiter and Saturn are believed to have four regions: a terrestrial rocky core, a liquid "ice" shell, a liquid metallic hydrogen shell, and a normal liquid hydrogen mantle. Their atmospheres are thin layers above the normal hydrogen, which boils upward, creating the belts and zones.

the terrestrial core down to a sphere only 10,000 km in radius (see Figure 5-44). By comparison, Earth's diameter is 12,756 km. At the same time, the pressure forced the lighter ices out of the rock and metal, thereby forming a shell of these compounds between the solid core and the liquid metallic hydrogen layer. Calculations reveal that the temperature and pressure inside Jupiter should make these "ices" liquid there! The pressure at Jupiter's very center is calculated to be about 70 million atm, and the temperature there is about 25,000 K, nearly four times hotter than the surface of the Sun. The interior of Jupiter is summarized in Figure 5-44a.

5-18 Impacts provide probes into Jupiter's atmosphere

On July 7, 1992, a comet nucleus (a clump of rock and ice a few kilometers across) passed so close to Jupiter that the planet's gravitational tidal force ripped it into at least 21 pieces. The debris from this comet was first observed in March 1993 by comet hunters Carolyn and Gene Shoemaker and David Levy. (Because it was the ninth comet they had found together, it was named Shoemaker-Levy 9 in their honor.) Shoemaker-Levy 9 was an unusual comet that actually orbited Jupiter, rather than just orbiting the Sun. Calculations of the comet's orbit showed that the pieces (Figure 5-46a) would return to strike Jupiter between July 16 and July 22, 1994.

Recall from Section 2-11 that impacts were extremely common in the first 800 million years of the solar system's existence. From more recent times, two chains of impact craters have been discovered on Earth, consistent with pieces of comets having hit our planet within the past 300 million years. However, it is very uncommon for pieces of space debris as large as several kilometers in diameter to collide with planets today. Therefore, the discovery that

Shoemaker-Levy 9 would hit Jupiter created great excitement in the astronomical community. Seeing how a planet and a comet respond to such an impact would allow astronomers to deduce information about the planet's atmosphere and interior and also about the striking body's properties.

Why is the terrestrial body inside Jupiter so small compared to what it would be in orbit without the outer layers around it?

The impacts occurred as predicted, with most of Earth's major telescopes—as well as those on several spacecraft—watching closely (Figure 5-46b). At least 20 fragments from Shoemaker-Levy 9 struck Jupiter, and 15 of them had detectable impact sites. The impacts resulted in fireballs some 10 km in diameter with temperatures of 7500 K, which is hotter than the surface of the Sun. Indeed, the largest fragment released as much energy as 600 million megatons of TNT, far more than the combined energy that could be released by all nuclear weapons remaining on Earth. Impacts were followed by crescent-shaped ejecta that contained a variety of chemical compounds. Ripples or waves spread out from the impact sites through Jupiter's clouds in splotches that lasted for months.

The observations suggested that the pieces of comet did not penetrate very far into Jupiter's upper cloud layer. This fact, in turn, supports the belief that the pieces were not much larger than a kilometer in diameter. The ejecta from each impact included a dark plume that rose high into Jupiter's atmosphere. The darkness was apparently due to carbon compounds that vaporized from the cometary bodies. Also detected from the comet were water, sulfur compounds, silicon, magnesium, and iron.

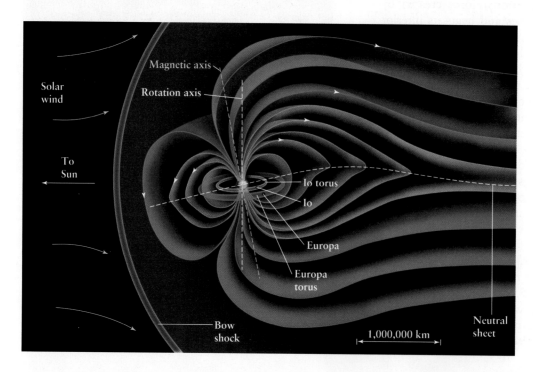

FIGURE 5-45
Jupiter's Magnetosphere Created by the planet's rotation, the ion-trapping regions of Jupiter's magnetosphere (in orange, analogous to the Van Allen belts) extend into the realm of the Galilean moons. Gases from Io and Europa form tori in the magnetosphere. Some of Io's particles are pulled by the field onto the planet. Pushed outward by the Sun, the magnetosphere reaches all the way to Saturn.

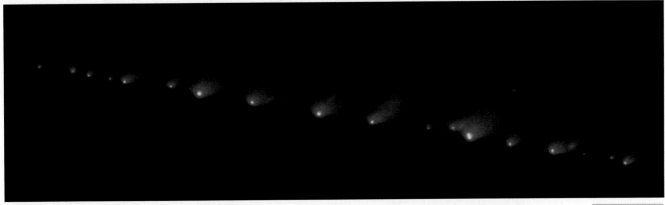

a   R I V U X G

b   R I V U X G   R I V U X G

FIGURE 5-46 **Comet Shoemaker-Levy 9 and Its Encounter with Jupiter** (a) The comet was torn apart by Jupiter's gravitational force on July 7, 1992, fracturing into at least 21 pieces. This comet originally orbited Jupiter, and its returning debris, shown here in May 1994, struck the planet between July 16 and July 22, 1994. (b) Shown here are visible (left) and ultraviolet (right) images of Jupiter taken by the Hubble Space Telescope after three pieces of Comet Shoemaker-Levy 9 struck the planet. Astronomers had expected white remnants (the color of condensing ammonia or water vapor); the darkness of the impact sites may have come from carbon compounds in the comet debris. Note the auroras in the ultraviolet image. Auroras and lightning are common on Jupiter, due in part to the planet's strong magnetic fields and dynamic cloud motions. *(a: H. A. Weaver, T. E. Smith, STScI and NASA; b: NASA)*

Why did the comet impacts of 1994 not create craters on Jupiter?

JUPITER'S MOONS AND RINGS

Jupiter hosts at least 63 moons. The four largest—Io, Europa, Ganymede, and Callisto—may have formed at the same time as Jupiter from debris orbiting the planet, just as the planets formed from debris orbiting the Sun (see Section 2-11). Jupiter's other moons, as we will see, are probably captured planetesimals and smaller pieces of space debris.

Galileo was the first person to observe the four largest moons, in 1610, seen through his meager telescope as pin-points of light. He called them the *Medicean stars* to attract the attention of the Medicis, rulers of Florence and wealthy patrons of the arts and sciences. To Galileo, the moons provided evidence supporting the then-controversial Copernican cosmology; at that time, Western theologians asserted that all cosmic bodies orbited Earth. The fact that the Medicean stars orbited Jupiter raised grave concerns in some circles.

To the modern astronomer, these moons are four extraordinary worlds, different both from the rocky terrestrial planets and from hydrogen-rich Jupiter. Now collectively called the **Galilean moons** or **Galilean satellites,** they are named after the mythical lovers and companions of the Greek god Zeus (called Jupiter by the early Romans). From the innermost moon outward, they are Io, Europa, Ganymede, and Callisto.

TABLE 5-2 The Galilean Moons: Vital Statistics

	Mean distance from Jupiter (km)	Sidereal period (day)	Diameter (km)	Mass (kg)	Mass (Moon = 1)	Mean density (kg/m³)
Io	421,600	1.77	3630	8.94×10^{22}	1.22	3570
Europa	670,900	3.55	3138	4.80×10^{22}	0.65	2970
Ganymede	1,070,000	7.16	5262	1.48×10^{23}	2.01	1940
Callisto	1,883,000	16.69	4800	1.08×10^{23}	1.47	1860
Mercury	—	—	4878	3.30×10^{23}	4.49	5430
Moon	—	—	3476	7.35×10^{22}	1.00	3340

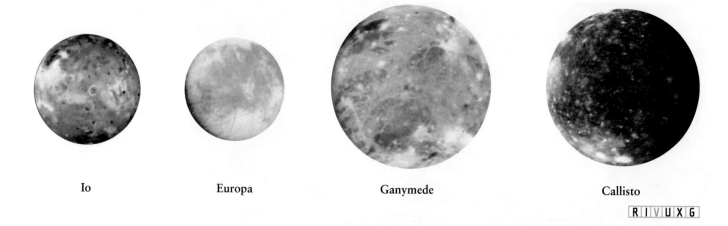

Io   Europa   Ganymede   Callisto

R I V U X G

GANYMEDE

Icy crust
Icy mantle
Rocky mantle

Iron core

IO

Molten mantle

Rocky crust   Iron core

CALLISTO

Icy crust

Ocean?

Mixed ice–rock interior

EUROPA

Rocky mantle
Ocean
Icy crust

Iron core

WEB LINK 5.21

FIGURE 5-47 **The Galilean Moons** The four Galilean satellites are shown here to the same scale. Io and Europa have diameters and densities comparable to our Moon and are composed primarily of rocky material. Ganymede and Callisto are roughly as big as Mercury, but their low average densities indicate that each contains a thick layer of water and ice. The cross-sectional diagrams of the interiors of the four Galilean moons show the probable internal structures of the moons, based on their average densities and on information from the *Galileo* mission. (*NASA and NASA/JPL*)

Prometheus
Volcano

a   R I V U X G

Pilan Patera

b   R I V U X G

These four worlds were photographed extensively by the *Voyager 1* and *Voyager 2* flybys and by the *Galileo* spacecraft (Figure 5-47 and Table 5-2). The two inner Galilean satellites, Io and Europa, are about the same size as our Moon. The two outer satellites, Ganymede and Callisto, are roughly the size of Mercury. Table 5-2 presents comparative information about these bodies.

5-19   Io's surface is sculpted by volcanic activity

Sulfurous Io is among the most exotic moons in our solar system (Figure 5-48a). At first glance, its density of 3570 kg/m^3 seems to place its chemical composition between that of the terrestrial and Jovian planets. However, its density has that value in large measure because its mass is so small that it cannot compress its interior nearly as much as can Earth or more massive planets. Allowing for its lower compression, astronomers calculate that Io is mostly rock and iron, like Earth, rather than being composed of lighter elements, as is Jupiter.

Io zooms through its orbit of Jupiter once every 1.8 days. Like our Moon, it is in synchronous rotation with its planet. Images of Io reveal giant plumes emitted by volcanoes (Figure 5-48b) and geysers, similar to Old Faithful on Earth. Most of this ejected material falls back onto Io's surface; the rest is moving fast enough to escape into space. These

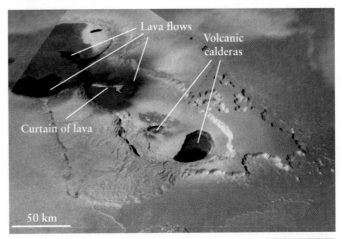

Lava flows

Volcanic
calderas

Curtain of lava

50 km

c   R I V U X G

FIGURE 5-48 **Io** (a) This true-color view was taken by the *Galileo* spacecraft in 1999. The range of colors results from surface deposits of sulfur ejected from Io's numerous volcanoes. Plumes from the volcano Prometheus rise up 100 km. Prometheus has been active in every image taken of Io since the *Voyager* flybys of 1979. (b) *Galileo* image of an eruption of Pilan Patera on Io. (c) Photographed in 1999 and then 2000 (shown here), the ongoing lava flow from this volcanic eruption at Tvashtar Catena has considerably altered this region of Io's surface. *(a: Moses Milazzo, PIRL, JPL, NASA; b: Galileo Project, JPL, NASA; c: University of Arizona/JPL/NASA)*

geysers are usually associated with Io's 300 or so active volcanoes. Observations indicate that these volcanoes emit 10 trillion tons of matter each year in plumes up to 500 km high. That is enough material to resurface Io to a depth of 1 m each century (Figure 5-48c). One eruption in 2002 occurred over an area the size of London, some 1600 sq km (620 sq mi). The volcanoes also emit basaltic lava flows rich in magnesium and iron. Io's volcanoes are named after gods and goddesses associated with fire in Greek, Norse, Hawaiian, and other mythologies. Io also has numerous black "dots" on its surface, which, apparently, are dormant volcanic vents. Old lava flows radiate from many of these locations, which are typically 10 to 50 km in diameter and cover 5% of Io's surface.

INSIGHT INTO SCIENCE

What's in a Name? Following up on the preconceptions that common words create (see Insight into Science: Imagine the Moon), objects with familiar names often have different characteristics than we expect. We tend to envision "moons" as inert, airless, lifeless, dry places similar to our Moon. However, as we will see throughout this chapter, starting with Io, some moons have active volcanoes, atmospheres, and vast, underground, liquid water oceans.

Just before their discovery, the existence of active volcanoes on Io was predicted from analysis of the gravitational forces to which that moon is subjected. As it rapidly orbits Jupiter, Io repeatedly passes between Jupiter and one or another of the remaining Galilean satellites. These moons pull on Io, causing it to slightly change its distance from Jupiter. This change in distance creates a change in the tidal forces (see Section 4-8) acting on it from the planet. As the distance between Io and Jupiter varies, the resulting tidal stresses alternately squeeze and flex the moon. This ongoing motion of Io's interior generates heat through friction, creating as much energy inside Io as the detonation of 2400 tons of TNT every second. Gas and molten rock from inside Io eventually make their way to the moon's surface, where they are ejected through the volcanoes and other vents.

Satellite instruments have identified sulfur and sulfur dioxide in the material erupting from Io's volcanoes. Sulfur is normally bright yellow. If heated and suddenly cooled, however, it forms molecules that assume a range of colors, from orange and red to black, which accounts for Io's tremendous range of colors (see Figure 5-48). Sulfur dioxide (SO_2) is an acrid gas commonly discharged from volcanic vents here on Earth and, apparently, on Venus. When eruptions on Io release this gas into the cold vacuum of space, it crystallizes into white flakes, which fall onto the surface and account for the moon's whitish deposits.

The *Galileo* spacecraft detected an atmosphere around Io. Composed of oxygen, sulfur, and sulfur dioxide, it is only one-billionth as dense as the air we breathe. Io's atmosphere can sometimes be seen to glow blue, red, or green, depending on the gases involved. Gases ejected from Io's volcanoes have also been observed extending out into space and forming a doughnut-shaped region around Jupiter called the *Io torus*, about which we will say more shortly.

What creates most of the heat inside Io that causes it to have volcanoes and geysers?

5-20   Europa apparently harbors liquid water below its surface

Images of Europa's ice and rock surface from *Voyager 2* and the *Galileo* spacecraft suggest that Jupiter's second-closest Galilean moon contains liquid water (Figure 5-49). Europa orbits Jupiter every 3½ days and, like Io, is in synchronous rotation. The changing gravitational tug of Io creates stress inside Europa similar to the magma-generating distortion that Io undergoes. These stresses apparently cause numerous cracks or fractures seen on Europa's surface (Figure 5-50a). Furthermore, astronomers think this stress creates enough heat inside Europa to keep the water in a liquid state just a few kilometers below the moon's ice and rock surface.

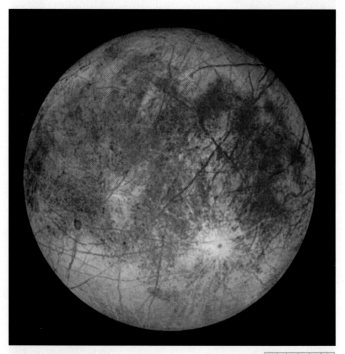

R I V U X G

FIGURE 5-49 **Europa** Imaged by the *Galileo* spacecraft, Europa's ice surface is covered by numerous streaks and cracks that give the satellite a fractured appearance. The streaks are typically 20 to 40 km wide. *(NASA/JPL)*

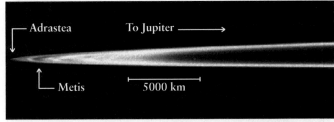

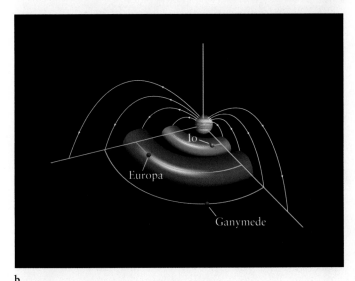

FIGURE 5-55 **Jupiter's Ring and Torus** (a) A portion of Jupiter's faint ring system, photographed by *Voyager 2.* The ring is probably composed of tiny rock fragments. The brightest portion of the ring is about 6000 km wide. The outer edge of the ring is sharply defined, but the inner edge is somewhat fuzzy. A tenuous sheet of material extends from the ring's inner edge all the way down to the planet's cloud tops. (b) Quarter images of Io's and Europa's tori (also called *plasma tori* because the gas particles in them are charged plasmas). Io is visible in its torus (green), while Europa is visible in its torus (blue). Some of Jupiter's magnetic field lines are also drawn in. Plasma from tori flow inward along these field lines toward Jupiter. *(a: NASA/JPL, Cornell University; b: NASA/JPL/Johns Hopkins University Applied Physics Laboratory)*

The other torus is in Europa's orbit (Figure 5-55b). It consists of hydrogen and oxygen ions and electrons created from water molecules kicked off Europa's surface by radiation from Jupiter. The mass of this ring of gas is calculated to be about 6×10^4 tons.

What features on our Moon and on Mercury are similar to Callisto's feature Valhalla?

What other objects that we have already studied in the solar system are similar to the smaller moons of Jupiter?

SATURN

Saturn, with its ethereal rings, presents the most spectacular image of the planets (Figure 5-56). Giant Saturn has 95 times as much mass as Earth, making it second in mass and size to Jupiter. Like Jupiter, Saturn has a thick, active atmosphere composed predominantly of hydrogen, and it radiates more energy than it receives from the Sun—in this case, three times as much. It also has a strong magnetic field. Ultraviolet images from the Hubble Space Telescope reveal auroras around Saturn differing from those of Jupiter in that Saturn's are spirals, while Jupiter's are rings (see Figure 5-55).

5-24 Saturn's atmosphere, surface, and interior are similar to those of Jupiter

Partly obscured by the thick, hazy atmosphere above them, Saturn's clouds lack the colorful contrast visible on Jupiter. Nevertheless, photographs do show faint stripes in Saturn's atmosphere, similar to Jupiter's belts and zones (Figure 5-57a). Their existence indicates that Saturn also has internal heat that transports the cloud gases by convection. Changing features in the atmosphere show that Saturn, too, has differential rotation—ranging from about 10 h and 14 min at the equator to 10 h and 40 min at high latitudes. As on Jupiter, some of the belts and zones move eastward, while others move westward. In 2006, the *Cassini* spacecraft revealed a hexagonal boundary surrounding Saturn's north pole (Figure 5-57b). While belts and zones flow around it, this feature is nearly stationary. The mechanism that creates and maintains it is still under investigation. Although Saturn lacks a long-lived spot like Jupiter's Great Red Spot, it does have storms, including a major one discovered by the Hubble Space Telescope in 1994 and a pair of storms that merged in 2004 (Figure 5-58). In 2004, the *Cassini* spacecraft detected lightning in Saturn's atmosphere.

Saturn's atmosphere is composed of the same basic gases as Jupiter. However, because its mass is smaller than Jupiter's, Saturn's gravitational force on its atmosphere is less. Therefore, Saturn's atmosphere is more spread out than that of its larger neighbor (see Figure 5-42). Saturn has impressive surface winds. At its equator, storms have been clocked moving at speeds of 1600 km/h (1000 mph), which is some 10 times faster than Earth's jet streams and three times faster than the fastest winds on Jupiter. The reason for Saturn's extremely high wind speeds is still under investigation.

Astronomers infer that Saturn's interior structure resembles Jupiter's. A layer of molecular hydrogen just below the clouds surrounds a layer of liquid metallic hydrogen, liquid "ices," and a rock and metal core (see Figure 5-44).

Because of its smaller mass, Saturn's interior is also less compressed than Jupiter's, and the pressure inside Saturn is insufficient to convert as much hydrogen into liquid metal. Saturn's rocky core is about 7400 km in radius, while its layer of liquid metallic hydrogen is 12,000 km thick (see

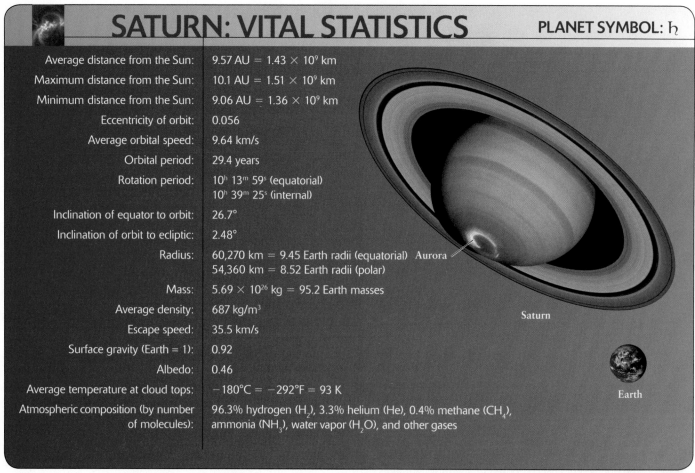

SATURN: VITAL STATISTICS

PLANET SYMBOL: ♄

Average distance from the Sun:	9.57 AU = 1.43 × 10⁹ km
Maximum distance from the Sun:	10.1 AU = 1.51 × 10⁹ km
Minimum distance from the Sun:	9.06 AU = 1.36 × 10⁹ km
Eccentricity of orbit:	0.056
Average orbital speed:	9.64 km/s
Orbital period:	29.4 years
Rotation period:	10ʰ 13ᵐ 59ˢ (equatorial) 10ʰ 39ᵐ 25ˢ (internal)
Inclination of equator to orbit:	26.7°
Inclination of orbit to ecliptic:	2.48°
Radius:	60,270 km = 9.45 Earth radii (equatorial) 54,360 km = 8.52 Earth radii (polar)
Mass:	5.69 × 10²⁶ kg = 95.2 Earth masses
Average density:	687 kg/m³
Escape speed:	35.5 km/s
Surface gravity (Earth = 1):	0.92
Albedo:	0.46
Average temperature at cloud tops:	−180°C = −292°F = 93 K
Atmospheric composition (by number of molecules):	96.3% hydrogen (H₂), 3.3% helium (He), 0.4% methane (CH₄), ammonia (NH₃), water vapor (H₂O), and other gases

Aurora

Saturn

Earth

R I V U X G

WEB LINK 5.28   VIDEO 5.1   VIDEO 5.2

FIGURE 5-56 Saturn: Vital Statistics Combined visible and ultraviolet images from the Hubble Space Telescope reveal spiral arcs of aurora near Saturn's south pole.

The image of Earth shows its size relative to Saturn. Note that there is much less contrast between Saturn's clouds than those of Jupiter. *(NASA; ESA; J. Clarke, Boston University; and Z. Levay, STScI; inset: NASA)*

Figure 5-44). To alchemists, Saturn was associated with the extremely dense element lead. This is wonderfully ironic in that at 687 kg/m³, Saturn is, on average, the least dense body in the entire solar system; an object with that density on Earth floats in water.

What two factors cause Saturn's belt and zone system?

5-25 Saturn's spectacular rings are composed of fragments of ice and ice-coated rock

WEB LINK 5.29   Even when viewed through a telescope from the vicinity of Earth, Saturn's magnificent rings are among the most interesting objects in the solar system. They are tilted 27° from the perpendicular to Saturn's plane of orbit, so sometimes they are nearly edge-on to us, making them virtually impossible to see (Figure 5-59). In

1675, Giovanni Cassini discovered an intriguing feature—a dark division in the rings. This 5000-km-wide gap, called the **Cassini division,** separates the dimmer **A ring** from the brighter **B ring,** which lies closer to the planet. By the mid-1800s, astronomers using improved telescopes detected a faint **C ring** just inside the B ring. Figure 5-60 shows a modern perspective of these rings.

The Cassini division occurs because the gravitational force from Saturn's moon Mimas combines with the gravitational force from the planet to keep the region clear of debris. Whenever matter drifts into the Cassini division, Mimas (orbiting at a different rate than the matter in the Cassini division) periodically exerts a force on this matter, eventually pulling it out of the division. This effect is called a **resonance** and is similar to what happens when you repeatedly push someone on a swing at the right time, enabling them to go higher and higher.

A second gap exists in the outer portion of the A ring, named the **Encke division** (Figure 5-60), after the German

a

R I V U X G

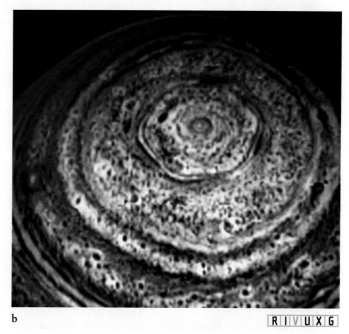

b

R I V U X G

FIGURE 5-57 **Belts and Zones on Saturn** (a) *Cassini* took this extremely high resolution image of Saturn in 2007. Details as small is 53 km (33 mi) across can be seen. There is less swirling structure between belts and zones on Saturn than on Jupiter. (b) In 2006,

Cassini's infrared imager took this view of a hexagonal pattern of clouds that rotates much more slowly than the surrounding belts and zones. Its origin is still under investigation. *(a: NASA/JPL/Space Science Institute; b: NASA/JPL/University of Arizona)*

astronomer Johann Franz Encke, who allegedly saw it in 1838. (Many astronomers have argued that Encke's report was erroneous, because his telescope was inadequate to resolve such a narrow gap.) The first undisputed observation

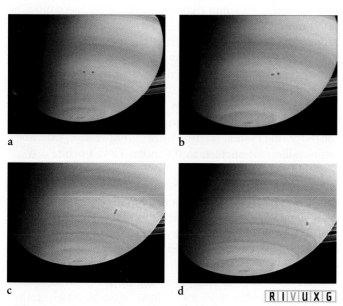

a

b

c

d

R I V U X G

FIGURE 5-58 **Merging Storms on Saturn** This sequence of *Cassini* images shows two hurricanelike storms merging into one on Saturn in 2004. Each storm is about 1000 km (600 mi) across. *(NASA/JPL/Space Science Institute)*

of the 270-km-wide division was made by the American astronomer James Keeler in the late 1880s, with the newly constructed 36-in refractor at the Lick Observatory in California. A 40-km space near the outer edge of the bright A ring is now named the *Keeler Gap* in honor of his work (Figure 5-60). Unlike the Cassini division, the Encke division is kept clear because a small moon, Pan, orbits within it. The gravitational tugs of Pan, Daphnis, Prometheus, and other moons also cause the rings to ripple, as shown in the inset of Figure 5-60.

The *Voyager* cameras sent back the first high-quality pictures of the **F ring**, a thin set of ringlets just beyond the outer edge of the A ring (Figure 5-61). Two tiny satellites, Prometheus and Pandora, have orbits on either side of the F ring and serve to keep it intact. The outer of the two satellites orbits Saturn at a slower speed than do the ice particles in the ring, as governed by Kepler's third law. As the ring particles pass near it, they receive a tiny, backward gravitational tug, which slows them down, causing them to fall into orbits a bit closer to Saturn. Meanwhile, the inner satellite orbits the planet faster than the F ring particles. Its gravitational force pulls the particles forward and nudges them back into a higher orbit. The combined effect of these two satellites is to focus the icy particles into a well-defined, narrow band about 100 km wide.

Because of their confining influence, Prometheus and Pandora are called **shepherd satellites** or **shepherd moons**. Among the most curious features of the F ring is that the ringlets are sometimes braided or intertwined and sometimes

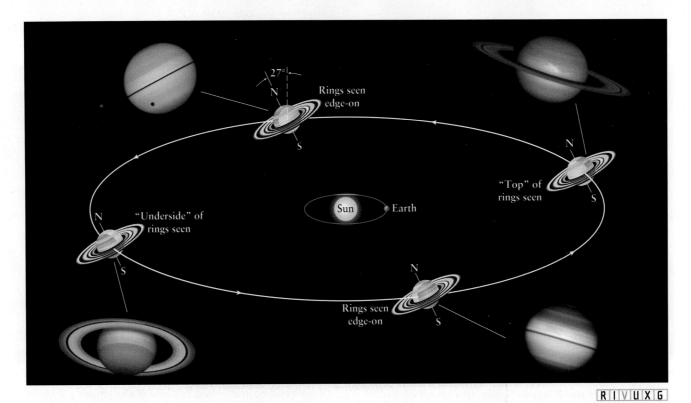

RIVUXG

FIGURE 5-59 **Saturn As Seen from Earth** Saturn's rings are aligned with its equator, which is tilted 27° from the plane of Saturn's orbit around the Sun. Therefore, Earth-based observers see the rings at various angles as Saturn moves around its orbit. The plane of Saturn's rings and equator keeps the same orientation in space as the planet goes around its orbit, just as Earth keeps its 23½° tilt as it orbits the Sun. The accompanying Earth-based photographs show how the rings seem to disappear entirely about every 15 years. *(Top left: E. Karkoschka/U. of Arizona Lunar and Planetary Lab and NASA; bottom left: AURA/STScI/NASA; bottom right and top right: A. Bosh/Lowell Obs. and NASA)*

separate. Although stabilized by the shepherd moons, the F ring does ripple and shows varying brightness. Astronomers theorize that the changes in brightness result from collisions between clumps of matter in the ring.

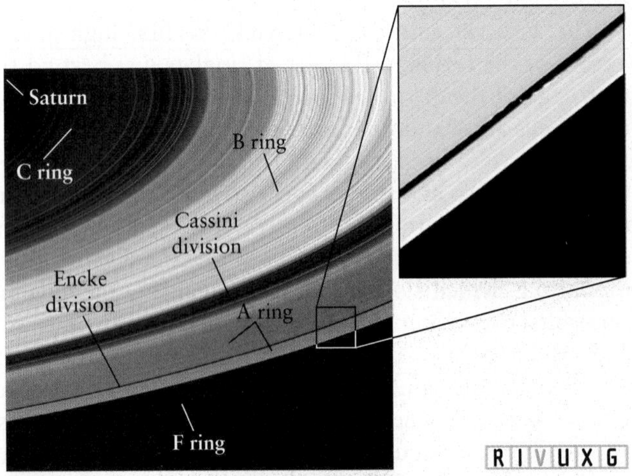

RIVUXG

FIGURE 5-60 **Numerous Thin Ringlets Constitute Saturn's Inner Rings** This *Cassini* image shows that Saturn's rings contain numerous ringlets. **Inset:** As moons orbit near or between rings, they cause the ring ices to develop ripples, often like the grooves in an old-fashioned record. *(NASA/JPL/ Space Science Institute)*

Beyond the major rings described above lie extensive rings of very fine dust particles, typically smaller than a millionth of a meter (a micron) across. Because such rings are very unstable, meaning that they drift away, there must be small moons that stabilize or renew this debris. The moons Janus, Epimetheus, and Pallene are now associated with these rings. Figure 5-62a shows a *Cassini* image of part of this system, including a ring discovered in the orbit of the moon Pallene. The *Cassini* image in Figure 5-62b gives us our most comprehensive view to date of Saturn's ring system. Nevertheless, the rings go out farther still. The scale of the outer ring system (Figure 5-62c) is impressive, extending out a million kilometers, to the moon Titan's orbit.

Pictures from the *Cassini* spacecraft and the Hubble Space Telescope show that Saturn's rings are remarkably thin—about 10 m thick according to recent estimates! This is amazing when you consider that the total ring system has a width (from inner edge to outer edge) of more than 380,000 km, more than the distance from Earth to our Moon.

Because Saturn's rings are very bright (their albedo is 0.80), the particles that form them must be highly reflective. Astronomers had long suspected that the rings consist primarily of water ice, along with ice-coated rocks. Spectra taken by Earth-based observatories and by the spacecraft visiting Saturn have confirmed this suspicion. Because the temperature of the rings ranges from 93 K (−290°F) in the

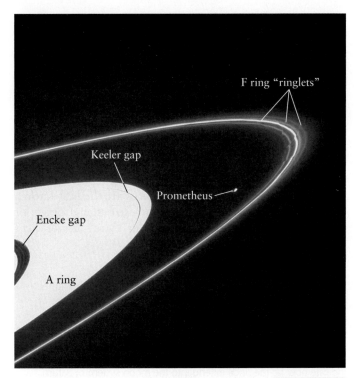

R I V U X G

FIGURE 5-61 **The F Ring and One of Its Shepherds** Two tiny satellites, Prometheus and Pandora, each measuring about 50 km across, orbit Saturn on either side of the F ring. Sometimes the ringlets are braided, sometimes parallel to each other. In any case, the passage of the shepherd moons causes ripples in the rings. The gravitational effects of these two shepherd satellites confine the particles in the F ring to a band about 100 km wide. *(NASA)*

sunshine to less than 73 K (−330°F) in Saturn's shadow, frozen water in them is in no danger of melting or evaporating.

Saturn's rings are slightly salmon-colored. This coloring suggests that they contain traces of organic molecules, which often have such hues. It appears that the rings have gained this material by being bombarded with debris from the outer solar system.

To determine the size of the particles in Saturn's rings, scientists measured the brightness of different rings from many angles as the *Voyager* and *Cassini* spacecraft flew past the planet. Astronomers also measured changes in radio signals received from these spacecraft as they passed behind the rings. Different rings are composed of debris with different sizes, ranging from particles of the sizes found in wood smoke, through those a few centimeters (or inches) across, to particles roughly 10 m in diameter. The outermost main ring, the A ring (see Figure 5-60), is composed of these larger clumps of rubble, which frequently collide with one another and thereby change sizes. This process is analogous to the planet-formation activity described in Chapter 2. Of the particles that would be visible to the naked eye (excluding dust particles), the centimeter-sized particles are most common.

High-resolution images reveal that the ring structure seen from Earth actually consists of thousands of closely spaced ringlets (see Figure 5-60). Even the spaces that we view from Earth as empty—namely Cassini and Encke's divisions—are filled with myriad smoke-sized dust particles. When spacecraft *Cassini* passed through a gap in the ring plane between the F and G rings (see Figure 5-62c), it was peppered with 100,000 impacts from these dust particles in less than 5 min.

Saturn's shell of liquid metallic hydrogen produces a planetwide magnetic field that apparently affects its rings. Saturn's much smaller volume of liquid metallic hydrogen produces a magnetic field at Saturn's surface that is only about two-thirds as strong as the magnetic field here on Earth. Data from spacecraft show that Saturn's magnetosphere contains radiation belts similar to Earth's Van Allen belts. Furthermore, dark **spokes** move around Saturn's rings (Figure 5-63); these are believed to be created by electric charges on the ring material interacting with the planet's magnetic field. The electric charge lifts charged particles out of the plane in which the rings orbit. Spreading the particles out decreases the light scattered from them and therefore makes the rings appear darker. The magnetic field causes the regions of spread particles to change, making the spokes appear to revolve around the planet.

Why are there gaps in Saturn's rings?

5-26   Titan has a thick atmosphere, clouds, and lakes filled with liquids

Only 7 of Saturn's 60 known moons are spherical. Several are shown in Figure 5-64a-e. The rest are oblong, suggesting that they are captured asteroids (Figure 5-64f, g). Twelve moons discovered in 2001 move in clumps, indicating that they may be pieces of a larger moon that was broken up by impacts. Saturn's largest moon, Titan, is second in size to Ganymede among the moons of the solar system and is the only moon to have a dense atmosphere. About 10 times more gas lie above each square meter of Titan's surface than lie above each square meter of Earth.

Christiaan Huygens discovered Titan in 1655, the same year he proposed that Saturn has rings. By the early 1900s, several scientists had begun to suspect that Titan might have an atmosphere. This moon is larger than (although less massive than) Mercury and far colder than that planet. Calculations reveal that Titan is cool enough and massive enough to retain heavy gases in its atmosphere. Because of this, Titan was a primary target for the *Voyager* missions (Figure 5-65a). Unexpectedly, its thick haze completely blocked any view of its surface. Observations also reveal that clouds of methane form and dissipate seasonally near both of Titan's poles and in the mid-southern latitudes. Although the causes for the appearance and disappearance of these clouds are not all known, volcanoes, geysers, and evaporation from Titan's lakes are all considered as factors.

Voyager data indicated that roughly 90% of Titan's atmosphere is nitrogen. Most of this gas probably formed from the breakdown of ammonia (NH_3) by the Sun's ultraviolet radiation into hydrogen and nitrogen atoms. Because Titan's gravity is too weak to retain hydrogen, this gas has escaped into space, leaving behind ample nitrogen. The unbreathable atmosphere is about 4 times as dense as that of Earth.

The second most abundant gas on Titan is methane, a major component of natural gas. Sunlight interacting with methane induces chemical reactions that produce a variety of other carbon–hydrogen compounds, or **hydrocarbons.** For example, spacecraft have detected small amounts of ethane (C_2H_6), acetylene (C_2H_2), ethylene (C_2H_4), and propane (C_3H_8) in Titan's atmosphere.

Methane and ethane condense into droplets in the cold air and fall as rain to Titan's surface. Nitrogen combines with various hydrocarbons to produce other compounds that can exist as liquids on Titan. Although one of these compounds, hydrogen cyanide (HCN), is a poison, some of the others are the building blocks of life's organic molecules. Indeed, biologists who study Titan's atmosphere now think that its chemistry is very similar to that of the young Earth.

Many of the hydrocarbons and carbon–nitrogen compounds in Titan's atmosphere can form complex compounds called **polymers.** These molecules are long repeating chains of atoms; rubber, cellulose, and plastics are the best-known examples on Earth. Scientists hypothesize that droplets of lighter polymers remain suspended in Titan's atmosphere

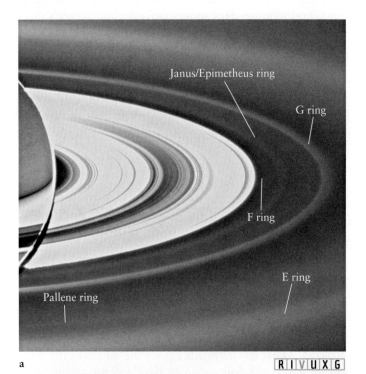

a R I V U X G

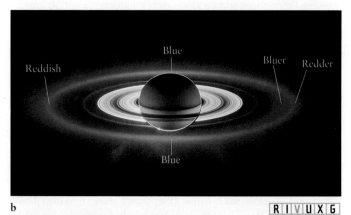

FIGURE 5-62 **Saturn's Outer Ring System** (a) Photographed with the Sun behind Saturn, the inner and intermediate regions of Saturn's ring system are seen to be very different from each other. Beyond the F ring, the particles are dust and pebble-sized. (b) Another view of the inner and intermediate rings, where subtle color differences are indicated. (c) Superimposed on this *Cassini* image are labels that indicate how far the rings extend into the moon system of Saturn. Titan (off image on right) is 1.2 million km (750, 000 mi) from the center of Saturn. *(a, b: NASA/JPL/Space Science Institute; c: NASA/JPL)*

b R I V U X G

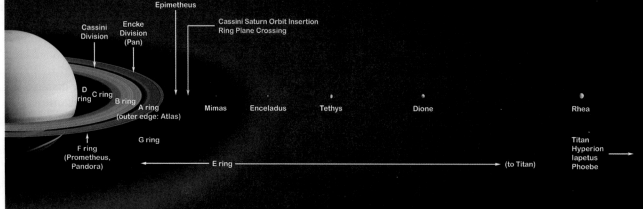

c R I V U X G

Crater Herschel

Mimas (diameter 392 km)

a

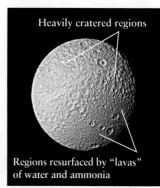

Heavily cratered regions

Regions resurfaced by "lavas" of water and ammonia

Tethys (diameter 1060 km)

b

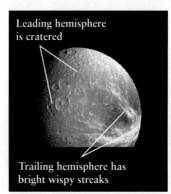

Leading hemisphere is cratered

Trailing hemisphere has bright wispy streaks

Dione (diameter 1120 km)

c

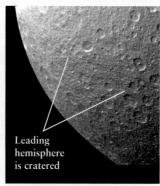

Leading hemisphere is cratered

Rhea (diameter 1530 km)

d

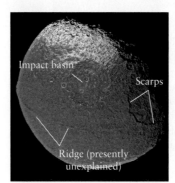

Impact basin

Scarps

Ridge (presently unexplained)

Iapetus (diameter 1460 km)

e

Hyperion (diameter 410 km)

f

R I V U X G

Phoebe (diameter 220 km)

g

R I V U X G

FIGURE 5-63 **Spokes in Saturn's Rings** Believed to be caused by Saturn's magnetic field moving electrically charged particles that are lifted out of the ring plane, these dark regions move around the rings like the spokes on a rotating wheel. *(NASA)*

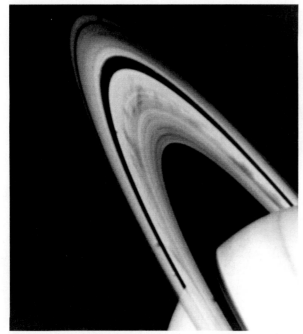

and form a mist, while heavier polymer particles settle down onto Titan's surface.

Images of the surface taken by the *Cassini* spacecraft (Figures 5-65b, c), which arrived in the Saturn system in 2004, show few craters, indicating that Titan's surface undergoes dynamic change. On the other hand, a mountain range with peaks as high as 1.5 km (1 mi) have been observed on it. They are covered with organic matter and possible methane snow. Some parts of Titan's surface have dunes 100 m high. Their composition is not yet clear, although they may be frozen organic molecules. Using the Keck I telescope in 1999, astronomers observed infrared emissions from Titan's surface consistent with

FIGURE 5-64 **Saturn's Diverse Moons** (a–e) These *Voyager 1, Voyager 2,* and *Cassini* images show the variety of surface features seen on five of Saturn's seven spherical moons. They are not shown to scale (refer to the diameters given below each image). The ridge running along the equator of Iapetus (e) has not yet been explained. (f) Perhaps the most bizarre object photographed in the solar system, Hyperion, shows innumerable impact craters. They are different from craters seen in other objects in that the crater walls here have not filled in the bottom of the craters. The moon's low gravity and the pull of nearby Titan may explain this unusual phenomenon. (g) This *Cassini* image of Phoebe, an irregular moon of Saturn almost as dark as coal, shows how the smaller moons have nonspherical shapes, along with many craters, landslides, grooves, and ridges. Phoebe is barely held in orbit by Saturn. Astronomers believe that it was captured after wandering in from beyond the orbit of Neptune. *(a, e–g: NASA/JPL/ Space Science Institute; b–d: JPL/NASA)*

FIGURE 5-65 Surface Features on Titan (a) *Voyager* images of Titan's smoggy atmosphere. (b) *Cassini* image of Titan of showing lighter highlands, called Xanadu, and dark, flat, lowlands that may be hydrocarbon seas. Resolution is 4.2 km (2.6 mi). (c) Riverbeds meandering across the Xanadu highlands of Titan. These are believed to have been formed by the flow of liquid methane and ethane. (d) Lake filled with liquid methane and ethane found at Titan's north pole, with Lake Superior, a Great Lake on Earth, for comparison. (e) The Huygens spacecraft took this image at Titan's surface on January 14, 2005. What appear like boulders here are actually pebbles strewn around the landscape. The biggest ones are about 15 cm (6 in.) and 4 cm (1.5 in.) across. *(a: NASA/JPL/Space Science Institute; b: NASA/JPL/University of Arizona; c: NASA/JPL; d: NASA/JPL/GSFC; e: NASA/JPL/ESA/University of Arizona)*

a partially liquid surface. Rain and rivers of methane have been observed (Figure 5-65c). In 2007, radar images from *Cassini* revealed a smooth dark surface, of more than 10^5 sq km (39,000 sq mi), the largest lake of liquid methane and ethane seen on Titan (Figure 5-65d). Other images reveal similar features elsewhere on that world, including islands in some lakes.

    Cassini launched a probe, named *Huygens,* that successfully parachuted onto Titan. It photographed rocks (Figure 5-65e), highlands, and channels. Upon impact, *Huygens* vaporized methane that had been just centimeters below the surface. The quantity of methane detected there implies that Titan has reservoirs of methane and other hydrocarbons, which are continually being broken down in its atmosphere by the Sun's ultraviolet radiation.

    We have little reason to suspect that life exists on Titan because its surface temperature of 95 K (−288°F) is prohibitively cold. Nevertheless, the material on Titan's surface

may be similar to what was on the young Earth, so a more detailed study of the chemistry of Titan may shed light on the origins of life here.

What liquids appear to exist on Titan?

5-27 Enceladus has an atmosphere and magnetic field

Saturn's sixth largest moon, Enceladus (500 km or 310 mi diameter), moved from the sidelines toward center stage in 2005, as *Cassini* began studying it (Figure 5-66a). This bright world has an icy, wrinkled surface similar in appearance to Jupiter's Europa and Ganymede. The fact that these latter two bodies contain liquid water suggests that Enceladus does, too. This theory is supported by observations that entire regions of Enceladus's southern hemisphere are

b. Zoom out to a field of view of about 30°. If the crosshair is not visible in the center of your screen, click the **View** tab and check the box labeled **Reticle/Crosshairs** in the **Overlays** pane to mark the position of Mars in the sky. Then click the **Fast Forward** button in the **Observing Time** panel until the time flow rate is at **X100000.** As the view tracks Mars, the background stars appear to move in the opposite direction. Remember, however, that it is Mars that is moving against the background of fixed stars. Is Mars moving toward the left (eastward) or toward the right (westward) relative to the background stars? Pause the flow before December, 2009. **c.** Click the **Fast Forward** button once to increase the time flow rate to **X1000000** and continue to watch the motion of Mars across the sky as indicated by the crosshair using a convenient field of view of 15° to 30°. Watch carefully as the date approaches mid-December, 2009 and moves forward to January, 2010. Describe the motion of Mars against the background stars during this period. **d.** Continue to observe the animation and pay particular attention to the time period between the end of February, 2010 and the end of March, 2010. Describe what happens to the motion of Mars against the background stars during this time period. **e. Pause** the time flow and reset the **Date** to October 1, 2009. Manipulate the time flow controls to determine i) the date of the first stationary point in Mars's motion, and ii) the date of the second stationary point in Mars's motion. For how long was Mars in retrograde motion?

PLANETARY SCIENCE

Dropping Acid

Mars may have needed acid rain to stay wet

BY SARAH SIMPSON

Sarah Simpson, "Dropping Acid," *Scientific American,* **April 2008, 27-28.**

On Mars, signs of wetness keep pouring in: deeply carved river valleys, vast deltas and widespread remnants of evaporating seas have convinced many experts that liquid water may have covered large parts of the Red Planet for a billion years or more. But most efforts to explain how Martian climate ever permitted such clement conditions come up dry. Bitterly cold and parched today, Mars needed a potent greenhouse atmosphere to sustain its watery past. A thick layer of heat-trapping carbon dioxide from volcanoes probably shrouded the young planet, but climate models indicate time and again that CO_2 alone could not have kept the surface above freezing.

Now, inspired by the surprising discovery that sulfur minerals are pervasive in the Martian soil, scientists are beginning to suspect that CO_2 had a warm-up partner: sulfur dioxide (SO_2).

Like CO_2, SO_2 is a common gas emitted when volcanoes erupt, a frequent occurrence on Mars when it was still young. A hundredth or even a thousandth of a percent SO_2 in Mars's early atmosphere could have provided the extra boost of greenhouse warming that the Red Planet needed to stay wet, explains geochemist Daniel P. Schrag of Harvard University.

That may not sound like much, but for many gases, even minuscule concentrations are hard to maintain. On our home planet, SO_2 provides no significant long-term warmth because it combines almost instantly with oxygen in the atmosphere to form sulfate, a type of salt. Early Mars would have been virtually free of atmospheric oxygen, though, so SO_2 would have stuck around much longer.

"When you take away oxygen, it's a profound change, and the atmosphere works really differently," Schrag remarks. According to Schrag and his colleagues, that difference also implies that SO_2 would have played a starring role in the Martian water cycle—thus resolving another climate conundrum, namely, a lack of certain rocks.

Schrag's team contends that on early Mars, much of the SO_2 would have combined with airborne water droplets and fallen as sulfurous acid rain, rather than transforming into a salt as on Earth. The resulting acidity would have inhibited the formation of thick layers of limestone and other carbonate rocks.

SULFUR MINERAL (*white*) that forms only in water was churned up by a Mars rover.

NASA/JPL/CORNELL

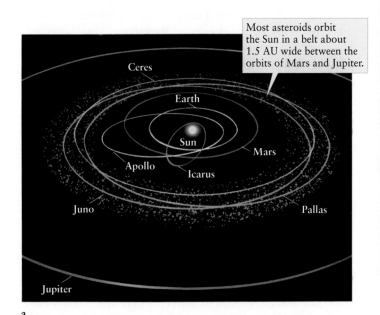

Most asteroids orbit the Sun in a belt about 1.5 AU wide between the orbits of Mars and Jupiter.

a

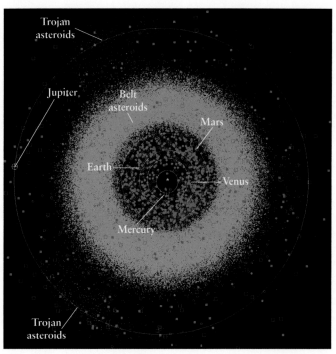

b

FIGURE 6-10 **Asteroid Orbits** (a) The orbits of belt asteroids Ceres, Pallas, and Juno are indicated to scale in this diagram. Some asteroids, such as Apollo and Icarus, have highly eccentric paths that cross Earth's orbit. Others, called the Trojan asteroids, follow the same orbit as Jupiter. (b) Actual positions of all known asteroids at Jupiter's orbit or closer. The locations of the belt asteroids are indicated by green dots. Objects passing closer than 1.3 AU to the Sun are shown by red circles. Objects observed at least twice are indicated by filled circles, and objects seen only once are indicated by outline circles. Jupiter's Trojan asteroids are deep blue squares. Comets are filled and unfilled light-blue squares. Although the asteroids appear packed together in this drawing, they are typically millions of kilometers apart. The small scale here is deceiving! *(Minor Planet Center)*

night to the next. With the advent of astrophotography, however, the floodgates were opened. Astronomers could simply aim a camera-equipped telescope at the stars and take long exposures. If an asteroid happened to be in the field of view, it left a distinctive trail on the photographic plate (Figure 6-11). Using this technique, Wolf alone discovered 228 asteroids.

INSIGHT INTO SCIENCE

Confirming Observations New findings in science must be confirmed or replicated by other competent scientists before discoveries or observations are accepted by the scientific community. Therefore, asteroid observations need to be repeated several times to determine exact orbits and eliminate the possibility that a sighting is of a known asteroid or other object.

Ceres, the largest asteroid, alone accounts for about 30% of the mass of all known asteroids combined. Only three asteroids—Ceres, Vesta, and Pallas—have diameters greater than 300 km. Records of all objects smaller than planets in our solar system are kept by the Minor Planet Center in Cambridge, Massachusetts. The center reports that as of September 2008, the existence of 189,407 asteroids had been confirmed and more than 51,272,383 other observations await confirmation as new asteroids. The number of asteroids increases dramatically with *decreasing size*: Only 41 asteroids have diameters between 200 and 300 km; 250 more are bigger than 100 km across; and there are thought to be tens of millions that are less than 1 km across.

You may have heard the common belief that the asteroids were formerly a single planet that was somehow destroyed. Indeed, it is plausible that the asteroid belt region once had many more objects in it—possibly several Earth masses worth of debris. However, Jupiter's gravitational force pulled many planetesimals out of this region before large numbers of them could meet, coalesce, and form a single planet-sized object.

If all of the present asteroids had once been part of a single body, it would have had a diameter of only 1500 km, or 12% of Earth's diameter. This is less than two-thirds the diameter of Pluto and half the diameter of our Moon. Considering that Pluto is not a planet, a single body that contains the mass of all of the asteroids also would not qualify as a planet. (We consider Jupiter's gravitational effect in more detail in the next section.)

Where is Eris located in the solar system?

R I V U X G

FIGURE 6-11 **Discovering Asteroids** In 1998, the Hubble Space Telescope found this asteroid while observing objects in the constellation Centaurus. The exposure, tracking stars, shows the asteroid as a 19-arcsec streak. This asteroid is about 2 km in diameter and was located about 140 million km (87 million mi) from Earth. (*R. Evans and K. Stapelfeldt, Jet Propulsion Laboratory and NASA*)

6-4   Jupiter's gravity creates gaps in the asteroid belt

In 1867, the American astronomer Daniel Kirkwood called attention to gaps in the asteroid belt. These features, called **Kirkwood gaps,** show the influence of Jupiter's gravitational attraction. Figure 6-12, a graph of asteroid orbital periods, has gaps at simple fractions (1/3, 2/5, 3/7, and 1/2) of Jupiter's orbital period. The gravitational pull of Jupiter created these empty regions by giving objects in them periodic reinforcing tugs that pulled the orbiting bodies to different orbits. The effect is analogous to the gravitational resonance of Mimas creating the Cassini division in Saturn's rings (see Section 5-25).

It is likely that the asteroid belt contains tens of millions of asteroids, yet their typical separation is an impressive 10 million km. This is quite unlike the image that has been created by innumerable popular movies and television shows of asteroids so close together that you must dodge them as you fly past.

INSIGHT INTO SCIENCE

Get Real Scientists continually apply the "laws of nature" to new situations, problems, observations, and experiments, even to popular culture. For example, applying Newton's law of gravity to the asteroids reveals that they could never swarm, as science fiction movies suggest. At those close quarters, their gravity would cause them either to collide or to pass so close together that they would subsequently fly rapidly and permanently apart.

Despite the large average separation between asteroids, the gravitational influences of Mars and Jupiter have sent some of them smashing into each other at various times over the past 4.6 billion years. A collision between kilometer-sized asteroids must be an awe-inspiring event. Typical collision velocities are estimated to be 3600 to 18,000 km/h (2000 to 11,000 mph), which is more than sufficient to shatter rock. In some collisions, the resulting fragments may not have enough speed to escape from one another's gravitational attraction, and they reassemble. The asteroid Toutatis appears to be composed of two comparably sized pieces connected to each other, as does the asteroid Castalia. These bodies are therefore likely to have been broken apart and reformed.

Alternatively, several large fragments, such as Ida and Dactyl (Figure 6-13), may end up orbiting each other. Dactyl is a pockmarked asteroid some 1.5 km in diameter that orbits Ida at a distance of 100 km. Petit-Prince is the satellite of Eugenia and is 13 km across, orbiting 1200 km from the larger body. To date, 151 asteroid/satellite systems have been observed.

In 1918, the Japanese astronomer Kiyotsugu Hirayama drew attention to groups of asteroids that share nearly identical orbits. These fragments are pieces of parent asteroids. Pursuing this line of research, astronomers in 2002 discovered that two large asteroids collided only a few million years ago (very recently in solar system history) and created about 20 separate families or clusters of smaller asteroids. Asteroids in each of these families orbit together. The largest

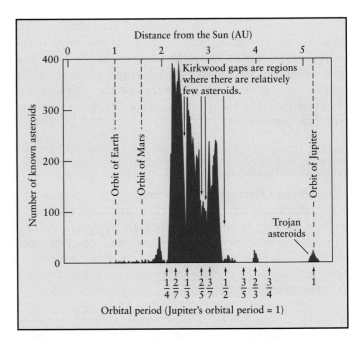

FIGURE 6-12 **The Kirkwood Gaps** This graph displays the number of asteroids at various distances from the Sun. Note that few asteroids have orbital periods that correspond to such simple fractions as 1/3, 2/5, 3/7, and 1/2 of Jupiter's orbital period. Repeated alignments with Jupiter have deflected asteroids away from these orbits. The Trojan asteroids accompany Jupiter as it orbits the Sun.

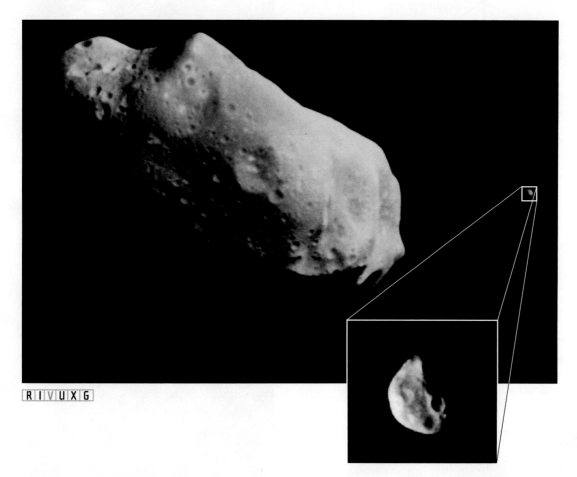

R I V U X G

FIGURE 6-13 **Ida and Its Satellite** The 55-km-long rocky asteroid Ida, shown here with its satellite Dactyl, is about twice the size of the younger asteroid Gaspra (see Figure 2-23). **Inset:** Dactyl is also heavily cratered. *(NASA)*

remnant cluster of this impact is named Karin, after its largest member, an asteroid some 20 km (12 mi) across.

In the early 1990s, the Jupiter-bound *Galileo* spacecraft passed near two asteroids—Gaspra (see Figure 2-23) and Ida (see Figure 6-13)—and sent back close-up views. Because Ida's surface is more heavily cratered than Gaspra's, we deduce that Ida is much older.

Why do astronomers doubt that the asteroid belt was once made up of a single planet?

6-5 Asteroids occur outside the asteroid belt

While Jupiter's gravitational pull clears out certain areas within the asteroid belt, it actually captures asteroids at two locations in the path of its own orbit. The gravitational forces of the Sun and Jupiter work together to hold asteroids in orbit at these locations, called **stable Lagrange points,** in honor of the French mathematician Joseph Lagrange, whose calculations explained them. One Lagrange point is located

60° ahead of Jupiter, and the other is 60° behind, as shown in Figure 6-14 (see also Figure 6-10b).

The asteroids trapped at Jupiter's Lagrange points are called **Trojan asteroids,** each named after a hero of the Trojan War. As of August 2008, 2543 Trojan asteroids orbiting with Jupiter have been catalogued. Neptune has 6 known Trojan asteroids, Mars has 1, and astronomers continue to search for Trojans orbiting with other planets.

Some asteroids have highly elliptical orbits that bring them into the inner regions of the solar system (see red dots in Figure 6-10b). Others have similarly elliptical orbits that extend from the asteroid belt out beyond the farthest reaches of Pluto's orbit. The Amor asteroids cross Mars's orbit, but do not get as close to the Sun as Earth's orbit, whereas the **Apollo asteroids** do cross Earth's orbit. At least 5525 *Earth-crossing* asteroids are known, several of which are pairs of asteroids that orbit each other. Not all of these Earth-crossing asteroids pose threats to Earth, but at least 1038 potentially hazardous asteroids (PHAs) may someday strike Earth.

Several close calls took place in recent years. In 1972, space debris was observed to skip off Earth's atmosphere

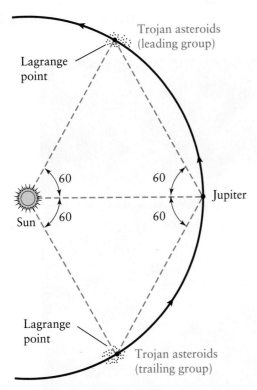

FIGURE 6-14 **Jupiter's Trojan Asteroids** Groups of asteroids orbit at the two stable Lagrange points along Jupiter's orbit, trapped by the combined gravitational forces of Jupiter and the Sun.

FIGURE 6-15 **Asteroid 1994 XM1** This image was obtained on December 9, 1994, shortly before the asteroid arrived in Earth's vicinity. When it passed by Earth just 12 h later, asteroid 1994 XM1 was less than half the distance from Earth to the Moon. *(Jim Scotti, Spacewatch on Kitt Peak)*

FIGURE 6-16 **Asteroids** (a) **Mathilde** Reflecting only half as much light as a charcoal briquette, Mathilde is half as dense as typical stony asteroids. Slightly larger than Ida, irregularly shaped Mathilde measures 66 km × 48 km × 46 km, rotates once every 17.4 days, and has a mass equivalent to 110 trillion tons. The part of the asteroid shown is about 59 km × 47 km. The large crater in shadow is about 20 km across. (b) **Itokawa** A near-Earth asteroid visited by the Japanese space probe *Hayabusa*. Samples were taken, but it is not known how much of the asteroid was captured by the spacecraft, which will return to Earth in 2010. *(a: Johns Hopkins University, Applied Physics Laboratory; b: ISAS, JAXA)*

and retreat back into space. On December 9, 1994, asteroid 1994 XM1 (10 m across—the size of a small bus; Figure 6-15) passed within 105,000 km (65,000 mi) of Earth. Two near misses occurred in 2002: one on January 8, when an asteroid passed within 375,000 km (233,000 mi), and the other on June 14, when an asteroid passed only 119,000 km (75,000 mi) away, which is less than a third the distance to the Moon.

During these close encounters, astronomers can examine the details of asteroids. For example, an asteroid's brightness often varies as it rotates because different surface features scatter different amounts of light. Such data show that typical rotation periods for asteroids are between 5 h and 20 h, although one asteroid, labeled 1998 KY26, rotates about once every 10.7 min. This is the fastest-rotating object known in the solar system.

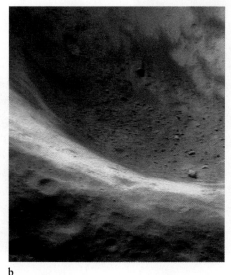

a   b   c   R I V U X G

FIGURE 6-17 Asteroid Eros (a) The *Near-Earth Asteroid Rendezvous (NEAR) Shoemaker* spacecraft took this image of asteroid Eros in February 1999. The top of the figure is the asteroid's north polar region. Eros's dimensions are 33 km × 13 km × 13 km (21 mi × 8 mi × 8 mi) and it rotates every 5¼ h. Its density is 2700 kg/m³, close to the average density of Earth's crust and twice as dense as asteroid Mathilde. (b) Looking into the large crater near the top of (a), which is 5.3 km (3.3 mi) across. (c) This is the penultimate image taken by *NEAR Shoemaker* before it gently landed on Eros. Taken from an altitude of 250 m (820 ft), the image is only 12 m across. You can see rocks and boulders buried to different depths in the regolith. *(Johns Hopkins Applied Physics Laboratory)*

The proximity of asteroids and the promise of learning more about these ancient and extremely varied members of our solar system prompted NASA to send the *Near Earth Asteroid Rendezvous (NEAR) Shoemaker* spacecraft to visit the asteroids Mathilde (Figure 6-16a) and Eros. The Japanese space agency sent the *Hayabusa* spacecraft to the asteroid Itokawa (Figure 6-16b), where the spacecraft attempted to take samples. It will return to Earth in 2010 with its cargo.

NEAR Shoemaker revealed that Mathilde is only 1.3 times denser than water and has an albedo of 0.04, making it darker than charcoal. This heavily cratered, carbon-rich body therefore has only half the density of the other asteroids astronomers have studied, such as Ida. Eros is about 3 times as dense as water and rotates once every 5¼ h. In February 2000, *NEAR Shoemaker* went into orbit around Eros, and for a year the spacecraft's cameras and other sensors sent back a wealth of information about it.

NEAR Shoemaker showed that Eros (Figure 6-17) is a chunk of rock and metal. Whether it is solid or a collection of loosely held pieces is still under investigation. Analyzing Eros's spectra reveals that it is probably much the same as it was when it coalesced 4.6 billion years ago. This means that it was never hot enough to differentiate (separate rock from metal). Infrared observations reveal that like our Moon, Eros has a regolith. It also has several substantial craters and is strewn with boulders (Figure 6-17c). Astronomers infer from the lack of high crater walls that recent impacts have caused the asteroid to vibrate, collapsing earlier cra-

ters. On February 12, 2001, NASA engineers landed *NEAR Shoemaker* on Eros so gently that the spacecraft continued to transmit data after landing. Details of Eros as small as 1.4 m across were imaged by *NEAR Shoemaker*.

The Sun is not the only star with an asteroid belt. In 2001, astronomers discovered that the star Zeta Leporis, 70 ly from Earth in the constellation Lepus (the Hare), has a disk of debris that appears to contain asteroids. This star system is less than 0.5 billion years old; astronomers hope it will provide insights into the early evolution of the asteroids and other objects in the disk of gas and dust that surrounded the early Sun.

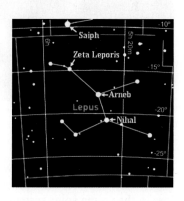

Describe the appearance of small asteroids.

COMETS

 While asteroids consist primarily of rock and metal, other space debris is composed of frozen water, along with rock, metal, and ices of other compounds. We have seen in earlier

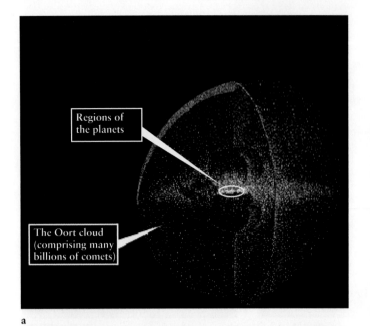

a

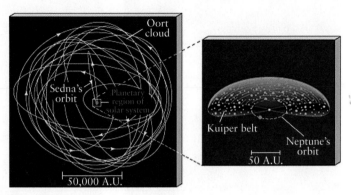

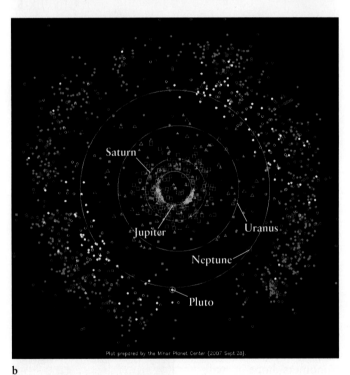

b

FIGURE 6-18 **The Kuiper Belt and Oort Comet Cloud** (a) The classical Kuiper belt of comets spreads from Neptune out 50 AU from the Sun. Most of the estimated 200 million belt comets are believed to orbit in or near the plane of the ecliptic. The spherical Oort comet cloud extends outward from the Kuiper belt. More detail of Sedna's orbit are given in Figure 6-20. (b) The current positions of the known dwarf planets and SSSBs in the outer solar system are shown in this diagram. Objects with unusually high eccentricity orbits are shown as cyan triangles. Objects roaming among the outer planets, Centaur objects, are orange triangles. Plutinos are white circles. Miscellaneous objects are magenta circles, and classical KBOs are red circles. Objects observed only once are denoted by open symbols; objects with two separate observations are denoted by filled symbols. Comets are filled and unfilled light-blue squares. (a: NASA and A. Field/Space Telescope Science Institute; b: Courtesy of Gareth Williams, Minor Planet Center)

chapters that water ice, along with carbon dioxide, methane, and ammonia ices, was locked up in planets and moons. Ices in the young solar system also condensed with roughly equal amounts of small rocky and metallic debris into bodies that still remain in orbit around the Sun. These dirty snowballs in space are the **comets.**

6-6   Comets come from far out in the solar system

In the first few hundred million years of the solar system's existence, comets formed in its outer reaches at roughly the distances of Saturn, Uranus, and Neptune. In that region, water was plentiful and the temperature was low enough for the ices to condense into chunks several kilometers across. After they formed, gravitational tugs from Uranus and Neptune flung the comets in every direction.

In 2001, astronomers were able to measure the temperature at which ammonia ice formed in Comet LINEAR. This temperature determines the ammonia's solid structure and so, by determining the structure of the ammonia in Comet LINEAR, they were able to calculate the temperature when it formed and how far the comet was from the Sun at that time. That comet coalesced between the orbits of Saturn and Uranus.

Today, the solar system is thought to contain two reservoirs of comets: the Kuiper belt and the Oort comet cloud. Most comets that eventually enter the inner solar system are believed to come from a doughnut-shaped region beyond the orbit of Pluto, called the **Kuiper belt.** Comets head inward as a result of collisions or near misses with other cometary bodies. Named after the American astronomer Gerard Kuiper, who first proposed its existence in 1951, the Kuiper belt is centered on the plane of the ecliptic. Its main body of objects extends out some 50 AU from the Sun (Figure 6-18).

The Kuiper belt consists of two groups of objects. Those with roughly circular orbits are called *classical KBOs* (Kuiper belt objects) and are found between 30 AU (Neptune's orbit) and 50 AU from the Sun. The second group, called *scattered KBOs,* have more elliptical orbits that range from 35 AU to at least 200 AU from the Sun. At least 1471 KBOs have been observed (Figure 6-18 and Figure 6-19a, b). Although most of these objects are tens of kilometers across, astronomers have also begun to observe some KBOs only 10 to 100 m in diameter. Based on these latter observations, astronomers estimate that 10^{15} such objects exist out there.

At least 1% of the KBOs are orbiting pairs (Figure 6-19c), including Quaoar, which is 1250 km across, some 400 km larger than Ceres or about half the size of Pluto. At least 95 of the KBOs orbit the Sun in the same region as Pluto. Because these latter bodies are smaller, they are called *Plutinos.* The periods of the orbits of both Pluto and the Plutinos are characterized by being 1½ (often written as 3:2) times as long as the period of Neptune's orbit. Indeed, it appears that Pluto and the Plutinos were forced into their present orbit by Neptune's gravity. Combining these last two points, we say that they are locked into a *3:2 resonance* with Neptune.

The vast majority of comets are believed to lie even farther from the Sun. Unlike the Kuiper belt comets and the rest of the solar system, these bodies are believed to have a spherical distribution around the Sun, called the **Oort comet cloud** (sometimes shortened to Oort cloud, see Figure 6-18a), named after the Dutch astronomer Jan Oort, who first proposed its existence in the 1950s.

In 2003, astronomers discovered an object larger than Quaoar. Called Sedna (see Section 2-12), this body is rough-ly 1800 km (1100 mi) across and is presently 86 AU from the Sun. It has a highly elliptical orbit (Figure 6-20) that takes it from the outer realm of the main Kuiper belt and possibly into the Oort comet cloud. Because the distance from the Sun to the inner edge of the Oort cloud is not known, astronomers are still debating whether it is another KBO or the first known Oort comet cloud object. Sedna has a sidereal rotation period of 10 h.

Astronomers calculate that the Oort comet cloud extends out at least 50,000 AU, one-fifth the distance to the nearest stars. Most of these comets have orbits so nearly circular that they never even get as close as Pluto is to the Sun. However, occasionally, a collision between two comet nuclei or the gravitational tug from a passing star or passing giant molecular cloud nudges a distant comet toward the inner solar system. As a result of its inward plunge, comets from the Oort comet cloud, such as Hale-Bopp and Hyakutake, have highly elliptical, parabolic, or even slightly hyperbolic orbits (see Section 2-8). Often comets also have orbits that are highly tilted, even perpendicular, to the plane of the ecliptic.

Because the comets in the Kuiper belt and Oort comet cloud are far from the Sun, they are completely frozen. Solid cometary bodies, called **nuclei** (*singular:* **nucleus**), are typically between 1 and 10 km across, although some, like that of Comet Halley, are 15 km or more in diameter. The first pictures of a comet's nucleus were obtained when a fleet of spacecraft flew past Comet Halley in 1986 (Figure 6-21a). Halley's potato-shaped nucleus is darker than coal, probably because of carbon-rich compounds left behind as ice evaporates. In 2001, the *Deep Space 1* spacecraft took high-resolution images of Comet Borrelly's nucleus (Figure 6-21b). The ends of the comet are very rugged, while the

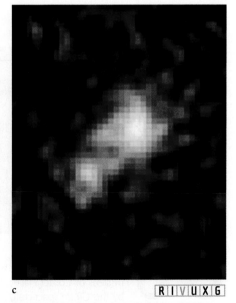

a   b   c   R I V U X G

FIGURE 6-19 **Kuiper Belt Objects** (a, b) These 1993 images show the discovery (white arrows) of one of at least 1277 known KBOs. These two images of KBO 1993 SC were taken 4.6 h apart, during which time the object moved against the background stars. (c) The KBO 1998 WW31 and its moon (lower left). *(a, b: Alan Fitzsimmons, Queen's University of Belfast; c: C. Veillet/CFHT)*

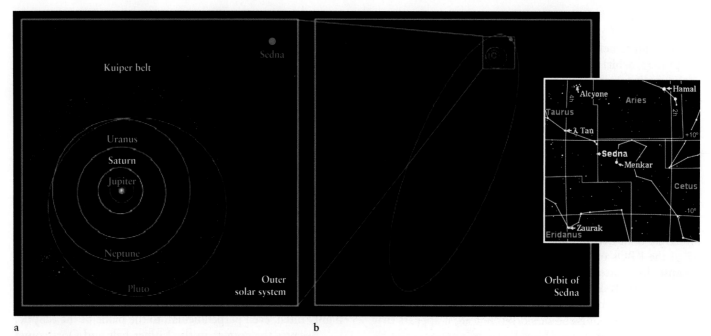

a   b

FIGURE 6-20 **Sedna's Orbit** (a) The farthest known body in the solar system is in a highly elliptical orbit (b) that ranges from the outer reaches of the Kuiper belt and possibly extends to the inner Oort comet cloud. *(NASA/Caltech)*

center has long, rolling plains, from which jets of gas appear to originate.

To better determine the composition of the gas emitted by comets, the *Stardust* probe visited Comet Wild (pronounced *Vilt*) 2 (Figure 6-22a), where it collected samples (Figure 6-22b, c) and returned them to Earth in 2006. Comet Wild 2 showed many unexpected features, including towering spires, many craters and cliffs, and more than 24 jets, rather than the 2 or 3 jets seen on other comets.

As a comet nucleus comes within 20 AU of the Sun, solar radiation begins to vaporize the ices on its surface. The liberated gases form an atmosphere, or **coma**, around the nucleus. Because the coma scatters sunlight, it appears as a fuzzy, luminous ball. The largest coma ever measured was more than a million kilometers across—nearly as large as the Sun. Not visible to the human eye is the **hydrogen envelope**, a sphere of tenuous gas surrounding the comet's nucleus and measuring as much as 20 million km in diameter (Figure 6-23).

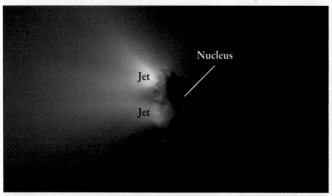

a   R I V U X G

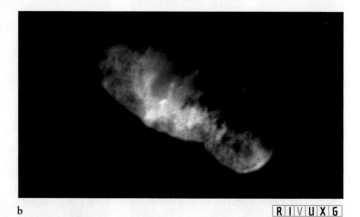

b   R I V U X G

FIGURE 6-21 **Comet Nuclei** (a) The nucleus of Comet Halley. This image, taken by the *Giotto* spacecraft, shows the potato-shaped nucleus of the comet. Its dark nucleus measures 15 km in its longest dimension and about 8 km in its shortest. The numerous bright areas on the nucleus are icy outcroppings that reflect more sunlight than surrounding areas of the comet. Two jets of gas can be seen emanating from the left side of the nucleus. (b) The nucleus of Comet Borrelly, in an image taken by *Deep Space 1*. The nucleus is 8 km (5 mi) long. *(a: Max-Planck-Institut für Aeronomie; b: Deep Space 1 Team, JPL, NASA)*

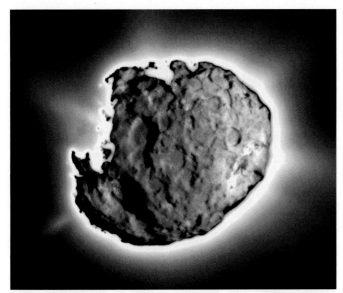

a   R I V U X G

Ejecta in aerogel

Tiny crater in aluminum
foil holding aerogel.

b   R I V U X G

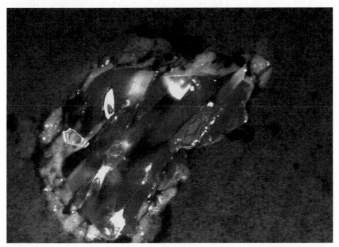

c   R I V U X G

FIGURE 6-22 **Comet Wild 2** (a) This picture shows two images combined. One is a high-resolution photograph showing the surprisingly heavily cratered comet. The other image is a longer photograph showing gas and dust jetting from the comet. Its tails are millions of kilometers long. (b) A substance called aerogel was used to capture particles from Comet Wild 2's dust tail. A piece of space debris pierced the aluminum foil holding the aerogel and embedded in it, along with pieces of the foil. (c) A 2-μm piece of comet dust, composed of a mineral called *forsterite*. On Earth this mineral is used to make gems called *peridot*. *(a, b: NASA/JPL; c: NASA/JPL-Caltech/ University of Washington.)*

6-7   Comet tails develop from gases and dust pushed outward by the Sun

The most visible and inspiring features of comets are their **3** long, flowing, diaphanous tails (Figure 6-24). Comet tails **4** develop from coma gases and dust pushed outward by radiation and particles from the Sun. This means that comet tails do not trail behind the nucleus, as the exhaust from a jet plane does in Earth's atmosphere. Rather, *at the comet's nucleus, the tails always point away from the Sun* (Figure 6-25), regardless of the direction of the comet's motion. The implication that something from the Sun "blows" the comet's gases radially outward led Ludwig Biermann to predict the existence of the solar wind (see Sections 4-4 and

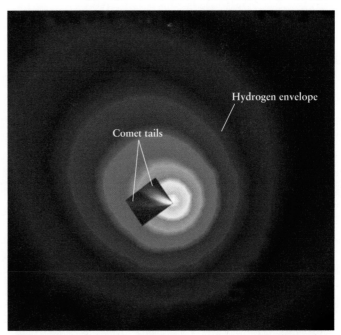

Hydrogen envelope

Comet tails

R I V U X G

FIGURE 6-23 **Comet Hale-Bopp** In 1997, Hale-Bopp had a hydrogen envelope 1 AU in diameter (blue ovals). This gas was observed in the ultraviolet. The visible-light inset shows the scale of the visible tails (see also the image at the beginning of the chapter). *(a: Johns Hopkins University and Naval Research Laboratory; b: Mike Combi)*

R I V U X G

FIGURE 6-24 **Comet West** Astronomer Michael M. West first noticed this comet on a photograph taken with a telescope in 1975. After passing near the Sun, Comet West became one of the brightest comets of the 1970s. This photograph shows the comet in the predawn sky in March 1976. *(Hans Vehrenberg)*

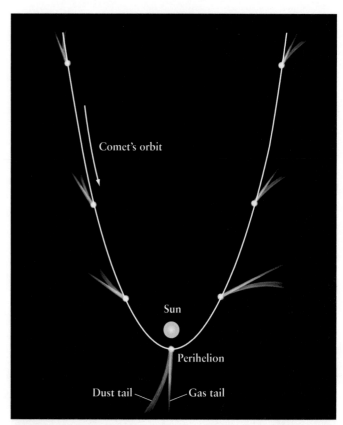

FIGURE 6-25 **The Orbit and Tails of a Comet** The sunlight and solar wind blow a comet's dust particles and ionized atoms away from the Sun. Consequently, comets' tails always point away from the Sun.

7-3). This stream of particles from the Sun was actually discovered in 1962, a full decade after it was predicted, by instruments on the spacecraft *Mariner 2*. In 2001, NASA launched the *Genesis* spacecraft, which collected pristine solar wind particles for 2½ years. Although it crash-landed back on Earth in September 2004, some of the particles that it collected were salvaged and are now being studied.

What is the solid part of a comet called?

INSIGHT INTO SCIENCE

Starting with Observations Science often advances by working from observations and experiments that require scientific explanations. For example, seeing comet tails and how they behave led Biermann to wonder what causes them. The simplest physical explanation is that matter from each cometary body is being evaporated and pushed upon by something from the Sun—the solar wind. Remember Occam's razor (Section 2-1).

Outflowing particles and radiation from the Sun produce two comet tails: a **gas** (or **ion** or **plasma**) **tail** and a **dust tail** (Figure 6-26a and the figure opening this chapter). Positively charged ions (atoms missing one or more electrons) from the coma are swept directly away from the Sun by the solar wind to form the gas tail. This tail often appears blue, because, like the gas in our air, the tail's gas strongly scatters blue light. The relatively straight gas tail can change dramatically from night to night (see Figure 6-26a). This occurs because the solar wind that pushes the gases changes rapidly and the gases themselves are very light and easily moved about, like the smoke from a fire (Figure 6-26b).

The dust tail is formed when sunlight strikes dust particles that have been freed from the comet's evaporating nucleus. Light exerts pressure on any object that absorbs or scatters it. While this pressure, called **radiation pressure** or **photon pressure,** is quite weak, fine-grained dust particles in a comet's coma are sufficiently light to be pushed from the vicinity of the comet, thus producing a dust tail. The dust tail is often white or light gray. The dust particles are massive enough not to flow straight away from the Sun; rather, the dust tail arches in a path that lies between the gas tail and the direction from which the comet came (Figure 6-25 and Figure 6-27).

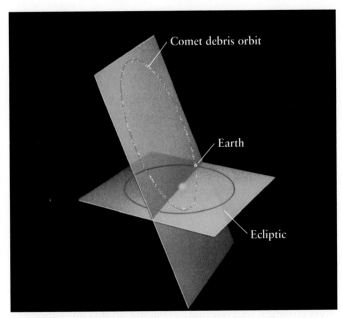

FIGURE 6-37 **The Origin of Meteor Showers** As comets dissipate (see Figure 6-31), they leave debris behind that spreads out along their orbits. When Earth plows through such material, many meteors can be seen emanating from the same place within a very short time—a meteor shower. As shown in this diagram, many comets have high orbital inclinations.

showers as the Perseids, which take place in the summertime and are among the most exciting light shows in astronomy. Except for the Lyrids, meteor showers are best seen after midnight.

The Moon also moves through the debris that creates meteor showers on Earth. Whereas our atmosphere vaporizes most of the meteors before they strike Earth, the Moon's thin atmosphere provides no protection, so the Moon is struck by numerous meteoroids during each shower. The Leonid meteor shower of 1999 provided astronomers with an excellent opportunity to observe contemporary impacts on the Moon. Some of these impacts were even bright enough to see with the naked eye (Figure 6-39). When a 10-kg meteorite strikes the Moon, the impact is as powerful as 10^4 pounds of TNT, and the resulting cloud is momentarily between 5×10^4 and 10×10^4 K, which is hotter than the surface of the Sun. Subsequent Leonids have provided further impact sightings on the Moon.

6-11 Meteorites are space debris that land intact

Although most meteors that enter Earth's atmosphere completely vaporize or are destroyed on impact, some land without totally disintegrating. Such pieces of debris are called meteorites, and they come from a variety of sources. Many are thought to have been broken off from asteroids by collisions; some are debris that were never part of larger bodies; still others come from the Moon, Mars, Vesta, or comets. We saw

TABLE 6-1 Prominent Yearly Meteor Showers

Shower	Date of maximum intensity	Typical hourly rate	Constellation
Quadrantids	January 3	40	Boötes
Lyrids	April 22	15	Lyra
Eta Aquarids	May 4	20	Aquarius
Delta Aquarids	July 30	20	Aquarius
Perseids	August 12	80	Perseus
Orionids	October 21	20	Orion
Taurids	November 4	15	Taurus
Leonids	November 16	15	Leo Major
Geminids	December 13	50	Gemini
Ursids	December 22	15	Ursa Minor

FIGURE 6-38 **Meteor Shower** The time exposure, taken in 1998, shows meteors streaking away from the constellation Leo Major. They are part of the Leonid meteor shower. This shower occurs because Earth is moving through debris left by Comet Temple-Tuttle. The table lists highly active annual meteor showers, which last for several days. *(Tony and Daphne Hallas/Photo Researchers, Inc.)*

in Section 5-12, for example, how SNC meteorites offer clues to the chemistry of Mars. By studying the chemistry and age of one meteorite from the Moon that was found on Earth, geologists have determined that it was ejected into space when the Mare Imbrium impact basin formed. Meteorites from the asteroid Vesta show signs of having been molten rock on that world. They were ejected when an impact occurred there, creating a 460 km (280 mi) wide, 13 km (8 mi) deep crater, which has been seen by the Hubble Space Telescope.

Why are the times when meteor showers occur so predictable?

Meteorites tell us the age of the solar system. Measuring the relative amounts of various radioactive elements in them allows astronomers to determine how long ago they formed (see Appendix H-3). The oldest known meteorites became solid bodies about 4.57 billion years ago, and impacts on Earth as early as 4 billion years ago have been identified, based on how they disturbed preexisting rock here. Therefore, the solar system is at least 4.57 billion years old, and we will use 4.6 billion years as its age throughout the book.

People have been picking up debris from space for thousands of years. The descriptions of meteorites in historical Chinese, Indian, Islamic, Greek, and Roman literature show that early peoples placed special significance on these "rocks from heaven." They also have a practical "impact": Infalling space debris is increasing Earth's mass by nearly 300 tons per day on average (Figure 6-40).

R I V U X G

FIGURE 6-39 **Recent Impacts on the Moon** The locations A–F are places on the Moon where impacts were observed from Earth in 1999 during the Leonid meteor shower. The impacting bodies hit the Moon at around 260,000 km/h (160,000 mph) and had masses of between 1 and 10 kg. Each impact created a short-lived cloud that momentarily heated to between 5×10^4 and 10×10^4 K, much hotter than the surface of the Sun. *(NASA)*

Stony Meteorites Often Look Like Ordinary Rocks Meteorites are classified as stones (or stony), stony-irons, and irons. Most **stony meteorites** look much like ordinary rocks, although some are covered with a dark *fusion crust* (Figure 6-41a), created when the meteorite's outer layer melts during its fiery descent through the atmosphere (see Figure 6-38). When a stony meteorite is cut in two and polished, tiny flecks of iron are sometimes found in the rock (Figure 6-41b).

Meteorites that have a high iron content can be located with a metal detector. They also look unusual and hence are more likely to be noticed. Consequently, the easily found iron and stony-iron meteorites dominate most museum collections.

Iron Meteorites Are Very Dense and Look Noticeably Different from Rocks Iron meteorites (Figure 6-42a) may also contain from 10% to 20% nickel by weight. Iron is moderately abundant in the universe as well as being one of the most common rock-forming elements, so it is not surprising that iron is an important constituent of asteroids and meteoroids. Another element, iridium, is common in the iron-rich minerals of meteorites but rare in ordinary rocks on Earth's surface because most of the iridium that collected here when Earth formed settled deep into Earth eons ago. Measurements of iridium in Earth's crust can thus tell us the rate at which meteoritic material has been deposited over the ages.

In 1808, Count Alois von Widmanstätten, director of the Imperial Porcelain Works in Vienna, discovered a conclusive test for the most common type of iron meteorite. Most iron meteorites have a unique structure of long

R I V U X G

FIGURE 6-40 **The Mass of Impacts on Earth** The Vatican Obelisk is about 300 tons, the amount of mass that strikes Earth daily. As a result, Earth's mass increases by this amount every day. *(John and Dallas Heaton/Corbis)*

a   R I V U X G

b   R I V U X G

FIGURE 6-41 **Stony Meteorites** (a) Most meteorites that fall to Earth are stones. Many freshly discovered specimens, like the one shown here, are coated with thin, dark crusts. This stony meteorite fell in Morocco. (b) Some stony meteorites contain tiny specks of iron, which can be seen when the stones are cut and polished. This specimen was discovered in Ohio. *(The R. N. Hartman Collection)*

nickel–iron crystals, called **Widmanstätten patterns**, which become visible when the meteorites are cut, polished, and briefly dipped into a dilute solution of acid (Figure 6-42b). Because nickel–iron crystals can grow to lengths of several centimeters only if the molten metal cools slowly over many millions of years, Widmanstätten patterns are never found in counterfeit meteorites, or "meteorwrongs."

Stony-Iron Meteorites Are the Most Exotic of All Space Debris on Earth The final category of meteorites are the **stony-iron meteorites**, which consist of roughly equal amounts of rock and iron. Figure 6-43, for example, shows the greenish mineral olivine suspended in a matrix of iron.

To understand why different types of meteorites exist, we consider their formation. Many meteorites were once pieces of asteroids. Heat from impacts and the rapid decay of radioactive isotopes melted newly formed asteroid

a   R I V U X G

b   R I V U X G

FIGURE 6-42 **Iron Meteorites** (a) Irons are composed almost entirely of iron–nickel minerals. The surface of a typical iron is covered with thumbprintlike depressions created as the meteorite's outer layers vaporized during its high-speed descent through the atmosphere. This specimen was found in Argentina. (b) When cut, polished, and etched with a weak acid solution, most iron meteorites exhibit interlocking crystals in designs called *Widmanstätten patterns*. This meteorite was found in Australia. *(a: The R. N. Hartman Collection; b: R. A. Oriti)*

R I V U X G

FIGURE 6-43 **Stony-Iron Meteorite** Stony-irons account for about 1% of all meteorites that fall to Earth. This specimen, a variety called a pallasite, was found in Antarctica. It has been thinly cut and appears to glow because of a light located behind it. *(James C. Hartman)*

interiors. Over the next few million years, differentiation occurred, just as in young Earth. Iron sank toward the asteroid's center, while lighter rock floated up to its surface. Iron meteorites are fragments of asteroid cores, and stony-irons come from the boundary regions between the iron cores and stony crusts.

Stony meteorites have a variety of origins. Some are from the outer layers of asteroids. Other stony meteorites, ordinary **chondrites,** show no evidence of ever having been melted as parts of asteroids. They may therefore be primordial material from which our solar system was created. **Carbonaceous chondrites** are rare chondrites that contain small glass-rich beads called *chondrules.* These meteorites also contain complex carbon compounds, including simple sugars and glycerin. They also have as much as 20% water bound into their minerals. The organic compounds would have been broken down and the water driven out if these meteorites had been significantly heated. Undifferentiated asteroid Mathilde, shown in Figure 6-16a, has a very dark gray color and virtually the same spectrum as a carbonaceous chondrite meteorite, so it is likely composed of primordial material.

Amino acids, the building blocks of proteins upon which terrestrial life is based, are among the organic compounds occasionally found inside carbonaceous chondrites. These amino acids may be contaminants acquired after the meteoroids entered Earth's atmosphere. However, space scientists have actually created them in the laboratory under the conditions found in deep space, showing that amino acids may exist out there. Indeed, some scientists suspect that amino-acid–rich carbonaceous chondrites may have played a role in the origin of life on Earth.

Not All Meteorite Impacts ("Falls") Lead to Meteorite "Finds" Stony meteorites account for about 95% of all meteoritic material that fall on Earth. Most stony meteorites are not identified as meteorites because, well, they look like stones. Indeed, the percentages of the stony, iron, and stony-iron meteorites that are discovered, called *finds,* are quite different from the percentages that actually land, called *falls.* Nevertheless, astronomers and geologists have a good idea about how many of each type strike land. They get the correct percentages of impacts by carefully surveying areas in which only meteorites land, namely snow and ice-covered regions, such as Antarctica, or on deserts. By counting all of the debris under the surface using metal detectors and other technologies, the actual number of impacts of each type of meteorite is determined.

6-12 The Allende meteorite and Tunguska mystery provide evidence of catastrophic explosions

A chance to study debris immediately after impact came shortly after midnight on February 8, 1969, when a brilliant, blue-white light shot across the night sky over Chihuahua, Mexico. Hundreds of people witnessed the dazzling display. The light disappeared in a spectacular and deafening explosion that dropped thousands of rocks and pebbles over the terrified onlookers. Within hours, teams of scientists were on th eir way to collect specimens of carbonaceous chondrites, collectively named the *Allende meteorite,* after the locality in which they fell (Figure 6-44).

One of the most significant discoveries to come from the Allende meteorite was evidence of the detonation of a nearby supernova 4.6 billion years ago. Among nature's most violent and spectacular phenomena, a *supernova explosion* occurs when a massive star blows apart in a cataclysm that hurls matter outward at tremendous speeds, as we will explore in Chapter 10. During this detonation, violent collisions between atomic nuclei produce a host of radioactive elements, including a short-lived radioactive isotope of aluminum. Based on its decay products, scientists found unmistakable evidence that this isotope once lay within the Allende meteorite. Some astronomers interpret this as evidence for a supernova in our vicinity at about the time the Sun was born. By compressing interstellar gas and dust, the supernova's shock wave may have helped stimulate the birth of our solar system.

Why are stony-iron meteorites so rare compared to stony or iron meteorites?

Sixty years before the meteoritic event in Chihuahua, Mexico, at 7:14 A.M. local time on June 30, 1908, a much more spectacular explosion occurred, this one over the Tunguska region of Siberia. The blast, comparable to a nuclear detonation of several megatons, knocked a man off his porch some 60 km away and was audible more than 1000 km away.

a   R I V U X G

b   R I V U X G

FIGURE 6-44 **Pieces of the Allende Meteorite** (a) This carbonaceous chondrite fell near Chihuahua, Mexico, in February 1969. Note the meteorite's dark color, caused by a high abundance of carbon. Geologists believe that this meteorite is a specimen of primitive planetary material. The ruler is 15 cm long. (b) Sliced open, the Allende meteorite shows round, rocky *inclusions* called *chondrules* in a matrix of dark rock. *(a: J. A. Wood; b: The R. N. Hartman Collection)*

Millions of tons of dust were injected into the atmosphere, darkening the air as far away as California.

Preoccupied with wars, along with political and economic upheaval, neither Russia nor its successor, the former Soviet Union, sent a scientific expedition to the site until 1927. At that time, Soviet researchers found that trees had been seared and felled radially outward in an area about 30 km in diameter (Figure 6-45). There was no clear evidence of a crater. In fact, the trees at "ground zero" were left standing upright, although they were completely stripped of branches and leaves. Because no significant meteorite

R I V U X G

FIGURE 6-45 **Aftermath of the Tunguska Event** In 1908, an object traveling at supersonic speed struck Earth's atmosphere and exploded over the Tunguska region of Siberia. Trees were blown down for many kilometers in all directions from the impact site. *(SOVFOTO)*

samples were found, for many years scientists assumed that a small comet had exploded in the atmosphere.

What evidence do astronomers have from the Allende meteorite that a supernova explosion occurred at about the time the solar system formed?

Recently, however, several teams of astronomers have argued that a small comet, composed of primarily light elements and ice, breaks up too high above the ground to cause significant damage. They maintain that the Tunguska explosion was actually caused by a small asteroid or large meteoroid traveling at supersonic speed. The Tunguska event is consistent with an explosion of an asteroid about 80 m (260 ft) in diameter entering Earth's atmosphere at 79,000 km/h (50,000 mph) and exploding in the air as a result of becoming exceedingly hot. The resulting shock wave, slamming into the ground, would cause the damage seen without creating a crater. The debate on the origin of the Tunguska event continues.

6-13 Asteroid impacts with Earth have caused mass extinctions

In the late 1970s, the geologist Walter Alvarez and his father, physicist Luis Alvarez, discovered the remnants of a far more powerful impact. Working at a site of exposed marine limestone in the Apennine Mountains in Italy that had been on Earth's surface 65 million years ago, the Alvarez team discovered an exceptionally high abundance of iridium in a dark-colored layer of clay between limestone strata (Figure 6-46).

RIVUXG

FIGURE 6-46 **Iridium-Rich Layer of Clay** This photograph of strata in the Apennine Mountains of Italy shows a dark-colored layer of iridium-rich clay sandwiched between white limestone (bottom) from the late Mesozoic era and grayish limestone (top) from the early Cenozoic era. The coin is the size of a U.S. quarter. *(W. Alvarez)*

Since 1979, when this discovery was announced, a comparable layer of iridium-rich material has been uncovered at numerous sites around the world. In every case, geologic dating reveals that this apparently worldwide iridium-rich layer was deposited about 65 million years ago. Paleontologists were quick to realize the significance of this date, because it was back then that all of the dinosaurs rather suddenly became extinct. In fact, two-thirds of the species on Earth disappeared within a brief span of time 65 million years ago.

The Alvarez discovery supported an astronomical explanation for the dramatic extinction of so much of the life that once inhabited our planet—an asteroid impact. There is no universal agreement on the disasters that befell Earth during this episode. One scenario has an asteroid 10 km in diameter slamming into Earth at high speed and throwing enough dust into the atmosphere to block out sunlight for several years. As the temperature dropped drastically and plants died for lack of sunshine, the dinosaurs would have perished, along with many other creatures in the food chain that were highly dependent on vegetation. The dust eventually settled, depositing an iridium-rich layer over Earth. Another theory posits that the impact created enough heat to cause planetwide fires, followed by changes in the oceans that killed many species of life in them. Whatever scenario turns out to be correct, small creatures capable of ferreting out seeds and nuts were among the animals that managed to survive this holocaust, setting the stage for the rise of mammals and, consequently, the evolution of humans.

In 1992, a team of geologists discovered that the hypothesized asteroid crashed into a site on the eastern edge of Mexico. They based this conclusion on glassy debris and violently shocked grains of rock ejected from the multiringed, 195-km-diameter Chicxulub (pronounced chih-chuh-lube) Crater buried under the Yucatán Peninsula and western Caribbean Sea in Mexico (Figure 6-47). From the known rate at which radioactive potassium decays, the scientists have pinpointed the date when the asteroid struck—64.98 million years ago. In 1998, geologists digging on the Pacific Ocean floor discovered a piece of meteoritic debris with precisely the same age, apparently a piece of the offending asteroid. While some geologists and paleontologists are not yet convinced that an asteroid impact led to the extinction of the dinosaurs, most agree that this hypothesis better fits the available evidence than any other explanation that has been offered so far.

Another Impact Led to an Earlier, More Devastating Mass Extinction The end of the dinosaurs' reign was not the only mass extinction caused by an impact. In 2001, evidence came to light concerning the Permian-Triassic boundary of 250 million years ago that suggests another devastating blow from space. That time was called the "Great Dying," a mass extinction during which some 80% of the species of life living on land and 90% of those living in the oceans perished. Rocks from that time discovered in places from Japan to Hungary show evidence, in the form of *fullerenes*—soccer ball–shaped molecules containing at least 60 carbon atoms—of an impact from space. Trapped inside these fullerenes were gases that could only have been forced into them by exploding stars. (This signature of gas-filled fullerenes from space has now also been discovered in the layer of rock that existed on Earth's surface 65 million years ago.)

A 250-million-year-old crater that may have been the relevant impact site was found in 2004 off the northwestern coast of modern-day Australia. Keep in mind that both this impact and the one that ended the reign of the dinosaurs occurred on a world whose surface is in continual motion. Therefore, their present locations are not the same as their initial impact sites. The Chicxulub impact actually occurred in the Pacific Ocean, whereas the Australian impact occurred before the continents were separated from Pangaea (see Section 4-2). The impact explanation of the Great Dying is not universally accepted. Another possible explanation is that a massive amount of volcanic eruption darkened the skies and changed the climate.

Could a catastrophic impact happen again? So many asteroids cross Earth's path that scientists agree that it is a matter of "when" rather than "if." The good news is that studies of craters show that larger asteroids strike Earth significantly less often than do smaller ones. Although asteroids large enough to create Meteor Crater strike Earth about once every 10,000 years, killer asteroids, like the one that killed off the dinosaurs, collide with Earth only once every 100 million years. The threat of a catastrophic impact by an asteroid or comet in our lifetimes, thankfully, is remote.

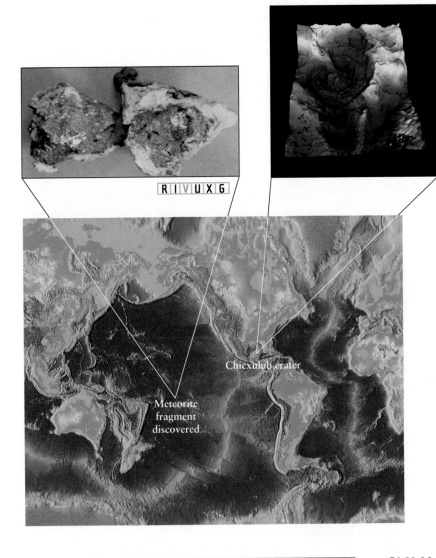

R I V U X G

Chicxulub crater

Meteorite
fragment
discovered

FIGURE 6-47 **Confirming an Extinction-Level Impact Site** By measuring slight variations in the gravitational attraction of different materials under Earth's surface, geologists create images of underground features. Concentric rings of the underground Chicxulub Crater (right inset) lie under a portion of the Yucatán Peninsula. This crater has been dated to 65 million years ago and is believed to be the site of the impact that led to the extinction of the dinosaurs. A piece of 65-million-year-old meteorite discovered in the middle of the Pacific Ocean in 1998 is believed to be a fragment of that meteorite. The fragment, about 0.3 cm (0.1 in) long, was cut into two pieces for study (left inset). *(Virgil L. Sharpton, Lunar and Planetary Institute; right inset: digital image by Peter W. Sloss, NOAA-NESDIS-NGDCD; left inset: Frank T. Kyte, UCLA)*

How did the Chicxulub crater move from the Pacific to the Caribbean?

6-14   Frontiers yet to be discovered

The years of the *Pioneer* and *Voyager* spacecraft were the first golden era of solar system research. We are now in another such period, with spacecraft either in development or already en route to planets, asteroids, comets, and even, as we will see shortly, out observing the weather in space. From studies of solar system debris, astronomers hope to learn whether life on Earth was brought here from elsewhere by asteroid or meteoroid impacts. They also hope to answer such questions as the evolutionary history of the space debris; how much of Earth's water came here during the planet's formation and how much landed afterward from comet impacts; whether the asteroids have sufficiently valuable compositions to justify mining them; whether comets can be harvested to supply water and other materials for people colonizing the solar system; and whether the Oort comet cloud really exists.

SUMMARY OF KEY IDEAS

• Astronomical objects smaller than the eight planets are classified as dwarf planets or small solar-system bodies (SSSBs).

• A variety of other names, including asteroids, comets, meteoroids, trans-Neptunian objects, Kuiper belt objects (KBOs), and Oort comet cloud objects, overlap with dwarf planet and SSSB.

• KBOs and Oort comet cloud objects are trans-Neptunian objects—they orbit farther from the Sun than the outermost planet.

• To date, five objects—Pluto, Ceres, Eris, Makemake, and Haumea—have been classified as dwarf planets.

• Pluto, a KBO and dwarf planet, is an icy world that may well resemble the moon Triton.

• Other objects orbit the Sun beyond Neptune. For example, at least 1471 KBOs have been observed. A few potential Oort comet cloud objects have also been identified.

Asteroids

• Tens of thousands of belt asteroids with diameters larger than a kilometer are known to orbit the Sun between the orbits of Mars and Jupiter. The gravitational attraction of Jupiter depletes certain orbits within the asteroid belt. The resulting Kirkwood gaps occur at simple fractions of Jupiter's orbital period.

• Jupiter's and the Sun's gravity combine to capture Trojan asteroids in two locations, called stable Lagrange points, along Jupiter's orbit.

• The Apollo asteroids move in highly elliptical orbits that cross the orbit of Earth. Many of these asteroids will eventually strike the inner planets.

Comets

• Many comet nuclei orbit the Sun in the Kuiper belt, a doughnut-shaped region beyond Pluto. Billions of cometary nuclei are also believed to exist in the spherical Oort comet cloud located far beyond Pluto.

• Comet nuclei are fragments of ice and rock often orbiting at a great inclination to the plane of the ecliptic. In the Kuiper belt and Oort comet cloud, they have fairly circular orbits. When close to the Sun, they generally move in highly elliptical orbits.

• As an icy comet nucleus approaches the Sun, it develops a luminous coma surrounded by a vast hydrogen envelope. A gas (or ion) tail and a dust tail extend from the comet, pushed away from the Sun by the solar wind and radiation pressure.

Meteoroids, Meteors, and Meteorites

• Boulder-sized and smaller pieces of rock and metal in space are called meteoroids. When a meteoroid enters Earth's atmosphere, it produces a fiery trail, and it is then called a meteor. If part of the object survives the fall, the fragment that reaches Earth's surface is called a meteorite.

• Meteorites are grouped in three major classes according to their composition: iron, stony-iron, and stony meteorites. Rare stony meteorites, called carbonaceous chondrites, may be relatively unmodified material from the primitive solar nebula. These meteorites often contain organic hydrocarbon compounds, including amino acids.

• Fragments of rock from "burned-out" comets produce meteor showers.

• An analysis of the Allende meteorite suggests that a nearby supernova explosion may have been involved in the formation of the solar system some 4.6 billion years ago.

• An asteroid that struck Earth 65 million years ago probably contributed to the extinction of the dinosaurs and many other species. Another impact caused the "Great Dying" of life 250 million years ago. Such devastating impacts occur on average every 100 million years.

WHAT DID YOU THINK?

1 *Are the asteroids a former planet that was somehow destroyed? Why or why not?* No. The gravitational pull from Jupiter prevented a planet from ever forming in the asteroid belt. Also, the total mass of the asteroids is much less than even the mass of tiny Pluto, a dwarf planet.

2 *How far apart are the asteroids on average?* The distance between asteroids averages 10 million km.

3 *How are comet tails formed? Of what are they made?* Ices in comet nuclei are turned into gas by absorbing energy from the Sun. Debris is released in this process. Sunlight and the solar wind push on the gas and dust, creating the tails.

4 *In which directions do a comet's tails point?* Comets' gas tails point directly away from the Sun; their dust tails make arcs pointing away from the Sun.

5 *What is a shooting star?* A shooting star is a piece of space debris plunging through Earth's atmosphere—a meteor. It is not a star.

Review Questions

1. A piece of space debris that you pick up from the ground is called a(n): **a.** asteroid, **b.** meteoroid, **c.** meteor, **d.** meteorite, **e.** comet

2. Space debris that is a roughly equal mix of rock and ice is called a(n): **a.** asteroid, **b.** comet, **c.** meteoroid, **d.** meteorite, **e.** meteor

3. Which is the rarest type of meteorite in space? **a.** iron, **b.** stony-iron, **c.** stony meteorite

4. Which part of a comet is solid? **a.** nucleus, b. halo, c. gas tail, d. dust tail, e. coma

5. Suppose you were standing on Pluto. Describe the motions of Charon relative to the horizon. Under what circumstances would you never see Charon?

6. Describe the circumstantial evidence that supports the idea that Pluto is one of a thousand similar icy worlds that once occupied the outer regions of the solar system.

7. What role did Charon play in enabling astronomers to determine Pluto's mass?

8. Why are asteroids, meteoroids, and comets of special interest to astronomers who want to understand the early history of the solar system?

9. Describe the objects in the asteroid belt, including their sizes, orbits, and separation.

10. Why are there many small asteroids but only a few very large ones?

11. Describe the different chemistries of the two tails of a comet.

12. In what directions do comet tails point, and why?

13. What are the Kirkwood gaps, and what causes them?

14. What are the Trojan asteroids, and where are they located?

15. Describe the three main classifications of meteorites. How do astronomers believe that these different types of meteorites originated?

16. Why do astronomers believe that the debris that creates many isolated meteors comes from asteroids, whereas the debris that creates meteor showers is related to comets?

17. Why is the phrase "dirty snowball" an appropriate characterization of a comet's nucleus?

18. What and where is the Kuiper belt, and how is it related to debris left over from the formation of the solar system?

19. What evidence in Figure 6-33 supports the labeling of the gas and dust tails?

Observing Projects

20. **Halley's Comet** In this exercise we will use *Starry Night Enthusiast*™ to locate Halley's comet. Select **Favourites > Solar System > Comets and Asteroids > Comet Halley's Orbit** from the menu. Click the **Stop** button to stop time flow. Explain the reason for the direction in which the comet's tail is pointed. From the direction of the tail, can you predict the direction of Halley's motion when this image was taken? If so, give its direction and explain how you deduced it. If not, explain why not. Click the **Play** button to start time flow and watch the orbital motion of Halley's comet. Does the comet orbit the Sun in the same or in the opposite sense to the planets visible in the view? Right-click over the Sun, click on **Centre** and then use the **Location Scroller** (hold down the **Shift** key as well as the mouse button while moving the mouse) to move around the Sun until Earth's orbital plane is flat on the screen. Run time back and forward to watch Halley's comet as it crosses this plane in its close approach to the Sun, noting particularly the direction of its tail.

21. You can use *WorldWide Telescope* (WWT) to explore several places where massive meteorites have impacted Earth and left a visible crater. The most famous of these, in Chicxulub, Mexico, is now thought to have been the site of a massive impact that devastated Earth and caused the extinction of three-quarters of all living creatures, including the dinosaurs. This impact produced a massive dust cloud that spread throughout Earth and settled to produce a distinct layer that can be traced by the abnormal amount of the rare metal iridium, found in meteorites. The crater is under the sea off the coast of Mexico and cannot be seen visibly but can be traced with seismic sounding. Erosion by wind and weather and, in the longer term, plate tectonic motions, have obliterated most of the impact craters, but we can be sure that many impacts have occurred over geological time, from the evidence seen upon our Moon.

Open WWT and click on **Earth** in the **Look At** drop-down list, then click on **Virtual Earth Aerial** in the **Imagery** drop-down list. Zoom in on Earth and use the values of latitude and longitude listed to locate several impact craters and to center them in the view. (Latitude and longitude of the center of the view are displayed at the bottom-right of the screen.) Four of these craters are in the rock of the Canadian Shield, while the fifth is in the outback of northern Australia. Zoom in further until detail can be seen in these craters. Write a short description for each of them, giving relevant details (for example, water-filled, central peak, and crater wall shape). **a. Barringer Crater:** age 49,000 years; latitude +35°, 2'; longitude −111°, 1'. **b. New Quebec Crater:** age 1.4 million years; latitude +61°, 16', longitude −73°, 40' **c. Deep Bay,** Saskatchewan: age 99 million years; latitude +56°, 24'; longitude −102°, 59' **d. Manicouagan,** Quebec: age 214 million years; latitude +51°, 23'; longitude −68°, 42' **e. Wolfe Creek Crater,** Western Australia: age 300,000 years; latitude −19°, 10'; longitude +127°, 47'.

The Sun: Our Extraordinary Ordinary Star

R I V U X G

Magnetic Fields Carry Gases above the Sun's Surface in Loops That Can Extend Hundreds of Thousands of Kilometers *(NASA)*

WHAT DO YOU THINK?

1 What percentage of the solar system's mass is in the Sun?

2 Does the Sun have a solid and liquid interior like Earth?

3 What is the surface of the Sun like?

4 Does the Sun rotate? If so, how fast?

5 What makes the Sun shine?

Answers to these questions appear in the text beside the corresponding numbers in the margins and at the end of the chapter.

Dawn. Light and heat from the Sun begin to arouse our senses and signal the start of daily activity. Earliest societies, realizing that the Sun is essential to the existence and maintenance of life on Earth, revered it. The same respect for the Sun's awesome power motivated astronomers in the nineteenth and twentieth centuries to figure out how it shines.

1 A quick glance (never longer) at the Sun shows a brilliant disk of light, apparently no larger than the Moon. Despite appearances, the Sun is a body of monumental size and mass compared to any object we have studied so far. If the Sun's surface were as close to Earth as the Moon's surface is now, the Sun would spread two-thirds of the way across the sky, not to mention that it would be so hot on Earth that our planet would quickly vaporize. The Sun contains as much mass as 333,000 Earths. Indeed, all of the planets and other objects orbiting the Sun *combined* have only 0.15% as much mass as the Sun.

The Sun's tremendous mass not only holds the planets in orbit, but also explains why it shines. The Sun's mass creates a crushing force at its center that literally fuses atoms together—a process that generates the energy that eventually leaves the Sun and lights up our days. Some of the energy rushing into space (namely the ultraviolet radiation, visible light, and infrared radiation emitted by it) is essential to maintaining Earth's habitability.

The Sun is the closest star to Earth. By studying it, astronomers have come to learn how most stars throughout the universe work. As powerful and majestic as it is in our sky, we are going to discover in the next few chapters that the Sun is an ordinary star, neither among the most massive nor among the least massive. Similarly, it is neither among the brightest nor among the dimmest of stars. Figure 7-1 lists the Sun's properties.

In this chapter you will discover

• why the Sun is a typical star

• how today's technology has led to a new understanding of solar phenomena, from sunspots to the powerful ejections of matter that sometimes enter our atmosphere

• that some features of the Sun generated by its varying magnetic field occur in cycles

• how the Sun generates the energy that makes it shine

• new insights into the nature of matter from solar neutrinos

THE SUN'S ATMOSPHERE

2 Although astronomers often speak of the solar "surface," the Sun is so hot that it has neither liquid nor solid matter anywhere inside it. Moving down through the Sun, we continually encounter ever denser and hotter gases.

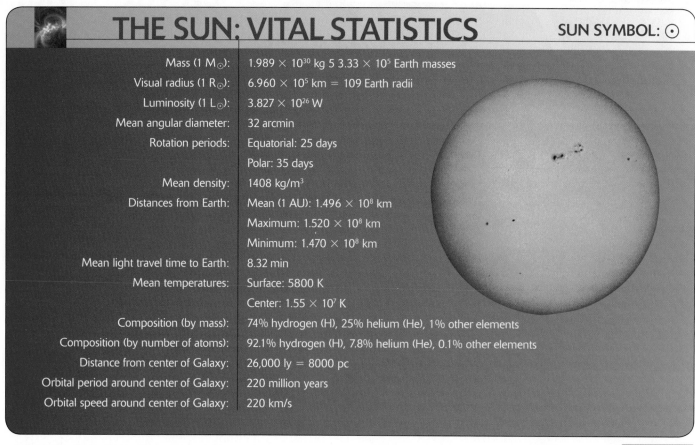

THE SUN: VITAL STATISTICS

SUN SYMBOL: ☉

Mass (1 M☉):	1.989×10^{30} kg 5 3.33×10^5 Earth masses
Visual radius (1 R☉):	6.960×10^5 km = 109 Earth radii
Luminosity (1 L☉):	3.827×10^{26} W
Mean angular diameter:	32 arcmin
Rotation periods:	Equatorial: 25 days
	Polar: 35 days
Mean density:	1408 kg/m³
Distances from Earth:	Mean (1 AU): 1.496×10^8 km
	Maximum: 1.520×10^8 km
	Minimum: 1.470×10^8 km
Mean light travel time to Earth:	8.32 min
Mean temperatures:	Surface: 5800 K
	Center: 1.55×10^7 K
Composition (by mass):	74% hydrogen (H), 25% helium (He), 1% other elements
Composition (by number of atoms):	92.1% hydrogen (H), 7.8% helium (He), 0.1% other elements
Distance from center of Galaxy:	26,000 ly = 8000 pc
Orbital period around center of Galaxy:	220 million years
Orbital speed around center of Galaxy:	220 km/s

R I V U X G

FIGURE 7-1 **Our Star, the Sun** The Sun emits most of its visible light from a thin layer of gas, called the photosphere, as shown. Although the Sun has no solid or even liquid region, we see the photosphere as its "surface." Astronomers always take great care when viewing the Sun by using extremely dark filters or by projecting the Sun's image onto a screen. (*Celestron International*)

7-1 The photosphere is the visible layer of the Sun

The Sun appears to have a surface only because most of its visible light comes from one specific gas layer (see Figure 7-1). This region, which is about 400 km thick, is appropriately called the **photosphere** ("sphere of light"). The density of the photosphere's gas is low by Earth standards, about 0.01% as thick as the air we breathe. The photosphere has a blackbody spectrum (recall Chapter 3) that corresponds to an average temperature of 5800 K (review Figure 3-41).

The photosphere is the innermost of the three layers that comprise the Sun's atmosphere. Because the upper two layers (discussed in the following two sections) are transparent to most wavelengths of visible light, we see through them down to the photosphere. We cannot, however, see through the shimmering gases of the photosphere, so everything below the photosphere is the Sun's interior.

As you can see in Figure 7-1, the photosphere appears darkest toward the edge, or **limb,** of the solar disk, a phenomenon called **limb darkening.** This occurs because we see regions of different temperatures at different depths in the photosphere. Here is how it works. Wherever we look on the Sun, we see light through roughly equal amounts of the Sun's atmosphere. Because the Sun is spherical, the light we see leaving the Sun at different places actually comes from different levels of the photosphere (Figure 7-2). Looking from Earth at the center of the Sun's disk (bottom black line, Figure 7-2), we see farther into the Sun's atmosphere than when we look toward the limb (top black line, Figure 7-2). The photosphere is hottest and brightest at its base, and so the center of the Sun's disk, where we see deepest into it, looks brighter and yellower than does its limb.

Under good observing conditions and using a telescope and special dark filters for protection, you can see a blotchy pattern, called *granulation,* on the photosphere (Figure 7-3). The lightly colored **granules** measure about 1000 km across and are surrounded by relatively dark boundaries. Time-lapse photography shows that granules form, disappear, and then reform in cycles that last several minutes. At any single moment, several million granules cover the solar surface.

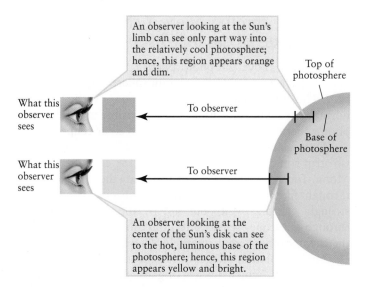

An observer looking at the Sun's limb can see only part way into the relatively cool photosphere; hence, this region appears orange and dim.

What this observer sees

To observer

Top of photosphere

Base of photosphere

What this observer sees

To observer

An observer looking at the center of the Sun's disk can see to the hot, luminous base of the photosphere; hence, this region appears yellow and bright.

FIGURE 7-2 **Limb Darkening** The Sun's edge, or limb, appears distinctly darker and more orange than does its center, as seen from Earth (see Figure 7-1). This occurs because we look through the same amount of solar atmosphere at all places. As a consequence, we see higher in the Sun's photosphere near its limb than when we look at its central regions. The higher photosphere is cooler and, because it is a blackbody, darker and more orange than the lower, hotter region of the photosphere.

Like liquid in a pot of simmering soup, the gases in granules rise and fall. In Section 3-23 we learned that radial motion of a light source affects the wavelengths of its spectral lines through the Doppler effect (see Figures 3-5 and 3-53). By carefully measuring the Doppler shifts of spectral lines in various parts of individual solar granules, astronomers have determined that hot gases move upward in the center of each granule and cooler gases cascade downward around its edges. This motion is caused by convection. We saw in Section 4-3 that convection moves the continents on Earth, and in Section 5-16 that it also helps form the belts and zones on the giant planets. Once the Sun's hot gases arrive at the photosphere, they radiate energy out into space. We see this energy as visible light and other electromagnetic radiation. Upon radiating, the gases cool, spill over the edges of the granule, and plunge back down into the Sun along the boundaries between granules (see Figure 7-3a inset).

According to the Stefan-Boltzmann law (see Section 3-18), hotter regions emit more photons per square meter than do cooler regions. Note in Figure 7-3a that the centers of the granules are brighter than their edges. Observations indicate that a granule's center is typically 100 K hotter than its edge, thus explaining the observed brightness difference.

Which part of the disk of the Sun appears brighter, the center or the edge? Why?

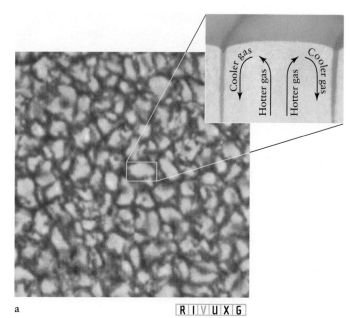

a   R I V U X G

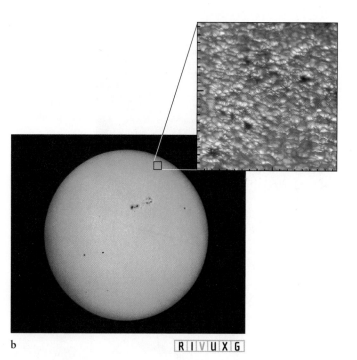

b   R I V U X G

FIGURE 7-3 **Solar Granulation** (a) High-resolution photographs of the Sun's surface reveal a blotchy pattern, called granulation. Granules, which measure about 1000 km across, are convection cells in the Sun's photosphere. Inset: Gas rising upward produces the bright granules. Cooler gas sinks downward along the darker, cooler boundaries between granules. This convective motion transports energy from the Sun's interior outward to the solar atmosphere. (b) At lower resolution, the Sun's surface appears relatively smooth (the dark regions will be discussed shortly). Inset: Seen near the Sun's limb, granules bulge upward at their centers as a result of the convection that creates them. *(a: MSFC/ NASA; inset: Goran Scharmer, Lund Observatory; b: Celestron International; inset: G. Scharmer [ISP, RSAS] et al., Lockheed-Martin Solar & Astrophysics)*

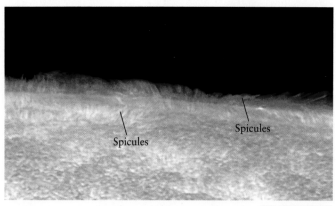

FIGURE 7-4 **The Chromosphere** This photograph of the chromosphere was taken by the *Hinode* (Japanese for "sunrise") satellite. The dark bumps are the tops of granules, and the light regions are hotter gases in spicules. The spicules on the edge or limb of the Sun give a sense of the height of these gas jets. *(Hinode JAXA/NASA)*

7-2 The chromosphere is characterized by spikes of gas called spicules

Immediately above the photosphere is a dim layer of less dense stellar gas, called the **chromosphere** ("sphere of color"). This unfortunate name suggests that it is the layer we normally see, but for centuries it was visible only when the photosphere was blocked during a total solar eclipse.

During an eclipse, the chromosphere is visible as a pinkish strip some 2000 km thick around the edge of the dark Moon. Today, astronomers can also study the chromosphere through filters that pass light with specific wavelengths strongly emitted by it—but not by the photosphere—or through telescopes sensitive to nonvisible wavelengths that the chromosphere emits intensely (Figure 7-4).

High-resolution images of the chromosphere reveal numerous spikes, which are jets of gas called **spicules** (Figure 7-4 and Figure 7-5a). A typical spicule rises for several minutes at the rate of 72,000 km/h (45,000 mph) to a height of nearly 10,000 km (Figure 7-5b). Then it collapses and fades away. At any one time, roughly a third of a million spicules cover a few percent of the Sun's chromosphere.

Spicules are generally located on the boundaries of enormous regions of rising and falling chromospheric gas called

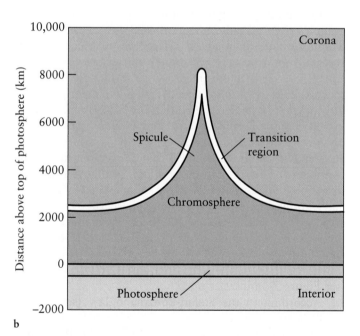

FIGURE 7-5 **Spicules and Supergranules** (a) Spicules appear in this photograph of the Sun's chromosphere. The Sun appears rose-colored in this image because it was taken through an H$_\alpha$ filter that passes red light from hydrogen and effectively blocks most of the photosphere's light. Surrounded by spicules, supergranules are regions of rising and falling gas in the chromosphere. Each supergranule spans hundreds of granules in the photosphere below. Inset: A view of spicules from above. (b) The spicules are jets of gas that surge upward into the Sun's outer atmosphere. This schematic diagram shows a spicule and its relationship to the solar atmosphere's layers. The photosphere is about 400 km thick. The chromosphere above it extends to an altitude of about 2000 km, with spicules jutting up to nearly 10,000 km above the photosphere. The outermost layer, the corona (discussed in Section 7-3), extends millions of kilometers above the photosphere. *(a: NASA; inset: Swedish Solar Telescope/Royal Swedish Academy of Sciences, La Palma, Spain, by Bert De Pontieu, Lockheed-Martin Solar and Astrophysics Lab)*

supergranules (Figure 7-5a). A typical supergranule has a diameter slightly larger than Earth's and contains about 900 granules.

7-3 Temperatures increase higher in the Sun's atmosphere

It seems plausible that the temperature should decrease as you rise through the Sun's atmosphere. After all, by moving upward, you move farther from the Sun's internal heat source. Indeed, starting in the photosphere at 5800 K, the temperature drops to around 4000 K in the lower chromosphere. Surprisingly, however, the temperature then begins to increase much more as you rise, reaching about 10,000 K at the top of the chromosphere. The outermost region of the Sun's atmosphere, the **corona**, extends several million kilometers from the top of the chromosphere (Figure 7-6). Between the chromosphere and corona is a **transition zone** in which the temperature skyrockets to about 1 million K (Figure 7-6c).

This unexpected increase in temperature was discovered around 1940 as a result of the high temperature's effect on the spectrum of the Sun's corona. The hotter a gas is, the more it is ionized (electrons are stripped off the atoms). Astronomers discovered that the corona contains the emission lines of several highly ionized (therefore, very hot) elements. For example, a prominent green line caused by the presence of Fe XIV, an iron atom stripped of 13 electrons, appears in the coronal spectrum. (Recall from Section 3-21 that Fe I is a neutral iron atom.) It is now known that coronal temperatures are typically in the range of 1 to 2 million K, although some regions are even hotter.

Which part of the Sun do we normally see?

a   R I V U X G

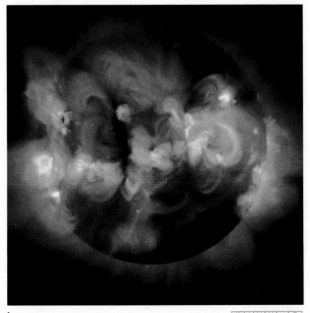

b   R I V U X G

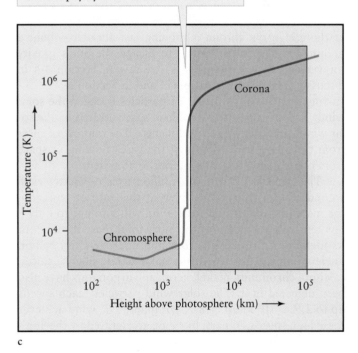

In this narrow transition region between the chromosphere and the corona, the temperature rises abruptly by about a factor of 100.

c

FIGURE 7-6 **The Solar Corona** (a) This visible-light photograph was taken during the total solar eclipse of July 11, 1991. Numerous streamers are visible, extending millions of kilometers above the solar surface. (b) This X-ray image of the Sun's corona, taken by the *Yohkoh* satellite in 1999, provides hints of the complex activity taking place on and in the Sun. The million-degree gases in the corona emit the X rays visible here. (c) This graph shows how temperature varies with altitude in the Sun's chromosphere and corona and in the transition region between them (white). Note that both the height and temperature scales are nonlinear. *(a: R. Christen and M. Christen, Astro-Physics Inc.; b: Greg Slater; c: adapted from A. Gabriel)*

Although its temperature is extremely high, the density of gas in the corona is very low, about 10 trillion times less dense than the air at sea level on Earth. The low density partly accounts for the dimness of the corona, which otherwise would outshine the photosphere. Astronomers have mounting evidence that the corona is heated by energy carried aloft and released there by the Sun's complex magnetic fields, discussed in Sections 7-5 and 7-6.

The total amount of visible light that we receive from the solar corona is comparable to the brightness of the full Moon—or only about one-millionth as bright as the photosphere. As with the chromosphere, the corona can be seen only when the photosphere is blocked out or through special filters or at nonvisible wavelengths (such as ultraviolet and X ray), at which the corona is especially bright compared to the photosphere. The photosphere is blocked naturally during a total eclipse or artificially with a specially designed telescope, called a *coronagraph*. Figure 7-6a is a photograph of the corona taken during a total eclipse (see also Figure 1-26). Figure 7-6b shows a stunning X-ray image of the corona.

Just as Earth's gravity prevents most of our atmosphere from escaping into space, so, too, does the Sun's gravity keep most of its outer layers from leaving. However, some of the gas in the corona is moving fast enough—about a million kilometers per hour—to escape the Sun's gravity forever and race into space. As we saw in Section 4-4 in discussing Earth's magnetosphere and in Section 6-7 in discussing comets, this outflow of particles is called the **solar wind.** The *heliosphere,* a bubble in space created by the solar wind, contains the Sun and planets. The outflow of gases from the Sun prevents most of the gases flowing in space from other stars from entering our solar system.

The Sun ejects around a million tons of matter each second as the solar wind. Even at this rate of emission, the mass loss due to the solar wind will amount to only a few tenths of a percent of the Sun's total mass throughout its lifetime. Although electrons and hydrogen and helium nuclei comprise 99.9% of the solar wind, silicon, sulfur, calcium, chromium, nickel, neon, and argon ions have also been detected in it. The solar wind particles reach speeds up to 2.9×10^6 km/h (1.8×10^6 mph). The wind achieves these high speeds, in part, by being accelerated by the Sun's magnetic field, discussed in Section 7-5. By the time the solar wind reaches Earth, a distance of 1 AU from the Sun, it has spread out so much that there are typically just 5 particles per cubic centimeter (cm^3), and they are moving around 1.1×10^6 to 1.4×10^6 km/h (6.8×10^5 to 8.5×10^5 mi/h). For comparison, there are 6×10^{19} molecules/cm^3 in the air we breathe, and the winds at Earth's surface are moving less than 240 km/h (150 mi/h).

Astronomers think that the Sun's surface chemical composition today is nearly identical to the composition of the solar nebula (Section 2-10). To better understand the chemistry of the early solar system, the *Genesis* spacecraft was launched to collect pristine solar wind particles far outside Earth's magnetosphere (Section 4-4) and to return them to Earth. By studying the amounts of the different chemicals in the solar wind, astronomers hope to better understand how the Sun works, as well as to learn the chemistries of the earlier generation of stars that exploded and whose gases eventually recondensed and formed the Sun and the other objects in the solar system. *Genesis's* parachute failed and the spacecraft crashed into Earth at 320 km/h (about 200 mph), creating a small crater. Nevertheless, the collected samples have been salvaged and are being studied.

An important result from the *Genesis* mission concerns the chemistry of neon in the Sun. The amounts of different neon isotopes (atoms of neon that have different numbers of neutrons in their nuclei) in the Moon rocks brought back by Apollo astronauts suggested that the Sun was more active (hotter) in the past than it is today. The amounts of neon isotopes collected by *Genesis* were different from those in the Moon rocks. Astronomers deduce that the neon at different depths of the Moon was chemically altered by different amounts due to the radiation from space over billions of years. Taking this into account, they conclude that the Sun has maintained a more nearly constant temperature than previously believed.

Which of the Sun's three atmospheric layers is coolest? Which is densest?

THE ACTIVE SUN

 Granules, supergranules, spicules, and the solar wind occur continuously. But the Sun's atmosphere is periodically disrupted by magnetic fields that stir things up, creating a group of phenomena known collectively as the *active* Sun. The Sun's most obvious transient features are **sunspots,** regions of the photosphere that appear dark because they are cooler than the rest of the Sun's lower atmosphere. Sometimes sunspots occur in isolation (Figure 7-7a), but often they arise in clusters, called *sunspot groups* (Figure 7-7b).

INSIGHT INTO SCIENCE

Perception versus Reality Our senses (and often our technology) are limited, so we must be careful about interpreting what we perceive. For example, the sunspots in Figure 7-7 certainly look like black spots. However, they appear black only in contrast to the bright light around them. As we will discuss shortly, sunspots are actually red and orange.

16 months long. One important discovery from helioseismology is that deep inside, the Sun rotates like a rigid body, rather than with the differential rotation we see on its surface (Figure 7-12b).

Observing the vibrations of the Sun around sunspots, astronomers in 2001 apparently found the answer to the question of how sunspots persist for months despite the repulsion of the magnetic fields inside them. They discovered that below the photosphere, the gases that surround each sunspot whip around like a hurricane as large across as Earth's diameter. The circulation of charged gases around the magnetic fields holds them in place.

Recall from Section 4-4 that magnetic field lines form complete loops, and each magnetic field has a north pole and a south pole. Likewise, when the Sun's magnetic field emerges through one sunspot or sunspot group, it forms a loop that reenters the Sun at another sunspot or sunspot group (Figure 7-13; insets). We associate the name *north pole* or *south pole* with each sunspot or sunspot group, depending on whether the magnetic field there is pointing outward (a south pole) or pointing inward (a north pole).

Sunspots and sunspot groups are connected in pairs—one where the magnetic field points out of the Sun, the other where it points into the Sun.

During his observations, Hale found that on one hemisphere of the Sun, sunspots with a magnetic north pole always come into view before the corresponding sunspots with a magnetic south pole. At the same time on the other hemisphere, the order is reversed (see Figure 7-13; insets). Hale also found that this pattern reverses itself about every 11 years. The hemisphere where north magnetic poles come first during one 11-year cycle has south magnetic poles coming first during the next. Astronomers, therefore, speak of the 22-year **solar cycle,** the time it takes the magnetic fields to return to the original orientation.

Based on these observations, we can now explain the Sun's magnetic field. This field is created as a result of the Sun's rotation and the resulting motion of the ionized particles found throughout it. This explanation was proposed in 1960 by another American astronomer, Horace Babcock, as the **magnetic dynamo** model, in an effort to explain the 22-year solar cycle. When the Sun is quiet, its magnetic field

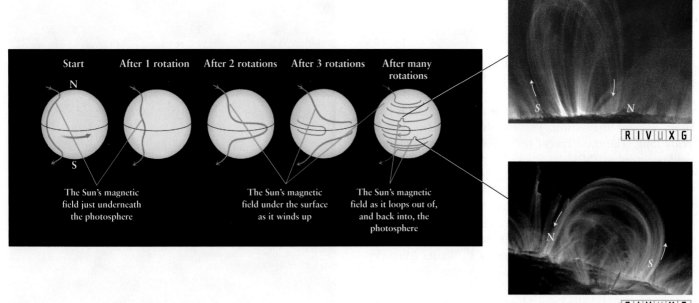

FIGURE 7-13 **Babcock's Magnetic Dynamo** In a plausible partial explanation for the sunspot cycle, differential rotation wraps a magnetic field around the Sun just under its surface. Convection under the photosphere tangles the field, which becomes buoyant and rises through the photosphere, creating sunspots and sunspot groups. Insets: In each group, the sunspot that appears first as the Sun rotates (on the right side of each loop as you view them) has the same polarity as the Sun's magnetic pole in that hemisphere (north in the upper hemisphere and south in the lower hemisphere). To understand why, note that magnetic fields make complete loops: If a north pole enters the Sun, it leads to a south pole emerging, and vice versa. Follow the magnetic field in the northern hemisphere from where it enters the Sun (near the top) in counterclockwise loops until you encounter the bump in that hemisphere. Where the field emerges in the bump is a south pole and where it reenters the Sun is a north pole, as drawn. Conversely, following the fields in the southern hemisphere, starting at the south pole, you encounter the southern bump from the left, emerging as a north pole and reentering the Sun as a south pole, as drawn. The Sun's magnetic fields are revealed by the radiation emitted from the gas they trap. These ultraviolet images show coronal loops up to 160,000 km (100,000 mi) high, with gases moving along the magnetic field lines at speeds of 100 km/s (60 mi/s). *(Top inset: TRACE, Stanford-Lockheed Institute for Space Research, and NASA Small Explorer Program; bottom inset: NASA)*

lies just below the surface in the highly conducting plasma located there, unlike Earth's field, which passes through our planet's center.

As shown in Figure 7-13, the Sun's differential rotation causes its magnetic field to become increasingly stretched. Like stretching a rubber band, stretching the field this way causes it to store energy. At the same time, the magnetic field becomes tangled like the jumble of computer wires behind my desk as convection of the gases under the photosphere also causes the fields to rise and fall. Unlike a rubber band, however, magnetic field lines cannot break to release the energy stored in them. Rather, the fields must untangle themselves.

This untangling process begins as the mixed-up regions of magnetic field trap gases. This gas expands, thereby becoming buoyant and floating up through the solar surface, carrying the magnetic fields with them. Sunspots, with magnetic fields typically 5000 times stronger than Earth's magnetic field, form where loops or tangles of solar magnetic field leave and reenter the Sun. The gas bottled up in the loops of magnetic field eventually leaks out, and the fields untangle, interact with other parts of the Sun's magnetic field, and gradually settle back under the photosphere, at which point the sunspots associated with them disappear. This process of magnetic fields piercing the Sun's surface begins at high latitudes, and over the next 11 years it occurs closer and closer to the Sun's equator (see Figure 7-10).

Because of how these fields interact and vanish, every 11 years the Sun's entire magnetic field is reversed—the Sun's north magnetic pole becomes its south magnetic pole, and vice versa. After another 11-year cycle, the field is back to its original orientation. This is why the solar cycle is 22 years long. The most recent reversal of the Sun's magnetic field occurred in 2001.

Notable irregularities occur in the solar cycle. For example, the overall reversal of the Sun's magnetic field is often piecemeal and haphazard. More intriguing still is the strong historical evidence that all traces of sunspots and the sunspot cycle have vanished for decades at a time. For example, in 1893 the British astronomer E. Walter Maunder used historical observations to conclude that virtually no sunspots occurred from 1645 through 1715 (see Figure 7-8a). This period, called the *Maunder minimum,* coincides with a period of cold so extreme in Europe that it was called the *Little Ice Age.* At the same time, western North America was subject to severe drought.

Similar sunspot-free periods apparently occurred at irregular intervals in earlier times as well. Conversely, periods of increased sunspot activity in the eleventh and twelfth centuries coincided with periods of warmer-than-average temperatures. It remains to be seen, however, whether scientists can find all of the links between the number of sunspots and periods of extreme temperatures on Earth.

INSIGHT INTO SCIENCE

Cause and Effect Although extremes of sunspot activity often occur at the same times as weather extremes on Earth, models of these phenomena must also take into account the possibility of purely terrestrial causes for these temperature changes. Remember, just because one event follows another does not mean that the first causes the second.

7-6 Solar magnetic fields also create other atmospheric phenomena

Figure 7-14 shows the active Sun's chromosphere and corona. The bright areas in this photograph are called **plages** (pronounced *plahzh,* from the French word for "beaches"). By studying the light emitted by calcium or hydrogen atoms in them, we know that they are hotter, and therefore brighter, than the surrounding chromosphere. Plages, which often appear just before nearby sunspots form, are thought

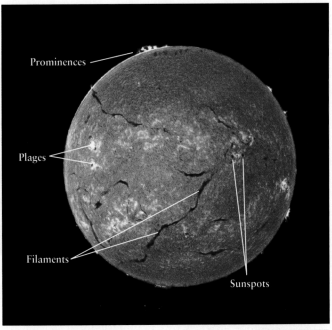

RIVUXG

FIGURE 7-14 Active Sun in H$_\alpha$ This photograph shows the chromosphere and corona during a solar maximum, when sunspots are abundant. The image was taken through a filter that allowed only light from H$_\alpha$ emission to pass through. The hot, upper layers of the Sun's atmosphere are strong emitters of H$_\alpha$ photons. (See Section 3-22 for details of the Balmer series, H$_\alpha$–H$_\infty$.) A few large sunspots are evident. Most notable are features that do not appear at the solar minimum, such as the snakelike features shown here and called filaments, bright areas called plages, and prominences (filaments seen edge-on) observed at the solar limb. *(NASA)*

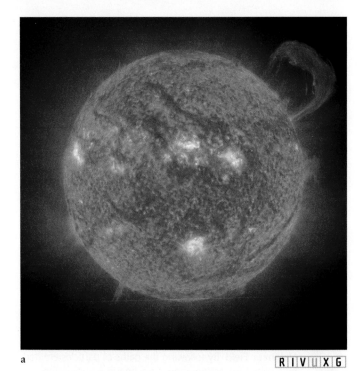

a RIVUXG

b RIVUXG

Approximate size of Earth for comparison

FIGURE 7-15 **Prominences** (a) A huge prominence arches above the solar surface in this *SOHO* image taken in 2001. The radiation that exposed this picture is from singly ionized helium at a wavelength of 30.4 nm, corresponding to a temperature of about 50,000 K. (b) The gas in some prominences is so energetic that it breaks free from the magnetic fields that shape and confine it. This eruptive prominence occurred in 1999 and did not strike Earth (shown only for size). *(a: ESA/NASA; b: Joseph B. Gurman, Solar Data Analysis Center, NASA)*

to be created by the magnetic fields under the photosphere crowding upward just before they emerge through the photosphere. In pushing upward, the fields compress the gases of the upper Sun, causing this gas to become hotter and therefore to glow more brightly.

The dark streaks in Figure 7-14 are features in the corona, called **filaments**. These are huge volumes of gas lofted upward from the photosphere by the Sun's magnetic field. When viewed from the side rather than from above, filaments form gigantic loops or arches, called **prominences** (see Figure 7-14 and Figure 7-15a). The temperature of the gas in prominences can reach 50,000 K. These features are almost always associated with sunspots. Some prominences last for only a few hours, whereas others persist for months. The gas in the most energetic prominences escape the magnetic fields that confine them and surge out into space (Figure 7-15b).

X-ray photographs also reveal numerous coronal bright and dark spots that are hotter or colder, respectively, than the surrounding corona (Figure 7-16). Temperatures in the bright regions occasionally reach 4 million K. Many of the bright coronal hot spots are located over sunspots. The darker, cooler **coronal holes** act as conduits for gases to flow out of the Sun. Therefore, when a coronal hole on the rotating Sun faces Earth, the solar wind in our direction increases dramatically.

Violent, eruptive events on the Sun, called **solar flares,** release vast quantities of high-energy particles, as well as

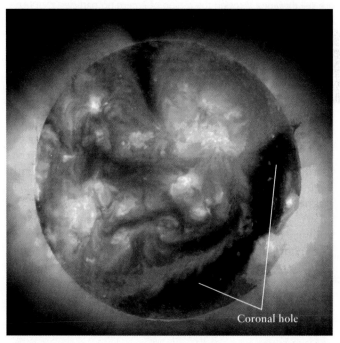

Coronal hole

RIVUXG

FIGURE 7-16 **A Coronal Hole** This X-ray picture of the Sun's corona was taken by the *SOHO* satellite. A huge coronal hole dominates the lower right side of the corona. The bright regions are emissions from sunspot groups. *(SOHO/EIT/ESA/NASA)*

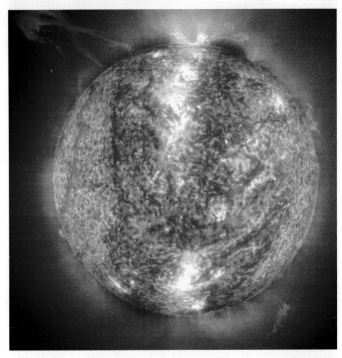

R I V U X G

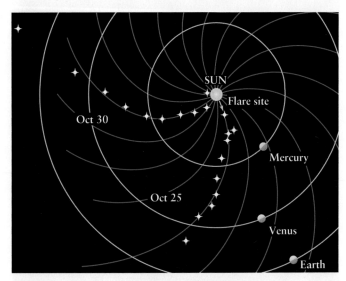

FIGURE 7-17 **A Flare** Solar flares, which are associated with sunspot groups, produce energetic emissions of particles from the Sun. This image, taken in 2000 by *SOHO*, shows a twisted flare in which the Sun's magnetic field lines are still threaded through the region of emerging particles. *(NASA)*

FIGURE 7-18 **A Snapshot of the Sun's Global Magnetic Field** By following the paths of particles emitted by a solar flare, astronomers have begun mapping the solar magnetic field outside the Sun. The field guides the outflowing particles, which, in turn, emit radio waves that indicate the position of the field. These data were collected by the *Ulysses* spacecraft in 1994. *Ulysses* was the first spacecraft to explore interplanetary space from high above the plane of the ecliptic.

X rays and ultraviolet radiation, from the Sun (Figure 7-17). Flares are so powerful that they leave the region of the surface of the Sun in their vicinity quaking for an hour or more. At the maximum of the sunspot cycle, about 1100 flares occur per year. Most of them last for less than an hour, but, during that time, temperatures soar to 5 million K. By following the paths of particles emitted by flares as they are guided outward by the Sun's magnetic field, astronomers have recently begun mapping that field as it extends out beyond Earth (Figure 7-18).

Astronomers have also observed huge, balloon-shaped volumes of high-energy gas being ejected from the corona. These **coronal mass ejections** typically expel 2 trillion tons of matter at 1.4×10^6 km/h, and each one lasts for up to a few hours (Figure 7-19).

Coronal mass ejections have enough energy to break through the Sun's magnetic fields that normally contain them or even to carry fields outward. Flares have been observed to create shock waves that initiate some of the coronal mass ejections. The origins of others are still under investigation. The numbers of plages, prominences, flares, and coronal mass ejections vary with the same 11-year cycle as the number of sunspots. Coronal mass ejections, the major source of hazardous particles from the Sun, occur with varying frequency throughout the sunspot cycle, but they never completely cease.

Some coronal mass ejections, solar flares, and filaments head toward Earth. It takes about 8 min for their electromagnetic radiation to get here; their high-energy particles arrive a few days later. At times when these surges of particles are *not* coming to us, the normal solar wind particles are trapped by Earth's magnetic fields in the Van Allen belts (see Section 4-4). The additional particles from a coronal mass ejection or other solar event that *do* come our way overwhelm the Van Allen belts, enabling matter in them to cascade Earthward. One of the most spectacular of such events in recent years occurred on and around October 31, 2003 (Halloween in the United States), filling the night skies with aurorae in many places around the world. Perhaps you saw them.

Using the technology described in Chapter 3, several spacecraft now monitor the Sun and the space between it and Earth. These include the *Solar and Heliospheric Observatory (SOHO)* launched in 1995 by the European Space Agency (ESA) and NASA, the *Transition Region and Coronal Explorer (TRACE)* launched in 1998, the *Reuven Ramaty High Energy Solar Spectroscopic Imager (RHESSI)* launched in 2002, and the *Solar Radiation and Climate Experiment (SORCE)* launched in 2003. These spacecraft observe the Sun in gamma ray, X ray, ultraviolet, and visible parts of the spectrum, and they provide both scientific and space weather information.

What holds the gas in loops above the Sun's surface?

Thus, scientists detect neutrinos by observing Cerenkov radiation flashes with light-sensitive devices, called *photomultipliers*, mounted in the water (see Figure 7-22). The three flavors of neutrinos have different interactions with water, all of which have been detected. This evidence that neutrinos can change *requires* that they have mass, and it explains the earlier low rate of solar neutrino observations. Astronomical research therefore led to a change in our understanding of some of the lowest-mass particles in existence.

Why did the earlier neutrino detectors not detect the predicted number of neutrinos from the Sun?

7-10   Frontiers yet to be discovered

Despite centuries of observations and decades of modeling it, the Sun holds innumerable secrets yet to be uncovered. We have launched an armada of spacecraft specifically designed to observe it in different parts of the spectrum. Astronomers soon hope to understand more about such questions as how the Sun's magnetic field is generated and how it changes throughout the solar cycle; how the Sun's atmosphere heats so dramatically in the transition zone; why the solar luminosity varies with time; just how the Sun's changing output affects the temperatures on Earth; the details of how flares and coronal mass ejections occur; and why the Sun rotates differentially.

R I V U X G

FIGURE 7-22 **The Solar Neutrino Experiment** Located 2073 m (6800 ft) underground in the Creighton nickel mine in Sudbury, Canada, the Sudbury Neutrino Observatory is centered around a tank that contains 1000 tons of water. Occasionally, a neutrino entering the tank interacts with one or another of the particles already there. Such interactions create flashes of light, called Cerenkov radiation. Some 9600 light detectors sense this light. The numerous silver protrusions are the back sides of the light detectors prior to their being wired and connected to electronics in the lab, seen at the bottom of the photograph. *(Ernest Orlando Lawrence/Berkeley National Laboratory)*

which consists of a proton and a neutron. When a solar neutrino is absorbed by a deuterium nucleus, the nucleus breaks apart into two protons and an electron. As this electron rushes through the water, it emits a flash of light, called **Cerenkov radiation**. As the Russian physicist Pavel A. Cerenkov (pronounced *Che-ren-kov*) first observed, the flash occurs whenever a particle moves through a medium, such as water, faster than light can. Such motion does not violate the tenet that the speed of light *in a vacuum* (3×10^5 km/s) is the ultimate speed limit in the universe. Light is slowed considerably as it passes through water, and high-energy particles can exceed this reduced speed of light in a medium without violating the laws of nature.

SUMMARY OF KEY IDEAS

The Sun's Atmosphere

• The thin shell of the Sun's gases we see are from its photosphere, the lowest level of its atmosphere. The gases in this layer shine nearly as a blackbody. The photosphere's base is at the top of the convective zone.

• Convection of gas from below the photosphere produces features called granules.

• Above the photosphere is a layer of hotter, but less dense, gas called the chromosphere. Jets of gas, called spicules, rise up into the chromosphere along the boundaries of supergranules.

• The outermost layer of thin gases in the solar atmosphere, called the corona, extends outward to become the solar wind at great distances from the Sun. The gases of the corona are very hot, but they have extremely low densities.

The Active Sun

• Some surface features on the Sun vary periodically in an 11-year cycle, and the magnetic fields that cause these changes actually vary over a 22-year cycle.

- Sunspots are relatively cool regions produced by local concentrations of the Sun's magnetic field protruding through the photosphere. The average number of sunspots and their average latitude vary in an 11-year cycle.

- A prominence is gas lifted into the Sun's corona by magnetic fields. A solar flare is a brief, but violent, eruption of hot, ionized gases from a sunspot group. Coronal mass ejections send out large quantities of gas from the Sun. Coronal mass ejections and flares that head our way affect satellites, communication, and electric power, and cause aurorae.

- The magnetic dynamo model suggests that many transient features of the solar cycle are caused by the effects of differential rotation and convection on the Sun's magnetic field.

The Sun's Interior

- The Sun's energy is produced by the thermonuclear process called hydrogen fusion, in which four hydrogen nuclei release energy when they fuse to produce a single helium nucleus.

- The energy released in a thermonuclear reaction comes from the conversion of matter into energy, according to Einstein's equation $E = mc^2$.

- The solar model is a theoretical description of the Sun's interior derived from calculations based on the laws of physics. The solar model reveals that hydrogen fusion occurs in a core that extends from the center to about a quarter of the Sun's visible radius.

- Throughout most of the Sun's interior, energy moves outward from the core by radiative diffusion. In the Sun's outer layers, energy is transported to the Sun's surface by convection.

- Neutrinos were originally believed to be massless. The electron neutrinos generated and emitted by the Sun were originally detected at a lower rate than is predicted by our model of thermonuclear fusion. The discrepancy occurred because electron neutrinos have mass, which causes many of them to change into other forms of neutrinos before they reach Earth. These alternative forms are now being detected.

WHAT DID YOU THINK?

1 *What percentage of the solar system's mass is in the Sun?* The Sun contains about 99.85% of the solar system's mass.

2 *Does the Sun have a solid and liquid interior like Earth?* No. The entire Sun is composed of hot gases.

3 *What is the surface of the Sun like?* The Sun has no solid surface. Indeed, it has no solids or liquids anywhere. The level we see, the photosphere, is composed of hot, churning gases.

4 *Does the Sun rotate? If so, how fast?* The Sun's surface rotates differentially, varying between once every 35 days near its poles and once every 25 days at its equator.

5 *What makes the Sun shine?* Thermonuclear fusion in the Sun's core is the source of the Sun's energy.

Review Questions

The answers to all computational problems, which are preceded by an asterisk (*), appear at the end of the book.

1. Describe the features of the Sun's atmosphere that are always present.

2. Describe the three main layers of the solar atmosphere and how you would best observe them.

3. Name and describe seven features of the active Sun. Which two are the same, seen from different angles? Which can have a direct impact on Earth? Explain.

4. Describe the three main layers of the Sun's interior.

*5. When will the next sunspot minimum and sunspot maximum occur after the maximum in 2001 and the minimum in 2007? Explain your reasoning.

6. Why is the solar cycle said to have a period of 22 years, even though the sunspot cycle is only 11 years long?

7. How do astronomers detect the presence of a magnetic field in hot gases, such as the field in the solar photosphere?

8. Describe the dangers in attempting to observe the Sun. How have astronomers learned to circumvent these hazards?

9. Give an everyday example of hydrostatic equilibrium not presented in the book.

10. Give some everyday examples of heat transfer by convection and radiative transport.

11. What do astronomers mean by a "model of the Sun"?

12. Why do thermonuclear reactions in the Sun take place only in its core?

13. What is hydrogen fusion? This process is sometimes called "hydrogen burning." How is hydrogen burning fundamentally unlike the burning of a log in a fireplace?

14. Describe the Sun's interior, including the main physical processes that occur at various depths within the Sun.

15. What is a neutrino, and why are astronomers so interested in detecting neutrinos from the Sun?

Observing Projects

16. Use *Starry Night Enthusiast*™ software to determine the Sun's rotation rate. Select **Favourites > Solar System > Sun > Flares and Prominences** from the menu. Note the locations of the features you see on the Sun so you can use them for reference as it rotates. Change the **Time Flow Rate** to **1 sidereal day** and use the single step forward button (❙▶) to observe the Sun's rotation. Step through enough timesteps to determine the rotation rate for the Sun. Compare the number of days for one rotation to the rotation information for the Sun presented in the text, and state which part of the Sun the program's rotation is portraying. (Note that the program does not show the Sun's differential rotation.)

17. You can use *WorldWide Telescope* (WWT) to examine several features on the surface of the Sun that have enhanced our knowledge of our star. This knowledge has been applied to the study of stars in general.

Open WWT and click on **Explore > Solar System > Sun** to display a full image of the Sun. **a. Limb darkening** Examine the solar image and note that the Sun appears brighter at the center than at its edge, a phenomenon known as limb darkening. Because the Sun is composed entirely of gas, when you look at the center of the Sun's image, you are looking deeper into the Sun than when you look close to the Sun's limb. The observed limb darkening indicates that the temperature of the gas is decreasing with height over the region just below the solar surface. **b. Granulation** Zoom in to examine the surface of the Sun in detail. You will see that the whole surface is covered by mottled patterns of light regions surrounded by narrower dark regions. This pattern is known as granulation and is evidence of convective motion under the solar surface, which is carrying heat from the hotter layers. The brighter centers of these cells indicate hotter gases that upwell and cool by radiating energy and then return to the solar interior down the darker cell granular boundaries. This type of convective flow is common in many areas of science, such as meteorology and oceanography. **c. Solar activity** The dark regions on this image of the Sun are sunspots. These are cooler regions where strong magnetic fields have inhibited the normal convective flow and have allowed the gas to cool to lower temperatures than those of the normal photosphere. Zoom in on several of these active regions in turn. Note that there are dark umbra surrounded by slightly less dark regions, the penumbra, in the larger of these sunspots and that they are in groups in general. Sunspots normally appear in pairs or clusters and are connected by arches of powerful magnetic fields. Areas surrounding sunspots are known as active regions. The gases above active regions are heated by energy released from magnetic fields. The detailed structure of these magnetic arches can be seen in images of the Sun taken in ultraviolet light and X-rays that are shown in the textbook.

You can use the Internet to examine the current appearance of the Sun. Open a Web browser and type in the URL http://sohowww.nascom.nasa.gov to go to the SOHO home page. Click the thumbnail beneath the label "The Sun Now" in the upper right of the page and select the MDI continuum image from the list of thumbnails. Draw a sketch of the Sun as it appears in this image, noting particularly the positions of any sunspots. Repeat this exercise every day for approximately a week. From your drawings, try to estimate the rotational period of the Sun.

8

Characterizing Stars

Stars Come in Many Colors *(Hubble Heritage Team/AURA/STScI/ NASA)*

R I V U X G

WHAT DO YOU THINK?

1 How near to us is the closest star other than the Sun?

2 How luminous is the Sun compared with other stars?

3 What colors are stars, and why do they have these colors?

4 Are brighter stars hotter than dimmer stars?

5 Compared to the Sun, what sizes are other stars?

6 Are most stars isolated from other stars, as the Sun is?

Answers to these questions appear in the text beside the corresponding numbers in the margins and at the end of the chapter.

The Sun is our stepping stone to the stars. The physics that describes why the Sun shines can be applied to the other stars, but we must first know how far they are from us, how bright and hot they are, and what their masses are. In this chapter we describe how this information is accquired. In Chapters 9 and 10, we examine stellar evolution, the process by which stars use up their nuclear fuel and therefore change.

In this chapter you will discover

• that the distances to many nearby stars can be measured directly, whereas the distances to farther ones are determined indirectly

• the observed properties of stars on which astronomers base their models of stellar evolution

• how astronomers analyze starlight to determine a star's temperature and chemical composition

• how the total energy emitted by stars and their surface temperatures are related

• the different classes of stars

• the variety and importance of binary star systems

• how astronomers calculate stellar masses

8-1 Distances to nearby stars are determined by stellar parallax

Nothing in human experience prepares us for understanding the distances between the stars. On Earth, the greatest distance we read about in our daily lives is perhaps halfway around the globe, 20,000 km, which is about the distance from New York City to Perth, Australia. By observing the Sun, the Moon, and planets, we have some comprehension of hundreds of thousands, or even millions, of kilometers.

How far away are the nearest stars? Ask 10 people and **1** you will probably hear answers that range from millions to billions of kilometers or miles. In fact, the closest star other than the Sun, Proxima Centauri in the constellation Centaurus, is about 40 *trillion* km (25 trillion mi) away. It takes light about 4 years to get here from there. Most of the stars you see in the night sky are many times farther away. Given such enormous separations between astronomical bodies, how do we measure the distance to a nearby star?

Because Proxima Centauri is closer to us than other stars, its position among the background stars changes as Earth orbits the Sun. This is precisely the same effect that we see when, for example, Mars appears to have retrograde motion as we pass between it and the Sun (see Figures 2-2 and 2-3) or that Tycho Brahe sought for the supernova of 1572 (see Figure 2-7). The apparent motion of nearby stars among the background of more distant stars, due to Earth's motion around the Sun, is called **stellar parallax.**

Parallax is an everyday phenomenon. We experience it when nearby objects appear to shift their positions against a distant background as we move (review Figure 2-6). We also experience it continuously when we are awake, because our eyes change angle when looking at objects at different distances. As you view a tree 10 m away, your eyes cross only slightly. Looking at something closer, say this book, your eyes cross much more in order for both of them to focus on the same word. The parallax angle formed between your eyes and the tree or the book lets your brain judge just how close these objects are to you. Working out distances of nearby stars is done similarly (Figure 8-1), but it requires painstaking measurements of tiny angles and a little geometry.

As the Earth moves from one side of its orbit around the Sun to the other, a nearby star's apparent position shifts among the more distant stars. Referring to Figure 8-1a, the parallax angle, p, is half the angle by which Earth shifts positions through the year as seen from that star, measured in arcseconds. The difference in parallax angles for stars at different distances can be seen by comparing Figures 8-1a and 8-1b. Appendix H-5 explores some details of how distances are determined from parallax angles.

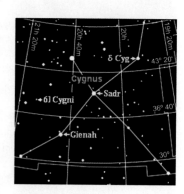

The first stellar parallax measurement was made in 1838 by Friedrich Wilhelm Bessel, a German astronomer and mathemati-

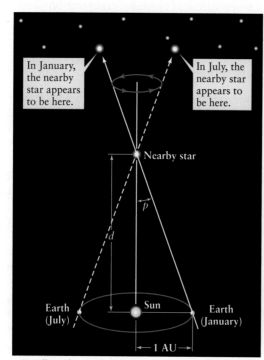

a Parallax of a nearby star

b Parallax of an even closer star

FIGURE 8-1 Using Parallax to Determine Distance (a) As Earth orbits the Sun, a nearby star appears to shift its position against the background of distant stars. The star's parallax angle (p) is equal to the angle between the Sun and Earth, as seen from the star. The stars on the scale of this drawing are shown much closer than they are in reality. If drawn to the correct scale, the closest star, other than the Sun, would be about 5 km (3.2 mi) away. (b) The closer the star is to us, the greater the parallax angle p. The distance to the star (in parsecs) is found by taking the inverse of the parallax angle p (in arcseconds), $d = 1/p$. (a, b: Mark Andersen/JupiterImages)

TABLE 8-1 The Spectral Sequence

Spectral class	Color	Temperature (K)	Spectral lines	Examples
O	Blue-violet	50,000–30,000	Ionized atoms, especially helium	Naos (ζ Puppis), Mintaka (δ Orionis)
B	Blue-white	30,000–11,000	Neutral helium, some hydrogen	Spica (α Virginis), Rigel (β Orionis)
A	White	11,000–7500	Strong hydrogen, some ionized metals	Sirius (α Canis Majoris), Vega (α Lyrae)
F	Yellow-white	7500–5900	Hydrogen and ionized metals, such as calcium and iron	Canopus (α Carinae), Procyon (α Canis Minoris)
G	Yellow	5900–5200	Both neutral and ionized metals, especially ionized calcium	Sun, Capella (α Aurigae)
K	Orange	5200–3900	Neutral metals	Arcturus (α Boötis), Aldebaran (α Tauri)
M	Red-orange	3900–2500	Strong titanium oxide and some neutral calcium	Antares (α Scorpii), Betelgeuse (α Orionis)

tral classification scheme we use today. Many of the early A through P categories were dropped because the Balmer lines in a star's spectrum can be weak whether the star is very cool or very hot. The remaining Balmer-based spectral types were thus reordered by stellar surface temperature into the **OBAFGKM sequence**. This sequence is most easily learned with the aid of a mnemonic, such as "Oh, **B**e **A** **F**ine **G**uy (or **G**irl), **K**iss **M**e!"

Why do very hot stars have weak hydrogen lines, even though they are composed primarily of hydrogen?

The hottest stars have surface temperatures of 30,000 to 50,000 K and are classified as O type; their spectra are dominated by He II and Si IV (triply ionized silicon). The coolest stars have surface temperatures of 2500 to 3000 K and are classified as M type. Table 8-1 includes representative examples of each spectral type.

The wide range of temperatures covered by each spectral type prompted astronomers to further subdivide the OBAFGKM temperature sequence. Each spectral type is now broken up into 10 temperature subranges. These 10 finer steps are indicated by adding an integer from 0 (hottest) through 9 (coolest). Thus, an A8 star is hotter than an A9 star, which is hotter than an F0 star, which is hotter than an F1 star, and so on. The Sun, whose spectrum is dominated by singly ionized metals (especially Fe II and Ca II), is a G2 star.

What class of star is just slightly cooler than a K9? (The caption for Figure 8-7 has the answer.)

Which star is hottest: F5, B6, or M3?

INSIGHT INTO SCIENCE

Tolerate Idiosyncrasies In astronomy, the word *metal* applies to all elements other than hydrogen or helium. In chemistry, elements are classified as metals or nonmetals according to their physical and chemical properties. To a chemist, sodium and iron are metals, whereas carbon and oxygen are not.

TYPES OF STARS

Keeping in mind that patterns of objects often reveal information about them, early twentieth-century astronomers began searching for relationships between different properties of stars. The relationship between luminosities and surface temperatures turned out to be a key to understanding types of stars. As often happens in such scientific quests, the significance of this relationship was discovered nearly simultaneously by two independent researchers.

8-7 The Hertzsprung-Russell diagram identifies distinct groups of stars

Around 1911, the Danish astronomer Ejnar Hertzsprung noticed that patterns emerge when the luminosities of stars (or their equivalent absolute magnitudes) are plotted against their surface temperatures (or their equivalent spectral types). Luminosity and absolute magnitude indicate the total energy emitted by a star or other body. They refer to what we loosely call "brightness." Within 2 years, the American astronomer Henry Norris Russell independently discovered the same result. Graphs of stellar luminosity or absolute magnitude against surface temperature or spectral type are now known as **Hertzsprung-Russell diagrams, or H-R diagrams.**

The H-R diagram is valuable because it shows that stars do not have random surface temperatures and luminosities; the two factors are correlated. Figure 8-7 is a typical H-R diagram. Each dot represents a star whose luminosity and spectral type have been determined. The surface temperatures are plotted along the top of this figure and the absolute magnitudes along the right side. You can therefore see the equivalence of the temperature and spectral type and the equivalence of the luminosity and absolute magnitude.

Bright stars are near the top of the diagram; dim stars are near the bottom. Contrary to intuition, hot (O and B) stars are toward the left side and cool (M) stars are toward the right. Hertzsprung and Russell made this choice because of the standard sequence OBAFGKM.

> ### INSIGHT INTO SCIENCE
>
> **From Patterns to Models** Scientists look at patterns of behavior in related objects as valuable clues to underlying properties and their causes. Guided by this data, they then create mathematical models, which are used to make fresh predictions. For example, astronomers analyze the relationship between the luminosities and surface temperatures of stars as displayed on an H-R diagram to gain insight into the internal activities of stars.

The band of stars in Figure 8-7 stretching diagonally across the H-R diagram and on which a red curve is superimposed represents most of the stars we see in the nighttime sky. Called the **main sequence,** it extends from the hot, bright, bluish stars in the upper left corner of the diagram down to the cool dim stars (called *red dwarfs*) in the lower right corner. Each star on this band is a **main-sequence star.** Just over 91% of the stars surrounding the solar system fall into this category.

Observations reveal that the number of main-sequence stars decreases with increasing surface temperature. Therefore, along the main sequence, the cooler M, K, and G stars are the most common ones and the hot O stars are the rarest. The Sun (spectral type G2, absolute magnitude +4.8) is a main-sequence star.

To the right of the main sequence on the H-R diagram is a second major grouping of stars. These stars are bright but cool. From the Stefan-Boltzmann law, we know that a cool object radiates much less light from each unit of surface area than does a hot object. To be so luminous, these cool stars must therefore be huge compared to main-sequence stars of the same temperature, so they are called **giant stars.**

Giants are typically 10 to 100 times the radius of the Sun and have surface temperatures between 2000 and 20,000 K. The cooler members of this class of stars (those with surface temperatures between 2000 and 4500 K) are often called **red giants** because they appear reddish in the nighttime sky. Aldebaran in the constellation Taurus and Arcturus in Boötes are examples of red giants that you can easily see with the naked eye.

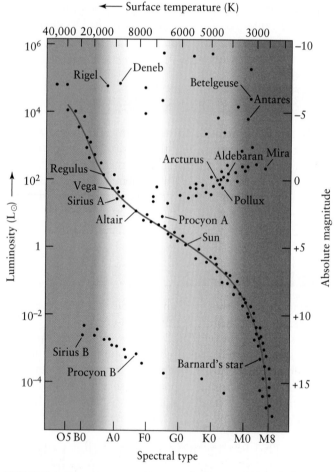

FIGURE 8-7 A Hertzsprung-Russell Diagram On an H-R diagram, the luminosities of stars are plotted against their spectral types. Each dot on this graph represents a star whose luminosity and spectral type have been determined. Some well-known stars are identified. The data points are grouped in just a few regions of the diagram, revealing that luminosity and spectral type are correlated: Main-sequence stars fall along the red curve, giants are to the right, supergiants are on the top, and white dwarfs are below the main sequence. The absolute magnitudes and surface temperatures are listed at the right and top of the graph, respectively. These are sometimes used on H-R diagrams instead of luminosities and spectral types. (*Answer to text question:* An M0 star is the next coolest after a K9.)

A few rare stars are considerably bigger and more luminous than typical giants. Located along the top of the H-R diagram, these superluminous stars are appropriately called **supergiants.** They can extend in radius up to about 1000 $R_\odot$.

work and the nature of the material between us and the stars. Therefore, a major goal of space-based observational astronomy is to get more and better parallax measurements for stars at greater distances. To this end, satellites will be sent far from Earth to increase the baseline used to make these measurements (see Figure 8-1) and hence the angular accuracy that is possible.

Classifying stars by luminosity and spectral type is an ongoing process providing an ever-growing body of data about stellar properties. This information yields continual surprises. For example, there is a growing number of individual "stars" whose properties astronomers thought they understood but which turned out to be binary systems, each star having different properties from that of the single "star" they were originally believed to be.

Another significant piece of information astronomers need to understand star formation better is how many stars of a given mass are formed. Observations strongly indicate that very-high-mass stars form quite infrequently, but is the same true of very-low-mass stars? Or do the numbers of stars formed increase with decreasing mass? Or perhaps there is a peak of star formation at some mass with decreasing numbers of stars that have higher and lower masses. The number of stars formed with different masses is expressed as the **initial mass function.** Although there are several theoretic initial mass functions, in the end, observations will determine which is correct.

SUMMARY OF KEY IDEAS

• Stars differ in size, luminosity, temperature, color, mass, and chemical composition—facts that help astronomers understand stellar structure and evolution.

Magnitude Scales

• Determining stellar distances from Earth is the first step to understanding the nature of the stars. Distances to the nearer stars can be determined by stellar parallax, which is the apparent shift of a star's location against the background stars while Earth moves along its orbit around the Sun. The distances to more remote stars are determined using spectroscopic parallax.

• The apparent magnitude of a star, denoted m, is a measure of how bright the star appears to Earth-based observers. The absolute magnitude of a star, denoted M, is a measure of the star's true brightness and is directly related to the star's energy output, or luminosity.

• The absolute magnitude of a star is the apparent magnitude it would have if viewed from a distance of 10 pc. Absolute magnitudes can be calculated from the star's apparent magnitude and distance from Earth.

• The luminosity of a star is the amount of energy emitted by it each second.

The Temperatures of Stars

• Stellar temperatures can be determined from stars' colors or stellar spectra.

• Stars are classified into spectral types (O, B, A, F, G, K, and M) based on their spectra or, equivalently, their surface temperatures.

Types of Stars

• The Hertzsprung-Russell (H-R) diagram is a graph on which luminosities of stars are plotted against their spectral types (or, equivalently, their absolute magnitudes are plotted against surface temperatures).

• The H-R diagram reveals the existence of four major groupings of stars: main-sequence stars, giants, supergiants, and white dwarfs.

• The mass–luminosity relation expresses a direct correlation between a main-sequence star's mass and the total energy it emits.

• Distances to stars can be determined using their spectral types and luminosity classes.

Stellar Masses

• Binary stars are surprisingly common. Those that can be resolved into two distinct star images (even if it takes a telescope to do this) are called visual binaries.

• The masses of the two stars in a binary system can be computed from measurements of the orbital period and orbital dimensions of the system.

• Some binaries can be detected and analyzed, even though the system may be so distant (or the two stars so close together) that the two star images cannot be resolved with a telescope.

• A spectroscopic binary is a system detected from the periodic shift of its spectral lines. This shift is caused by the Doppler effect as the orbits of the stars carry them alternately toward and away from Earth.

• An eclipsing binary is a system whose orbits are viewed nearly edge-on from Earth, so that one star periodically eclipses the other. Detailed information about the stars in an eclipsing binary can be obtained by studying the binary's light curve.

• Mass transfer occurs between binary stars that are close together.

WHAT DID YOU THINK?

1 *How near to us is the closest star other than the Sun?* Proxima Centauri is about 40 trillion km (25 trillion mi) away. Light from there will take about 4 years to reach Earth.

2 *How luminous is the Sun compared with other stars?* The most luminous stars are about a million times brighter, and the least luminous stars are about a hundred thousand times dimmer than the Sun.

3 *What colors are stars, and why do they have these colors?* Stars are found in a wide range of colors, from red through violet as well as white. They have these colors because they have different surface temperatures.

4 *Are brighter stars hotter than dimmer stars?* Not necessarily. Many brighter stars, such as red giants, are cooler but larger than hotter, dimmer stars, such as white dwarfs.

5 *Compared to the Sun, what sizes are other stars?* Stars range from more than 1000 times the Sun's diameter to less than 1/100 the Sun's diameter.

6 *Are most stars isolated from other stars, as the Sun is?* In the vicinity of the Sun, one-third of the stars are found in pairs or larger groups.

Review Questions

The answers to all computational problems, which are preceded by an asterisk (*), appear at the end of the book.

1. Stellar parallax measurements are used in astronomy to determine which of the following properties of stars? **a.** speeds, **b.** rotation rates, **c.** distances, **d.** colors, **e.** temperatures

2. The brightness a star would have if it were at 10 pc from Earth is called its **a.** absolute magnitude, **b.** apparent magnitude, **c.** luminosity, **d.** spectral type, **e.** center of mass.

3. Measurements of a binary star system are required to determine what property of stars? **a.** luminosity, **b.** apparent magnitude, **c.** distance, **d.** mass, **e.** temperature

4. A star with which of the following apparent magnitudes appears brightest from Earth? **a.** 6.8, **b.** 3.2, **c.** 0.41, **d.** −0.44, **e.** −1.5

5. A star of what spectral class has the strongest (darkest) H_α line? (Hint: See Figure 8-5.) **a.** B2, **b.** A0, **c.** F5, **d.** G5, **e.** M0

6. Describe how the parallax method of finding a star's distance is similar to the binocular (two-eye) vision of animals.

7. What is stellar parallax?

8. How do astronomers use stellar parallax to measure the distances to stars?

9. Why do stellar parallax measurements work only with relatively nearby stars?

10. What is the difference between apparent magnitude and absolute magnitude?

11. Briefly describe how you would determine the absolute magnitude of a nearby star.

12. What does a star's luminosity measure?

13. Why is the magnitude scale "backward" from what common sense dictates?

14. Does the star Betelgeuse, whose apparent magnitude is $m = +0.5$, look brighter or dimmer than the star Pollux, whose apparent magnitude is $m = 11.1$?

*15. Consider two identical stars, with one star 5 times farther away than the other. How much brighter will the closer star appear than the more distant one?

16. How and why is the spectrum of a star related to its surface temperature?

17. What is the primary chemical component of most stars?

18. A star of which spectral type has the strongest Na I absorption lines? At approximately what wavelength is this line normally found? (Hint: See Figure 8-5.)

19. Why does a G2 star have many more absorption lines than a B0 star?

20. Draw an H-R diagram and sketch the regions occupied by main-sequence stars, giants, supergiants, and white dwarfs. Briefly discuss the different ways you could have labeled the axes of your graph.

21. How can observations of a visual binary lead to information about the masses of its stars?

22. What is a radial-velocity curve? What kinds of stellar systems exhibit such curves?

23. What is the difference between a single-line and a double-line spectroscopic binary?

24. What is meant by the light curve of an eclipsing binary? What sorts of information can be determined from such a light curve?

25. What is the mass–luminosity relation? To what kind of stars does it apply?

26. Refer to Figure 8-7: **a.** What are the hottest and coolest named stars on the diagram? **b.** What are the brightest and dimmest named stars on the diagram? **c.** What are the hottest and coolest named main-sequence stars on the diagram? **d.** What named stars are white dwarfs? Giants? Supergiants?

Observing Projects

27. Mizar and Alcor In this exercise, you can use *Starry Night Enthusiast™* to explore two stars in the Big Dipper. Click the **Home** button in the toolbar. Open the **Find** pane and type "Mizar" in the search box, then click the **Enter** key. If the program indicates that Mizar is not visible at the current time, click the **Best Time** button. As seen from Earth, Mizar has a companion star, Alcor. Use the **Zoom** button in the toolbar to zoom in until a second star moves away from Mizar. Why does it appear to move away? Use the heads-up display feature of the **Hand Tool** to identify this star as Alcor. Where is the smaller star Alcor located with reference to Mizar? Continue zooming in until the field of view is about 2° wide and then use the angular measurement feature of the **Hand Tool** to measure the angular separation and actual physical distance between Mizar and Alcor. The closest star to Earth is Proxima Centauri, about 4.3 ly away. Comparing the data collected on Mizar and Alcor with this information, consider whether Mizar and Alcor are a visible binary, a spectroscopic binary, or an optical double, and explain your reasoning to support this conclusion.

28. Astronomers have studied the stars in our sky for several centuries. To understand how stars generate the energy that makes them shine, painstaking measurements are made of their properties, such as their brightnesses, distances from Earth, surface temperatures, energy outputs, masses, physical sizes, and the chemical compositions of their interiors. From these data and mathematical models, the ages of stars and their stages of evolution are determined.

The study of clusters of stars has been particularly fruitful in this continuing effort because the stars in a cluster are all at the same distance from Earth, they were probably all formed at about the same time, and all the stars in a given cluster were formed from the same gas cloud, and hence they started with the same chemical compositions.

You can use the *WorldWide Telescope* (WWT) program to explore some of the properties of stars and look at important types of stars that have acted as signposts in our quest to understand them. The stars in two open clusters of stars in Perseus show some of these properties.

Open WWT and center the view on the position RA = 02 h 21 min and Dec = 57° 10 min, using the display of the coordinates of the view center shown at the bottom right of the screen. Zoom in on this double cluster.

You can see a wide variety of apparent star brightnesses. Because most of these stars are members of the clusters and are therefore at the same distance from Earth, they show a wide variation in intrinsic brightness or luminosity.

What is the predominant color of these cluster stars? (A few of the stars are reddish in color. These are nearby stars that are not members of the clusters.) The observed color is an indication of the surface temperature of the stars, the blue stars being much hotter than the red stars.

One more thing can be inferred from these observations. Because most stars are blue and therefore hot, they are burning their nuclear fuel at a high rate and cannot continue to do so for long before depleting this fuel. Thus, these stars are young in age. Because they were born together, these clusters have only existed for a short time, astronomically speaking.

The masses of stars are determined by their motion in binary systems. You can learn about these types of stars by opening **Guided Tours > Star Clusters** and clicking on the **Binary Stars** thumbnail. There is also a relationship between the mass of most stars and their luminosities. The higher the mass of a star, the faster it burns its fuel and the brighter it becomes. Thus, the brighter stars in the Perseus clusters are the most massive and they will burn out more quickly and have shorter lives than lower mass stars.

The sizes or diameters of some stars can be measured in binary star systems where we see one star periodically move in front of its companion. These are eclipsing binary stars.

The Hubble Space Telescope has enhanced our knowledge of stars in clusters. Open **Explore > Hubble Images** and locate the thumbnail for **NGC 6397**. Use the **Image Fader** to alternate between the Hubble image and the Digitized Sky Survey image taken from the Earth's surface to show how the improvement in image quality has led to the detection of many faint stars not visible in Earth-based images. Click on the **Too Close for Comfort** thumbnail to see the center of this cluster and note the different brightnesses and colors of stars in this region of the cluster.

The high-resolution imagery that has resulted from the Hubble telescope has revealed many white dwarf stars that otherwise were too faint to detect. Click on the **Faint White Dwarf** thumbnail to see one of these stars. These very hot but very small stars are at a late stage in their evolution and are important in stellar evolution studies.

With this short survey, you have seen how astronomers have been able to collect and organize observational data for stars, which they can use for comparison with the predictions from theoretical modeling.

ASTRONOMY

Red Star Rising

SMALL, COOL STARS MAY BE HOT SPOTS FOR LIFE   BY MARK ALPERT

Mark Alpert, "Red Star Rising," *Scientific American*, November 2005, 28.

As every comic-book fan knows, Superman was born on the planet Krypton, which orbited a red star. Scientists are now learning that the Superman legend may contain a kernel of truth: the best places to find life in our galaxy could be on planets that circle the small but common stars known as red dwarfs.

Last June astronomers reported the discovery of Krypton's real-life counterpart, a planet orbiting a red dwarf called Gliese 876, about 15 light-years from our sun. Although researchers had previously identified two other planets orbiting Gliese 876, the detection of the third body made headlines because it is so much like Earth. Nearly all the planets found outside our solar system to date are gas giants comparable in size to Jupiter or Neptune. The newly discovered world, in contrast, is most likely a rocky body only about twice as large as Earth.

ROCKY PLANET **found near red dwarf star Gliese 876 is twice Earth's size** (*artist's rendering*).

Red dwarfs, also known as M dwarfs, have less than half the mass of the sun and are hundreds of times dimmer. Scientists had long doubted that life could arise near such faint stars because the "Goldilocks zone"— the area around the star that is neither too hot nor too cold for liquid water—would extend no more than about 40 million kilometers from the M dwarf, less than the distance between the sun and Mercury. At such close quarters, the planet could become tidally locked: one side would always face the star while the other would be turned away. Researchers had previously assumed that the planet would inevitably lose its atmosphere because of freezing on the bitterly cold night side, but computer models have shown that an atmosphere containing modest amounts of carbon dioxide would spread enough heat to the night side to prevent freezing.

M dwarfs are by far the most abundant stars in the Milky Way galaxy, outnumbering sunlike G stars more than 10 to one, so the possibility that they could harbor habitable worlds has excited researchers involved in the search for extraterrestrial intelligence (SETI). In July the SETI Institute in Mountain View, Calif., held a workshop in which scientists debated whether life could emerge on an M-dwarf planet. "No one found any showstoppers to habitability," says Gibor Basri of the University of California, Berkeley. One concern was that because M dwarfs frequently produce flares, the resulting torrents of charged particles could strip the atmosphere off any nearby planet. If the planet had a magnetic field, though, it would deflect the particles from the atmosphere. And even the slow rotation of a tidally locked M-dwarf planet—it spins once for every time it orbits its star—would be enough to generate a magnetic field as long as part of the planet's interior remained molten.

Unfortunately, the new planet found near Gliese 876 is only three million kilometers from the star, so its surface would be far too hot for life as we know it. But the sheer number of M stars in the Milky Way—an estimated 300 billion—has made them a prime target for SETI investigators, who are now building the Allen Telescope Array in northern California to scan the galaxy for radio signals. Besides their abundance, M dwarfs have another advantage over G stars as possible sites for intelligent life: because they shine longer before exhausting their hydrogen fuel and do not brighten with age, they provide a more stable long-term environment for life-bearing planets. One billion years from now, intensifying solar radiation will make Earth uninhabitable, but the galaxy's M dwarfs will burn steadily for hundreds of billions of years. It's enough to make Superman homesick.

COUNTING OUR DIM NEIGHBORS

Before scientists can look for habitable planets or radio signals near red dwarfs, they must overcome the challenge of finding the stars. Because red dwarfs are so faint—even the closest ones cannot be seen with the naked eye—they are seriously underrepresented in the astronomical catalogues. To correct this deficiency, researchers are measuring the parallax of faint red stars to estimate their distance from Earth, which helps to determine whether they are nearby red dwarfs or more luminous red giants that only appear dim because they are farther away. In recent years a team led by Todd J. Henry of Georgia State University has used this method to add about two dozen red dwarfs to the list of stars within 10 parsecs (32.6 light-years) of our solar system.

The Lives of Stars from Birth through Middle Age

Cosmic Mountains of Creation in the Star-Forming Region W5 in the Constellation Cassiopeia *(Lori Allen of Harvard-Smithsonian CfA, et al., JPL-Caltech, NASA)*

R I V U X G

WHAT DO YOU THINK?

1 How do stars form?

2 Are stars still forming today? If so, where?

3 Do more massive stars shine longer than less massive ones? How is this reasoned?

Answers to these questions appear in the text beside the corresponding numbers in the margins and at the end of the chapter.

Gravity provides the energy that enables stars to shine. As we saw in studying the Sun (Chapter 7), gravity compresses stars, thereby heating them and causing fusion to occur. As a result, stars emit huge amounts of radiation. Over time, their chemical compositions, masses, and brightnesses all vary—they age. Major stages in the life of each star can last from days to hundreds of billions of years. Cumulative lifetimes of stars range from millions to hundreds of billions of years. We will begin exploring their evolution in this chapter.

INSIGHT INTO SCIENCE

Beware Poetic License All too often, the names and descriptions we use for nonliving objects seem to give them human qualities. Anthropomorphism—assigning human attributes to nonhuman creatures or even to nonliving objects, such as stars—is descriptive but not scientific. In fact, scientists avoid this practice where possible, because it creates unjustified expectations. Nevertheless, anthropomorphism does creep into astronomy. For lack of better terms, astronomers describe stellar change with words related to living things, such as *evolution, birth, aging, maturing, growing old,* and *dying,* when talking about stars. This is poetic license of a sort: It piques our interest but muddles our science. As you study stellar evolution, always be mindful of that.

In this chapter you will discover

• how stars form

• what a stellar "nursery" looks like

• how astronomers use the physical properties of stars to learn about stellar evolution

• the remarkable transformations of older stars into giants

• how the Hertzsprung-Russell (H-R) diagram is your guide to the stellar life cycle

• how pairs of orbiting stars change each other

PROTOSTARS AND PRE–MAIN-SEQUENCE STARS

We saw in Chapter 2 that the solar system is thought to have formed from a collapsing, rotating cloud of gas and dust some 4.6 billion years ago. This scientific theory requires that interstellar matter existed before then so that the solar system had something from which to form. Taking this reasoning one step further, if such interstellar matter exists today, perhaps star formation continues.

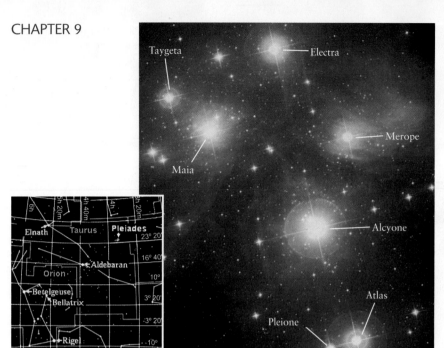

R I V U X G

a

R I V U X G   b

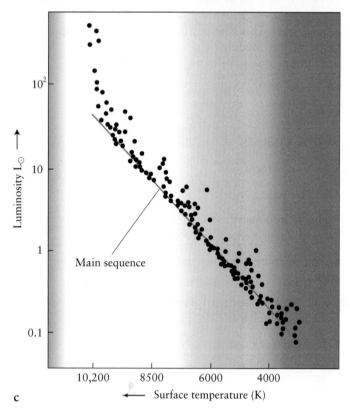

c

FIGURE 9-1 **Stars and the Interstellar Medium** (a) This open cluster, called the Pleiades, can easily be seen with the naked eye in the constellation Taurus (the Bull). It lies about 375 light-years (116 pc) from Earth. The stars are not shedding mass, unlike the star in Figures 9-12a and 9-20. The blue glow surrounding the stars of the Pleiades is a reflection nebula created as some of the stars' radiation scatters off preexisting dust grains in their vicinity. (b) The same region of the sky in a false-color infrared image taken by the Spitzer Space Telescope. Gases are seen here to exist in more areas than can be detected in visible light. (c) Each dot plotted on this H-R diagram represents a star in the Pleiades, whose luminosity and surface temperature have been determined. Note that most of the cool, low-mass stars have arrived at the main sequence, indicating that hydrogen fusion has begun in their cores. The cluster has a diameter of about 5 light-years, is about 100 million years old, and contains about 500 stars. *(a: Anglo-Australian Observatory; b: NASA/JPL-Caltech/J. Stauffer, SSC/Caltech)*

9-1   Gas and dust exist between the stars

Matter does indeed exist between the stars. We can see a relatively small amount of it through visible-light telescopes (Figure 9-1a); however, the bulk of it is too cold to be seen optically and requires the use of infrared (Figure 9-1b) and radio telescopes. We call the material between stars the **interstellar medium,** and it contains at least 10% of all the observed mass in our Galaxy. Observations of the spectra of the interstellar medium reveal that it is composed of gas containing isolated atoms and molecules and tiny pieces of dust. Table 9-1 summarizes the composition of the interstellar medium.

A variety of molecules is found in interstellar space, including molecular hydrogen (H_2), carbon monoxide (CO), carbon dioxide (CO_2), water (H_2O), ammonia (NH_3), formaldehyde (H_2CO), and simple sugars, among many others. We can identify these molecules by their unique spectral emissions, just as we can identify elements from atomic spectra.

The dust found in space also has a variety of structures. Some carbon-based (that is, organic) particles out there are typically 0.005 micrometers (μm) across (10,000 times smaller than the diameter of a typical human hair). Among these are collections of molecules similar to those found in engine exhaust and burnt meat called *polycyclic aromatic hydrocarbons* (PAHs), which are composed of only carbon and hydrogen atoms. Much larger pieces of space dust have cores of carbon or silicon compounds surrounded with mantles of ice and other materials. These grow to more than

TABLE 9-1	Composition of the Interstellar Medium	
	Particle number	
	(%)	**Mass (%)**
Hydrogen (atoms and molecules)	90	74
Helium	9	25
Metals*	1	1

*Metals are all elements except hydrogen and helium.

0.30 μm in diameter (nearly 200 times smaller than the diameter of a human hair).

Working with the knowledge that new stars form from the gas and dust of the interstellar medium, astronomers map this matter to identify places to look for young stars. When it can be seen in the visible part of the spectrum, the gas and dust in the interstellar medium glow as a result of scattered light from stars in its vicinity. For example, the interstellar medium is dramatically highlighted by stars in the Pleiades star cluster (Figure 9-1a) located in the constellation Taurus. The bluish haze, starlight scattered by the interstellar gas and dust, is called a **reflection nebula**. A **nebula** (plural: nebulae) is a dense region of interstellar gas and dust. Nebulae are often embedded in much larger bodies of gas and dust, called **molecular clouds.** The largest molecular clouds, called **giant molecular clouds,** contain upward of a million solar masses of matter and extend up to about 300 light-years across.

Astronomers identify several other types of nebulae. Most important among these are dark and emission nebu-

lae. **Dark nebulae** are sufficiently dense regions of interstellar gas and dust that they prevent most of the visible light from behind them getting to us. They look like regions of empty space, whereas they are actually among the densest of interstellar nebulae. **Emission nebulae** are regions of interstellar gas and dust that glow from energy they receive from nearby stars, from exploding stars, and from collisions between nebulae.

Finding more distant gas and dust is difficult because the nearby interstellar medium dims, or even blocks, light from behind it, in just the same way that thick clouds or several thin cloud layers in our sky can prevent us from seeing the Sun or stars behind them. This darkening of light by intervening gas and dust in space is called **interstellar extinction.** Dark nebulae are extreme examples of interstellar extinction.

Even when we can see a star or other object at visible wavelengths through the interstellar medium, it appears redder than it actually is, a phenomenon called **interstellar reddening.** This effect occurs because short-wavelength starlight is scattered by dust grains more than is the long-wavelength light. (Violet is scattered most, followed by blue, green, yellow, and orange, with red scattered least.) Therefore, when we observe a star through gas and dust, we are seeing less short-wavelength light from it than we would if the interstellar medium were not there (Figure 9-2a). The more gas and dust between the objects and us, the redder the objects appear

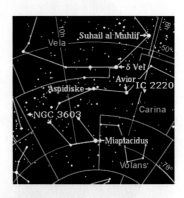

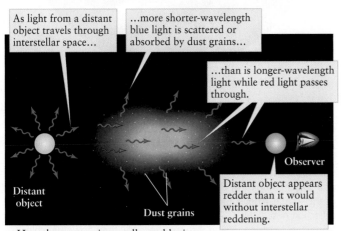

As light from a distant object travels through interstellar space...

...more shorter-wavelength blue light is scattered or absorbed by dust grains...

...than is longer-wavelength light while red light passes through.

Distant object

Dust grains

Observer

Distant object appears redder than it would without interstellar reddening.

a How dust causes interstellar reddening

NGC 3576: A closer nebula (9,000 ly from Earth)

NGC 3603: A more distant nebula (20,000 ly from Earth)

b Reddening depends on distance   R I V U X G

FIGURE 9-2 **Interstellar Reddening** (a) Dust in interstellar space scatters more short-wavelength (blue) light passing through it than longer-wavelength colors. Therefore, stars and other objects seen through interstellar clouds appear redder than they would otherwise. (b) Light from these two nebulae pass through different amounts of interstellar dust and therefore they appear to have different colors. Because NGC 3603 is farther away, it appears a ruddier shade of red than does NGC 3576. *(Anglo-Australian Observatory)*

(Figure 9-2b). Interstellar reddening is different from reddening due to the Doppler shift (see Section 3-2). The Doppler shift causes all wavelengths of electromagnetic radiation to lengthen equally, whereas interstellar reddening, due to the stronger scattering of shorter wavelengths, does not change the wavelengths of the starlight we receive—only their intensities.

We have been able to discover distant stars and clouds of interstellar gas and dust whose visible light is obscured by the nearby interstellar medium because the distant objects emit radio or infrared photons that are scattered relatively little on their way to us. Likewise, we use radio and infrared telescopes to find nearby interstellar gas and dust that are too cold to emit much visible light.

Stars form from gas and dust that are sufficiently cool so that they can collapse together. Most of this matter is hydrogen in the form of molecules (rather than atoms).

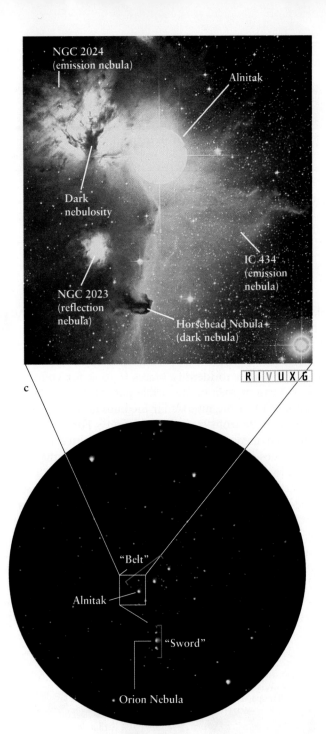

c

RIVUXG

Radio image

a RIVUXG

b Visible image RIVUXG

FIGURE 9-3 **A Gas-Rich and Dust-Rich Region of Orion** (a) This color-coded radio map of a large section of the sky shows the extent of giant molecular clouds in Orion and Monoceros as seen in the radio part of the spectrum. The intensity of carbon monoxide (CO) emission is displayed by colors in the order of the rainbow, from violet for the weakest to red for the strongest. Black indicates no detectable emission. The locations of four prominent star-forming nebulae are indicated on the star chart overlay. Note that the Orion and Horsehead nebulae are sites of intense CO emission, indicating that stars are forming in these regions. (b, c) A variety of nebulae appear in the sky around Alnitak, also called ζ (zeta) Orionis, the easternmost star in the belt of Orion. To the left of Alnitak is a bright, red emission nebula, called NGC 2024. The glowing gases in emission nebulae are excited by

ultraviolet radiation from young, massive stars. Dust grains obscure part of NGC 2024, giving the appearance of black streaks, while the distinctively shaped dust cloud, called the Horsehead Nebula, blocks the light from the background nebula IC 434. The Horsehead is part of a larger complex of dark interstellar matter, seen in the lower left of this image. Above and to the left of the Horsehead Nebula is the reflection nebula NGC 2023, whose dust grains scatter blue light from stars between us and it more effectively than any other color. All of this nebulosity lies about 1600 light-years from Earth, while the star Alnitak is only 815 light-years away from us. NGC refers to the New General Catalog of stars and IC stands for Index Catalogs, two supplements to the NGC. *(a: R. Maddalena, M. Morris, J. Moscowitz, and P. Thaddeus; b: Royal Observatory, Edinburgh; c: R. C. Mitchell, Central Washington University)*

However, molecular hydrogen is relatively hard to detect in space. Therefore, radio astronomers often search instead for carbon monoxide, which emits lots of photons at short radio (microwave) wavelengths of 2.6 and 1.3 mm. Calculations based on the known abundances of elements reveal that there are about 10,000 hydrogen molecules (H_2) for every CO molecule in the interstellar medium. Consequently, wherever astronomers detect strong emission of CO, they deduce that an enormous amount of hydrogen gas must also be present.

In mapping the locations of CO emission (Figure 9-3a), astronomers came to realize that interstellar gas and dust are often concentrated in giant molecular clouds. In some cases, these clouds appear as dark areas silhouetted against a glowing background light, such as Orion's famous Horsehead Nebula (Figure 9-3c). In other cases, the clouds appear as dark nebulae that obscure the background stars (Figure 9-4). Some 6000 of these molecular clouds are estimated to exist in our Milky Way Galaxy and have masses that range from 10^5 to 2×10^6 $M_\odot$ and diameters that range from 50 to 300 light-years. The density inside each of these clouds ranges from 10^2 to 10^5 hydrogen molecules per cubic centimeter—thousands of times greater than the average density of the gas and dust dispersed throughout interstellar space, but some 10^{15} times less dense than the air we breathe. Having located interstellar matter, astronomers turned to explaining why some of this gas and dust collapses to form new stars (and planets).

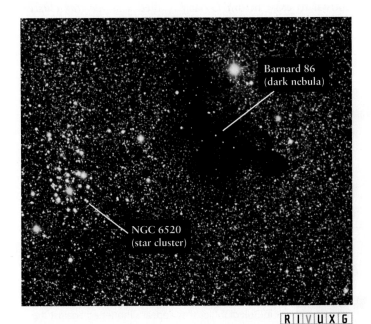

FIGURE 9-4 A Dark Nebula The dark nebula Barnard 86 is located in Sagittarius. It is visible in this photograph simply because it blocks out light from the stars beyond it. The bluish stars to the left of the dark nebula are members of a star cluster called NGC 6520. *(Anglo-Australian Observatory)*

9-2 Supernovae, collisions of interstellar clouds, and starlight trigger new star formation

We will see in detail in Chapter 10 that a supernova is a violent detonation that ends the life cycle of a massive star. The core of the doomed star collapses in a matter of seconds, releasing vast quantities of particles and energy that blow the star apart. The star's outer layers are blasted into space at speeds of several thousand kilometers per second.

The interstellar medium is composed primarily of what kinds of things?

Astronomers have found the ashes (more properly, the gas and dust) of many such dead stars scattered across the sky. These **supernova remnants** are another type of nebula. Supernova remnants, such as the Cygnus Loop shown in Figure 9-5, have a distinctly arched appearance, as would be expected for a shell of gas expanding at supersonic speeds. As it passes through the surrounding interstellar medium, the supernova remnant slams into preexisting matter, exciting the electrons in the atoms and molecules there, causing the gases to glow. If the expanding shell of a supernova remnant rams into a giant molecular cloud, it can compress the cloud, thus stimulating star birth in it. As we learned in Chapter 2, there is evidence that such an event happened around the time the solar system formed.

A simple collision between two interstellar clouds can also create regions sufficiently dense and cool to collapse and form new stars. Likewise, radiation from O and B stars, which are especially bright and hot, will ionize (remove electrons from their atomic orbits) the gas that surrounds them, which then moves away and compresses the nearby interstellar medium (Figure 9-6).

Once a giant molecular cloud contracts and cools enough, gravitational attraction causes small regions of gas and dust in it to collapse. As these regions become denser, they block light from behind and inside them, becoming dark **Bok globules** (very small dark nebulae; see Figure 9-6), named after astronomer Bart Bok, who studied them. Infrared observations show compact regions of gas and dust, called **dense cores**, inside Bok globules. These cores are destined to become stars.

To collapse, dark cores must be cool, because the hotter a gas, the more its particles collide with one another. The collisions between atoms and molecules create pressure in the cloud. If the temperature is too high, this pressure overcomes the gravitational attraction between the particles, preventing them from drawing close enough together to form stars.

Dense cores have temperatures of around 10 K, which is sufficiently cold so that their own gravitational attraction overwhelms the pressure they feel and thus pulls the gas together to form new stars. The conditions of temperature and density in the gas that ensure this collapse are called a

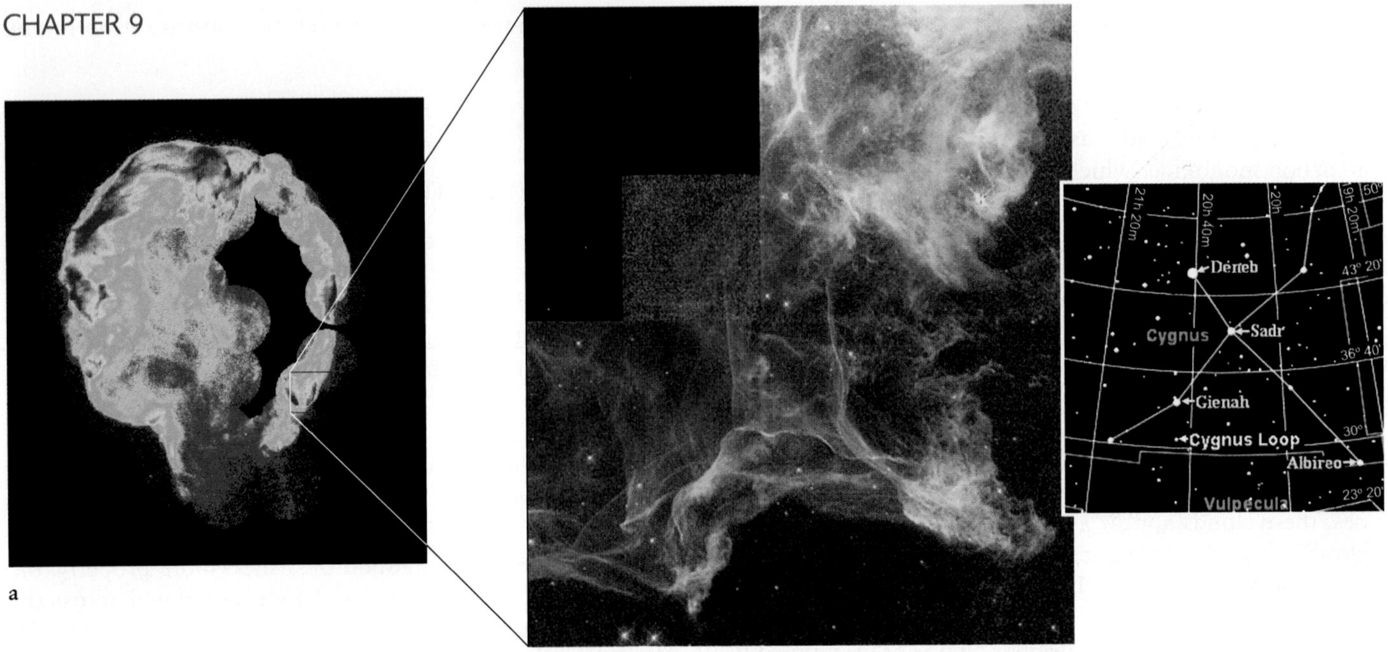

R I V U X G

b

R I V U X G

R I V U X G

FIGURE 9-5 **A Supernova Remnant** (a) X-ray image of the Cygnus Loop, the remnant of a supernova that occurred nearly 20,000 years ago. The expanding spherical shell of gas now has a diameter of about 120 light-years. The entire Cygnus Loop has an angular diameter in our sky 6 times wider than the Moon. (b) This visible-light Hubble Space Telescope image of part of the Cygnus Loop shows emission from different atoms false-color coded with blue from oxygen, red from sulfur, and green from hydrogen. *(a: Nancy Levenson/NASA; b: Jeff Hester, Arizona State University and NASA)*

Jeans instability, after the British physicist Sir James Jeans, who calculated them in 1902.

Often a giant molecular cloud has several hundred or even thousands of dense cores. In that case, hundreds or thousands of stars form together. Such stellar nurseries will become **open clusters** of stars like the Pleiades (see Figure 9-1). Open clusters are gravitationally unbound systems, meaning that the stars in them eventually drift apart. It is likely that the solar system formed in an open cluster. In 2004, astronomers detected an open cluster in our Galaxy that contains roughly half a million stars in a volume about 6 light-years across. It includes at least 200 O and B stars. This cluster, Westerlund 1, is the most massive open cluster known to exist in our Galaxy. Called a *super star cluster,* such systems have been observed in distant galaxies, but never before in the Milky Way.

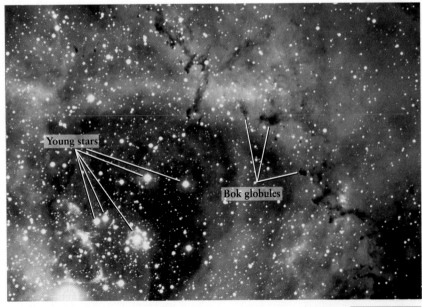

R I V U X G

FIGURE 9-6 **The Core of the Rosette Nebula** The large, circular Rosette Nebula (NGC 2237) is near one end of a sprawling giant molecular cloud in the constellation Monoceros (the Unicorn). Radiation from young, hot stars has blown gas away from the center of this nebula. Some of this gas has become clumped in Bok globules that appear silhouetted against the glowing background gases. New star formation is taking place within these globules. The entire Rosette Nebula has an angular diameter on the sky nearly 3 times that of the Moon, and it lies some 3000 light-years from Earth. *(Anglo-Australian Observatory)*

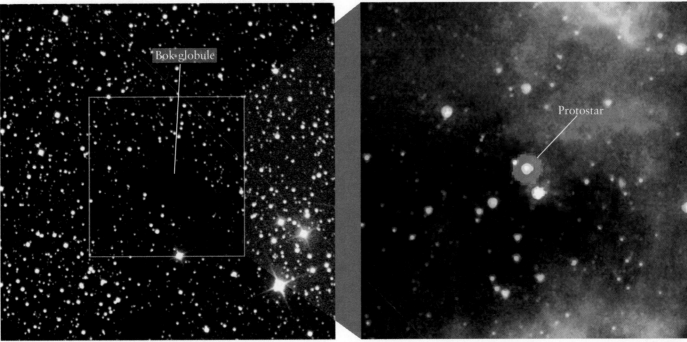

a A dark nebula R I V U X G

b A hidden protostar within the dark nebula R I V U X G

FIGURE 9-7 **Protostar in a Bok Globule** (a) This visible-light image shows a small dark nebula (equivalently, Bok globule) called L1014 located in the constellation Cygnus. (b) When viewed in the infrared, a protostar is visible within the nebula. *(a: Deep Sky Survey; b: NASA/JPL-Caltech/N. Evans, University of Texas at Austin)*

At first, a collapsing dense core is just a cool, dusty region thousands of times larger than our solar system. The dense core actually collapses from the inside out. The inner region falls in rapidly, leaving the outer layers of the dense core to drift in at a more leisurely rate. This process of increasing mass in the central region is called *accretion,* and the newly forming object at the center is called a **protostar** (Figure 9-7). Although fusion has not begun, a protostar emits energy, some of which comes from the compression and heating of its interior caused by the gravitational force from its growing mass of hot gas. However, most of the energy it releases comes from infalling gases colliding with the surface of the protostar. Figure 9-8 is an infrared image showing the locations of myriad protostars in and around a nebula in Centaurus.

What types of events can initiate the process of star formation?

If a dense core is not spinning, it collapses into a sphere, which, ultimately, becomes an isolated star. If it is spinning, it collapses into a disk, which may then condense into two or three stars. Or, if the disk has a low enough mass, it may become a single star with orbiting protoplanets, as we discussed in Chapter 2.

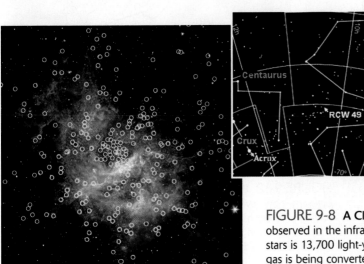

R I V U X G

FIGURE 9-8 **A Cluster of Protostars** More than 300 protostars (yellow circles) were observed in the infrared by the Spitzer Space Telescope. This cluster of newly forming stars is 13,700 light-years away in the constellation Centaurus. The nebula, some of whose gas is being converted into stars, is called RCW 49 and contains more than 2200 stars and protostars. Most of the interior of this nebula is hidden from our eyes by the dust it contains. *(NASA/JPL-Caltech/E. Churchwell, University of Wisconsin)*

9-3   When a protostar ceases to accumulate mass, it becomes a pre–main-sequence star

Protostars are physically larger than the main-sequence stars into which they are evolving. A protostar of 1 $M_\odot$, for example, has about 5 times the diameter of the Sun. Because of their large sizes, protostars emit great quantities of radiation and gas (analogous to the solar wind), and they can be observed as sources of infrared radiation. At this point, they cannot be seen in visible light because they are enshrouded by their outer layer of gas and dust (see Figure 9-7). Much matter is still slowly falling inward from the dense core's outer shell. Eventually, the radiation and particles that flow off the protostar exert enough outward force to halt the infall of this gas and dust. As a result, mass accretion stops, and the protostar becomes a **pre–main-sequence star.**

A pre–main-sequence star contracts slowly, unlike the rapid collapse of a protostar. When the temperature at its core reaches 10^7 K, hydrogen fusion begins. As we saw in Chapter 7, this thermonuclear process releases enormous amounts of energy. The outpouring of energy from hydrogen fusion creates enough pressure inside the pre–main-sequence star to stop its contraction. In the final stages of pre–main-sequence evolution, the outer shell of gas and dust finally dissipates (Figure 9-9; see also Figure 9-6). For the first time, the star is revealed via visible light to the outside universe.

9-4   The evolutionary track of a pre–main-sequence star depends on its mass

The more massive a pre–main-sequence star is, the more rapidly it begins hydrogen fusion in its core. For example, calculations indicate that a 5-$M_\odot$ pre–main-sequence star starts fusing less than a million years after it first forms from a protostar, whereas a 1-$M_\odot$ pre–main-sequence star takes a few tens of millions of years to begin fusion.

Astrophysicists use computers and the equations of stellar structure (described in Section 7-8) to model the evolution of a pre–main-sequence star. By calculating changes in the energy that the contracting star emits, computer simulations can follow its changing position on a Hertzsprung-Russell (H-R) diagram (Figure 9-10). Keep in mind that such an **evolutionary track** represents changes in a star's temperature and luminosity, not its motion in space.

R I V U X G

FIGURE 9-9 Pre–Main-Sequence Stars Seen in infrared, the two large bright objects in the center of this image are pre–main-sequence stars. They have recently shed their cocoons of gas and dust, but still have strong stellar winds that create their irregular shapes. The two stars are an optical double, that is, they are not orbiting each other. *(Atlas image courtesy of 2MASS/UMASS/IPAC-Caltech/NASA/NSF)*

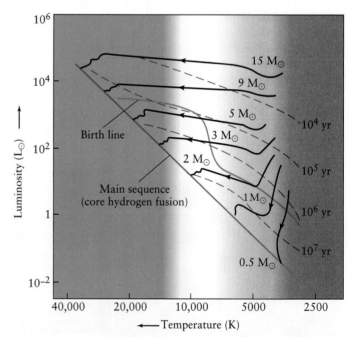

FIGURE 9-10 Pre–Main-Sequence Evolutionary Tracks This H-R diagram shows evolutionary tracks based on models of seven stars having different masses. The dashed lines indicate the stage reached after the indicated number of years of evolution. The birth line, shown in blue, is the location where each protostar stops accreting matter and becomes a pre–main-sequence star. Note that all tracks terminate on the main sequence at points that agree with the mass–luminosity relation (see Figure 8-14a).

Protostars transform into pre–main-sequence stars as they cross a curve called the **birth line** (see the blue line on Figure 9-10). A star's exact location on this curve depends primarily on its mass and to a much smaller extent, on the amount of metal it contains. (Recall from Section 8-6 that all elements other than hydrogen and helium are considered metals by astronomers.)

Spectroscopic observations of pre–main-sequence stars show that many are vigorously ejecting gas just before they reach the main sequence. Gas-ejecting stars in spectral classes G and cooler (that is, G, K, and M) are called **T Tauri stars,** after the first example discovered in the constellation of Taurus. Some astronomers propose that the onset of hydrogen fusion is preceded by vigorous chromospheric activity marked by enormous spicules (see Section 7-2) and flares (see Section 7-6) that propel the star's outermost layers back into space. In fact, an infant star going through its T Tauri stage can lose as much as 0.4 $M_\odot$ of matter and also shed its cocoon while still a pre–main-sequence star. In light of all this activity, it is not surprising that observations reveal T Tauri stars to be variable, meaning that they change brightness much more than, say, the Sun does today. We will discuss variable stars further in Sections 9-12 and 9-13.

Why can't protostars be observed with visible-light telescopes?

Pre–main-sequence stars more massive than 2 $M_\odot$ become hotter without much change in overall luminosity. The evolutionary tracks of these pre–main-sequence stars thus traverse the H-R diagram nearly horizontally, from right to left. A star more massive than about 7 $M_\odot$ has no pre–main-sequence phase at all. Its gravitational compression is so great that it begins to fuse hydrogen in its protostellar phase.

How are T Tauri stars different from main-sequence stars, like the Sun?

Calculations reveal that the minimum temperature required to start normal hydrogen fusion (see Appendix H-4) is 10 million K. However, pre–main-sequence stars less massive than 0.08 $M_\odot$ do not have enough gravitational force compressing and heating their cores to ever get this hot. As a result, these small bodies contract to become planetlike orbs of hydrogen and helium, called **brown dwarfs** (Figure 9-11).

Until 2004, the upper limit to main-sequence stellar mass was thought to be around 120 $M_\odot$. This number was based on calculations that show that above this mass, protostars rapidly develop extremely high fusion rates in their cores, which lead to extremely high surface temperatures—temperatures so great that their outer layers are superheated and thereby expelled into interstellar space. This, in turn, decreases their masses and their temperatures. An example

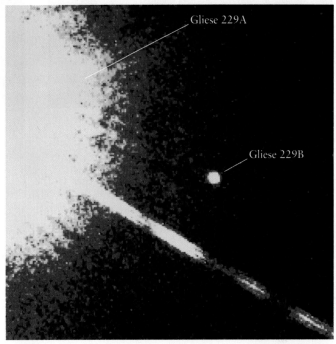

R I V U X G

FIGURE 9-11 **A Brown Dwarf** Located 18 light-years (6 pc) from Earth in the constellation Lepus (the Hare), Gliese 229B was the first confirmed brown dwarf ever observed. With a surface temperature of about 1000 K, its spectrum is similar to that of Jupiter. Gliese 229B is in orbit around a star. The overexposed image of part of its companion, Gliese 229A, appears on the left. The two bodies are separated by about 43 AU. Gliese 229B has from 20 to 50 times the mass of Jupiter, but the brown dwarf is compressed to the same size as Jupiter. The spike of light was produced when Gliese 229A overloaded part of the Hubble Space Telescope's electronics. *(S. Kularni, California Institute of Technology; D. Golimowski, Johns Hopkins University; NASA)*

of a very massive star in the process of shedding mass as it settles down onto the main sequence is the Pistol Star (Figure 9-12a). One of the most luminous stars in our Galaxy, the Pistol Star may have started as a protostar with 200 $M_\odot$. Observations reveal that every few thousand years it expels shells of gas, and it may have less than 10 $M_\odot$ left when the expulsion of matter stops. The entire process of mass loss by very massive stars takes only a few million years.

In 2004, astronomers observed a main-sequence star that apparently has between 130 and 150 $M_\odot$ (Figure 9-12b). Astronomers are trying to reconcile the theory of massive star formation with this observation.

9-5   H II regions harbor young star clusters

We can detect young open clusters of stars from the magnificent glows they create in the nebulae in which they form. Figure 9-3 and Figure 9-13 show examples of these emission nebulae. Because these nebulae are predominantly ionized hydrogen, they are also called **H II regions.** To see why

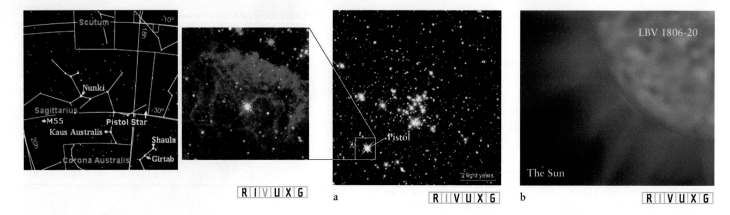

R I V U X G   a   R I V U X G   b   R I V U X G

FIGURE 9-12 Mass Loss from a Supermassive Star (a) The Quintuplet Cluster is 25,000 light-years from Earth. Inset: Within the cluster is one of the brightest known stars, called the Pistol. Astronomers calculate that the Pistol formed nearly 3 million years ago and originally had 100–200 $M_\odot$. The structure of the gas cloud suggests the star ejected the gas we see in two episodes, 6000 and 4000 years ago. The gas from any previous ejections is so thinly spread now that we cannot see it. The nebula shown in the inset is more than 4 ly (1.25 pc) across—it would stretch from the Sun nearly to the closest star, Proxima Centauri. The name "Pistol" was given to the star based on early, low-resolution radio images of its gas, which initially looked like an old-fashioned pistol aimed to the left near the top of the inset. (b) The largest, most massive known star, LBV 1806-20, is 5 million times brighter and apparently some 150 times more massive than the Sun. This drawing shows the star's color and its size compared to the Sun. *(a: D. Filger, NASA; b: Image by Dr. Stephen Eikenberry, Meghan Kennedy/University of Florida)*

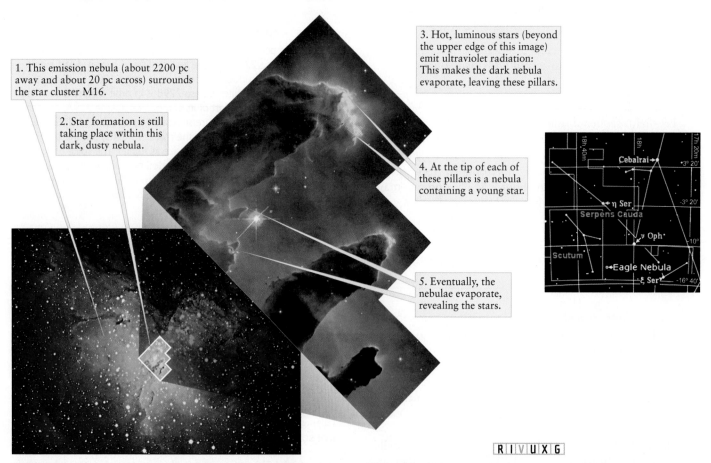

1. This emission nebula (about 2200 pc away and about 20 pc across) surrounds the star cluster M16.

2. Star formation is still taking place within this dark, dusty nebula.

3. Hot, luminous stars (beyond the upper edge of this image) emit ultraviolet radiation: This makes the dark nebula evaporate, leaving these pillars.

4. At the tip of each of these pillars is a nebula containing a young star.

5. Eventually, the nebulae evaporate, revealing the stars.

R I V U X G

FIGURE 9-13 An H II Region This emission nebula, M16, called the Eagle Nebula because of its shape, surrounds a star cluster. It is so numbered because it was the sixteenth object in the Messier Catalogue of astronomical objects. Star formation is presently occurring in M16, which is located 7000 light-years from Earth in the constellation of Serpens Cauda (the Serpent's Tail). Several bright, hot O and B stars are responsible for the ionizing radiation that causes the gases to glow. Inset: Star formation is occurring inside these dark pillars of gas and dust. Intense ultraviolet radiation from existing massive stars off to the right of this image is evaporating the dense cores in the pillars, thereby prematurely terminating star formation there. Newly revealed stars are visible at the tips of the columns. *(Anglo-Australian Observatory; J. Hester and P. Scowen, Arizona State University; NASA)*

H II regions occur, remember that the most massive pre–main-sequence stars, those of spectral types O and B, are exceptionally hot. Their surface temperatures are typically 15,000 to 35,000 K, causing them to emit vast quantities of ultraviolet radiation. This energetic radiation easily ionizes any surrounding hydrogen gas, thereby creating an H II region. Photons from an O5 star can ionize hydrogen atoms up to 500 light-years away.

H II denotes ionized hydrogen, which has no electrons in orbit; therefore, how do we observe it? While some hydrogen atoms in the H II regions are being knocked apart by ultraviolet photons, some of the free protons and electrons manage to get back together. As these new hydrogen atoms assemble, their electrons return to their ground states ($n = 1$). This downward cascade through each atom's energy levels, releasing photons with each jump between levels, is what makes the nebula glow. Particularly prominent is the transition from $n = 3$ to $n = 2$, which produces H_α photons at 656 nm in the red portion of the visible spectrum (review the emission line spectrum in Figure 3-46c, d). Thus, the nebula around a newborn star cluster often shines with a distinctive reddish hue (see Figure 9-13). Nebulae that have oxygen often look green when seen with the eye through a telescope because this gas has a green emission line at 501 nm. Because the eye is more sensitive to green light than red, the dimmer oxygen-emission line appears brighter in our brains than does the H_α line, which shows up well on CCD images, which usually appear red.

An H II region is a small, bright "hot spot" in a giant molecular cloud. The collection of hot, bright O and B stars that produces the ionizing ultraviolet radiation is called an **OB association.** The famous Orion Nebula (Figure 9-14) is an example. Four O and B stars in an open cluster called the Trapezium, at the heart of the Orion Nebula, are the primary sources of the ionizing radiation that causes the surrounding gases to glow. The Orion Nebula is embedded in a giant molecular cloud whose mass is estimated at 5.0×10^5 $M_\odot$.

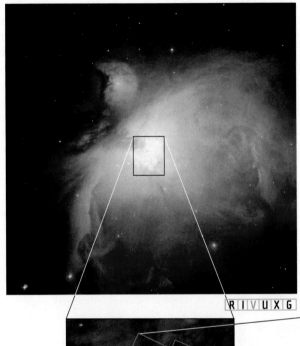

Trapezium stars

Young stars

Shock waves

FIGURE 9-14 The Orion Nebula The middle "star" in Orion's sword is actually the Orion Nebula, part of a huge system of interstellar gas and dust in which new stars are now forming. The Orion Nebula is a region visible to the naked eye. It is 1600 ly (490 pc) from Earth and has a diameter of roughly 16 ly (5 pc). This nebula's mass is about 300 $M_\odot$. Left Inset: This view at visible wavelengths shows the inner regions of the Orion Nebula. At the lower left are four massive stars, the brightest members of the Trapezium star cluster, which cause the nebula to glow. Right Inset: This view shows that infrared radiation penetrates interstellar dust that absorbs visible photons. Numerous infrared objects, many of which are stars in the early stages of formation, can be seen, along with shock waves caused by matter flowing out of protostars faster than the speed of sound waves in the nebula. Shock waves from the Trapezium stars may have helped trigger the formation of the protostars in this view. *(European Southern Observatory; left inset: C. R. O'Dell, S. K. Wong, and NASA; right inset: R. Thompson, M. Rieke, G. Schneider, S. Stolovy, E. Erickson, D. Axon, and NASA)*

The OB association that creates an H II region also affects the rest of the giant molecular cloud in which it is imbedded (see Figure 9-14 insets). Detailed models indicate that vigorous stellar winds, along with ionizing ultraviolet radiation from these stars, carve out a cavity in the cloud. Where this outflow is supersonic, it creates a shock wave, such as the sonic boom created by fast-flying aircraft (see the shock waves shown in the right inset of Figure 9-14). The shock wave forms along the outer edge of the expanding H II region, compressing hydrogen gas as it passes and thereby stimulating a new round of star birth. As more O and B stars form, they power the expansion of the H II region still farther into the giant molecular cloud. Meanwhile, the older O and B stars left behind begin to disperse (Figure 9-15). In this way, an OB association "eats into" a giant molecular cloud, creating stars in its wake. The inset in Figure 9-15 shows star formation around a single O star.

What are groups of high-mass stars called?

9-6   Plotting a star cluster on an H-R diagram reveals its age

As noted in Section 9-2, stars are often observed to form in open clusters, such as seen in the nearby Orion Nebula (see Figure 9-14). Numerous other star-forming regions have been identified, and the young open clusters in them offer astronomers a rich source of information about stars in their infancy. By measuring each star's apparent magnitude, color, and distance, an astronomer can deduce its luminosity and surface temperature. The data for all the stars in the cluster can then be plotted on an H-R diagram, as shown for the Pleiades in Figure 9-1 or for the cluster NGC 2264 in Figure 9-16.

Because all of the stars in a cluster begin forming at the same time, and stars with different masses arrive on the main sequence at different times, astronomers can use the H-R diagram to determine the age of a cluster. For example, note that the hottest stars in NGC 2264 lie on the main sequence. These hot stars have surface temperatures around 20,000 K and are extremely bright and massive. Their radiation also causes the surrounding gases to glow. Most of the stars cooler than about 10,000 K have not yet arrived at the main sequence. These less massive stars, which are in the final stages of pre–main-sequence contraction, are just now beginning to ignite thermonuclear reactions at their centers.

From the H-R diagram of a young cluster we can see which are the lowest-mass stars that have already entered the main sequence. Using this information, along with theories of stellar evolution, reveals the cluster's age. The cluster NGC 2264, for example, is roughly 2 million years old. In contrast, nearly all of the stars in the Pleiades (see Figure 9-1c) have completed their pre–main-sequence stage. That

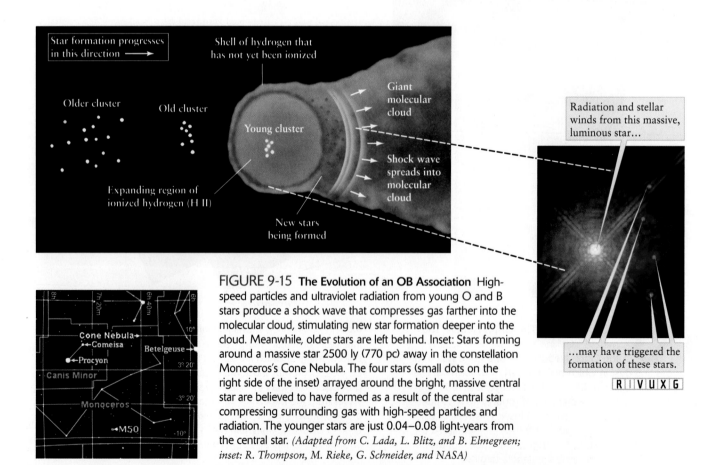

FIGURE 9-15 **The Evolution of an OB Association** High-speed particles and ultraviolet radiation from young O and B stars produce a shock wave that compresses gas farther into the molecular cloud, stimulating new star formation deeper into the cloud. Meanwhile, older stars are left behind. Inset: Stars forming around a massive star 2500 ly (770 pc) away in the constellation Monoceros's Cone Nebula. The four stars (small dots on the right side of the inset) arrayed around the bright, massive central star are believed to have formed as a result of the central star compressing surrounding gas with high-speed particles and radiation. The younger stars are just 0.04–0.08 light-years from the central star. (*Adapted from C. Lada, L. Blitz, and B. Elmegreen; inset: R. Thompson, M. Rieke, G. Schneider, and NASA*)

NGC 2264

R I V U X G

a

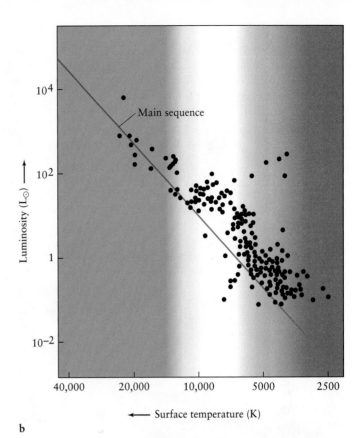

b

FIGURE 9-16 **Plotting the Ages of Stars** (a) This photograph shows a region of ionized hydrogen and the young star cluster NGC 2264 in the constellation Monoceros. The red nebulosity is located about 2600 light-years from Earth and contains numerous stars that are about to begin hydrogen fusion in their cores. (b) Each dot plotted on this H-R diagram represents a star in this cluster whose luminosity and surface temperature have been measured. Note that most of the cool, low-mass stars have not yet arrived at the main sequence. Calculations of stellar evolution indicate that this star cluster started forming about 2 million years ago. *(a: David Malin/Anglo-Australian Observatory)*

cluster's age is calculated to be about 100 million years, which is how long it takes for the least massive stars to finally begin hydrogen fusion in their cores.

As noted earlier, open clusters, such as the Pleiades and NGC 2264, possess barely enough mass to hold themselves together. A star moving faster than the average speed for the cluster occasionally escapes. This lowers the total gravitational force of the cluster, making it easier for other stars to leave. Astronomers predict that a few hundred million years after they begin forming, the stars in an open cluster have separated from each other and mixed with the rest of the stars in the Galaxy. At that time, most open clusters have completely dispersed.

Which star arrives on the main sequence first, one that is 0.5 M$_\odot$ or one that is 2 M$_\odot$?

MAIN-SEQUENCE AND GIANT STARS

In the first part of this chapter, we saw how stars form from dense cores of gas and dust inside giant molecular clouds. As these cores collapse, the gas and dust in their centers quickly form protostars. Eventually, protostars have pulled in (accreted) most of the mass available to them and they become pre–main-sequence stars that slowly contract. As a pre–main-sequence star evolves, fusion begins in its core.

Recall from Chapter 7 that in our Sun, thermal pressure created by fusion in the core pushes outward, everywhere balancing the inward force of gravity so the Sun neither expands nor collapses. The Sun is said to be in **hydrostatic equilibrium** (see Figure 7-19). When each layer of a pre–main-sequence star can finally support all of the layers above it (meaning that collapse ceases), that star also comes into hydrostatic equilibrium, and a main-sequence star is born (see Figure 9-10). *Main-sequence stars are those stars in hydrostatic equilibrium, in which nuclear reactions fuse hydrogen into helium in their cores at nearly constant rates.* Figure 9-17 summarizes the star-formation process.

9-7 Stars spend most of their lives on the main sequence

The **zero-age main sequence** (**ZAMS**) is the set of locations on the H-R diagram where pre–main-sequence stars of different masses first become stable objects, neither shrinking nor expanding. This is the solid red line that appears on several of the H-R diagrams in this chapter. Note that the theoretical evolutionary tracks in Figure 9-10 end at locations along the main sequence that agree with the observed

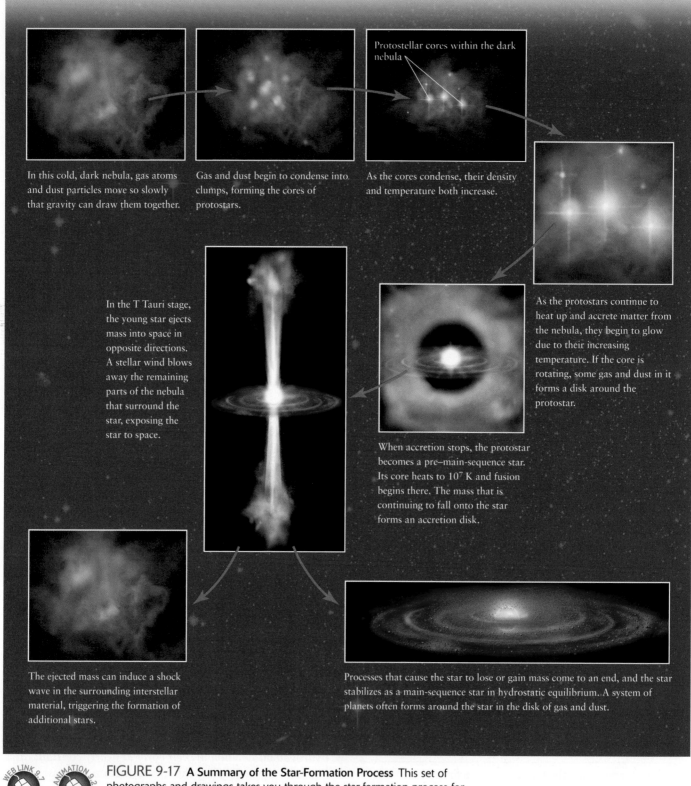

In this cold, dark nebula, gas atoms and dust particles move so slowly that gravity can draw them together.

Gas and dust begin to condense into clumps, forming the cores of protostars.

Protostellar cores within the dark nebula

As the cores condense, their density and temperature both increase.

As the protostars continue to heat up and accrete matter from the nebula, they begin to glow due to their increasing temperature. If the core is rotating, some gas and dust in it forms a disk around the protostar.

In the T Tauri stage, the young star ejects mass into space in opposite directions. A stellar wind blows away the remaining parts of the nebula that surround the star, exposing the star to space.

When accretion stops, the protostar becomes a pre–main-sequence star. Its core heats to 10^7 K and fusion begins there. The mass that is continuing to fall onto the star forms an accretion disk.

The ejected mass can induce a shock wave in the surrounding interstellar material, triggering the formation of additional stars.

Processes that cause the star to lose or gain mass come to an end, and the star stabilizes as a main-sequence star in hydrostatic equilibrium. A system of planets often forms around the star in the disk of gas and dust.

FIGURE 9-17 **A Summary of the Star-Formation Process** This set of photographs and drawings takes you through the star-formation process for stars with less than about 1.5 M$_\odot$.

mass–luminosity relation (recall Figure 8-14): The most massive main-sequence stars are the most luminous, whereas the least massive stars are the least luminous.

The energy emitted by stars is primarily generated by thermonuclear fusion (see Appendix H-4). More luminous (brighter) stars generate and radiate more energy each sec-

ond than do dimmer stars. Put another way, higher-mass stars are consuming more fuel than are lower-mass stars. This raises the interesting question of whether higher-mass stars have enough extra nuclear fuel to shine longer than lower-mass stars. The answer is *no*. The equations of nuclear fusion reveal that higher-mass stars consume fuel so

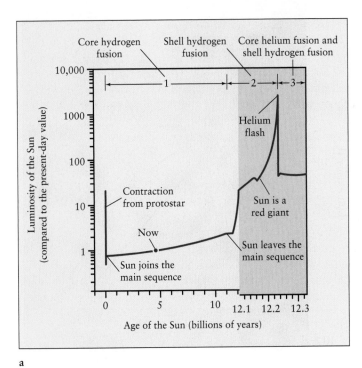

a

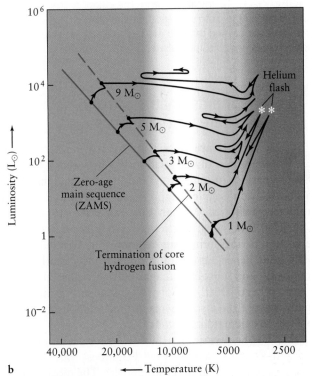

b

FIGURE 9-22 **Post–Main-Sequence Evolution** (a) The luminosity of the Sun changes as our star evolves. It began as a protostar with decreasing luminosity. On the main sequence today, it gradually brightens. Giant-phase evolution occurs more rapidly, with faster and larger changes of luminosity. Note the change in scale of the horizontal axis at 12 billion years. (b) Model-based evolutionary tracks of five stars are shown on this H-R diagram. In the high-mass stars, core helium fusion ignites smoothly where the evolutionary tracks make a sharp turn upward into the giant region of the diagram.

called alpha particles. They were later identified as helium nuclei.) The triple alpha process works in two steps. First, two helium nuclei fuse to create beryllium. Within 10^{-8} s, a third helium nucleus collides with the short-lived beryllium to create carbon. This process also releases energy and can be summarized as follows:

$$^4He + {}^4He + {}^4He \rightarrow {}^{12}C + \gamma$$

where γ denotes energy emitted as photons. The fusion in giants does not stop there. Some of the carbon created can then fuse with another helium nucleus to produce oxygen:

$$^{12}C + {}^4He \rightarrow {}^{16}O + \gamma$$

As in main-sequence stars, energy is released in the form of gamma rays. Giant stars fuse helium in their cores for about 10% as long as they spend fusing hydrogen while on the main sequence. For example, the Sun will spend a total of about 10 billion years on the main sequence, after which it will be a giant for about 1 billion years. While helium fusion is occurring in a giant's core, further gravitational contraction of the star's core ceases.

Despite the fact that all stars spend most of their lives on the main sequence, most of the stars visible to the naked eye are giants! This occurs because the relatively few giants in our neighborhood are so luminous that they outshine their neighbors in the night sky.

VARIABLE STARS

After core helium fusion begins, the evolutionary tracks of mature stars move partway back toward the main sequence (right to left on an H-R diagram; see Figure 9-22). Their pressure-temperature safety valves are not perfect, so as they shrink, they overheat, and core fusion causes them to expand to larger sizes than they would be at equilibrium. Becoming too large, their pressure and temperature decrease and they collapse, overcompressing themselves and overheating themselves again. This is analogous to dropping a tennis ball, which bounces repeatedly after striking the ground. In other words, a star under these conditions becomes unstable and pulsates. The region on the H-R diagram between the main sequence and the right side of the giant branch, where this is going on, is called the **instability strip** (Figure 9-23). These so-called **variable stars** can be easily identified by their changes in brightness amid a field of stars of constant luminosity.

Lower-mass, post–helium-flash stars pass through the lower end of the instability strip as they move horizontally on the H-R diagram along their

evolutionary tracks. These stars become **RR Lyrae variables,** named after the prototype in the constellation of Lyra (the Lyre). RR Lyrae variables all have periods shorter than 1 day. Higher-mass stars pass back and forth through the upper end of the instability strip on the H-R diagram. These stars become Cepheid variables, often simply called Cepheids.

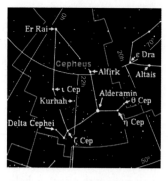

How many helium atoms does it take to make one oxygen atom?

9-12 A Cepheid pulsates because it is alternately expanding and contracting

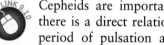 A **Cepheid variable** is characterized by the way in which its light output varies—rapid brightening followed by gradual dimming. A Cepheid variable brightens and fades because the star's outer layers cyclically expand and contract. Lines in the spectrum of δ (delta) Cephei shift back and forth with the same 5.4-day period that characterizes its variations in magnitude. According to the Doppler effect measured from it, these shifts mean that the star's surface is alternately approaching and receding from us.

When a Cepheid variable pulsates, its gases alternately heat up and cool down; the surface temperature cycles between about 6300 K and 5000 K. Thus, the characteristic light curve (luminosity versus time curve, as in Figure 8-13) of a Cepheid variable results from changes in both size and surface temperature.

Just as a bouncing ball eventually comes to rest, a pulsating star would soon stop pulsating without some mechanism to maintain its oscillations. A variable star keeps going because it receives an extra push from inside compared to stars that are stable. Variable stars have a layer of gas rich in partially ionized helium (helium with one electron stripped off). This gas, absorbing a lot of photons leaving the core, heats, expands, and pushes the outer layers outward, just as steam raises the lid of a pot. Eventually, the ionized helium layer is spread so thin that photons pass through it. The ionized helium cools and contracts, and the star's outer layers collapse, compressing the interior gases until the process repeats itself. In our analogy with the pot of boiling water, when enough of the steam inside escapes, the lid falls back and the steam inside builds up again. Variability only occurs when the conditions of temperature and pressure inside the star are appropriate. As a result, giants are variable only when the stars are in the instability strip of an H-R diagram.

How do astronomers observe that Cepheids are changing size?

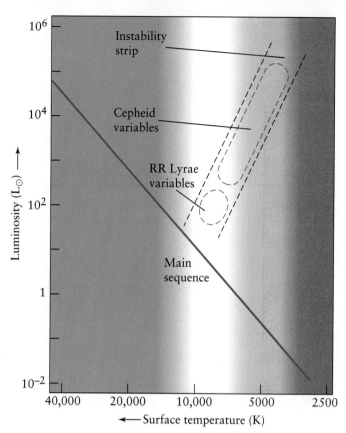

FIGURE 9-23 **The Instability Strip** The instability strip occupies a region between the main sequence and the giant branch on the H-R diagram. A star passing through this region along its evolutionary track becomes unstable and pulsates.

9-13 Cepheids enable astronomers to estimate vast distances

Cepheids are important to astronomers because there is a direct relationship between a Cepheid's period of pulsation and its average luminosity. This correlation is called, appropriately enough, the **period–luminosity relation.** We can determine the distance to a Cepheid in four steps. First, we observe its period. Second, we use the period–luminosity relation to determine its luminosity and, hence, its absolute magnitude (recall that the luminosity and absolute magnitudes are directly related to each other). Third, we observe its apparent magnitude. Finally, we use the distance–magnitude relationship (Appendix H-6) to calculate its distance.

Dim Cepheid variables pulsate rapidly, with periods of 1 to 2 days, and have average brightness variations of a few hundred times our Sun's luminosity. The most luminous Cepheids have the longest periods of all Cepheids, with variations occurring over 100 days and average brightness variations equaling 10,000 $L_\odot$. Being so bright, we can see Cepheids far beyond the boundaries of our Milky Way Galaxy. Because the changes in brightness of Cepheids can be seen at distances where other techniques for measuring distance, such as stellar parallax, fail, the period–luminosity relation plays an important role in determining the overall size and structure of the universe. We will explore this application further in Chapter 11.

16. Why do astronomers believe that globular clusters are made of old stars?

17. What are Cepheid variables, and how are they related to the instability strip?

18. What occurs in Cepheid stars that is analogous to the vapor raising the lid on a pot of boiling water?

19. What are RR Lyrae variables, and how are they related to the instability strip?

20. What is a Roche lobe, and what is its significance in close binary systems?

21. What are the differences between detached, semidetached, contact, and overcontact binaries?

Observing Projects

 22. Use the *Starry Night Enthusiast*™ program to examine a star-forming region. Select **Favourites > Deep Space > Clusters and Nebulas > Nebulas in Sagittarius** from the menu. The view is centered on the constellation Sagittarius with several prominent clusters and nebulae labeled. Time is passing at 1 sidereal day per frame. You can temporarily stop motion by left-clicking (or clicking on a Mac) on the screen, as necessary. **a.** In which month is M20, the Trifid Nebula, highest in the sky at noon? Explain how you determined this. **b.** In which month is M20 highest in the sky at midnight, so that it is best placed for observing with a telescope? Explain how you determined this. **c.** Click the right mouse button (Ctrl+Click on a Mac) with the cursor over M20, the Trifid Nebula, and select **Magnify** from the Object Contextual menu. Explain why different areas of the nebula have the colors that they have.

23. Use *WorldWide Telescope* (WWT) to explore M42, The Orion Nebula, a vast star-forming region in the Milky Way. Click the **Guided Tours** tab followed by the thumbnail of the **Nebula** folder. Then select the thumbnail labeled **Orion Nebula–Hubble's Universe** to begin the tour. You may wish to click the **Watch Again** button at the conclusion of the tour to help answer the following questions: **a.** What is the name of the group of stars that are at the core of the Orion Nebula? **b.** What are the three noteworthy properties of the stars in this group? **c.** What is the nature of the radiation that causes the gas and dust of the Orion nebula to glow and what is the source of this radiation? **d.** What are proplyds? **e.** What causes the proplyds near the center of the Orion nebula to appear tadpole-shaped?

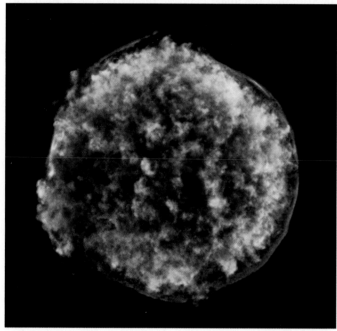

R I V U X G

Tycho's Supernova First seen as a visible light object in 1572, this supernova remnant reveals some of its secrets in an X-ray image taken in 2003 by the Chandra X-ray telescope. Multimillion-degree gas and dust (red and green) are expanding outward at about 10 million km/h following a shell of high energy elections (blue). *(NASA/CXC/Rutgers/J. Warren & J. Hughes et al.)*

WHAT DO YOU THINK?

1 Will the Sun someday cease to shine brightly? If so, how will this occur?

2 What is a nova? How does it differ from a supernova?

3 What are the origins of the carbon, silicon, oxygen, iron, uranium, and other heavy elements on Earth?

4 What are cosmic rays? Where do they come from?

5 What is a pulsar?

6 Are black holes empty holes in space? If not, what are they?

7 Does a black hole have a solid surface? If not, what is at its surface?

8 What power or force enables black holes to draw things into themselves?

9 How close to a black hole do you have to be for its special effects to be apparent?

10 Can you use black holes to travel to different places in the universe?

11 Do black holes last forever? If not, what happens to them?

Answers to these questions appear in the text beside the corresponding numbers in the margins and at the end of the chapter.

Most video and computer games have one thing in common: Players move through different levels, each progressively more difficult than the last. The biggest challenge comes at the final level, when you have to face the most powerful or evil entity. Stellar evolution is similar, in which stars pass through different stages of stellar activity before they can move on to the next stage. There is one big difference—a game player can win the final level, but a star at that stage of evolution always ceases to shine with the vigor that it had previously. In the end, stars eject vast quantities of gas and dust into interstellar space. In human terms, they die. In this chapter we learn how the later stages of stellar evolution are significantly different for stars with different masses. Some of them stop evolving by relatively mild emissions of their outer layers, while others have spectacular finales.

In this chapter you will discover

- what happens to stars when core helium fusion ceases

- how heavy elements are created

- the characteristics of the end of stellar evolution

- why some stars go out relatively gently, and others go out with a bang

- the incredible density of the matter in neutron stars and how these objects are observed

- that Einstein's theory of general relativity predicts the existence of regions of space and time that are severely distorted by the extremely dense matter they contain

- that space and time are not separate entities

- how black holes arise

- the surprisingly simple theoretical properties of black holes

- that X rays and jets of gas are created near many black holes

- the fate of black holes

- the unsurpassed energy emitted by gamma-ray bursts

LOW-MASS STARS AND PLANETARY NEBULAE

We have seen that red dwarf stars, with masses less than $0.4 M_\odot$, never get to the giant phase of stellar evolution. Rather, they convert all their mass to helium, stop fusing, and thereafter just cool off. We will now explore the fates of stars greater than $0.4 M_\odot$ as they proceed along through the giant phase and beyond.

In Chapter 9, we discovered that when hydrogen fusion in the shell surrounding the core first begins, the energy from it causes a star to expand and become a giant. During this process, low-mass stars move over to, and ascend, the giant branch on the H-R diagram for the first time (Figure 10-1a; also, review Figure 9-22b). As this happens, mass is expelled into space in the form of stellar winds, which reduce the masses of these stars. Then core helium fusion begins, with stars of less than about 2 $M_\odot$ undergoing a core helium flash in the giant stage (Figure 10-1a). Stars with more than 2 $M_\odot$ begin helium fusion more gradually. In either case, after core fusion begins again, stars shrink and move onto the *horizontal branch* (Figure 10-1b). Their cores are eventually converted into carbon and oxygen, helium fusion ceases, and these stars undergo another stage that closely parallels the end of core hydrogen fusion (Figure 10-1c). The destinies of stars depend on their masses, and we have two mass ranges to consider: 0.4 to 8 $M_\odot$ (hereafter, *low-mass* stars) and more than 8 $M_\odot$ (hereafter, *high-mass* stars).

10-1 Low-mass stars become supergiants before expanding into planetary nebulae

Calculations reveal that carbon and oxygen atoms require a temperature of at least 600 million K to fuse. Because the core of a low-mass giant on the horizontal branch only reaches about 200 million K, fusion of these elements in the core does not occur. Photon production in the core therefore decreases when its helium fuel is nearly used up, and the inner regions of the star again contract, compressing and heating the shell of helium-rich gas just outside the core. As a result, **helium shell fusion** begins outside of the core; this shell is itself surrounded by a hydrogen-fusing shell (Figure 10-2). All of this takes place within a volume roughly the size of Earth.

Once helium shell fusion commences, the new outpouring of energy again pushes the outer envelope of the star out. A low-mass star thus leaves the horizontal branch and ascends the H-R diagram for a second, and final, time. Powered by the fusion from two shells, such an **asymptotic giant branch star**, or **AGB star**, becomes brighter than ever before. When the Sun moves along the asymptotic giant branch, it will expand to a radius of about 1 AU, enveloping Earth in its tenuous outer layers (Figure 10-2). An 8 $M_\odot$ star will have a diameter as big as the orbit of Mars and shine with the luminosity of 10^4 $L_\odot$. At the top of the asymptotic giant branch, these stars are now so bright that they are classified as low-temperature, red supergiants (see Figure 10-1c).

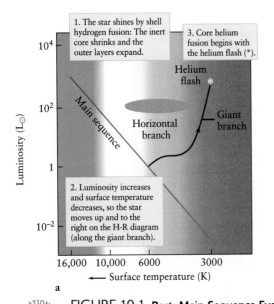

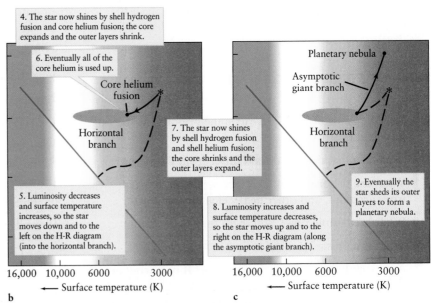

FIGURE 10-1 **Post–Main-Sequence Evolution of Low-Mass Stars** (a) A typical evolutionary track on the H-R diagram as a star makes the transition from the main sequence to the giant phase. The asterisk (*) shows the helium flash occurring in a low-mass star. (b) After the helium flash, the star converts its helium core into carbon and oxygen. While doing so, its core re-expands, decreasing shell fusion. As a result, the star's outer layers re-contract. (c) After the helium core is completely transformed into carbon and oxygen, the core recollapses, and the outer layers re-expand, powered up the asymptotic giant branch by hydrogen shell fusion and helium shell fusion.

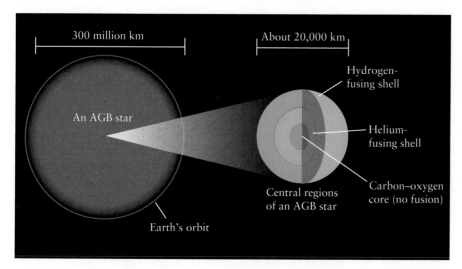

FIGURE 10-2 **The Structure of an Old Low-Mass Star** Near the end of its life, a low-mass star, like the Sun, travels up the AGB and becomes a supergiant. (The Sun will be about as large as the diameter of Earth's orbit.) The star's inert core, the hydrogen-fusing shell, and the helium-fusing shell are contained within a volume roughly the size of Earth. The inner layers are not shown to scale here.

AGB stars are destined to self-destruct. Giant stars of all spectral types emit strong stellar winds. Low-mass stars expel their outer layers more slowly than higher-mass stars, but they all reduce their masses significantly from what they were on the main sequence. During their initial ascent up the giant branch, stars lose mass. On the asymptotic giant branch they lose even more, often surrounding themselves with thickening cocoons of gas and dust. A star of the Sun's mass loses about 10^{-5} $M_\odot$ per year at this stage, eventually dumping half of its mass back into space.

This mass loss limits the force of gravity available to compress the star's core and the regions of shell fusion. A fine balance is struck: Our low-mass AGB star is compressed until its core of carbon and oxygen becomes degenerate, meaning that the electrons provide a growing repulsive force that stops its contraction (as discussed in Section 9-10). However, even for the most massive of these low-mass stars (now much less than their original 8 $M_\odot$ due to their stellar winds), the core temperature cannot reach the 600 million K necessary to fuse carbon or oxygen. Therefore, no further core fusion occurs.

The final stage through which a low-mass star passes begins with a "thermal runaway" (meaning a rapid rise in temperature) in the helium shell, just like the helium flash in its core earlier in its life (described in Section 9-10). This occurs because the triple alpha process (see Section 9-11) is extremely sensitive to temperature, and, as the temperature goes up slightly, the fusion rate skyrockets. The increase in energy output from the thin helium-fusing shell, called a **helium shell flash,** expands the star, thereby briefly decreasing the temperature in the shell and slowing the rate of fusion.

A star undergoes several helium shell flashes as its helium shell thickens. During each flash, the helium shell's energy output jumps a thousandfold. These brief outbursts are separated by relatively quiet intervals lasting about 100,000 years, during which the helium shell gradually becomes thicker. The energy from the increased fusion during the helium shell flashes causes the star's outer layers to expand and, therefore, cool further. Eventually, the outer gases are sufficiently cool so that electrons and ions in them can recombine. This is exactly the opposite of ionization (discussed in Section 3-21). Whereas ionization requires an electron to absorb a photon, recombination forces an electron to emit a photon.

Just as the collisions of particles create pressure (force acting over an area), the collisions of photons with matter also create pressure. The impacts of photons emitted during recombination, along with the photons from the helium flashes and the photons normally flowing outward from fusion, generate enough pressure to eject more and more of the star's outer layers into space. Eventually, the mass of the star decreases so much that there is not enough pressure to sustain any shell fusion. As the ejected material expands and cools, some of it condenses to form dust grains. When enough of the gas has left the star for the core to be visible, the expanding dust and gases are considered to be a **planetary nebula** (see Figure 10-1c). As the outer layers of the star are shed, an increasingly hot interior is revealed, and the star moves to the left across the H-R diagram (Figure 10-3). Including the mass lost prior to the planetary nebula phase, low-mass stars lose as much as 80% of their masses by shedding their outer layers.

INSIGHT INTO SCIENCE

Names Often Don't Change As Understanding Improves Planetary nebulae have *nothing* to do with planets. This term was coined by the astronomer William Herschel in the eighteenth century. When viewed through the telescopes of the day, these glowing objects, often green in appearance, looked like planets similar to Uranus. Astronomers have continued to use this term, but do not be fooled—we now know that planetary nebulae indicate the presence of dying stars, not planets.

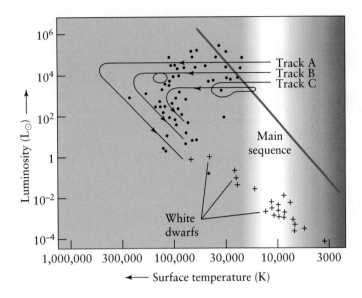

	Mass ($M_\odot$)		
Evolutionary track	Red supergiant	Ejected nebula	White dwarf
A	3.0	1.8	1.2
B	1.5	0.7	0.8
C	0.8	0.2	0.6

FIGURE 10-3 **Evolution from Supergiants to White Dwarfs** The evolutionary tracks of three low-mass supergiants are shown as they eject planetary nebulae. The table gives their masses as supergiants, the amount of mass they lose as planetary nebulae, and their remaining (white dwarf) masses. The loops indicate periods of instability and adjustment for the white dwarfs. The dots on this graph represent the central stars of planetary nebulae whose surface temperatures and luminosities have been determined. The crosses are white dwarfs for which similar data exist.

1   Planetary nebulae are quite common in our Galaxy. More than 1800 have been identified, and astronomers estimate that 20,000 to 100,000 exist in the Milky Way. Indeed, the fate of the Sun is to become a planetary nebula. The rate at which the gases are shed in a planetary nebula is so slow that they really are not explosive, compared either to our common conception of an explosion on Earth or to the supernova explosions that we will examine shortly.

Because stellar winds from red giants and subsequent planetary nebulae are so plentiful, astronomers estimate that they return a total of about 5 $M_\odot$ per year of gas and dust to the disk of our Galaxy. This mass amounts to about 85% of all matter expelled by all types of stars. This material goes into the formation of new, metal-rich Population I stars and associated planets. Thus, planetary nebulae play an important role in the chemical and physical evolution of the Galaxy.

The outflowing gases ejected in a planetary nebula take a breathtaking variety of shapes when they interact with gases surrounding their stars, with companion stars, and with the stars' magnetic fields, which are often 10 to 100 times stronger than the Sun's field (Figure 10-4 and Figure 10-5). The Hourglass Nebula (Figure 10-5c) appears initially to have shed mass in a doughnut shape around itself. In the star's final death throes, this gas and dust forced the outflow to go in two directions perpendicular to the plane of the doughnut, creating what is called a *bipolar planetary nebula*.

Spectroscopic observations of planetary nebulae show bright emission lines of hydrogen, carbon, neon, magnesium, oxygen, and nitrogen. From the Doppler shifts of these lines, we conclude that the gas and dust are moving outward from the dying low-mass star at speeds of 10 to 30 km/s. A typical planetary nebula moving outward for 10,000 years has a diameter of a few light-years.

By astronomical standards, a planetary nebula is short-lived. After about 50,000 years, its gases spread over distances so far from the cooling central star that its nebulosity simply fades from view. The gases then mingle and mix with the surrounding interstellar medium, enriching it with moderately heavy elements.

10-2   The burned-out core of a low-mass star becomes a white dwarf

The exposed carbon–oxygen cores of low-mass stars are stable objects supported by electron degeneracy pressure. Such burned-out hulks are called **white dwarfs,** and this is the fate of the Sun's core. Typical white dwarfs are roughly the size of Earth and covered with a thin coating of hydrogen and helium. The densities of these stellar "corpses" are typically 10^9 kg/m^3. In other words, a teaspoonful of white dwarf matter brought to Earth would weigh 5 tons.

What luminosity class will the Sun be in just prior to when it becomes a planetary nebula?

Theoretical evolutionary tracks of three burned-out stellar cores are shown in Figure 10-3. These particular white dwarfs evolved from main-sequence stars of between 0.8 and 3.0 $M_\odot$. During the ejection phase, the appearance of these remnants changes rapidly. They appear to race along their evolutionary tracks on the H-R diagram, sometimes executing loops called thermal pulses as they temporarily reheat (tracks B and C in Figure 10-3). Finally, as the ejected planetary nebulae fade and the stellar cores cool, the evolutionary tracks of the dying stars take a sharp turn downward toward the white dwarf region of the diagram.

As billions of years pass, an isolated white dwarf radiates its stored energy into space, cooling and decreasing in luminosity. It moves down and to the right on the H-R diagram. The Sun will become a cold, dark, dense sphere of

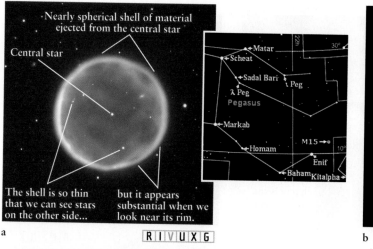

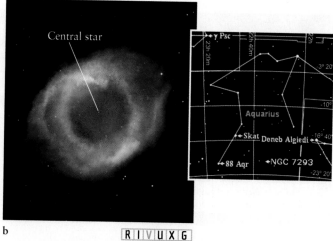

a

b

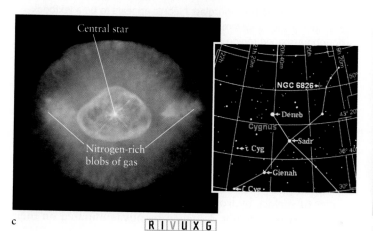

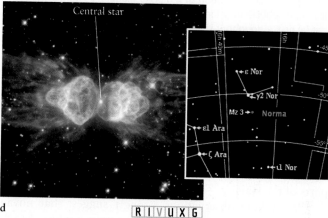

c

d

FIGURE 10-4 **Some Shapes of Planetary Nebulae**
The outer shells of dying low-mass stars are ejected in a wonderful variety of patterns. (a) An exceptionally spherical remnant, this shell of expanding gas, in the globular cluster M15 in the constellation Pegasus, is about 7000 ly (2150 pc) away from Earth. (b) The Helix Nebula, NGC 7293, located in the constellation Aquarius, about 700 ly (215 pc) from Earth, has an angular diameter equal to about half that of the full Moon. Red gas is mostly hydrogen and nitrogen, whereas the blue gas is rich in oxygen. (c) NGC 6826 shows, among other features, lobes of nitrogen-rich gas (red). The process by which they were ejected is as yet unknown. This planetary nebula is located in Cygnus. (d) MZ 3 (Menzel 3), in the constellation Norma (the Carpenter's Square), is 3000 ly (900 pc) from Earth. The dying star, creating these bubbles of gas, may be part of a binary system. *(a: WIYN/NOAL/NSF; b: Howard Bons, STScI; Robin Ciardullo, Pennsylvania State University; NASA; c: AURA/STScI/ NASA; d: R.Sahai and J. Trauser, JPL, WFPC-2 Science Team, and NASA)*

degenerate gases rich in oxygen and carbon, about the size of Earth. We can actually follow the theory a little further. It predicts that after billions of years of radiating away their heat, the interior temperatures of white dwarfs will decrease to about 4000 K, and the carbon and oxygen in them will solidify, transforming them into giant crystals.

Many white dwarfs are found in our solar neighborhood, but all are too faint to be seen with the naked eye. The first white dwarf to be discovered is a companion to the bright star Sirius. The binary nature of Sirius was first deduced in 1844 by the German astronomer Friedrich Bessel, who noticed that the star was moving back and forth, as if orbited by an unseen object. This companion, Sirius B (Figure 10-6) was first glimpsed in 1862. Recent satel-lite observations at ultraviolet wavelengths—where white dwarfs emit most of their light—demonstrate that the surface temperature of Sirius B is about 30,000 K.

Of what two elements is a white dwarf composed?

10-3   White dwarfs in close binary systems can create powerful explosions

Occasionally, a star in the sky suddenly becomes between 10^4 and 10^6 times brighter than it is normally. This phenomenon is called a **nova.** Its abrupt rise in brightness is followed by a gradual decline that may stretch over several months or

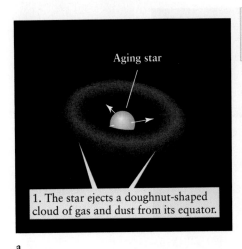

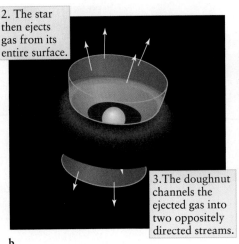

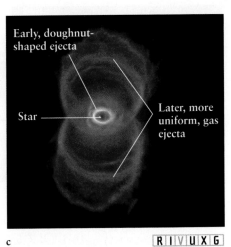

a   b   c   R I V U X G

FIGURE 10-5 **Formation of a Bipolar Planetary Nebula** Bipolar planetary nebulae may form in two steps. Astronomers hypothesize that (a) first, a doughnut-shaped cloud of gas and dust is emitted from the star's equator, (b) followed by outflow that is channeled by the original gas to squirt out perpendicularly to the plane of the doughnut. (c) The Hourglass Nebula appears to be a textbook example of such a system. The bright ring is believed to be the doughnut-shaped region of gas lit by energy from the planetary nebula. The Hourglass is located about 8000 ly (2500 pc) from Earth. *(c: R. Sahai and J. Trauer, JPL; WFPC-2 Science Team; and NASA)*

more (Figure 10-7 and Figure 10-8). Astronomers typically observe two or three novae in our Galaxy each year. Novae are fairly common, with about 20 estimated to occur in the Milky Way Galaxy annually. We do not see all of them because of the interstellar gas and dust that obscures many parts of the Galaxy.

Painstaking observations strongly suggest that novae occur in close binary systems that contain a white dwarf. The ordinary companion star fills its Roche lobe (recall Figures 9-29, 9-30, and 9-31), so it gradually deposits fresh hydrogen onto the white dwarf. This new mass becomes a dense layer covering the hot surface of the white dwarf. As more gas is deposited and compressed, the temperature in the hydrogen layer continues to increase. Finally, at about 10^7 K, hydrogen fusion ignites throughout the layer, blowing it into interstellar space. This explosion is the nova. After a nova, fusion ceases on the white dwarf. The companion star, however, may retain enough mass to supply a new layer of surface hydrogen, enabling some novae to reoccur. White dwarfs are not damaged by novae initiated on their surfaces.

Recent visible observations by the Hubble Space Telescope and X-ray observations by the Chandra X-ray Observatory reveal that novae are much more complex than astronomers had thought. Rather than smooth shells of expanding gas, they are composed of thousands of clumps

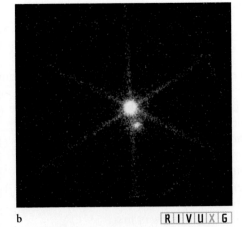

a   R I V U X G   b   R I V U X G

FIGURE 10-6 **Sirius and Its White Dwarf Companion** (a) Sirius, the brightest-appearing star in the night sky, is actually a double star. The smaller star, Sirius B, is a white dwarf, seen here at the 5 o'clock position in the glare of Sirius. The spikes and rays around the bright star, Sirius A, are created by optical effects within the telescope. (b) Since Sirius A (11,000 K) and Sirius B (30,000 K) are hot blackbodies, they are strong emitters of X rays. *(a: R. B. Minton; b: NASA/SAO/ CXC)*

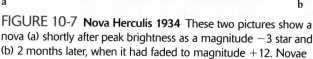

a
b

FIGURE 10-7 Nova Herculis 1934 These two pictures show a nova (a) shortly after peak brightness as a magnitude −3 star and (b) 2 months later, when it had faded to magnitude +12. Novae are named after the constellation and year in which they appear. (*UCO/Lick Observatory*)

of gas. These observations are stimulating astrophysicists to reexamine the details of the nova mechanism described above to try to explain these newly observed effects.

Theoretical models predict that degenerate matter, such as the electrons in a white dwarf, can withstand only a finite amount of pressure, above which the degenerate matter cannot exert enough outward force to prevent the core from collapsing further. This limits the mass of a white dwarf to less than 1.4 M$_\odot$. This mass is called the **Chandrasekhar limit,** after the astrophysicist Subrahmanyan Chandrasekhar, who received a Nobel prize in 1983 for his pioneering theoretical studies of white dwarfs a half-century earlier.

Why are novae not associated with isolated white dwarfs?

10-4 Accreting white dwarfs in close binary systems can also explode as Type Ia supernovae

Whereas a nova just removes hydrogen and helium that build up on the surface of a white dwarf, some of these stellar remnants actually blow apart completely. This event is called a **Type Ia supernova** that occurs with a white dwarf in a semidetached binary system. To trigger a Type Ia supernova, a swollen giant companion star dumps gas onto a white dwarf that is close to the Chandrasekhar limit. When the added gas causes the white dwarf's mass to cross that limit, the increased pressure deep inside enables carbon fusion to begin in the core. In a cata-

strophic runaway process reminiscent of the helium flash, the rate of carbon fusion skyrockets, and, with no outer layers to absorb the energy, the star blows up.

A Type Ia supernova is powered by nuclear energy. What we see is the fallout from a gigantic thermonuclear explosion, which produces a wide array of radioactive isotopes. Especially abundant is an unstable isotope of nickel that decays into a radioactive isotope of cobalt. Most of the electromagnetic display of a Type Ia supernova

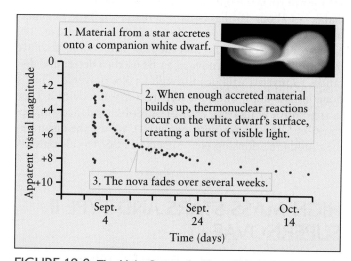

1. Material from a star accretes onto a companion white dwarf.

2. When enough accreted material builds up, thermonuclear reactions occur on the white dwarf's surface, creating a burst of visible light.

3. The nova fades over several weeks.

FIGURE 10-8 The Light Curve of a Nova This graph shows the history of Nova Cygni 1975, a nova that was observed to blaze forth in the constellation of Cygnus in September 1975. The rapid rise in magnitude followed by a gradual decline is characteristic of many novae, although some oscillate in intensity as they become dimmer.

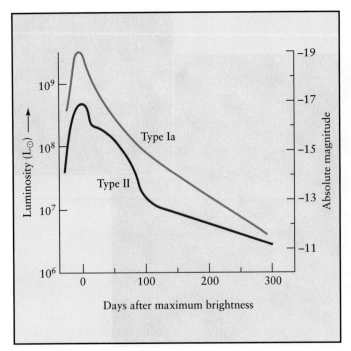

FIGURE 10-9 **Supernova Light Curves** A Type Ia supernova, which gradually declines in brightness, is caused by an exploding white dwarf in a close binary system. A Type II supernova is caused by the explosive death of a massive star and usually has alternating intervals of steep and gradual declines in brightness.

results directly from the radioactive decay of this nickel into cobalt.

A Type Ia supernova's spectrum lacks hydrogen lines because the star from which it came was virtually free of this element. The light curve of the supernova, which shows its change of brightness over time, begins with a sudden rise (Figure 10-9). It typically reaches an absolute magnitude of −19 at peak brightness and then declines gradually for more than a year.

Because these supernovae are so bright, they can be seen far beyond the bounds of the Milky Way. Whereas our Galaxy is about 10^5 light-years in diameter, Type Ia supernovae can be seen in other galaxies at distances of over 10^9 light-years away. Because they all peak at the same absolute magnitude, these supernovae can be used to determine distances to myriad galaxies throughout the universe. This is done in two steps. First, the apparent magnitude of a Type Ia supernova at its peak brightness is observed. Second, because we know its absolute magnitude, we use the distance–magnitude relationship (see Appendix H-6) to calculate its distance.

HIGH-MASS STARS AND TYPE II SUPERNOVAE

In main-sequence stars of at least 8 $M_\odot$, sufficient gravitational force compresses and heats the core enough to enable its carbon and oxygen to fuse. This leads to the creation of a host of other elements that are eventually ejected into interstellar space.

10-5   A series of fusion reactions in high-mass stars leads to luminous supergiants

When helium fusion ends in the core of a star more massive than 8 $M_\odot$, gravitational compression collapses the carbon–oxygen core, driving the star's central temperature above 600 million K. Now, carbon fusion begins, producing such elements as neon and magnesium. When a star compresses itself enough to elevate its central temperature to 1.2 billion K, neon fusion occurs, creating oxygen and magnesium. When the central temperature of the star then reaches 1.5 billion K, oxygen fusion begins, producing sulfur, silicon, and phosphorus. The process of converting lower-mass elements into higher-mass ones is called **nucleosynthesis.**

As the core's temperature increases, the core's fusion rate soars. Also, each succeeding core stage has fewer and fewer nuclei than the previous one. For example, because each helium atom is made up of four hydrogen atoms, only one-quarter as many helium atoms exist in a star's core as there were hydrogen atoms. Combining the higher fusion rate and fewer fusible atoms causes each cycle of core thermonuclear fusion to occur more rapidly than the previous stage. For example, detailed calculations for a star that was initially 25 $M_\odot$ on the main sequence reveal that carbon fusion occurs for 600 years, neon fusion for 1 year, and oxygen fusion for only 6 months. After half a year of core oxygen fusion, gravitational compression forces the central temperature up to 2.7 billion K, and silicon fusion begins. This thermonuclear process proceeds so furiously that the entire core supply of silicon in a 25-$M_\odot$ star is converted into iron in 1 day.

Why do type Ia supernovae not have any hydrogen lines in their spectra?

Each stage of fusion adds a new shell of matter outside of the core (Figure 10-10), creating a structure that resembles the layers of an enormous onion. Together, the tremendous numbers of fusion-generated photons created in all of these shells, as well as in the core, push the outer layers of the star farther and farther outward. The star expands to become a luminous supergiant, almost as wide across as Jupiter's orbit around the Sun.

High-mass main-sequence stars evolve into luminous supergiant stars that are brighter than 10^5 $L_\odot$, emit winds throughout most of their existence, and have mass-loss rates that exceed those of giants (see Section 9-9). Figure 10-11 shows a supergiant star losing mass. Betelgeuse (see Figure 8-2a), 470 light-years away in the constellation of Orion, is another good example of a supergiant experiencing mass loss. Spectroscopic observations show that Betelgeuse is losing mass at the rate of 1.7×10^{-7} $M_\odot$ per year and is surrounded by ejected gas in a *circumstellar shell.* This huge shell is expanding at 10 km/s, and escaping gases have been detected at distances of 10,000 AU from the star. The expanding circumstellar shell has an overall diameter of one-third of a light-year.

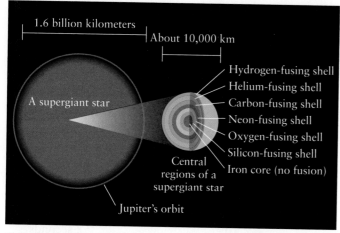

FIGURE 10-10 **The Structure of an Old High-Mass Star** Near the end of its life, a high-mass star becomes a supergiant with a diameter almost as wide as the orbit of Jupiter. The star's energy comes from six concentric fusing shells, all contained within a volume roughly the same size as Earth.

Silicon fusion in high-mass stars involves many types of nuclear reactions, but its major final product is iron. Eventually, the core is converted entirely into iron, which is surrounded by layers of shell fusion that consume the star's remaining reserves of fuel (see Figure 10-10). Whereas the star's enormously bloated atmosphere is nearly as big as the orbit of Jupiter, its entire energy-producing region is again contained in a volume the size of Earth. The buildup of an inert core of iron nuclei and electrons signals the impending violent death of a massive star.

R I V U X G

FIGURE 10-11 **Mass Loss by a Supermassive Star** Gas and dust ejected by the massive star HD 148937 in the constellation Norma. Orange is H$_\alpha$, blue is oxygen, and red is sulfur. Located about 4200 ly (1300 pc) away, this star has 40 M$_\odot$ and is more than halfway through its 6-million-year lifespan. *(Gemini Observatory/AURA)*

10-6 High-mass stars blow apart in violent supernova explosions

Unlike lighter elements, iron cannot fuel further thermonuclear reactions. The protons and neutrons inside iron nuclei are already so tightly bound together that no further energy can be extracted by fusing still more particles with them. The sequence of fusion stages in the cores of high-mass stars therefore ends.

Because iron atoms do not fuse and emit energy, the electrons in the core must now support the star's outer layers by the strength of electron degeneracy pressure alone. Soon, however, the continued deposition of fresh iron from the silicon-fusing shell causes the core's mass to exceed the Chandrasekhar limit. Electron degeneracy suddenly fails to support the star's enormous weight, and the core collapses.

Any isolated main-sequence star of more than 8 M$_\odot$ develops an iron core as a luminous supergiant. When the core collapses, a rapid series of cataclysms is triggered that tears the star apart in a few seconds. Let us see how this happens in the death of a 25-M$_\odot$ star, according to computer simulations (Table 10-1).

Where in massive stars are the higher-mass elements formed?

In a 25-M$_\odot$ star, electron degeneracy pressure fails when the density is sufficiently high inside the iron core. The core, only some 3000 km in diameter, then collapses immediately. In roughly one-tenth of a second, the central temperature exceeds 5 billion K. Gamma-ray photons associated with this intense heat have so much energy that they begin to break apart the iron nuclei, a process called **photodisintegration**. Although millions of years passed from the time the star arrived on the main sequence with a hydrogen- and helium-filled core until its core became iron, it takes less than a second to convert the core back into elemental protons, neutrons, and electrons!

Within another tenth of a second, as the density continues to climb, the electrons in the core are forced to combine with the core's protons to produce neutrons, and the process releases a flood of neutrinos. As we saw in Section 7-9, neutrinos have no electric charge and very little mass, and most pass through Earth or the Sun without interaction. Nevertheless, the matter deep inside a collapsing high-mass star is so fantastically dense that its newly created neutrinos cannot immediately escape from the star's core. Slamming into nearby particles, the neutrinos provide a pressure pushing outward from the core.

According to the calculations, about a quarter of a second after the collapse begins, the density of the entire core reaches 4×10^{17} kg/m^3, which is *nuclear density*, the density at which neutrons and protons are normally packed together inside atomic nuclei. (Compare this with the density of water, 10^3 kg/m^3.) When the neutron-rich material of the core reaches nuclear density, it suddenly stiffens. This means that it resists the collapse much more than when it was less dense. The collapse of the core halts so abruptly that it rebounds and begins rushing back out in a process called *core bounce*.

TABLE 10-1 Evolutionary Stages of a 25-$M_\odot$ Star

Stage	Central temperature (K)	Central density(kg/m³)	Duration of stage
Hydrogen fusion	4×10^7	5×10^3	7×10^6 years
Helium fusion	2×10^8	7×10^5	5×10^5 years
Carbon fusion	6×10^8	2×10^8	600 years
Neon fusion	1.2×10^9	4×10^9	1 year
Oxygen fusion	1.5×10^9	1×10^{10}	6 months
Silicon fusion	2.7×10^9	3×10^{10}	1 day
Core collapse	5.4×10^9	3×10^{12}	0.2 seconds
Core bounce	2.3×10^{10}	4×10^{17}	milliseconds
Supernova explosion	about 10^9	varies	hours

The model predicts that, during this critical stage, the star's unsupported layers of shell-fusing matter are plunging inward at up to 15% of the speed of light. The outward-flowing neutrinos and rebounding core slam into this matter (the red regions in Figure 10-12a). The impact stops the core's rebound while causing the infalling matter to reverse course. In just a fraction of a second, a tremendous volume of matter begins to move back up toward the star's surface. This matter accelerates rapidly as it encounters less and less resistance, and soon it forms an outgoing shock wave. After a few hours, this shock wave reaches the star's surface, lifting the star's outer layers away from the core in a mighty blast. The star becomes a **Type II supernova.**

As the supernova expands, the star's luminosity, which was already some 10^4 to 10^5 $L_\odot$, suddenly increases by another factor of between 10^3 and 10^6. The brightest supernova can therefore emit nearly as much energy as a hundred billion stars as luminous as our Sun! In other words, a supernova is often 10% as bright as all of the other stars in its galaxy combined.

As the layers of a massive dying star are blasted into space, they are compressed so much by the neutrinos and the shock wave that fusion actually occurs during the explosion, creating a broad assortment of heavy elements and most of the larger interstellar dust particles (see Sections 2-9 and 2-10). Indeed, the products of shell fusion and this final burst of fusion during supernovae are where many of the elements on Earth and other terrestrial planets come from, including many of the atoms in our bodies. We really are made of stardust.

Computer simulations indicate that the onion skins of different elements do not just move out of a star "going supernova" in lockstep (Figures 10-12a, b). Therefore, the ejected gases are not expected to be uniform shells of matter. Indeed, observations reveal that the material emerging in a supernova comes out quite irregularly. Figures 10-12c, d, and e show the distributions of some of the elements in the supernova Cassiopeia A (Cas A). Other observations show that some of the dense elements, such as iron, ejected in supernovae are actually created deep inside the innermost silicon and oxygen layers of the supernovae and then shot out like bullets or geysers. Other elements are created throughout the expanding stars. However, calculations and observations reveal that supernovae do not create enough of the heaviest elements, like gold, silver, and platinum, to account for the amounts of these elements found on Earth. We will explore their likely origin in Section 10-14.

Over its lifetime, our model 25-$M_\odot$ star ejects more than 20 $M_\odot$ of its mass back into space by stellar winds, by gas ejected during unstable periods, and during its supernova phase. Less massive stars return proportionately less mass to the interstellar medium.

What role do neutrinos play in Type II supernovae?

INSIGHT INTO SCIENCE

The Process of Science The best scientific theories and models also provide explanations about matters that they were not initially designed to study. In modeling stellar evolution and supernovae, the processes that create and eject metals (that is, elements heavier than hydrogen and helium) from stars provide the elements necessary for life to exist.

The spectrum and light curve of a Type II supernova are both significantly different from those of a Type Ia supernova. Whereas hydrogen lines are absent from a Type Ia's spectrum, they are strong in the spectra of Type II supernovae. Both types begin with a sudden rise in brightness (see Figure 10-9). However, a Type II supernova usually peaks at −17, a hundred times dimmer than a Type Ia. Whereas Type Ia supernovae decline gradually for more than a year, Type II supernovae alternate between periods of steep and gradual declines in brightness. Type II light curves, therefore, have a steplike appearance.

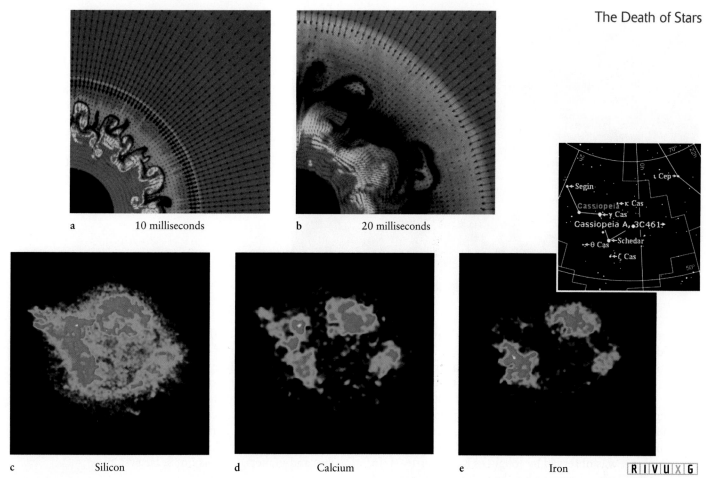

a   10 milliseconds

b   20 milliseconds

c   Silicon   d   Calcium   e   Iron   R I V U X G

FIGURE 10-12 Supernovae Proceed Irregularly Images (a) and (b) are computer simulations showing the chaotic flow of gas deep inside the star as it begins to explode as a supernova. This helps account for the globs of iron and other heavy elements emitted from deep inside, as well as the lopsided distribution of all elements in the supernova remnant, as shown in (c), (d), and (e). These three pictures are X-ray images of supernova remnant Cassiopeia A taken by Chandra at different wavelengths. *(a and b: Courtesy of Adam Burrows, University of Arizona and Bruce Fryxell, NASA/GSFC; c, d, and e: U. Hwang et al., NASA/GSFC)*

10-7   Supernova remnants are observed in many places

Astronomers have seen more than 1000 supernovae. These observations suggest that in a typical galaxy like the Milky Way, Type Ia supernovae occur approximately once every 36 years, whereas Type II supernovae occur about once every 44 years. Thus, there should be about 5 supernovae exploding in our Milky Way Galaxy each century. Vigorous stellar evolution in our Milky Way occurs primarily in the Galaxy's disk, because stars there continue to form in giant molecular clouds. The disk (which we are in and which we see edge-on as the milky-white band of stars and glowing gas and dust across the sky) is also where supernovae explode.

Astronomers find the debris of local supernova explosions in the Milky Way scattered across the Galaxy's disk. A beautiful example is the Cygnus Loop (see Figure 9-5). The star's outer layers were blasted into space 20,000 years ago so violently that they are still traveling at supersonic speeds. As this expanding shell of gas plows through the interstellar medium, it collides with atoms and molecules, making the gases glow.

Many supernova remnants cover sizable fractions of the sky. The largest is the Gum Nebula, with an angular diameter of 60° (its central region is shown in Figure 10-13). This nebula looks so big because it is so close: Its near side is only about 300 light-years from Earth. Studies of the Gum Nebula's expansion rate suggest that the supernova exploded about 11,000 years ago. At maximum brilliance, the exploding star was probably as bright as the Moon at first quarter.

The Milky Way's disk is so filled with interstellar gas and dust that we cannot see very far into it. Supernovae are thought to erupt every few decades in remote parts of our Galaxy, but their detonations are hidden from optical telescopes by the interstellar medium. Many supernova remnants in our Galaxy can be detected only at nonvisible wavelengths, ranging from X rays through radio waves. For example, Figure 10-14 shows images of the supernova remnant Cassiopeia A as seen in X-ray (Figure 10-14a) and radio (Figure 10-14b) images. Visible-light photographs of this part of the sky reveal only a few small, faint wisps. Thus, radio searches for supernova remnants are more fruitful than visual searches. Only 24 supernova remnants have been found in visible-light photographs, but more than

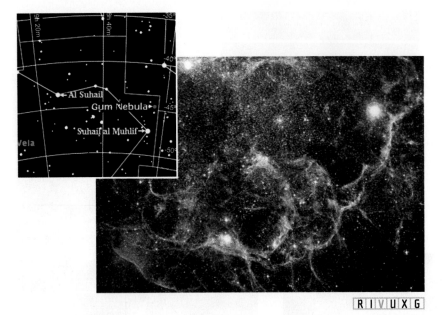

FIGURE 10-13 **The Gum Nebula** The Gum Nebula is the largest known supernova remnant. It spans 60° across the sky and is centered roughly on the southern constellation of Vela. The nearest portions of this expanding nebula are only 300 light-years from Earth. The supernova explosion occurred about 11,000 years ago, and the remnant now has a diameter of about 2300 light-years. Only the central regions of the nebula are shown here. *(Royal Observatory, Edinburgh)*

RIVUXG

3100 remnants in our Galaxy and in others have been discovered by radio astronomers.

From the expansion rate of the nebulosity in Cassiopeia A, astronomers calculate that the light from that supernova explosion first reached Earth about 300 years ago. Although telescopes were in wide use by the late 1600s, no record of the event is known. In fact, the last supernova of a star near its maximum brightness in our Galaxy, seen by the naked eye, was observed by Johannes Kepler in 1604. In 1572, Tycho Brahe also recorded the sudden appearance of an exceptionally bright star in the sky.

In 2002, astronomers discovered that 2 million years ago a group of bright O and B stars, called the Scorpius-Centaurus OB association, passed within 150 light-years of Earth, and that one or more supernovae in this group may have occurred at that time. Such a close explosion could have damaged Earth's ozone layer and caused the extinction of some ocean life at the end of the Pliocene epoch.

Besides the gas and dust enriched with heavy elements, supernovae leave a heritage long after their remnants have dissipated. Much of this gas and dust is recycled into new stars and planets, such as our solar system.

Why might we not be able to see all the supernovae in our Galaxy?

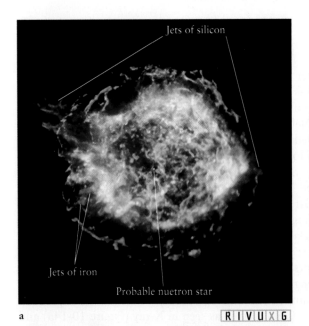

a

Jets of silicon

Jets of iron

Probable nuetron star

RIVUXG

b

RIVUXG

FIGURE 10-14 **Cassiopeia A** Supernova remnants, such as Cassiopeia A, are typically strong sources of X rays and radio waves.
(a) An X-ray picture of Cassiopeia A taken by Chandra. The opposing jets of silicon, probably guided by powerful magnetic fields, were ejected early in the supernova, before the iron-rich jets were released.

(b) A corresponding radio image produced by the Very Large Array (VLA). Radiation from the supernova that produced this nebula first reached Earth 300 years ago. The explosion occurred about 10,000 light-years from here. *(a: NASA/ CXC/GSFC/U. Hwang et al.; b: Very Large Array)*

WEB LINK 10.5   ANIMATION 10.4

10-8   Cosmic rays are not rays at all

4 In observations connected with supernovae, astronomers have detected particles flying through space at speeds that exceed 90% of the speed of light. These high-speed particles are called **cosmic rays** or **primary cosmic rays**. They were named *rays* before their true identity as particles was known, and, as often happens, this misleading name has stuck. Cosmic rays are primarily hydrogen nuclei, along with less than 1% of the nuclei of all the more massive elements and a few percent of electrons and positrons (positively charged electrons).

The origins of cosmic rays are just beginning to become clear. At first, many astronomers thought that moderate-energy cosmic rays were emitted directly by supernovae. This made sense because these explosions are the best-known source of the energy necessary to give such cosmic rays their tremendous speeds. However, observations by the *Advanced Composition Explorer* satellite reveal that the isotopes of nickel coming to Earth in the form of cosmic rays are not the isotopes emitted by supernovae. Instead, evidence is mounting that many cosmic rays are created when supernova debris slams into gas already in space, causing some of this preexisting matter to accelerate and become cosmic rays.

Much less common than cosmic rays from supernova remnants are ultrahigh-energy cosmic rays that come from outside of our Galaxy. Each of these particles has as much energy as a baseball thrown by a major league pitcher. Observations of their paths suggest that many of them come from galaxies containing supermassive black holes, which we will discuss in Section 10-25. Whether these black holes themselves generate the ultrahigh-energy cosmic rays is still under investigation.

Most primary cosmic rays coming in our direction collide with gas in the atmosphere some 15 km above Earth's surface. This is fortunate, because each cosmic ray packs an enormous amount of energy that potentially could harm living tissue. The collision in the air divides a cosmic ray's energy among several gas particles, which are shoved Earthward. These, in turn, often hit other gas atoms, creating a cascade of lower-energy particles that eventually reach Earth's surface as a **cosmic ray shower** (Figure 10-15). These cosmic rays created from atoms in our atmosphere are called **secondary cosmic rays**.

10-9   Supernova 1987A offered a detailed look at a massive star's death

Some supernovae in other galaxies can be seen without a telescope. In 1885, a supernova in the Andromeda Galaxy was just barely visible to the unaided eye. On February 23, 1987, a supernova was observed in the Large Magellanic Cloud, a galaxy near our own Milky Way. The supernova, designated SN 1987A, was the first to be observed that year (Figure 10-16), and it was so bright that it could easily be seen with the naked eye. What made SN 1987A such an exciting discovery was that it gave astronomers a rare opportunity to study the death of a nearby massive star using modern equipment that provided data with which to verify the theory of supernovae described earlier.

At first, SN 1987A reached only a tenth of the luminosity predicted for an exploding massive star. For the next 85 days, it gradually brightened. Then it began dimming, as is characteristic of an ordinary supernova. Fortunately, the star had been observed before it became a supernova (Figure 10-16 insets). The Large Magellanic Cloud is about 160,000 light-years from Earth—near enough to us that many of its stars have been individually observed and catalogued. The star had been identified as a B3 I supergiant. The theory of the evolution of such stars provided an explanation for SN 1987A's unusually slow brightening.

When this star was on the main sequence, its mass was about 20 $M_\odot$, although by the time it exploded, it had shed many solar masses. The evolutionary track for an aging 20-$M_\odot$ star wanders back and forth across the top of the H-R diagram, so the star alternates between being a hot (blue) supergiant and a cool (red) supergiant. The star's size changes significantly as its surface temperature changes. A blue supergiant is only 10 times larger in diameter than the Sun, but a red supergiant of the same luminosity is 1000 times larger. Because the doomed star was a relatively small blue supergiant when it exploded, it reached only a tenth of the brightness that it would have attained had it been a red supergiant at that time.

Ordinary telescopic observations of a supernova explosion can only show us the expanding outer layers of the dying star. No photons created in the core can penetrate through the supernova gases directly to our telescopes. As the outer layers of the supernova thin, however, neutrinos escape directly from the core into space.

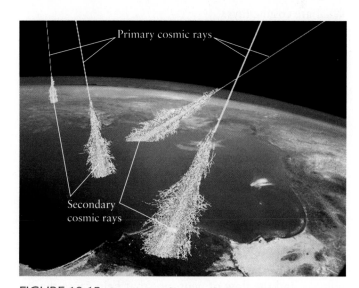

FIGURE 10-15 Cosmic Ray Shower Cosmic rays from space slam into particles in the atmosphere, breaking them up and sending them Earthward. These debris are called secondary cosmic rays. This process of impact and breaking up continues as secondary cosmic rays travel downward, creating a cosmic ray shower, as depicted in this artist's conception of four such events. *(Simon Swordy, U. Chicago; NASA)*

R I V U X G

WEB LINK 10.6 **FIGURE 10-16 Supernova 1987A** A supernova was discovered in a nearby galaxy called the Large Magellanic Cloud (LMC) in 1987. This photograph shows a portion of the LMC that includes the supernova and a huge H II region, called the Tarantula Nebula. At its maximum brightness, observers at southern latitudes saw the supernova without a telescope. **Insets:** The star before and after it exploded. *(European Southern Observatory; insets: Anglo-Australian Observatory/David Malin Images)*

By detecting these neutrinos and measuring their properties, astronomers can learn many details about the star's collapsing core, especially about core bounce. Several neutrino detectors were operating when the neutrinos from SN 1987A reached Earth.

What are cosmic rays?

Nearly a day before SN 1987A was first observed in the sky, teams of scientists at neutrino detectors in Japan and the United States reported observing Cerenkov flashes (see Section 7-9) from a burst of neutrinos. The Kamiokande II detector in Japan detected 12 neutrinos at about the same time that 8 were found by the IMB (Irvine-Michigan-Brookhaven) detector in a salt mine under Lake Erie. Neutrinos preceded the visible supernova outburst because they escaped from the dying star before the shock wave from the collapsing core reached the star's surface. They were detected in Earth's northern hemisphere, where the supernova is always below the horizon, after having passed through Earth. The discovery of these neutrinos provided strong support for the theory that supernovae are caused by the collapse of the star's core, its subsequent bounce, and the flow of neutrinos that help cause the supernova, as described in Section 10-6.

About 3 and a half years after SN 1987A's detonation was first seen from Earth, astronomers using the Hubble Space Telescope obtained a picture showing several rings of glowing gas around the exploded star (Figure 10-17). This gas had been emitted by the star 20,000 years before the supernova, when a hydrogen-rich stellar envelope was ejected by stellar winds from the doomed star, as discussed earlier in this chapter. This gas expanded in an hourglass shape, because it was blocked from expanding around the star's equator, either by a preexisting ring of gas or by the orbit of an as-yet-unseen companion star.

While emitting these gases, the doomed star was a red supergiant. It then shrank and in its final blue giant phase, the same star emitted higher-speed gas that compressed the hourglass gas into three rings—a narrow one around the equator and wider ones at the top and bottom of the hourglass. This relic gas is now being illuminated by photons from the supernova. In 1998, astronomers began to observe gas ejected from the supernova striking the preexisting circumstellar gas. This collision has caused the ring of gas to brighten, and it should eventually illuminate more of the earlier gas that had been ejected.

SN 1987A provided astronomers with invaluable information, in part, because the doomed star had been studied before it exploded and its distance from Earth was known. The supernova was also located in an unobscured part of the sky, and neutrino detectors happened to be operating at the time of the outburst. Astronomers will be monitoring the progress of this supernova for years to come.

Why were neutrinos from SN 1987A observed before the light from this event?

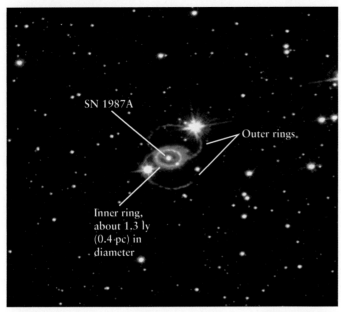

a Supernova 1987A seen in 1996   [R][I][V][U][X][G]

b An explanation of the rings

FIGURE 10-17 **Shells of Gas around SN 1987A** (a) Intense radiation from the supernova explosion caused three rings of gas surrounding SN 1987A to glow in this Hubble Space Telescope image. This gas was ejected from the star 20,000 years before the star detonated. All three rings lie in parallel planes. The inner ring is about 1.3 light-years across. The white and colored spots are unrelated stars. (b) When the progenitor star of SN 1987A was still a red supergiant, a slowly moving wind from the star filled the surrounding space with a thin gas. When the star contracted into a blue supergiant, it produced a faster-moving stellar wind. The interaction between the fast and slow winds somehow caused gases to pile up along an hourglass-shaped shell surrounding the star. The burst of ultraviolet radiation from the supernova ionized the gas in the rings, causing the rings to glow. The supernova itself, at the center of the hourglass, glows because of energy released from radioactive decay. *(Robert P. Kirshner and Peter Challis, Harvard-Smithsonian Center for Astrophysics; STScI)*

NEUTRON STARS AND PULSARS

When stars that initially had between 8 and 25 $M_\odot$ explode as supernovae, their expanding cores are stopped by the outer layers into which they slam. The cores recontract and remain intact as highly compressed clumps of neutrons, called **neutron stars**. Isolated neutron stars, resulting from these explosions, are stable stellar remnants with masses between 1.4 $M_\odot$ and about 3 $M_\odot$.

10-10 The cores of many Type II supernovae become neutron stars

The neutron was discovered during laboratory experiments in 1932. Inspired by the realization that white dwarfs are supported by electron degeneracy pressure, Fritz Zwicky and Walter Baade proposed, within a year, that a similar repulsion between neutrons could support remnant neutron cores. They used the fact that the Pauli exclusion principle also prevents neutrons with identical properties from packing too closely together. When neutrons are pressed together, many of them have to move rapidly (so as not to be identical to their neighbors). This motion provides a pressure, called **neutron degeneracy pressure,** that equations predict is even greater than electron degeneracy pressure (see Section 9-10) and thus allows stellar remnants with masses beyond the Chandrasekhar limit to be stable.

Most scientists ignored Zwicky and Baade's theory for years. After all, a neutron star seemed to be a rather unlikely object. To transform protons and electrons into neutrons, the density in the star would have to be equal to nuclear density, about 10^{17} kg/m³. Thus, a teaspoon full of neutron star matter brought back to Earth would weigh about 1 billion tons. Put another way, neutron star matter is 200 million times denser than that found in a white dwarf. Furthermore, an object compacted to nuclear density would be very small. A 2-$M_\odot$ neutron star would have a diameter of only 20 km (12 mi) and would fit inside any major city on Earth. The surface gravity on one of these neutron stars would be so strong that the escape velocity would equal one-half the speed of light. All of these conditions appeared to be so outrageous that few astronomers paid any serious attention to the subject of neutron stars—until 1968.

In that year, Jocelyn Bell, then a graduate student in radio astronomy at Cambridge University, noticed that the radio telescope she was using was detecting regular pulses from one particular location in the sky. Careful repetition of the observations demonstrated that the radio pulses were arriving with a regular period of 1.3373011 s. The regularity of this pulsating radio source was so striking that the Cambridge team suspected that they might be detecting signals from an advanced alien civilization, which accounts for the name first assigned to it: LGM1 (short for Little Green Men 1). This possibility was soon discarded as several more of these pulsating radio sources, which soon came to be known as **pulsars,** were discovered across the sky. The broad distribution of pulsars would require incredibly widespread civilizations, inconsistent with what we have found in our searches for alien intelligence (see Chapter 13). In all cases, the periods were extremely regular, ranging between 0.2 and 1.5 s (Figure 10-18).

When the discovery of pulsars was officially announced in early 1968, astronomers around the world began proposing all sorts of explanations. Many of these theories were bizarre, and arguments raged for months. We have already ruled out alien civilizations. Astronomers therefore looked for some recurring event in a star's life. We know that a star sheds matter as it swells to a giant. Perhaps some stars alternately expand and contract, emitting energy as they pulsate. However, this explanation also fails, because any star expanding and contracting as fast as a pulsar emits pulses would explode.

Late in 1968, all controversy was laid to rest with the discovery of a pulsar in the middle of the Crab Nebula. In A.D. 1054, Native American, Chinese, and possibly other peoples saw and recorded the appearance of a "guest star" in the constellation of Taurus. When we turn a telescope toward this location, we find the Crab Nebula, shown in Figure 10-19. It looks like the residue of an explosion and is, in fact, a supernova remnant. At the center of the Crab Nebula is the Crab pulsar (Figure 10-19b).

With expansion and contraction ruled out, the likely scenario is that pulsars are spinning objects. If the Crab pulsar is spinning, it is spinning too fast to be a white dwarf. Its period is 0.033 s, which means that it rotates 30 times each second. At that speed, something as wide as a white dwarf would immediately fly apart. Strengthening the case against pulsars being white dwarfs, astronomers discovered numerous other pulsars rotating even faster, including PSR 1937+21 (PSR for pulsar and 1937+21 for the object's approximate right ascension and declination), a stellar remnant that pulses 640 times each second. Staying with the belief that they are spinning, astronomers concluded that pulsars must be incredibly compact. Calculations revealed that, if neutron stars exist, they have a sufficiently small diameter to remain intact while rotating as fast as pulsars pulsate. Indeed, this was the *only* explanation that withstood the process of scientific scrutiny, and, as a result, the theory of neutron stars described earlier was quickly developed.

How big is a neutron star?

10-11 A rotating magnetic field explains the pulses from a neutron star

Why are neutron stars rotating so fast, and how do they emit pulses of electromagnetic energy? Based on observations of the rotation of the Sun, Betelgeuse, and other nearby stars, astronomers theorize that many, perhaps most, stars on the main sequence have some angular momentum (rotation). The Sun, for example, takes nearly a full month to rotate once around its axis. But the rotation rate of collapsing stars increases, just as pirouetting ice skaters speed up when they pull in their arms, as shown in Figure 2-12. Recall from Sec-

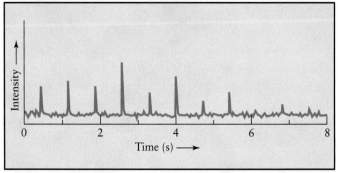

FIGURE 10-18 **A Recording of a Pulsar** This chart recording shows the intensity of radio emissions from one of the first pulsars to be discovered, PSR 0329+54. Note that some of the pulses are weak and others are strong. Nevertheless, the spacing between pulses is so regular (0.714 s) that it is more precise than most clocks on Earth.

tion 2-7 that this occurs because the total amount of angular momentum in an isolated system always remains constant. To conserve angular momentum, the core of a high-mass main-sequence star rotating once a month spins faster than once a second when it collapses to the size of a neutron star.

In addition to rapid rotation, most neutron stars are believed to have intense magnetic fields. In an average star such as our Sun, the magnetic field is spread out over millions upon millions of square kilometers just under the star's surface (see Chapter 7). However, when a star of solar dimensions collapses down to a neutron star, its magnetic field becomes very concentrated, in the same way as stalks of wheat in a field become concentrated when you gather them in a bundle (Figure 10-20). The strength of a collapsing star's magnetic field increases to as much as 10^{15} times Earth's magnetic field. A magnet that strong located halfway to the Moon would lift iron rods off Earth.

The axis of rotation of a typical neutron star is not the same as the axis connecting its north and south magnetic poles (Figure 10-21), just as the magnetic and rotation axes of the Sun and planets (except Saturn) are different from each other. As the neutron star rotates, its powerful magnetic field therefore rapidly changes direction. Like a giant electric generator, the star creates intense electric fields, which act on protons and electrons near its surface. The powerful electric fields channel these charged particles, causing them to flow out from the neutron star's polar regions, as sketched in Figure 10-21. As the particles stream along the field, they accelerate and emit energy. The result is two very thin beams of radiation that pour out of the neutron star's north and south magnetic polar regions and sweep through space—a pulsar.

This explanation for pulsars is often called the **lighthouse model.** A rotating, magnetized neutron star is somewhat like a lighthouse beacon. The magnetic field causes nearby gases to accelerate and radiate. Depending on the conditions, this radiation can range from radio waves to gamma rays, and many pulsars emit in more than one range. Calculations indicate that this radiation is beamed, and, as the star rotates, it sweeps around the sky. If Earth happens to be located in the right direction, a brief flash can be observed each time a beam whips past our line of sight. As a

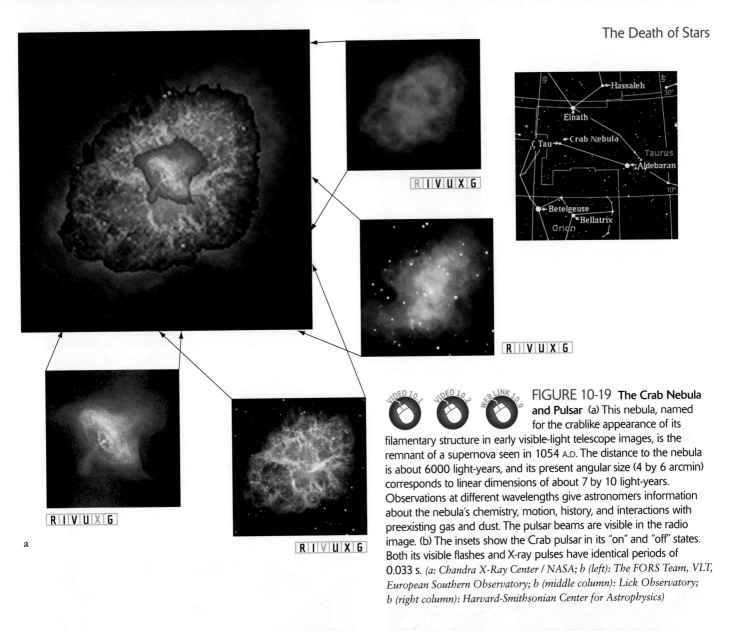

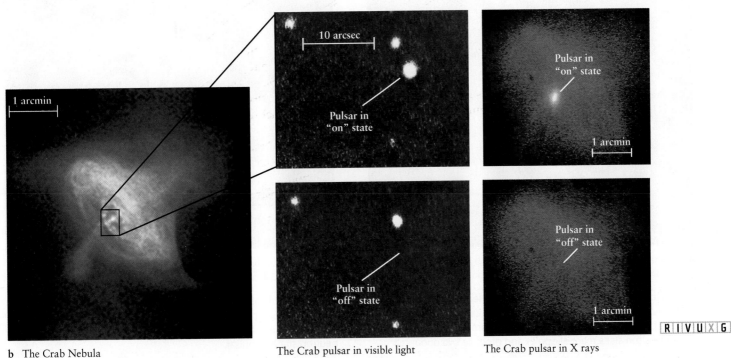

FIGURE 10-19 **The Crab Nebula and Pulsar** (a) This nebula, named for the crablike appearance of its filamentary structure in early visible-light telescope images, is the remnant of a supernova seen in 1054 A.D. The distance to the nebula is about 6000 light-years, and its present angular size (4 by 6 arcmin) corresponds to linear dimensions of about 7 by 10 light-years. Observations at different wavelengths give astronomers information about the nebula's chemistry, motion, history, and interactions with preexisting gas and dust. The pulsar beams are visible in the radio image. (b) The insets show the Crab pulsar in its "on" and "off" states. Both its visible flashes and X-ray pulses have identical periods of 0.033 s. (a: *Chandra X-Ray Center / NASA; b (left): The FORS Team, VLT, European Southern Observatory; b (middle column): Lick Observatory; b (right column): Harvard-Smithsonian Center for Astrophysics*)

a

b The Crab Nebula

The Crab pulsar in visible light

The Crab pulsar in X rays

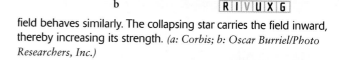

FIGURE 10-20 **Analogy about How Magnetic Field Strengths Increase** When growing, these wheat stalks cover a much larger area than when they are harvested and bound together. A star's magnetic field behaves similarly. The collapsing star carries the field inward, thereby increasing its strength. (*a: Corbis; b: Oscar Burriel/Photo Researchers, Inc.*)

result of its neutron star's rotation, the Crab Nebula is flashing on and off 30 times each second (see Figure 10-19).

The Crab pulsar is one of the youngest known pulsars, its creation having been observed some 950 years ago. Also visibly flashing is the Vela pulsar at the core of the Gum Nebula (see Figure 10-13). The Vela pulsar, with a period of 0.089 s, is the slowest pulsar ever detected at visual wavelengths. Because they use some of their energy of rotation to create the pulses they emit, *isolated pulsars slow down as they get older*. Only the very youngest pulsars are ener-

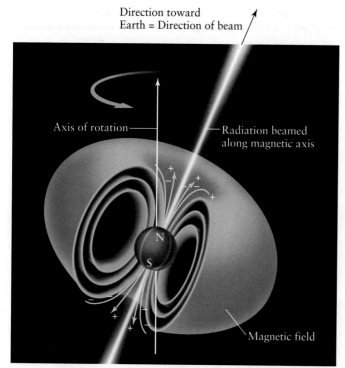

a One of the beams from the rotating neutron star is aimed toward Earth: We detect a pulse of radiation.

FIGURE 10-21 **A Rotating, Magnetized Neutron Star**
Calculations reveal that many neutron stars rotate rapidly and possess powerful magnetic fields. Charged particles are accelerated near a neutron star's magnetic poles and produce two oppositely

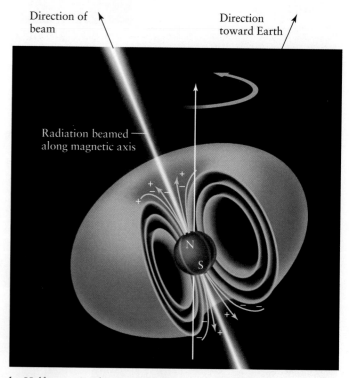

b Half a rotation later, neither beam is aimed toward Earth: We detect that the radiation is "off."

directed beams of radiation. As the star rotates (going from a to b is half a rotation period), the beams sweep around the sky. If Earth happens to lie in the path of a beam (a, but not b), we see the neutron star as a pulsar.

getic enough to emit visible flashes along with their radio pulses, and the Vela pulsar was created just 11,000 years ago, which is recent in astronomical terms.

If the lighthouse model of pulsars is correct, do we see all nearby pulsars? Why or why not?

10-12 Rotating neutron stars create other phenomena besides normal pulsars

Pulsars typically have magnetic fields about 10^{12} times stronger than the magnetic field at the surface of Earth. Exceptional circumstances can conspire to raise the field strength a thousand times higher in some pulsars. If the neutron star is very hot (around 10^{11} K) and spinning very fast (100 times a second) when it forms, convection occurs throughout the neutron star, just as it does for red dwarfs (see Section 9-8). This inward and outward motion, combined with the spin, generates extra magnetic fields by the dynamo effect (see Section 4-4). Combining the neutron star's two magnetic fields yields a field 10^{15} times greater than the one at Earth's surface. The resulting object is called a **magnetar**, and its effects are dramatic.

The magnetic fields of a magnetar are so strong that they cause the neutron star's surface to buckle and emit stupendous bursts of X rays and gamma rays, along with photons at other wavelengths, that last for fractions of a second. Such magnetars are called *soft gamma-ray repeaters,* or SGRs. ("Soft" means "low energy." Even so, soft gamma rays are very powerful photons.) Interacting with surrounding gases, the magnetic fields of a magnetar rapidly slow the neutron star's rotation rate. After 10^4 years, the body is spinning only once every 10 s, and the SGR activity subsides. Fourteen magnetars have been observed since 1998, including one in 2004 that released the brightest flash ever seen from outside our solar system. Its photons scattered off the Moon and lit up Earth's atmosphere.

In 2006, radio astronomers discovered yet another type of emission from rotating neutron stars. Called *rotating radio transients,* or RRATs, these objects give off bursts of radio and other emissions that last a few thousandths of a second, turn off for minutes or hours, and then spring briefly back to life. The neutron stars involved rotate slowly compared to most pulsars, spinning once every 0.4 to about 10 s. The most promising explanation, at present, is that these are older pulsars or magnetars that have slowed down.

How do the rotation rates of magnetars change with time?

10-13 Neutron stars have internal structure

Neutron stars are not composed solely of neutrons. Astronomers have developed a theoretical model for the surface and internal structure of a neutron star (Figure 10-22). Its interior has a radius of about 10 km, with a core of supercon-

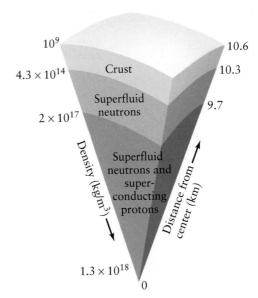

FIGURE 10-22 **Model of a Neutron Star's Interior** This drawing shows the theoretical model of a 1.4-$M_\odot$ neutron star. The neutron star has a superconducting, superfluid core 9.7 km in radius, surrounded by a 0.6-km-thick mantle of superfluid neutrons. The neutron star's crust is only 0.3-km thick (the length of four football fields) and is composed of heavy nuclei and free electrons. The thicknesses of the layers are not shown to scale.

ducting protons and superfluid neutrons. A *superconductor* is a material in which electricity and heat flow without the system losing energy, whereas a *superfluid* has the strange property that it flows without any friction. Both superconductors and superfluids have been created in the laboratory. Surrounding a neutron star's core is a layer of superfluid neutrons. The surface of the neutron star is a solid, brittle crust of dense nuclei and electrons about ⅓ km thick. The gravitational force of the neutron star is so great at its surface that climbing a bump there just 1 mm high would take more energy than it takes to climb Mount Everest.

Neutron stars may also have atmospheres, as indicated by absorption lines in the spectrum of at least one of them. The absorption lines are believed to be due to scattering of radiation from the neutron star by gases in its atmosphere, just as scattering in the Sun's atmosphere causes absorption lines in its spectrum. In 2005, astronomers found indications that some isolated neutron stars expel energy long after they form. In particular, the Cassiopeia A remnant has regions glowing from radiation deposited on them as recently as 50 years ago (some 275 years after the supernova itself). The cause of this activity is still being determined.

Just as Earth's crust has earthquakes, calculations show that the crusts of rotating neutron stars have equivalent events. When a rotating neutron star (pulsar) undergoes such a quake, its rotation rate suddenly changes to conserve its angular momentum. This creates a **glitch** in the rate at which the pulsar is slowing down. Such events have been observed in a number of pulsars (Figure 10-23).

While most pulsars appear to be isolated bodies, more than 90 have been discovered with companions (not necessarily other neutron stars). These pulsars are called *binary*

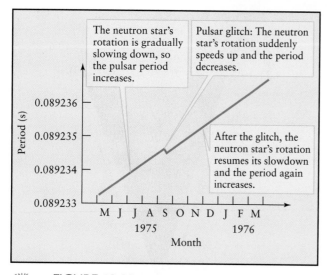

FIGURE 10-23 **A Glitch Interrupts the Vela Pulsar's Spindown Rate** An isolated pulsar radiates energy, which causes it to slow down. This "spin down" is not always smooth. As it slows down, it becomes more spherical, and so its spinning, solid surface must readjust its shape. Because the surface is brittle, this readjustment is often sudden, like the cracking of glass, which causes the angular momentum of the pulsar to suddenly jump. Such an event, called a glitch, shown here for the Vela pulsar in 1975, changes the pulsar's rotation period.

FIGURE 10-24 **Double Pulsar** This artist's conception shows two pulsars that orbit their center of mass. The double pulsar they represent is called PSR J0737-3039, which is about 1500 light-years from Earth in the constellation Puppis. One pulsar has a 23-m/s period and the other has a 2.8-s period. The two orbit once every 2.4 h. (*Michael Kramer/Jodrell Bank Observatory, University of Manchester*)

pulsars and are of interest for several reasons. First, many binary pulsars pull mass onto themselves from their companions. This infall causes the pulsar to speed up (like the skater in Figure 2-12 pulling in her arms). As a result of this effect, some pulsars spin nearly a thousand times per second and are called *millisecond pulsars*. Second, some pulsars, like the one labeled PSR 1916+13, have compact massive companions, which are often also neutron stars. Observations of binary pulsars with neutron star companions reveal that the two bodies are spiraling toward each other, as predicted by Einstein's gravitational theory, called *general relativity* (see Sections 2-1 and 10-21 for details).

In 2004, astronomers observed, for the first time, a pair of pulsars in orbit around each other (Figure 10-24). This discovery provides an exciting opportunity to further test the predictions of general relativity. For example, this theory predicts that these two pulsars should be precessing (the directions that their rotation axes point should be changing) and that, like the binary pulsars just discussed, they should be spiraling in toward each other. Observations of this system in the next few years will test these predictions.

Why do glitches change the rotation rates of pulsars?

10-14   Colliding neutron stars may provide some of the heavy elements in the universe

In Sections 10-5 and 10-6 we discussed nucleosynthesis, the process of creating the heavier elements. We saw how a variety of elements up to iron are created inside massive stars before the supernova occurs and are then ejected into space during the explosion. We also saw that even heavier elements are created and ejected during the supernovae themselves. However, models indicate that supernovae do not create enough of the elements heavier than iron to account for the amounts of these elements found in the universe.

Computer simulations of what happens during collisions of two neutron stars predict that, in these impacts, some of the neutrons are splashed into space, where many of them decay back into protons and help form various elements. The interactions that follow create a variety of heavy elements in amounts consistent with observations.

Why does most of the iron on Earth not come from the iron cores of high-mass stars that explode as supernovae?

10-15   Binary neutron stars create pulsating X-ray sources

Neutron stars in binary systems may also hold the key to another regular pulse from the sky—pulses from X-ray sources. During the 1960s, astronomers obtained tantalizing X-ray views of the sky during short rocket and balloon flights that briefly lifted X-ray detectors above Earth's atmosphere. Several strong X-ray sources were discovered, and each was named after the constellation in which it was located.

Astronomers were so intrigued by these preliminary discoveries that NASA built and launched *Explorer 42*, an X-ray-detecting satellite that could make observations 24 h a day. The satellite was launched from Kenya (to place it in an orbit above Earth's equator) in 1970, on the seventh anniversary of Kenyan independence. To commemorate the occasion, *Explorer 42* was renamed *Uhuru*, which means "freedom" in Swahili.

Uhuru gave us our first comprehensive look at the X-ray sky. As the satellite slowly rotated, its X-ray detectors swept across the heavens. Each time an X-ray source came into view, signals were transmitted to receiving stations on the ground. Before its battery and transmitter failed in early 1973, *Uhuru* had succeeded in locating 339 X-ray sources.

The discovery of pulsars was still fresh in everyone's mind when the *Uhuru* team discovered X-ray pulses coming from Centaurus X-3 in early 1971. Figure 10-25 shows data from one sweep of *Uhuru*'s detectors across Centaurus X-3. The pulses have a regular period of 4.84 s. A few months later, similar pulses were discovered coming from a source called Hercules X-1, which has a period of 1.24 s. Because the periods of these two X-ray sources are so short, astronomers suspected that they had found rapidly rotating neutron stars.

It soon became clear, however, that systems such as Centaurus X-3 and Hercules X-1 are not ordinary pulsars like the Crab or Vela pulsars. Every 2087 days, Centaurus X-3 completely turns off for almost 12 h. This fact suggests that Centaurus X-3 is an eclipsing binary (see Figure 8-13), and that the X-ray source takes nearly 12 h to pass behind its companion star.

The case for the binary nature of Hercules X-1 is even more compelling. It has an off state corresponding to a 6-h eclipse every 1.7 days, and careful timing of the X-ray pulses shows a periodic Doppler shift every 1.7 days. This information provides direct evidence of orbital motion around a companion star. When the X-ray source is approaching us, its pulses are separated by slightly less than 1.24 s. When the source is receding from us, slightly more than 1.24 s elapse between the pulses.

Astronomers now realize that systems such as Centaurus X-3 and Hercules X-1 are examples of binary systems in which one object is a neutron star. Because all of these binaries have very short orbital periods, the distance between the ordinary star and the neutron star must be small. This proximity enables the neutron star to capture gas escaping from the companion star (Figure 10-26).

To explain the pulsations of X-ray sources such as Centaurus X-3 or Hercules X-1, astronomers propose that the ordinary star either fills or nearly fills its Roche lobe (Figure 10-26). Either way, matter escapes from the star. If the star fills its lobe, as in the case of Hercules X-1, mass loss results from direct overflow through the Roche lobe. If the star's surface lies just inside its lobe, as with Centaurus X-3, a stellar wind carries off the mass. A typical rate of mass loss for the ordinary star in such a pair is approximately 10^{-9} M$_\odot$ per year.

The neutron star in a pulsating X-ray source, like an ordinary pulsar, rotates rapidly and has a powerful magnetic field inclined to the axis of rotation (recall Figure 10-21). As the gas from the companion star falls toward the neutron star, the magnetic field funnels the incoming matter down onto its magnetic polar regions. The neutron star's gravity is so strong that the gas is traveling at nearly half the speed of light by the time it crashes onto the star's surface. This violent

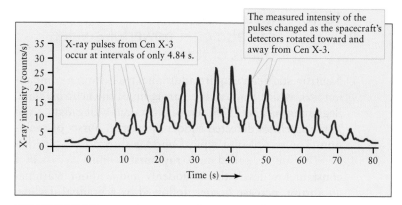

FIGURE 10-25 **X-Ray Pulses from Centaurus X-3** This graph shows the intensity of X rays detected by *Uhuru* as Centaurus X-3 moved across the satellite's field of view. The variation in the height of the pulses was a result of the changing orientation of *Uhuru*'s X-ray detectors toward the source as the satellite rotated. The short pulse period suggests that the source is a rotating neutron star.

impact creates hot spots at both poles, with temperatures of about 10^8 K, that emit abundant X rays with a luminosity nearly 100,000 times brighter than the Sun. Observations of three such X-ray sources in 2005 revealed the impact sites on the neutron stars to range in size from 100 m to 4 km across. As the neutron star rotates, two beams of X rays from the polar caps sweep around the sky. If Earth happens to be in the path of one of the beams, we can observe this pulsating X-ray source. The pulse period is thus equal to the neutron star's rotation period. For example, the neutron star in Hercules X-1 completes one rotation every 1.24 s.

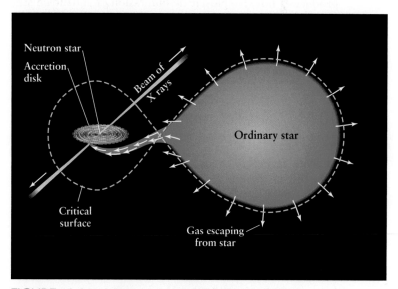

FIGURE 10-26 **A Model of a Pulsating X-Ray Source** Gas transfers from an ordinary star to the neutron star. The infalling gas is funneled down onto the neutron star's magnetic poles, where it strikes the star with enough energy to create two X-ray-emitting hot spots. As the neutron star spins, beams of X rays from the hot spots sweep around the sky.

10-16 Neutron stars in binary systems can also emit powerful isolated bursts of X rays

Neutron stars can acquire additional mass from a companion star, which leads to the flaring up of the neutron stars. Beginning in late 1975, astronomers analyzing data from X-ray satellites detected sudden, powerful bursts of radiation. The record of a typical burst is shown in Figure 10-27. The source, called an **X-ray burster**, emits X rays at a constant low level; then, suddenly and without warning, an abrupt increase occurs, followed by a gradual decline. A typical burst lasts for 20 s. Several dozen X-ray bursters have been located, most lying in the plane of the Milky Way Galaxy.

X-ray bursters, like novae, are believed to arise from mass transfer in binary star systems. With a burster, however, the stellar remnant is a neutron star rather than a white dwarf. Gases escaping from the ordinary companion star fall onto the neutron star. The energy released as this gas crashes down onto the neutron star's surface produces the low-level X rays that are continuously emitted by the burster.

Most of the gas falling onto the neutron star is hydrogen, which becomes compressed against the hot surface of the star by the star's powerful surface gravity. In fact, temperatures and pressures in this accreting layer are so high that the arriving hydrogen is promptly converted by fusion into helium. Constant hydrogen fusion soon produces a layer of helium that covers the entire neutron star. When the helium layer is about 1-m thick, helium fusion ignites explosively, and we observe a sudden burst of X rays. In other words, whereas explosive hydrogen fusion on a white dwarf produces a nova, explosive helium fusion on a neutron star produces an X-ray burster.

Is an X-ray burster similar to a supernova in the evolution of its source?

10-17 Smaller, more exotic stellar remnants composed of quarks may exist

In death as well as in life, the mass of a star determines its fate. Just as there is an upper limit to the mass of a white dwarf (the Chandrasekhar limit), there is also an upper limit to the mass of a neutron star, called the *Oppenheimer-Volkov limit*. Above this limit, neutron degeneracy pressure cannot support the overbearing weight of the star's matter pressing in from all sides. The Chandrasekhar limit for a white dwarf is 1.4 $M_\odot$, and the Oppenheimer-Volkov limit for a neutron star is about 3 $M_\odot$.

According to calculations first done in 1964 and experiments verifying the theory done in 1968, many elementary particles, such as neutrons (and protons, but not electrons), are actually composed of smaller particles, called **quarks**. Normally, quarks cannot exist as separate

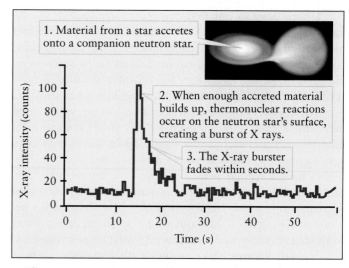

1. Material from a star accretes onto a companion neutron star.

2. When enough accreted material builds up, thermonuclear reactions occur on the neutron star's surface, creating a burst of X rays.

3. The X-ray burster fades within seconds.

FIGURE 10-27 X Rays from an X-Ray Burster A burster emits X rays with a constant low intensity interspersed with occasional powerful bursts. This burst was recorded in September 1975 by an X-ray telescope that was pointed toward the globular cluster NGC 6624.

objects—in the cases of neutrons and protons, they clump together in groups of three. However, when the pressure on them is great enough, neutrons can become so compressed that they dissolve into their constituent quarks. As free particles, quarks obey the Pauli exclusion principle, meaning that identical quarks cannot occupy the same place at the same time. As with electrons and neutrons, quarks therefore create a pressure that resists collapse, thereby creating quark stars. The upper limit for the mass of quark stars is unknown, but it could not be much greater than the Oppenheimer-Volkov limit.

If the remnant core from a supernova is above the mass limit of neutron stars (and quark stars, if their existence is verified), then there is another fate in store for it. As noted earlier, the gravitational attraction near a neutron star is so strong that the particles escaping from its surface must be moving at roughly half the speed of light. If the stellar corpse has slightly greater than 3 $M_\odot$, so much matter becomes crushed into such a small volume that the repulsion created by the Pauli exclusion principle on its neutrons is quickly overcome. The remnant therefore collapses further and the escape velocity exceeds the speed of light. Because nothing can travel faster than light, nothing—not even light—can leave such a collapsed object. The discovery of neutron stars and the calculations regarding the limit to the Pauli exclusion principle inspired astrophysicists to consider seriously one of the most bizarre and fantastic objects ever predicted by modern science—the black hole, which we examine next. Figure 10-28 summarizes the evolutionary history of stars, as discussed in Chapters 9 and 10.

What is the Oppenheimer-Volkov limit?

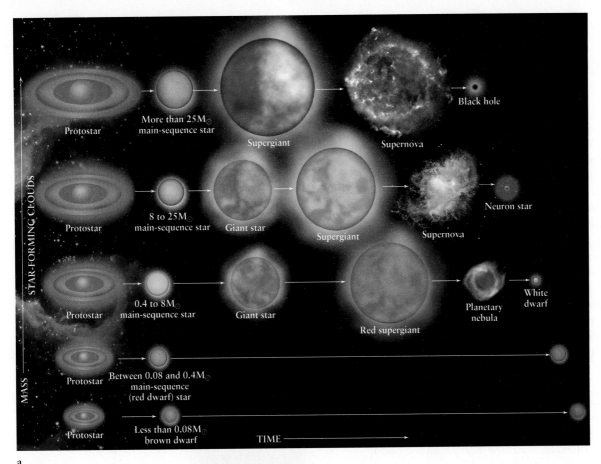

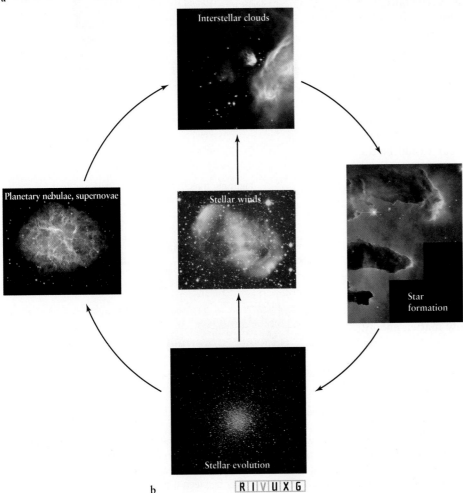

a

b

RIVUXG

FIGURE 10-28 **A Summary of Stellar Evolution** (a) The evolution of isolated stars depends primarily on their masses. The higher the mass, the shorter the lifetime. Stars less massive than about 8 $M_\odot$ can eject enough mass to become white dwarfs. High-mass stars can produce Type II supernovae and become neutron stars or black holes. The horizontal (time) axis is not to scale, but the relative lifetimes are accurate. (b) The cycle of stellar evolution is summarized in this figure. *(Top inset: Infrared Space Observatory, NASA; right inset: Anglo-Australian Observatory/ J. Hester and P. Scowen, Arizona State University/NASA; bottom inset: NASA; left inset: NASA; middle inset: Anglo-Australian Observatory)*

BLACK HOLES

Stars are formed and held together by gravity. At each stage that we have considered, they are prevented from collapsing further by the outward forces generated by fusion and electromagnetism. We now explore the fate of stellar remnants with more than 3 $M_\odot$ in which gravitation "wins" the battle, but eventually loses the war between the inward and outward forces.

When the gravitational force of an object is so great that it overcomes all opposing repulsive forces or pressures (such as neutron degeneracy pressure), the object collapses in on itself. Its gravitational attraction then becomes so strong that nothing—not even light—can escape from it. When this happens, the matter and the space around it become a *black hole*.

Black holes inspire awe, fear, and uncertainty. Many people harbor the mistaken belief that black holes are giant vacuum cleaners destined to "suck up" all matter in the universe. Happily, the equations that describe them reveal that black holes are more benign than that, but they are still truly strange.

The idea that gravity could prevent matter from escaping was first put forward in the eighteenth century, but it was not until the twentieth century, when Albert Einstein presented his ideas of *relativity* (how matter and motion affect mass, distance, and time), that the modern concept of a black hole developed. We must first explore the relativity theories in order to understand these exotic objects.

THE RELATIVITY THEORIES

It seems reasonable that if I watch you traveling at, say, 100 km/h (60 mph), my measurements of your mass, the rate your watch runs, and the length of your car would be the same as the measurements you would make of these things as you are traveling. In fact, they will not be the same. At the beginning of the twentieth century, Einstein began a revolution in science by disregarding his common sense, making a few physical assumptions about nature, and using mathematics to explore their consequences. The resulting equations revealed profound, albeit highly counterintuitive, insights into how nature works. Armed with the equations that Einstein derived, scientists have been exploring and using the consequences of his theories to develop new technology and a deeper understanding of how matter, energy, space, and time interact.

10-18   Special relativity changes our conception of space and time

In 1905, Albert Einstein derived his **theory of special relativity**, a description of how motion affects our measurements of distance, time, and mass. He was guided by two innovative ideas. Although the implications of the theory proved revolutionary, the first notion seems simple:

Your description of physical reality is the same regardless of the (constant) velocity at which you move.

In other words, as long as you are moving in a straight line at a constant speed, you experience the same laws of physics as anyone else moving at any other constant speed and in any other direction. To illustrate this first idea, suppose you were inside a closed boxcar moving smoothly in a straight line at 100 km/h and you dropped a pen from a height of 2 m. You time how long it takes the pen to reach the floor (about 0.64 s). Then the train is stopped and you repeat the same experiment while in the station. The time it would take the pen to fall would be identical.

The second idea seems more bizarre:

Regardless of your speed or direction, you always measure the speed of light to be the same.

Suppose that you are in a (very powerful) car moving toward a distant street lamp at 150,000 km/s (93,000 mi/s). In simpler terms, this is 50% the speed of light, denoted $0.5c$ (Figure 10-29a). A friend of yours leaning on the lamppost

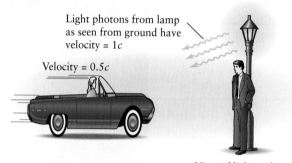

View of light and car from the ground

a

View of light and ground from the car

b

FIGURE 10-29 **The Speed of Light Is Constant** (a) As seen from the ground, photons of light from the lamp are traveling toward the car with a speed $v = c$ (ignoring the slight decrease in speed due to the presence of the air). (b) As seen from the car, moving toward the lamp at $v = 0.5c$, the photons are also traveling at the speed $v = c$. The difference in color between the light in the two figures is due to the Doppler shift.

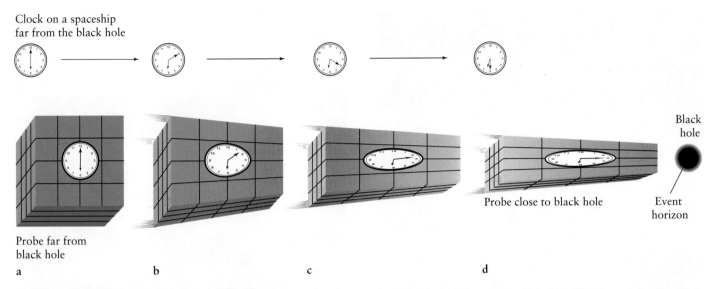

Clock on a spaceship far from the black hole

Black hole

Probe close to black hole

Event horizon

Probe far from black hole

a   b   c   d

FIGURE 10-41 Effect of a Black Hole's Tidal Force on Infalling Matter (a) A cube-shaped probe 1500 km from a 5-$M_\odot$ black hole. (b, c, d) Near the Schwarzschild radius, the probe is pulled long and thin by the difference in the gravitational forces felt by its different sides. This tidal effect is a greatly magnified version of the Moon's gravitational force on Earth. The probe changes color as its photons undergo extreme gravitational redshift and time slows down on the probe, as seen from far away.

As the probe nears the black hole, the blue photons leaving it must give up more and more energy to escape the increasing gravitational force. However, unlike a projectile fired upward, photons cannot slow down. Rather, they lose energy by increasing their wavelengths. This is another example of the gravitational redshift predicted by general relativity. The closer the probe gets to the event horizon, the more its light is redshifted—first to green, then yellow, then orange, then red, then infrared, and, finally, to radio waves (see Figure 10-41b, c, d).

How would a 1 $M_\odot$ black hole 1 AU from Earth affect our planet?

Stranger still is the black hole's effect on time. General relativity predicts that when the probe approaches within a few Schwarzschild radii of the black hole, its infall rate will slow down as seen from far away. Also, signals from the probe show you that its clocks are running much more slowly than they did when it left your spacecraft. Time dilation becomes so great near the event horizon that the probe will appear to hover above it and its clocks will stop. Anyone in the probe would observe something else altogether: They see the probe actually cross the event horizon and continue falling toward the black hole's singularity in a normal period of time, according to their own watches. Pulled apart by tidal effects, the probe disintegrates as it falls inward. Contrary to the science fiction concept of traveling great distances quickly by passing through a black hole, calculations indicate that objects entering them could not survive passage through, even if there were a way to come out somewhere else.

Could a black hole be connected to another part of spacetime or even some other universe? General relativity predicts such connections, called **wormholes,** for Kerr black holes, but astrophysicists are skeptical that the equations are correct in this regard. Their conviction is called **cosmic censorship:** Nothing can leave a local region of space that contains a singularity (that is, a black hole).

EVIDENCE FOR BLACK HOLES

10-24   Several binary star systems contain black holes

Black holes are more than fine points of relativity theory: They are real. Their presence has been observed by their effects on the orbits of other stars and on gas and dust near them, and more of them are being located all of the time. To find evidence for black holes, we look first to binary star systems.

The technique for detecting black holes formed from collapsing stars is based on the interaction between the black hole and its binary companion. When one star in a close binary becomes a black hole, its gravitational attraction pulls off some of its companion's atmosphere. However, such black holes have diameters of only a few kilometers, so there is not enough room for all that gas to fall straight in. Rather, the infalling gas swirls into the black hole like water going down a bathtub drain (Figure 10-42). The gas waiting to fall in forms an **accretion disk,** a disk of gas and dust spiraling in toward the black hole. Magnetic fields in the disk help pull the debris inward.

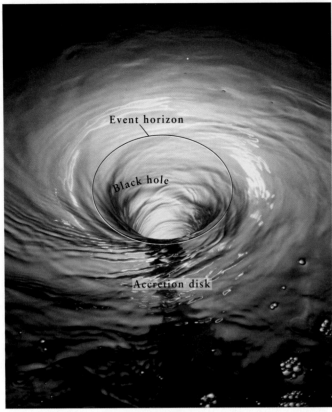

RIVUXG

FIGURE 10-42 Formation of an Accretion Disk Just as the water in this photograph swirls around waiting to get down the drain, the matter pulled toward a black hole spirals inward. Angular momentum of the infalling gas and dust causes them to form an accretion disk around the hole. *(Chris Collins/Corbis)*

2. As gases spiral toward the black hole, they are heated by friction and magnetic fields: Just outside the black hole, they are hot enough to emit X rays.

HDE 226868 (blue-white supergiant)

X rays

Black hole

1. Gases from the supergiant are captured into an accretion disk around the black hole.

FIGURE 10-43 X Rays Generated by Accretion of Matter Near a Black Hole Stellar-remnant black holes, such as Cygnus X-1, LMC X-3, V404 Cygni, and probably A0620-00, are detected in close binary star systems. This drawing (of the Cygnus X-1 system) shows how gas from the 30 $M_\odot$ companion star, HDE 226868, transfers to the black hole, which has at least 11 $M_\odot$. This process creates an accretion disk. As the gas spirals inward, friction and compression heat it so much that the gas emits X rays, which astronomers can detect. *(Courtesy of D. Norton, Science Graphics)*

Calculations reveal that this gas is compressed and heated so much from the collisions of its particles that it releases X rays (Figure 10-43). Thus, if a visible star has a sufficiently tiny, sufficiently massive X-ray-emitting companion, we have located a black hole.

To date, in the Milky Way Galaxy alone, at least 10 stellar-remnant black holes in binary star systems have been identified (Table 10-2). There is also growing evidence that black holes can collide with each other or with different types of objects, such as neutron stars. Such collisions would lead to the emission of tremendous amounts of gravitational radiation, which astronomers are trying to detect.

10-25 Other black holes range in mass up to billions of solar masses

Based on the equations of general relativity, the fate of massive neutron stars led as early as 1939 to the idea of black holes. Since then, calculations have supported at least three other types of black holes:

1. Early in the life of the universe, black holes could have formed from the condensation of vast amounts of gas, as well as the collisions of stars, during the process of galaxy formation, thereby creating **supermassive black holes,** each with millions or billions of solar masses.

2. We saw in Section 9-2 that stars form in clusters. If a cluster forms with enough stars concentrated in a sufficiently small volume of space, collisions between stars in the center of the cluster can lead to the formation of a black hole with between a few hundred and a few thousand solar masses. These are called **intermediate-mass black holes.**

3. Black holes could have formed during the explosive beginning of the universe as tiny amounts of matter were compressed sufficiently to form **primordial black holes.** These bodies would have masses from grams to the mass of a planet.

Why is the gas in an accretion disk heated?

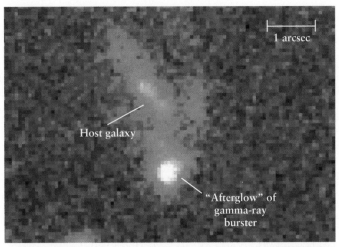

FIGURE 10-50 **The Host Galaxy of a Gamma-Ray Burst**
The Hubble Space Telescope recorded this visible-light image in 1999, 16 days after a gamma-ray burst was observed at this location. The image shows a faint galaxy that is presumed to be the home of the gamma-ray burst. The galaxy has a bright blue color, indicating the presence of many recently formed stars. *(Andrew Fruchter, STScI; NASA)*

are looking down the axis from Earth, the gamma rays create the burst, and the gas, traveling more slowly, creates the subsequent afterglow. Observations reveal that hypernovae and the associated gamma-ray bursts are extremely messy affairs, with powerful aftershocks, rather than a single, smooth outburst. The details of the explosions and the mechanism by which they generate gamma-ray bursts are still under investigation.

Other gamma-ray bursts are thought to occur when black holes collide; when a black hole swallows a neutron star; or when a pair of neutron stars collide and form a black hole. Calculations confirm that the collision of such dense objects should indeed lead to a rapid, intense burst of gamma rays.

To better understand gamma-ray bursts, astronomers have constructed telescopes both on Earth and in orbit dedicated to rapidly shifting to look in the direction in which these events occur in space. These telescopes, including the *Swift Gamma-Ray Burst Mission* and the *High Energy Transient Explorer-2*, see these bursts and provide new insights into their origins and dynamics. *Swift* carries not only a gamma-ray telescope to locate the bursts, but also separate X-ray and optical telescopes to study their afterglows.

10-28   Black holes evaporate

In exploring the fate of black holes, astrophysicists find that, once again, these objects confound common sense—they evaporate! With a black hole cloaked behind the event horizon and its mass collapsed into a singularity, it must seem that there is no way of getting mass from the black hole back out into the universe again. No way, that is, until you recall that mass and energy are two sides of the same coin. What if there was a way of converting the black hole's mass into a form of energy that *could* get out, such as gravitational energy? The transformation does occur, according to an idea first proposed by Stephen Hawking.

Black holes convert their mass into energy by a process called **virtual particle** production. If things in this chapter aren't weird enough, consider this: The laws of nature allow pairs of particles, called virtual particles, to spontaneously appear. Each pair consists of a particle and its antiparticle—such as a proton and an antiproton, an electron and an antielectron (positron), or a pair of photons. The particles in each pair are identical, except for having opposite electric charges, and they annihilate each other and normally disappear without a trace within a very short period of time, typically 10^{-21} s.

We know that virtual particles actually flash into existence spontaneously throughout the universe because some have been made into real particles in the laboratory. This is done using high-speed particles in particle accelerators that slam into the pair of virtual particles and push them apart before they annihilate each other. The mass or energy given to the virtual particles in making them real comes from the laboratory particles, which lose an equal amount of energy. Calculations reveal that billions and billions of pairs of virtual particles are seething in and out of existence in your body every second.

Now, suppose two virtual particles form *just outside* the event horizon of a black hole. They would experience an exceptionally powerful gravitational field. If one particle is created slightly farther from the hole than its companion, the two virtual particles feel a tidal force from the black hole's tremendous gravitational pull. If the gravitational tidal force is strong enough, it can pull the two virtual particles apart before they can annihilate each other. This separation makes them real (Figure 10-51).

To conserve momentum, at least one of the newly formed real particles always falls into the black hole. But the other particle will sometimes have enough energy to escape from the vicinity of the black hole. This latter particle flies free into space, and the black hole has effectively emitted energy equal to $E = mc^2$, where m is the mass of the freed particle and c is the speed of light. Alternatively, the black hole may lose the energy due to one of two photons created, $E = hc/\lambda$, where λ is the wavelength of the photon and h is Planck's constant.

How long do gamma-ray bursts last?

Where the virtual particles become real is a temporary void in space outside the event horizon. The black hole fills the void with energy that comes from its mass. This energy is transmitted outside the event horizon as gravitational radiation to replace the energy taken away by the newly formed particles. This is called the **Hawking process**, and

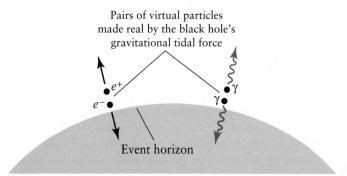

Pairs of virtual particles
made real by the black hole's
gravitational tidal force

Event horizon

FIGURE 10-51 **Evaporation of a Black Hole** Throughout the universe, pairs of virtual particles spontaneously appear and disappear so quickly that they do not violate any laws of nature. The tidal force just outside of the event horizon of a black hole is strong enough to tear apart two virtual particles that appear there before they destroy each other. The gravitational energy that goes into separating them makes them real and, therefore, permanent. At least one of each pair of newly created particles falls into the black hole. Sometimes the other particle escapes into the universe. Because the gravitational energy used to create the particles came from the black hole, the hole loses mass and shrinks, eventually evaporating completely. Here we see just a few particles in the making: an electron (e^-) and a positron (e^+) and a pair of photons (γ).

the particles leaving the vicinity of the black hole are called *Hawking radiation* (even though many of the particles are not electromagnetic radiation).

The time it takes a black hole to completely evaporate by the Hawking process increases with the mass of the hole. More massive black holes have lower tidal forces at their event horizons than do less massive and, therefore, smaller-diameter black holes. Therefore, the *rate* at which virtual particles are converted into real particles around bigger black holes is actually slower than it is around smaller ones. The faster the real particles are produced, the faster they wear down the mass of the black hole; thus, lower-mass black holes evaporate more rapidly. Indeed, calculations indicate that in its final moments of evaporation, a black hole should create particles so quickly that it appears to explode.

A stellar black hole of 5 $M_\odot$ would take more than 10^{62} years to evaporate (much longer than the present age of the universe), whereas a 10^{10}-kg primordial black hole (equivalent to the mass of Mount Everest; Figure 10-52) would take only about 15 billion years to evaporate. Depending on the precise age of the universe, a black hole of this size and age should be in the final throes of evaporating. Astronomers are trying to identify events in space that correspond to the violent, particle-generating deaths of primordial black holes.

Why does at least one particle in the Hawking process always fall into the black hole?

10-29 Frontiers yet to be discovered

The incredible telescopes now in existence or scheduled to go on line soon are revolutionizing our understanding of the final stages of star formation. In coming years, astronomers hope to pin down the details of how planetary nebulae, novae, supernovae, and stellar remnants behave. Among the specific questions to be answered—What are the physical processes that cause planetary nebulae? Why do planetary nebulae have such a variety of shapes? What causes the hourglass shape of so many stellar remnants? Why are novae composed of many blobs of gas? Do quark stars really exist, and, if so, what are their properties?

Black holes take us to the limits of science. They represent a state of matter that we are not yet fully capable of explaining. This state of affairs makes black holes an especially intriguing field of study. Understanding the nature of black hole singularities is a major research goal. Numerous other frontiers exist in this realm. For example, by directly observing gravitational radiation, we will have a new tool for exploring violent activity in the universe. We still need to understand the details of black hole formation, both from stars and by other means. We have yet to understand in detail why mid-mass and high-mass black holes have the masses they have.

If black holes merge, as is now expected, the details of these events and their effects on the rest of the universe have still to be understood. It also remains to be shown whether or not the cosmic censorship theorem is correct. If not, the implications of the existence of wormholes need to be examined in more depth. Among the more intriguing issues surrounding black holes is whether primordial black holes

R I V U X G

FIGURE 10-52 **Mount Everest** Primordial black holes, formed at the beginning of time, may have had masses similar to that of Mount Everest, as shown here. Calculations indicate that the universe is old enough for such black holes to have evaporated. Their final particle production rate is so high that they should look as though they are exploding. *(Galen Rowell/Corbis)*

exist. Although the evaporation process has not yet been seen, we should soon have the technology to discover it, if it is occurring.

Gamma-ray bursts still require much explanation. How does so much energy get transformed so quickly into electromagnetic radiation in these events? Why do some supernovae emit most of their energy as gamma rays, while others emit most as less energetic electromagnetic photons?

SUMMARY OF KEY IDEAS

- Stars with higher masses fuse more elements than do stars with lower masses.

- Stars lose mass via stellar winds throughout their lives.

Low-Mass Stars and Planetary Nebulae
- A low-mass (below 8 $M_\odot$) main-sequence star becomes a giant when hydrogen shell fusion begins. It becomes a horizontal-branch star when core helium fusion begins. It enters the asymptotic giant branch and becomes a supergiant when helium shell fusion starts.

- Stellar winds during the thermal-pulse phase eject mass from the star's outer layers.

- The burned-out core of a low-mass star becomes a dense carbon–oxygen body, called a white dwarf, with about the same diameter as that of Earth. The maximum mass of a white dwarf is 1.4 $M_\odot$ (the Chandrasekhar limit).

- Explosive hydrogen fusion may occur in the surface layer of a white dwarf in some close binary systems, producing sudden increases in luminosity that we call novae.

- An accreting white dwarf in a close binary system can also become a Type Ia supernova when carbon fusion ignites explosively throughout such a degenerate star.

High-Mass Stars and Type II Supernovae
- After exhausting its central supply of hydrogen and helium, the core of a high-mass (above 8 $M_\odot$) star undergoes a sequence of other thermonuclear reactions. These include carbon fusion, neon fusion, oxygen fusion, and silicon fusion. This last fusion eventually produces an iron core.

- A high-mass star dies in a supernova explosion that ejects most of the star's matter into space at very high speeds. This Type II supernova is triggered by the gravitational collapse and subsequent bounce of the doomed star's core.

- Neutrinos were detected from Supernova 1987A, which was visible to the naked eye. Its development supported theories of Type II supernovae.

Neutron Stars and Pulsars
- The core of a high-mass main-sequence star containing between 8 and 25 $M_\odot$ becomes a neutron star. The cores of slightly more massive stars may become quark stars. A neutron star is a very dense stellar corpse consisting of closely packed neutrons in a sphere roughly 20 km in diameter. The maximum mass of a neutron star, called the Oppenheimer-Volkov limit, is about 3 $M_\odot$.

- A pulsar is a rapidly rotating neutron star with a powerful magnetic field that makes it a source of periodic radio and other electromagnetic pulses. Energy pours out of the polar regions of the neutron star in intense beams that sweep across the sky.

- Some X-ray sources exhibit regular pulses. These objects are believed to be neutron stars in close binary systems with ordinary stars.

- Explosive helium fusion may occur in the surface layer of a companion neutron star, producing the sudden increase in X-ray radiation called an X-ray burster.

The Relativity Theories
- Special relativity reveals that space and time are intimately connected and change with an observer's motion.

- As seen by observers moving more slowly, the faster an object moves, the slower time passes for it (time dilation) and the shorter it becomes (length contraction).

- According to general relativity, mass causes space to curve and time to slow down. These effects are significant only near large masses or compact objects.

Black Holes
- If a stellar corpse is more massive than about 3 $M_\odot$, gravitational compression overcomes neutron degeneracy (or the theorized quark degeneracy pressure) and forces it to collapse further and become a black hole.

- A black hole is an object so dense that the escape velocity from it exceeds the speed of light.

- Observations indicate that some binary star systems harbor black holes. In such systems, gases captured by the black hole from the companion star heat up and emit detectable X rays and jets of gas.

- Supermassive black holes exist in the centers of many galaxies. Intermediate-mass black holes exist in clusters of stars. Very low mass (primordial) black holes may have formed at the beginning of the universe.

- The event horizon of a black hole is a spherical boundary where the escape velocity equals the speed of light. No matter or electromagnetic radiation can escape from inside the event horizon. The distance from the center of the black hole to the event horizon is called the Schwarzschild radius.

- The matter inside a black hole collapses to a singularity. The singularity for nonrotating matter is a point at the center of the black hole. For rotating matter, the singularity is a ring inside the event horizon.

- Matter inside a black hole has only three physical properties: mass, angular momentum, and electric charge.

• Nonrotating black holes are called Schwarzschild black holes. Rotating black holes are called Kerr black holes. The event horizon of a Kerr black hole is surrounded by an ergoregion in which all matter must constantly move to avoid being pulled into the black hole.

• Matter that approaches a black hole's event horizon is stretched and torn by the extreme tidal forces generated by the black hole; light from the matter is redshifted; and time slows down.

• Black holes can evaporate by the Hawking process, in which virtual particles near the black hole become real. These transformations of virtual particles into real ones decrease the mass of a black hole until, eventually, it disappears.

Gamma-Ray Bursts
• Gamma-ray bursts are events believed to be caused by some supernovae and by the collisions of dense objects, such as neutron stars or black holes. Some occur in the Milky Way and nearby galaxies, whereas many occur billions of light-years away from Earth.

• Typical gamma-ray bursts occur for a few tens of seconds and emit more energy than the Sun will radiate over its entire 10-billion-year lifetime.

WHAT DID YOU THINK?

1 *Will the Sun someday cease to shine brightly? If so, how will this occur?* Yes. The Sun will shed matter as a planetary nebula in about 6 billion years and then cease nuclear fusion. Its remnant white dwarf will dim over the succeeding billions of years.

2 *What is a nova? How does it differ from a supernova?* A nova is a relatively gentle explosion of hydrogen gas on the surface of a white dwarf in a binary star system. Supernovae, on the other hand, are explosions that cause the nearly complete destruction of massive stars.

3 *What are the origins of the carbon, silicon, oxygen, iron, uranium, and other heavy elements on Earth?* These elements are created during stellar evolution by supernovae and by colliding neutron stars.

4 *What are cosmic rays? Where do they come from?* Cosmic rays are high-speed particles (mostly hydrogen and other atomic nuclei) in space. Many of them are thought to have been created as a result of supernovae.

5 *What is a pulsar?* A pulsar is a rotating neutron star in which the magnetic field's axis does not coincide with the rotation axis. The beam of radiation it emits periodically sweeps across our region of space.

6 *Are black holes empty holes in space? If not, what are they?* No, black holes contain highly compressed matter—they are not empty.

7 *Does a black hole have a solid surface? If not, what is at its surface?* No. The surface of a black hole, called the event horizon, is empty space. No stationary matter exists there.

8 *What power or force enables black holes to draw things into themselves?* The only force that pulls things in is the gravitational attraction of the matter in the black hole.

9 *How close to a black hole do you have to be for its special effects to be apparent?* About 100 times the Schwarzschild radius.

10 *Can you use black holes to travel to different places in the universe?* No, most astronomers believe that the wormholes predicted by general relativity do not exist.

11 *Do black holes last forever? If not, what happens to them?* No, black holes evaporate.

Review Questions

1. Will the Sun shed most of its mass, and, if so, what is that event called? a. yes, as a planetary nebula, b. yes, as a supernova, c. yes, as a white dwarf, d. yes, as a neutron blast, e. no

2. A white dwarf is composed primarily of a. neutrons, b. hydrogen and helium, c. iron, d. cosmic rays, e. carbon and oxygen.

3. What prevents a neutron star from collapsing? a. hydrogen fusion, b. friction, c. electron degeneracy pressure, d. neutron degeneracy pressure, e. helium fusion

4. What is the difference between a giant star and a supergiant star?

5. Why is knowing the temperature in a star's core so important in determining which nuclear reactions can occur there?

6. What determines the temperature in the core of a star?

7. What is a planetary nebula, and how does it form?

8. What is the Chandrasekhar limit?

9. What is a neutron star?

10. Compare a white dwarf and a neutron star. Which of these stellar corpses is more common? Why?

11. What is the Oppenheimer-Volkov limit?

12. On an H-R diagram, sketch the evolutionary track that the Sun will follow between the time it leaves the

main sequence and when it becomes a white dwarf. Approximately how much mass will the Sun have when it becomes a white dwarf? Where will the rest of the mass have gone?

13. Why have searches for supernova remnants at visible wavelengths been less fruitful than searches at other wavelengths?

14. Why do astronomers believe that pulsars are rapidly rotating neutron stars?

15. What is the difference between Type Ia and Type II supernovae?

16. Compare a nova with a Type Ia supernova. What do they have in common? How are they different?

17. Compare a nova and an X-ray burster. What do they have in common? How are they different?

18. Describe what X-ray pulsars, pulsating X-ray sources, and X-ray bursters have in common. How are they different manifestations of the same type of astronomical object?

19. Which property, if any, of normal matter ceases to exist in a black hole? **a.** mass, **b.** chemical composition, **c.** angular momentum, **d.** electric charge, **e.** all of these properties exist in a black hole

20. Supermassive black holes are found in which of the following locations? **a.** in the centers of galaxies, **b.** in globular clusters, **c.** in open (or galactic) clusters, **d.** between galaxies, **e.** in orbit with a single star

21. Which feature is found with Kerr black holes but not Schwarzschild black holes? **a.** a singularity, **b.** an event horizon, **c.** gravitational redshift of photons outside of the black hole, **d.** an ergoregion, **e.** warping of nearby spacetime

22. Under what conditions do all outward pressures on a collapsing star fail to stop its inward motion?

23. In what way is a black hole blacker than black ink or a black piece of paper?

24. If the Sun suddenly became a black hole, how would Earth's orbit be affected?

25. What is cosmic censorship?

26. What are the differences between rotating and nonrotating black holes?

27. Why are all of the observed stellar-remnant black hole candidates members of close binary systems?

28. If light cannot escape from a black hole, how can we detect X rays from such objects?

Observing Projects

29. Planetary nebulae are found all across the sky. Some of the brightest are listed in Chapter 10 of the *Discovering the Essential Universe* Web site. With the help of star maps or *Starry Night Enthusiast*™ software, locate and observe as many planetary nebulae as possible on clear, Moonless nights, using the largest telescope at your disposal. Some of the more notable planetary nebulae include: Little Dumbbell (M76), NGC 1535, Eskimo, Ghost of Jupiter, Owl (M97), Ring (M57), Blinking Planetary, Dumbbell, Saturn Nebula, and NGC 7662. Note and compare the various shapes of the different nebulae. In which cases can you see the central star?

30. The ultimate fate of a star depends upon its mass. For stars with mass less than about 8 $M_\odot$, nuclear fusion in the star's core results in hydrogen conversion to helium and a loss of gravitational force from the core as mass is converted to radiative energy. Late in the life of such a star, the outer layers expand and cool to form a red-giant star. At some point in this process, core readjustment occurs, as helium "burning" becomes violent. The resulting intense bursts of radiation produce a relatively rapid expansion of the outer layers of this giant star. For a short time, astronomically speaking, we see the results of this expansion, and the interaction of radiation with matter surrounding the star, as a planetary nebula. These diffuse gas clouds take many forms and shapes as a result of this interaction. Left behind at the center of this expansion is a small, hot core of the star, a white dwarf star. Energy generation by nuclear fusion reactions has ceased in this star and the star is cooling off as it radiates energy into space.

You can use *WorldWide Telescope* (WWT) to examine the fine details of this expansion in some of these transient clouds at the resolution of the Hubble Space Telescope.

Open WWT and ensure that **Sky** is displayed in the **Look At** drop-down list and **DDS: Digital Sky Survey** is displayed in the **Imagery** drop-down list. Click on the **Explore** menu and click on the **Hubble Images** thumbnail. There are several superb images of planetary nebulae in this collection. Locate and click on the thumbnails for the following planetary nebulae to examine the different shapes that can result from the interaction between expanding gas clouds and their surrounding environment and magnetic fields in these late phases of low-mass stellar evolution: Helix Nebula, Ring Nebula, Retina Nebula, Butterfly wing-shaped planetary nebula, and Eskimo Nebula. Answer the following questions for each of these "stellar corpses."

a. Is the white dwarf star visible in this image? **b.** What is the approximate shape of this gas cloud? **c.** Is there evidence for outward pressure upon the gases in this cloud? Describe the features that lead to your conclusion about this outward flow of gas. **d.** What colors are visible in this gas cloud? These colors are produced by fluorescence in atoms excited by the UV light from the central white dwarf star. **e.** Are these colors separated into different spherical shells or are the colors mixed within the cloud?

In stars whose mass exceeds about 8 $M_\odot$, nuclear "burning" of a sequence of heavier elements takes place with increasing violence as gravitational compression heats the star's core, until the star explodes to release enormous energy that scatters the remaining matter into space. This explosion will also implode matter inward to produce compact objects with bizarre properties. The implosion in an intermediate-mass object with mass between about 8 and 25 $M_\odot$ will produce a neutron star, composed entirely of neutrons. This object has the typical size of a small city, with a mass of between 1.4 and 3 $M_\odot$. Many of these compact objects emit narrow beams of radiation that sweep across the Earth as they rotate, producing precisely spaced pulses, leading to their name, the pulsars.

The force of implosion within a supernova for stars with masses greater than 25 $M_\odot$ is so large that no force in nature can prevent the infall of matter and a black hole is produced. Nothing, not even light, can escape the gravitational force of this singularity in space.

Click on the thumbnails for the following objects in the **Hubble Images** sequence to see regions in space where these tremendous explosions have disrupted stars in past history: Crab Nebula, Giant Hubble mosaic of the Crab Nebula, Supernova remnant Cassiopeia A, Supernova 1987A mosaic (zoom in to see rings of light caused by illumination of shells of gas by UV light from the supernova), and Gum Nebula (open **Explore > Named Objects > Gum Nebula** and zoom out to a field of view of about 4° to see the full extent of this supernova remnant). Answer the following questions for each of them.

f. What shape is the supernova remnant? **g.** Does the gas cloud have structure within it? **h.** Are supernova remnants all about the same size in our sky?

Combined Infrared and Ultraviolet Images of the Andromeda Galaxy, 2.5 Million Light-Years Away and 260,000 Light-Years in Diameter (Red Indicates Cool, Dusty Regions; Blue Are Hot, Young Stars; Older Stars Are Depicted in Green; and Pinkish Regions Are Where Old and Young Stars Coexist) *(NASA/JPL-Caltech/K. Gordon [Univ. of Ariz.] & GALEX Science)*

R I V U X G

WHAT DO YOU THINK?

1 What is the shape of the Milky Way Galaxy?

2 Where is our solar system located in the Milky Way Galaxy?

3 Is the Sun moving through the Milky Way Galaxy and, if so, about how fast?

4 Are most of the stars in spiral galaxies located in their spiral arms?

5 Do all galaxies have spiral arms?

6 Are galaxies isolated objects?

7 Is the universe contracting, unchanging in size, or expanding?

Answers to these questions appear in the text beside the corresponding numbers in the margins and at the end of the chapter.

Centuries of observations have firmly established that our solar system is part of an enormous assemblage of hundreds of billions of stars, along with gas, dust, and other matter, all held together by their mutual gravitational attraction. This is our **Milky Way Galaxy**, which we study first in this chapter.

Most of the stars in our Galaxy are located in a disk that looks from its edge like a flying saucer in an old science fic-

tion movie (Figure 11-1a). The inner part of the disk is filled with stars, some of which form a bar, as shown (Figure 11-1b). Spiral arms swirl out from the ends of the bar, and, within these arms, new stars form from the debris of earlier generations of stars. The Galaxy's remaining stars are located in a spherical halo that surrounds the disk.

After considering our Galaxy, we explore the other galaxies, the clusters of galaxies, the superclusters of galaxies, and the nature of the space that connects them, along with what little we know about the vast bulk of the matter in the universe called dark matter. We end this chapter by examining the supermassive black holes that lie at the hearts of most galaxies, and the effects these black holes have on the surrounding matter.

In this chapter you will discover

• the Milky Way Galaxy—billions of stars, along with gas and dust, bound together by mutual gravitational attraction

• the structure of our Milky Way Galaxy

• Earth's location in the Milky Way

• how interstellar gas and dust enable star formation to continue

• that observations reveal the presence of significant mass in the Milky Way that astronomers have yet to identify

• that there is a massive black hole at the center of our Galaxy

• how galaxies are categorized by their shapes

• the processes that produce galaxies of different shapes

• that galaxies are found in clusters that contain huge amounts of dark matter

• why clusters of galaxies form in superclusters

• how some galaxies merge and others devour their neighbors

• that the universe is changing size

• bright and unusual objects, called active galaxies

• distant, luminous quasars

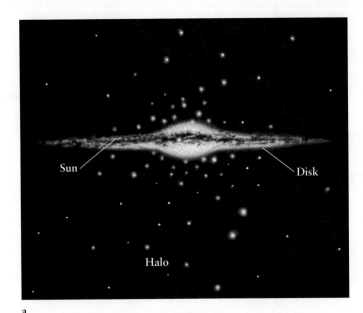

a

b

FIGURE 11-1 **Schematic Diagrams of the Milky Way** (a) This edge-on view shows the Milky Way's disk, containing most of its stars, gas, and dust, and its halo, containing many old stars. Individual stars in the halo are too dim to show, so the bright regions in the halo represent clusters of stars. (b) Our Galaxy has two major arms and several shorter arm segments, all spiraling out from the ends of a bar of stars and gas that passes through the Galaxy's center. The bar's existence and the presence of two major arms were both confirmed by the Spitzer Space Telescope. (b: *NASA/JPL-Caltech/R. Hurt [SSC]*)

- the unusual spectra and small volumes of quasars

- the extremely powerful BL Lac objects

- supermassive black holes that serve as central engines for radio galaxies, quasars, Seyfert galaxies, and BL Lac objects

DEFINING THE MILKY WAY

Prior to the twentieth century, astronomers did not know the large-scale distribution of stars and other matter in the universe. Throughout history, most people, including many astronomers, believed that the Milky Way contains all of the stars in the cosmos. In other words, they thought that the "Galaxy" and the "universe" were the same thing. We begin the study of galaxies by learning how that belief changed.

11-1 Studies of Cepheid variable stars revealed that the Milky Way is only one of many galaxies

The belief that our Galaxy is but one of many was put forth in 1755, when the German philosopher Immanuel Kant suggested that vast collections of stars lie far beyond the confines of the Milky Way. Less than a century later, the Irish astronomer William Parsons observed the structure of some of those "island universes" proposed by Kant. Parsons was the third Earl of Rosse in Ireland. He was rich, he liked machines, and he was fascinated by astronomy. Accordingly, he set about building gigantic telescopes. In February 1845, his pièce de résistance was finished. This telescope's massive mirror measured 1.8 m (6 ft) in diameter and was mounted at one end of an 18-m (60-ft) tube controlled by cables, straps, pulleys, and cranes (Figure 11-2a). For many years, this triumph of nineteenth-century engineering enjoyed distinction as the largest telescope in the world.

Using this new telescope, Lord Rosse examined many of the glowing interstellar clouds previously discovered and catalogued by the Herschels. William Herschel, his sister Caroline, and his son John, among others, discovered and recorded details of fuzzy-looking astronomical objects, called **nebulae** (singular: **nebula**). Using the high resolution provided by his telescope, Lord Rosse observed that some of these nebulae have a distinct spiral structure. A particularly good example is M51, also called the Whirlpool Galaxy or NGC 5194.

Lord Rosse had no photographic equipment in 1845, so he made drawings of what he observed. Figure 11-2b is his drawing of M51. Views such as this inspired him to echo Kant's proposal of island universes. Figure 11-2c shows a modern photograph of M51. It is interesting to note the differences between the perception and interpretation of astronomical objects and a camera's recording of them.

Most astronomers of Rosse's day did not agree with the notion of island universes outside of our Galaxy. They thought that the Milky Way contained all of the stars in the universe—that the Milky Way *was* the universe. In April 1920, a formal discussion, now known as the **Shapley–Curtis debate,** was held at the National Academy of Sciences in Washington, D.C. Harlow Shapley argued that the spiral nebulae are relatively small, nearby objects scattered around our Galaxy. Heber D. Curtis championed the island universe theory, arguing that each of these spiral nebulae is a

a

R I V U X G

b

c

R I V U X G

FIGURE 11-2 **High-Tech Telescope of the Mid-Nineteenth Century** (a) Built in 1845, this structure houses a 1.8 m-diameter telescope, the largest of its day. The improved resolution it provided over other telescopes was similar to the improvement that the Hubble Space Telescope provided over Earthbound optical instruments when it was launched. The telescope, as shown here, was restored to its original state in 1996–1998. (b) Using his telescope, Lord Rosse made this sketch of the spiral structure of the galaxy M51 and its companion galaxy, NGC 5195. (c) A modern photograph of M51 (also called NGC 5194) and NGC 5195. The spiral galaxy M51 in the constellation of Canes Venatici is known as the Whirlpool Galaxy because of its distinctive appearance. The two galaxies are about 20 million ly from Earth. *(a: Birr Castle Demesne; b: Lund Humphries; c: NOAO)*

separate rotating system of stars, much like our own Galaxy. Although the Shapley–Curtis debate focused scientific attention on the size of the universe, nothing was decided, because no one had any firm evidence to demonstrate exactly how far away the spiral nebulae were. Astronomers desperately needed to devise a way to measure the distances to them. A young teacher and former basketball player from Kentucky who moved to Chicago to study astronomy finally achieved this goal. His name was Edwin Hubble.

In 1923, Hubble took historic photograph of M31, then called the Andromeda Nebula. It was one of the spiral nebulae around which controversy raged. On the photographic plate he discovered what first appeared to be a nova. Referring to previous plates of that region, he soon realized that the object was actually a Cepheid variable star. As we saw in Section 9-12, these pulsating stars vary in brightness periodically. Further scrutiny over the next several months revealed many other Cepheids. Figure 11-3 shows a Cepheid in the galaxy M100, at different stages of brightness.

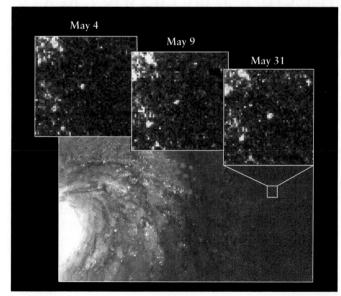

R I V U X G

FIGURE 11-3 **A Cepheid Variable Star in Galaxy M100** One of the most reliable ways to determine the distance to moderately remote galaxies is to locate Cepheid variable stars in them, as discussed in the text. At 56 million light-years (17 Mpc) from Earth, the galaxy M100 in the constellation Coma Berenices is among the most remote objects whose distances have been determined using Cepheids. Insets: The Cepheid in this view, one of 20 located to date in M100, is shown at different stages in its brightness cycle, which recurs over several weeks. *(Dr. Wendy L. Freedman, Observatories of the Carnegie Institution of Washington; NASA)*

Only a decade before, in 1912, the American astronomer Henrietta Leavitt had published an important study of Cepheid variables. Leavitt examined many of them in the Small Magellanic Cloud, then also believed to be a nebula, but now known to be a galaxy very close to the Milky Way. Leavitt's study led her to the period–luminosity relation for Type I Cepheids (see Section 9-13). Leavitt established that a direct relationship exists between a Cepheid's luminosity (or absolute magnitude) and its period of oscillation, which we saw in Figure 9-14, and which is presented in more detail in Figure 11-4. By observing the star's period and apparent magnitude, its distance can be calculated.

Referring to Figure 11-4, how would the brightness of a Cepheid variable with peak luminosity of 1000 L$_\odot$ change if it were observed every 5 days?

Thanks to this work, Hubble knew that he could use the characteristics of the fluctuating light of Cepheid variable stars to help him calculate the distance to M31 and put the Shapley–Curtis debate to rest once and for all. His observations of Type I Cepheids led him to determine that M31 is some 2.2 million light-years *beyond* the Milky Way. This proves that M31 is not an open or globular cluster in our Galaxy, but rather an enormous separate stellar system—a separate galaxy. M31, now called the *Andromeda Galaxy*, is the most distant object in the universe that can be seen with the naked eye. Similar calculations have been done for all galaxies in which Cepheids can be observed. The distances to even more remote galaxies have been determined from observations of Type Ia supernovae in them.

Hubble's results, which he presented at the end of 1924, did settle the Shapley–Curtis debate. The universe was recognized to be far larger and populated with far bigger objects than most astronomers had imagined. Hubble had discovered the realm of the galaxies. Today, we know that the universe contains myriad galaxies, of which the Milky Way is just one. Like the Milky Way, each **galaxy** is a grouping of millions, billions, or even trillions of stars, along with gas, dust, and matter in other forms, all gravitationally bound together.

We now apply Hubble's method to find Earth's place in the Milky Way Galaxy. Recall that the Sun's proximity to us makes it our best-understood star. It might seem that the nearness of the stars and clouds in the Milky Way would make it the best-understood galaxy. However, the clouds of gas and dust that surround the solar system make it very challenging for astronomers to survey completely the distant parts of the Galaxy, which are only now coming into focus.

THE STRUCTURE OF OUR GALAXY

Because the band of the Milky Way completely encircles us, astronomers long ago suspected that the Sun and all of the stars that we see are part of it. In the 1780s, William Herschel took the first steps toward mapping its structure. He attempted to deduce the Sun's location in the Galaxy by counting the number of stars in 683 regions of the sky. He reasoned that the greatest density of stars should be seen toward the Galaxy's center and a lesser density seen toward the edge. However, Herschel found roughly the same density of stars all along the Milky Way. He therefore concluded that we are at the center of the Galaxy.

Herschel was wrong: Earth has no privileged place in the Milky Way. The Sun is about 26,000 ly (8000 pc) from the Galaxy's center, the **galactic nucleus.** Herschel's physical understanding of the cosmos was incomplete, so he misinterpreted his observations and thus came to an incorrect conclusion.

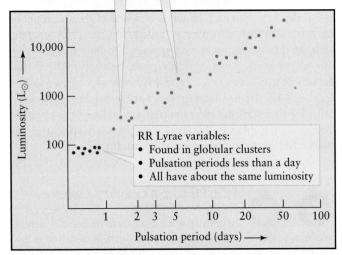

FIGURE 11-4 The Period–Luminosity Relation This graph shows the relationship between the periods and average luminosities of classical (Type I) Cepheid variables and the closely related RR Lyrae stars (discussed in Chapter 9). Each dot represents a Cepheid or RR Lyrae whose luminosity and period have been measured.

11-2   Cepheid variables help us locate our Galaxy's center

While studying star clusters in the 1930s, R. J. Trumpler discovered the reason for Herschel's mistake. Herschel did not know about interstellar gas and dust, which affected his counts of the stars. Trumpler noticed that remote star clusters appear dimmer than would be expected just from their distance alone. Something must be blocking starlight on its way toward Earth. He correctly concluded that interstellar space is not a perfect vacuum. Instead, it contains dust that absorbs light from distant stars. Great patches of this dust are clearly visible in wide-angle photographs (Figure 11-5). Like the stars, this dust is concentrated in the plane of the Galaxy.

This interstellar dust almost completely obscures from view visible light emanating from the center of our Galaxy. Visual photons from there are mostly absorbed or scattered before they reach us. Therefore, Herschel was seeing only nearby stars, and he measured apparent magnitudes that were dimmer than they would have been had there been no interstellar dust. Without adjusting for its effects, he concluded that the stars were farther away than they really are. He also had no idea of the true size of the Galaxy and could not see the vast number of stars located in the general direction of the galactic center that are hidden by the dust.

R I V U X G

FIGURE 11-5 **Our Galaxy** This wide-angle photograph spans half the Milky Way. The Northern Cross is at the left, and the Southern Cross is at the right. The center of the Galaxy is in the constellation Sagittarius, in the middle of this photograph. The dark lines and blotches are caused by hundreds of interstellar clouds of gas and dust that obscure the light from background stars, rather than by a lack of stars. *(Dirk Hoppe)*

INSIGHT INTO SCIENCE

A Little Knowledge Incomplete information often leads to incorrect interpretation of data and, therefore, to incorrect conclusions. Herschel's lack of knowledge about the matter in interstellar space prevented him from correctly interpreting the distribution of stars that surround Earth and, thus, led to his inaccurate conclusion about the position of the Sun within the Galaxy.

Because interstellar dust is concentrated in the plane of the Galaxy, the absorption of starlight is strongest in those parts of the sky covered by the Milky Way. Above or below the plane of the Galaxy, our view is relatively unobscured. Knowledge of our true position in the Galaxy eventually came from observations of globular clusters (see Section 9-14). Shapley used the period–luminosity relationship for variable stars to determine the distances to the then-known 93 globular clusters in the sky. (More than 150 are known today.) From their directions and distances, he mapped out the distribution of these clusters in three-dimensional space. By 1917, Shapley had discovered that the globular clusters are located in a spherical distribution centered not on Earth but on a point in the Milky Way toward the constellation

Sagittarius. Figure 11-6 shows two globular clusters in a relatively clear part of the sky in that direction. Shapley then made a bold conjecture: *The globular clusters orbit the center of the Milky Way, which is located in Sagittarius.* His pioneering research has since been observationally verified. Earth is not at the center of the Galaxy.

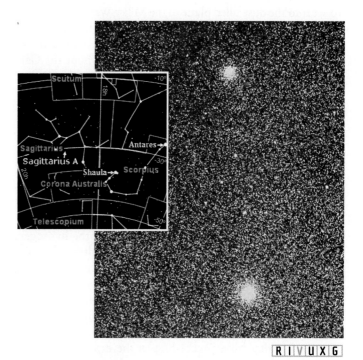

R I V U X G

FIGURE 11-6 **A View Toward the Galactic Center** More than a million stars in the disk of our Galaxy fill this view, which covers a relatively clear window just 4° south of the galactic nucleus in Sagittarius. Beyond the disk stars you can see two prominent globular clusters. Although most regions of the sky toward Sagittarius are thick with dust, very little obscuring matter appears in this tiny section of the sky. *(Harvard Observatory)*

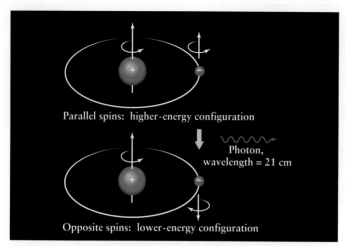

FIGURE 11-7 **Electron Spin and the Hydrogen Atom** Due to their spins, electrons and protons both act as tiny magnets. When an electron and the proton it orbits are spinning in the same direction, their energy is higher than when they are spinning in opposite directions. When the electron flips from the higher-energy to the lower-energy configuration, the atom loses a tiny amount of energy that is radiated as a radio photon with a wavelength of 21 cm.

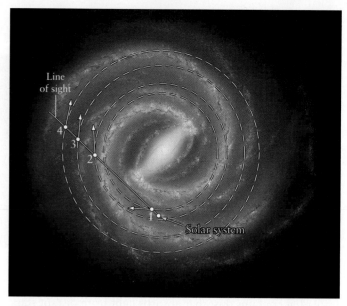

FIGURE 11-8 **A Technique for Mapping the Galaxy** Hydrogen clouds at different locations along our line of sight are moving around the center of the Galaxy at different speeds. Radio waves from the various gas clouds, therefore, exhibit slightly different Doppler shifts, permitting astronomers to sort out the gas clouds and map the Galaxy. *(NASA/JPL-Caltech/R. Hurt [SSC])*

11-3   Nonvisible observations help map the galactic disk

To see into the dust-filled plane of the Milky Way, astronomers use radio wave, infrared, X-ray, and gamma-ray telescopes (see Figure 3-38). These wavelengths are scattered much less by the interstellar gas and dust located throughout the Galaxy's disk than are visible or ultraviolet wavelengths. Observations of the distant parts of the Galaxy were first made using radio telescopes. Radio waves penetrate Earth's atmosphere, so we can observe them anywhere that we can build a radio telescope. (Recall that infrared observations must be made at high altitudes or in space, and that X-ray and gamma-ray observations are almost always made from space.)

Detecting the radio emission directly from interstellar hydrogen—by far the most abundant element in the universe—is a primary means of mapping the Galaxy. Unfortunately, the major transitions of electrons in the hydrogen atom (see Figure 3-49) produce photons at ultraviolet and visible wavelengths that do not penetrate the interstellar medium. How, then, can radio telescopes directly detect all of this hydrogen? The answer lies in atomic physics.

In addition to mass and charge, particles such as protons and electrons possess a tiny amount of angular momentum, commonly called **spin**. According to the laws of quantum mechanics, the electron and proton in a hydrogen atom can spin only in either parallel or opposite directions (Figure 11-7); they can have no other spin orientations. If the electron in a hydrogen atom flips from one orientation to the other, the atom must gain or lose a tiny amount of energy. In particular, when flipping from parallel to opposite spins,

the atom simultaneously emits a low-energy radio photon whose wavelength is 21 cm. This flip happens rarely in each atom, so it is only because the Galaxy has vast quantities of interstellar hydrogen gas that it can be detected at all. In 1951, a team of astronomers first succeeded in detecting the faint hiss of 21-cm radio static from spin flips.

The detection of **21-cm radio radiation** was a major breakthrough in mapping the **disk** of the Galaxy. To see why, suppose that you aim your radio telescope across the Galaxy, as sketched in Figure 11-8. Your radio receiver picks up 21-cm emission from hydrogen clouds at points 1, 2, 3, and 4. However, the radio waves from these various clouds are Doppler shifted (see Section 3-2) by slightly different amounts, because they have different radial velocities (motion toward or away from Earth). Because these radio waves from gas clouds in different parts of the Galaxy arrive at our radio telescopes with slightly different wavelengths as a result of being Doppler shifted, it is possible to identify which radio signals come from which gas clouds and thus to produce an initial map of the Galaxy, such as that shown in Figure 11-9a.

What are the two reasons that the gas and dust at points 1, 2, 3, and 4 (Figure 11-8) have different radial velocities, as measured from Earth?

Our radio map reveals numerous arched lanes of neutral hydrogen gas. If this were the overall structure of the Galaxy, then the Milky Way would appear unlike any other observed galaxy (see typical images of other galaxies later

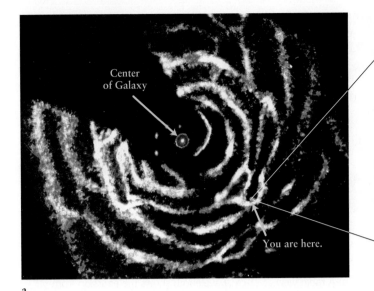

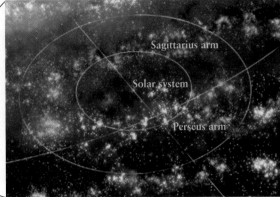

a

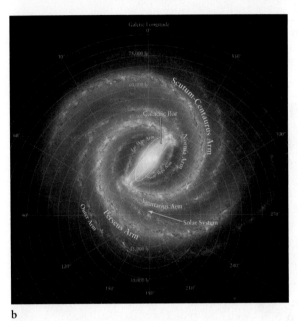

b

FIGURE 11-9 **A Map of the Galaxy** (a) This map, based on radio telescope surveys of 21-cm radiation, shows the distribution of hydrogen gas in a face-on view of the Galaxy. This view just hints at spiral structure. The galactic nucleus is marked with a dot surrounded by a circle. Details in the large, blank, wedge-shaped region toward the upper left of the map are unknown, because gas in this part of the sky is moving perpendicular to our line of sight and thus does not exhibit a detectable Doppler shift. Inset: This drawing, based on visible-light data, shows that our solar system lies between two arms of the Milky Way Galaxy. (b) This drawing labels the two major, and several minor, spiral arms in the Milky Way. *(a: Courtesy of G. Westerhout; inset: National Geographic; b: NASA/JPL-Caltech/ R. Hart [SSC])*

in this chapter). Indeed, the other disk-shaped galaxies we observe have spiral arms. We need different observations to improve our understanding of the Galaxy's disk features. Note that photographs of a barred spiral galaxy (Figure 11-10) show arms outlined by "spiral tracers"—bright, Population I stars and emission nebulae. As we saw in Chapter 9, these features indicate active star formation. If the Milky Way is spiral, then another useful way to further chart its structure and show that it has **spiral arms** is to map the locations of star-forming complexes filled with H II regions, giant molecular clouds, and massive, hot, young stars in groups, called *OB associations*.

What situation(s) on Earth are analogous to the obscuration caused by interstellar gas and dust?

---

1 Dust absorption limits the range of visual observations in the plane of the Galaxy to less than 10,000 light-years from

Earth. However, astronomers can use visible observations of nearby bright OB associations and associated H II regions to plot the spiral arms near the Sun. Radio observations of hydrogen and carbon monoxide molecules (discussed in Section 9-1) and infrared images of stars taken by the Spitzer Space Telescope have been used to chart more remote star-forming regions of the Galaxy. Taken together, all of these observations indicate that our Galaxy has about 200 billion stars located in and between two major spiral arms and several short arm segments (see Figure 11-9b). (Each arm is named after the constellation in which it is centered, as seen from Earth.) Recent observations by the Spitzer Space Telescope also confirm previous observations that a bar of stars and gas crosses the center of the Galaxy. We will explore the origin of the spiral arms in Section 11-9, when we present more evidence about the cause of spiral structure from observations of other galaxies.

The observable disk of our Galaxy is about 100,000 **2** light-years in diameter and about 2000 light-years thick

H II regions (red)
in the spiral arms

Hot luminous, young stars (blue)
in the spiral arms

a Visible-light view of M83

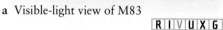

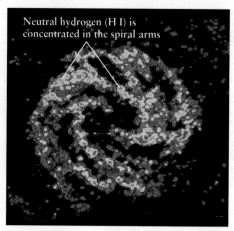

Neutral hydrogen (H I) is
concentrated in the spiral arms

b 21-cm radio view of M83

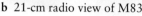

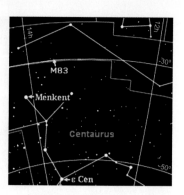

FIGURE 11-10 **Two Views of a Barred Spiral Galaxy** The galaxy M83 is in the southern constellation of Centaurus, about 12 million light-years from Earth. (a) At visible wavelengths, spiral arms are clearly illuminated by young stars and glowing H II regions.

(b) A radio view at 21-cm wavelength shows the emission from neutral hydrogen gas. Note that the spiral arms are more clearly demarcated by visible stars and H II regions than by 21-cm radio emission. *(a: S. Van Dyk/IPAC; b: VLA, NRAO)*

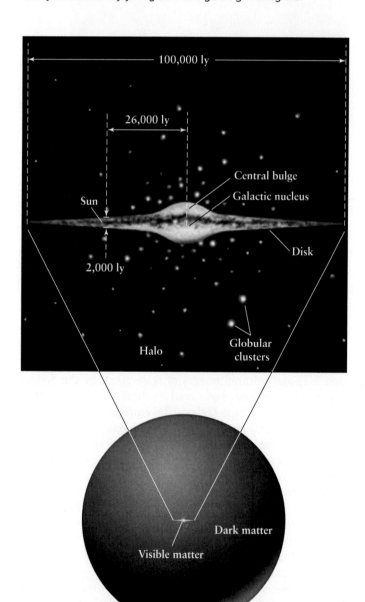

FIGURE 11-11 **Our Galaxy** As seen from the side, three major visible components of our Galaxy are a thin disk, a central bulge, and a halo. As noted earlier, there is also a central bar. The visible Galaxy's diameter is about 100,000 light-years, and the Sun is about 26,000 light-years from the galactic center. The disk contains gas and dust along with Population I (young, metal-rich) stars. The halo is composed almost exclusively of Population II (old, metal-poor) stars. Inset: The visible matter in our Galaxy fills only a small volume compared to the distribution of dark matter, whose composition is presently unknown. Its presence is felt by its gravitational effect on visible matter.

(Figure 11-11). Two spiral arms border either side of the Sun's position. On the side toward the galactic center is the Sagittarius arm, which stargazers in the northern hemisphere see in the summer when they look at the portion of the Milky Way stretching across Scorpius and Sagittarius (see Figures 11-5 and 11-9b). Directed away from the galactic center is the Perseus arm, one of the two major arms of the Galaxy. The Perseus arm is visible in the northern hemisphere in the winter. The other major spiral arm is the Scutum-Centaurus arm, which cannot be seen at visible wavelengths because of obscuring dust in the interstellar medium.

Observations reveal that the Galaxy's arms spiral out from a flattened sphere of stars, called the **central bulge** (sometimes called the *nuclear bulge*), that is about 20,000 light-years in diameter. This feature is also seen in Figure 11-12, a wide-angle infrared image of the Galaxy taken by the *COBE* satellite. The central bulge is centered on the galactic nucleus 26,000 light-years away from us. The bulge is pierced by a bar of stars, gas, and dust that move one way down the bar and then back the other way (see Figure 11-9b).

Sagittarius (center of Milky Way galaxy)

Solar system (view of Milky Way from Earth)

R I V U X G

a

b

FIGURE 11-12 **Infrared View of the Milky Way** Taken by the *COBE* satellite in 1997, this infrared image shows the disk and central bulge of our Galaxy. Most of the sources scattered above and below the disk are nearby stars. Stars appear white, whereas interstellar dust appears orange. Note that the dust that obscures light from more distant stars in Figure 11-5 is quite bright in this infrared image. *(The COBE Project, DIRBE, NASA)*

11-4 The galactic nucleus is an active, crowded place

If you lived on a planet near the center of the Galaxy, which is called the galactic nucleus, you would see a million stars as bright as Betelgeuse. The total intensity of starlight from all those nearby stars would be equivalent to 200 of our full Moons. Night would never really fall. Stranger still, the Galaxy around you would be filled with intense activity.

Figure 11-13 shows three infrared views that look toward the nucleus of the Galaxy. Figure 11-13a is a wide-angle view covering a 50° segment of the Milky Way through Sagittarius and Scorpius. The prominent band across this image is a thin layer of dust in the plane of the Galaxy. The numerous knots and blobs along the dust layer are interstellar clouds heated by young O and B stars. Figure 11-13b is an IRAS (Infrared Astronomical Satellite) view of the galactic center. Numerous streamers of dust (blue) surround it. The strongest infrared emission (white) comes from **Sagittarius A** (often abbreviated Sgr A), which is also a grouping of several powerful sources of radio waves. One of these sources, called Sagittarius A* (pronounced "A-star"), is believed to be the galactic nucleus. Figure 11-13c shows stars within 1 light-year of Sagittarius A*, with resolution of 0.02 light-year.

Radio observations add more to the picture of the galactic center. Some of the most detailed radio images of it come from the Very Large Array (VLA). Figure 11-14a is a wide-angle view of Sagittarius A and surrounding features, including arcs of gas, at least three supernovae remnants, and localized radio sources. Huge filaments, such as the one labeled "Arc," lie perpendicular to the plane of the Galaxy and stretch 200 light-years northward of the galactic disk, then abruptly arch southward toward Sagittarius A. The orderly arrangement of these filaments suggests that a magnetic field may be controlling the distribution and flow of ionized gas,

just as magnetic fields on the Sun funnel such gas to create solar prominences. This resulting radio emission, produced by high-speed electrons that spiral around these magnetic fields, is called **synchrotron radiation.** Despite its small size, Sagittarius A is one of the brightest sources of synchrotron radiation in the entire sky.

X-ray observations taken by the Chandra telescope in 2004 reveal that the galactic nucleus is also bathed in ultra-hot gas with a temperature of 100 million K. If this gas is more than a single outburst from an as-yet-unknown source, it must continually be replenished to remain as hot as it is. Astronomers are still trying to account for the source of this energetic gas.

If the center of our Galaxy is not active and bizarre enough, recent gamma-ray observations reveal positrons being ejected from that region. (Recall that positrons have positive electric charges but are otherwise identical to electrons.) The source of these positrons is still unknown. Positrons can be detected because when a positron and an electron collide, they annihilate each other. Their mass is converted into energy as gamma rays with well-defined wavelengths. These special gamma rays have also been observed emanating from the galactic center.

Infrared observations reveal stars and gas in very rapid orbits around Sagittarius A* (Figure 11-14b). Something massive must be holding this high-speed matter in such tight orbits around the galactic nucleus. Using Kepler's third law, astronomers calculate that 4×10^6 $M_\odot$ is needed to prevent the stars and gas from flying off into interstellar space. The observed broadening of spectral lines further suggests that an object with the mass of 4 million Suns in a volume only the size of the solar system is concentrated at Sagittarius A*. Astronomers believe that the object is a supermassive black hole. In 2001, the Chandra X-ray Telescope observed a burst of X rays from this black hole. The X rays were generated

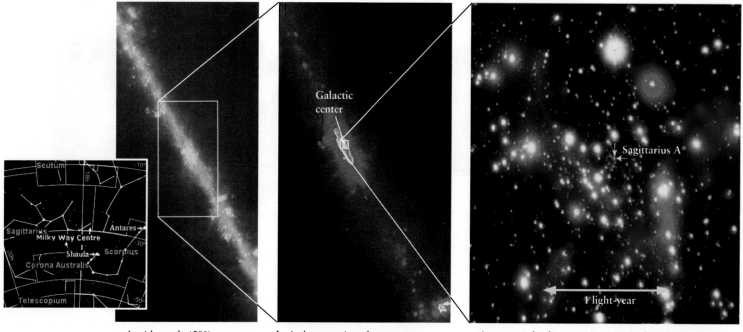

a A wide-angle (50°) infrared view

b A close-up view shows a more luminous region at the galactic center.

c An extremely close-up view centered on Sagittarius A*, a radio source at the very center of the Milky Way Galaxy, shows hundreds of stars within 1 ly (0.3 pc).

R I V U X G   R I V U X G   R I V U X G

FIGURE 11-13 **The Galactic Center** (a) This wide-angle view at infrared wavelengths shows a 50° segment of the Milky Way centered on the nucleus of the Galaxy. Black represents the dimmest regions of infrared emission, with blue the next strongest, followed by yellow and red; white represents the strongest emission. The prominent band diagonally across this photograph is a layer of dust in the plane of the Galaxy. Numerous knots and blobs along the plane of the Galaxy are interstellar clouds of gas and dust heated by nearby stars. (b) This close-up infrared view of the galactic center covers the area outlined by the white rectangle in (a). (c) This infrared image shows about 300 of the brightest stars less than 1 ly from Sagittarius A*, which is at the center of the picture. The distribution of stars and their observed motions around the galactic center imply a very high density (about a million solar masses per cubic light-year) of less luminous stars. *(a, b: NASA; c: R. Schödel et al., MPE/ESO)*

by gas heating up as it fell into it. The bigger a black hole, the longer a burst of X rays can last. The duration of the observed blast indicated that the black hole is no wider than 1 AU across, consistent with theoretical calculations.

The presence of supernova remnants at the center of our Galaxy implies what other activity is occurring in that region?

As we saw in Chapter 10, extraordinary activity is also occurring in the nuclei of many other galaxies, implying the presence of supermassive black holes at their centers as well. Astronomers are actively studying these regions in an effort to understand the complex, intriguing events that are happening there.

11-5 Our Galaxy's disk is surrounded by a spherical halo of stars and other matter

As mentioned at the beginning of this chapter, stars have been observed in a spherical distribution, called the **halo**, centered on the galactic nucleus and extended out far beyond the disk (see Figure 11-11). The halo was originally discovered because of the globular clusters that it contains. Looking out of the plane of the Milky Way's disk and between globular clusters, we see apparently unobstructed views of distant galaxies, such as the Whirlpool Galaxy observed by Lord Rosse (see Section 11-1). Appearances can be deceiving. Although the brightness of the globular clusters suggests that they contain most of the stars in the halo, it turns out that about 99% of the halo's stars are isolated *halo field stars* spread all through the halo. The localized concentrations of stars in globular clusters account for only 1% of the halo's stars. It is worth noting, however, that some globular clusters are observed to contain intermediate-mass black holes, which contribute significantly to their total mass.

The various components of the Milky Way Galaxy overlap and interpenetrate each other. For example, the disk slices through the central bulge, and globular clusters and halo field stars periodically pass through the plane of the disk. Figure 11-15 shows the shapes of the orbits of typical central bulge, disk, and halo stars and clusters. Our solar

Small Magellanic Cloud (SMC) are examples of Irr I. Both can be seen with the naked eye from southern latitudes.

Besides being an E0, what other classification could M105 in Figure 11-31 have?

Occasionally, irregulars are observed that appear highly distorted and completely asymmetrical, as though created by collisions between galaxies or by violent activity in their nuclei. These are denoted *Irr II*, as shown by NGC 4485 (Figure 11-32b). Irregular galaxies are typically smaller and less massive than spirals, containing between 10^8 and about 3×10^{10} $M_\odot$.

The Large Magellanic Cloud is visible to the naked eye in the southern hemisphere. From Figure 11-32, what two things do you think it could be mistaken for?

11-13 Hubble presented spiral and elliptical galaxies in a tuning fork–shaped diagram

Edwin Hubble connected the three regularly shaped types of galaxies—spirals, barred spirals, and ellipticals—in a diagram shaped like a tuning fork (Figure 11-33). According to his scheme, S0 or SB0 galaxies, called **lenticulars** (lens-shaped), are an intermediate type between ellipticals and the two kinds of spirals. Although they often look somewhat like ellipticals, lenticular galaxies have both a central bulge and a disk, like spiral galaxies, but they lack spiral arms. For want of any better scheme, the irregular galaxies are sometimes placed between the ends of the tuning fork tines of the Hubble diagram.

The idea that one type of galaxy may change into another type has been in and out of favor with astronomers for decades. Indeed, such transformations were part of Hubble's motivation in relating the different types of galaxies as he did. As we will see later in this chapter and in Chapter 12, extensive observations by the Hubble Space Telescope and by ground-based telescopes reveal that interactions between galaxies sometimes do lead to changes in their structures. For example, when two disk galaxies merge, they often morph into a giant elliptical galaxy. However, it appears that most galaxies remain unchanged in overall shape once they are fully formed. The properties of the different types of galaxies are summarized in Table 11-1.

11-14 Galaxies built up in size over time

In Chapter 12, we will explore the observational evidence for the formation and early evolution of galaxies. Nevertheless, it is worth summarizing here what astronomers know about why the galaxies have the structures we observe. With the exception of dwarf ellipticals, galaxies formed from smaller ensembles of stars, gas, and dust. The underlying dark matter that clumped together in the young universe drew this matter together, often assisted by a supermassive black hole formed early in the life of the universe. This black hole became the nucleus of the galaxy. The more massive the black hole, the larger a disk galaxy's central bulge. Once formed, about three-quarters of all galaxies have maintained their structures, whereas the remaining quarter have changed (for example, from spiral to elliptical).

Normal (unbarred) spiral galaxies formed when swirling eddies of gas-rich matter collided and formed a disk centered on an especially massive black hole and lots of dark matter. When the amount of dark matter and the mass of the central black hole were too low, the matter in the central regions of the disk had bar-shaped orbits, creating barred-spiral galaxies. Disk galaxies formed when the gases that created them had time to interact with each other

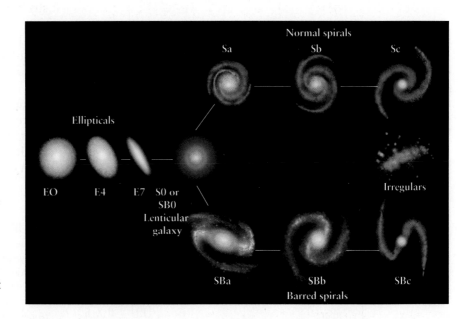

FIGURE 11-33 **Hubble's Tuning Fork Diagram** Hubble summarized his classification scheme for galaxies with this tuning fork diagram. Elliptical galaxies are classified by how oval they appear, whereas spirals and barred spirals are classified by the sizes of their central bulges and the correlated winding of their spiral arms. An S0 or SB0 galaxy, also called a lenticular galaxy, is an intermediate type between ellipticals and spirals. It has a disk but no spiral arms.

TABLE 11-1 Some Properties of Galaxies

	Spiral (S) and barred spiral (Sb) galaxies	Elliptical galaxies (E)	Irregular galaxies (Irr)
Mass ($M_\odot$)	10^9 to 4×10^{11}	10^5 to 10^{13}	10^8 to 3×10^{10}
Luminosity ($L_\odot$)	10^8 to 2×10^{10}	3×10^5 to 10^{11}	10^7 to 10^9
Diameter (ly)	1.6×10^4 to 8×10^5	3×10^3 to 6.5×10^5	3×10^3 to 3×10^4
Stellar populations	Disk: young Population I central bulge; halo: Population II and old Population I	Population II and old Population I	Mostly Population I
Percentage of observed galaxies	77%	*20%	3%

*This percentage does not include dwarf elliptical galaxies that are as yet too dim and distant to detect. Hence, the actual percentage of galaxies that are ellipticals is likely to be higher than shown here.

without forming too many stars. If the collision of the gases led to especially rapid star formation, the resulting galaxies were elliptical or irregular. Giant elliptical galaxies were (and still are) formed from the collisions of two spiral galaxies (either type).

What physical property of matter discussed in Chapter 2, besides gravitation, is essential in forming all spiral galaxies?

CLUSTERS AND SUPERCLUSTERS

As we consider the vastness of galaxies, it is hard to imagine that structures this huge orbit each other in groups. But they do.

R I V U X G

11-15 Galaxies occur in clusters, which occur in larger clumps called superclusters

Galaxies are not scattered randomly throughout the universe, but rather they are grouped together in **clusters.** The Fornax cluster (Figure 11-34), so called because it is located in the constellation of Fornax (the Furnace), is typical. Observations of galactic motion indicate that members of a cluster of galaxies are gravitationally bound together: The galaxies in a cluster orbit each other and, occasionally, even collide.

Astronomers observe very hot (10^7 K) gas between clusters of galaxies. Why does this gas not appear in Figure 11-34?

Clusters of galaxies are themselves grouped together in huge associations, called **superclusters.** A typical supercluster contains dozens of individual clusters spread over a volume 150 million light-years across. Figure 11-35 shows the distribution of superclusters in our vicinity of the universe. The nearer ones, out to the Virgo cluster, are members of our *Local Supercluster.* Observations indicate that most superclusters are not gravitationally bound units—most clusters in each supercluster are drifting away from most of the other clusters in that same supercluster. Furthermore, the superclusters are all moving away from one another.

FIGURE 11-34 **A Cluster of Galaxies** This group of galaxies, called the Fornax cluster, is about 60 million light-years from Earth. Both elliptical and spiral galaxies are easily identified. The barred spiral galaxy at the lower left is NGC 1365, the largest and most impressive member of the cluster. For a closer view of NGC 1365, see Figure 11-28. *(Anglo-Australian Observatory/Royal Observatory, Edinburgh)*

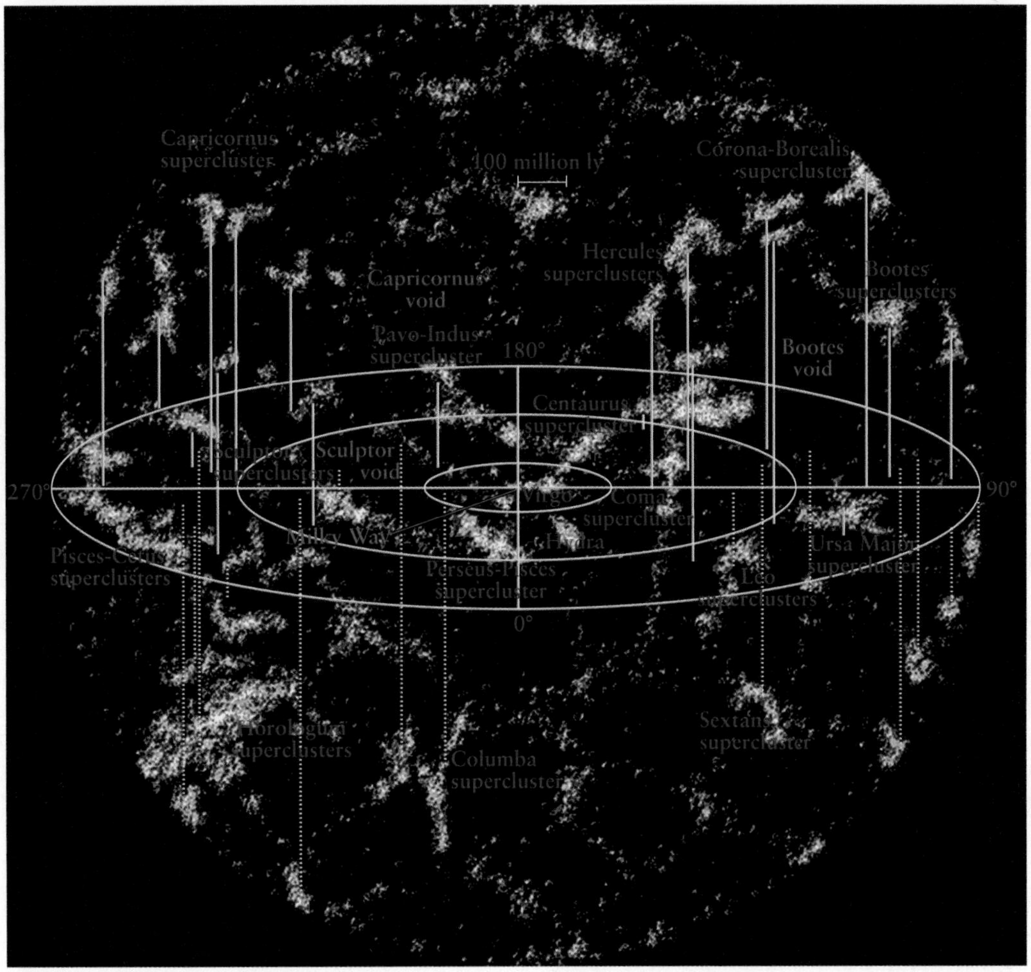

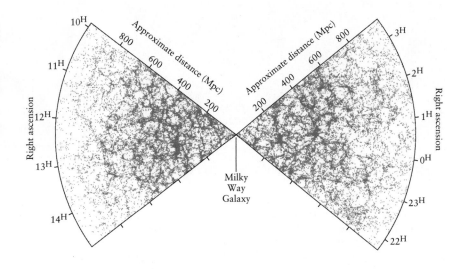

FIGURE 11-35 **Superclusters in Our Neighborhood** This diagram shows the distances and relative positions of superclusters within 950 million light-years of Earth. Note also the labeling of some of the voids, which are large, relatively empty regions between superclusters. *(Kirk Korista)*

How superclusters are arranged in space became clearer, beginning in the early 1980s, when astronomers began systematic surveys of the positions of observable objects in the nearby universe. Completed in 2005, the first Sloan Digital Sky Survey observed nearly 200 million galaxies, stars, and other celestial objects, and measured the redshifts of more than a million of them. This survey, made in the northern hemisphere, covered about one-quarter of the sky and gives us a three-dimensional map through 100 times the volume of space as we had mapped before. Another set of observations, called the Two Degree Field (2dF) Survey, studied regions of both the northern and southern hemispheres.

These data reveal that most superclusters are located on the boundaries between enormous bubblelike *voids* in which few galaxies are observed (Figure 11-36). These voids are roughly spherical and measure between 100 million and 400 million light-years in diameter. They are not

FIGURE 11-36 **Structure in the Universe** This map shows the distribution of 62,559 galaxies in two wedges extending in opposite directions from Earth out to distances of 3.5 billion light-years. For an explanation of right ascension (r.a.), see Section 1-3. Note the prominent voids surrounded by thin areas full of galaxies. *(Courtesy of the 2dF Galaxy Redshift Survey Team)*

The 2dF Galaxy Survey

completely empty, however. Observations reveal hydrogen clouds in some of them, while others may be subdivided by strings of dim galaxies.

The distribution of clusters of galaxies throughout the universe appears to have the structure of a sponge (Figure 11-37) or soap bubbles, with the galaxies located where the sponge or soapy material is, surrounded by voids. Astronomers believe that this spongy pattern contains important clues about conditions in the early universe that have not yet been fully fathomed.

11-16 Clusters of galaxies may appear densely or sparsely populated and regular or irregular in shape

A cluster of galaxies is said to be either a **poor cluster** or a **rich cluster,** depending on whether it contains less than or more than a thousand galaxies. For example, the Milky Way Galaxy, the Andromeda Galaxy, and the Large and Small Magellanic Clouds belong to a poor cluster, called the **Local Group.** The Local Group contains roughly 40 galaxies, more than a third of which are dwarf ellipticals. Figure 11-38 shows a map covering most of the Local Group. The Virgo cluster is the nearest rich cluster, with over 2000 galaxies.

New galaxies in the Local Group continue to be found. This happens primarily because astronomers are developing techniques to observe objects that lie in the same plane as, but beyond, the Milky Way. In addition to the 1994 discovery of the Sagittarius Dwarf Galaxy, the dwarf galaxy Antlia (Figure 11-39) was first observed in 1997 and the Canis Major Dwarf (see Figure 11-16) was found in 2003.

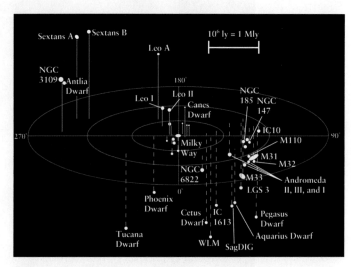

FIGURE 11-38 **The Local Group** Our Galaxy belongs to a poor, irregular cluster that consists of about 40 galaxies, called the Local Group. This map shows the distribution of about three-quarters of the galaxies. The Andromeda Galaxy (M31) is the largest and most massive galaxy in the Local Group. The second largest is the Milky Way itself. M31 and the Milky Way are each surrounded by a dozen satellite galaxies. The recently discovered Canis Major Dwarf Galaxy is the Milky Way's nearest known neighbor.

R I V U X G

FIGURE 11-39 **A Recently Discovered Member of the Local Group** The galaxy Antlia was first detected in 1997. It lies about 3 million light-years away, outside the region depicted in Figure 11-38. This galaxy contains only about a million stars. (*M. J. Irwin, Royal Greenwich Observatory, and A. B. Whiting and G. K. T. Hau, Institute of Astronomy, Cambridge University*)

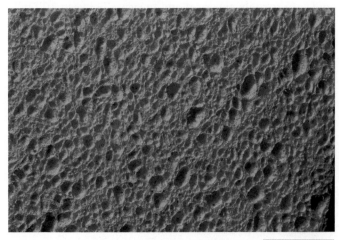

R I V U X G

FIGURE 11-37 **Foamy Structure of the Universe** A sponge that recreates the distribution of bright clusters of galaxies throughout the universe. The empty spaces in the foam are analogous to the voids found throughout the universe. The spongy regions are analogous to the locations of most of the galaxies. (*Image Source/Super Stock*)

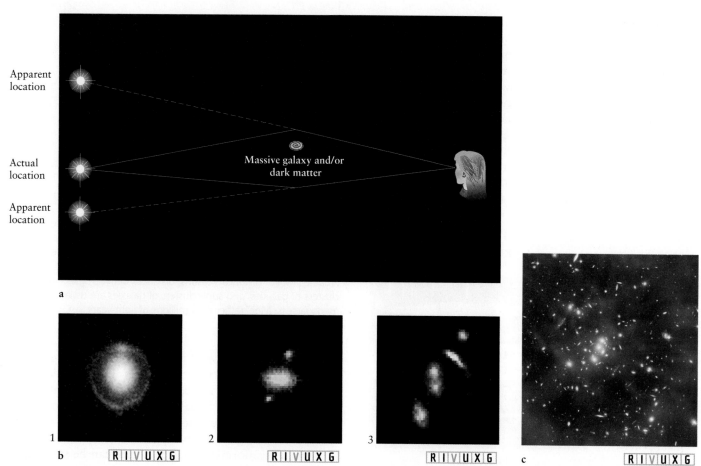

FIGURE 11-49 Gravitational Lensing of Extremely Distant Galaxies (a) Schematic of how a gravitational lens works. Light from the distant object changes direction due to the gravitational attraction of the intervening galaxy and underlying dark matter. The more distant galaxy appears in different places than it actually is to the observer on the right. (b) Three examples of gravitational lensing: (1) The bluer arc is a galaxy that has been lensed by the redder elliptical galaxy. (2) A pair of bluish images of the same object lensed symmetrically by the brighter, redder galaxy between them. (3) The lensed object appears as a blue arc under the gravitational influence of the group of four galaxies. (c) Superimposed in blue on this image of the galaxy cluster CL 0024+17 is the location of dark matter that is gravitationally lensing the galaxies behind it. *(b: NASA, ESA, A. Bloton [Harvard-Smithsonian CfA] and the SLACS Team; c: NASA, ESA, and M.J. Jee [Johns Hopkins University])*

SUPERCLUSTERS IN MOTION

Whenever an astronomer finds an object in the sky, one of the first tasks is to determine its composition. As we saw in Chapter 3, that means attaching a spectrograph to a telescope and recording the object's spectrum. As long ago as 1914, V. M. Slipher, working at the Lowell Observatory in Arizona, took spectra of "spiral nebulae," now known to be spiral galaxies. He was surprised to discover that the spectral lines of 11 of the 15 spiral nebulae that he studied were substantially redshifted. These redshifts indicate that they are moving away from the Milky Way at significant speeds. This left scientists with a major puzzle: Is the entire universe expanding?

If the dark matter in Figure 11-49c suddenly vanished, what would we see of the clusters of galaxies behind it?

11-19 The redshifts of superclusters indicate that the universe is expanding

During the 1920s, Edwin Hubble and Milton Humason recorded the spectra of many galaxies with the 100-in telescope on Mount Wilson, confirming that most galaxies are rapidly receding from the Milky Way. This implies that the universe *is* expanding, although, as we will see in a moment, it does not imply that the Milky Way is at the center of it!

Using the Doppler effect (recall Figure 3-7), Hubble calculated the speed at which the galaxies are moving away from us. Using techniques such as the brightness of Cepheid variables (see Section 11-1), he also estimated the distances to several of these galaxies. He found a direct correlation between the distance to a galaxy and the size of its redshift: *Galaxies in distant clusters and superclusters are moving away from us more rapidly than galaxies in nearby clusters and superclusters.* Figure 11-50 shows this correlation for five elliptical galaxies. This recessional motion pervades the universe and is now called the **Hubble flow.**

The separation of clusters in the same supercluster is slower than the general Hubble flow of the universe because these clusters are relatively close to one another, so their gravitational attractions slow each other down. Keep in mind that the Hubble flow does not occur for galaxies in any given cluster, because all of those galaxies are gravitationally bound to each other. To recap: The Hubble flow applies to the separation of all superclusters from one another and many clusters of galaxies in the same supercluster from one another.

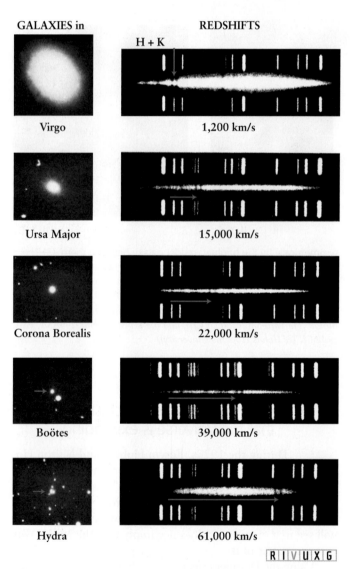

GALAXIES in   **REDSHIFTS**

H + K

Virgo — 1,200 km/s

Ursa Major — 15,000 km/s

Corona Borealis — 22,000 km/s

Boötes — 39,000 km/s

Hydra — 61,000 km/s

R I V U X G

FIGURE 11-50 Five Galaxies and Their Spectra The photographs of these five elliptical galaxies were all taken at the same magnification. They are labeled according to the constellation in which each galaxy is located. The spectrum of each galaxy is the hazy band between the comparison spectra at the top and bottom of each plate. In all five cases, the so-called H and K lines of calcium are seen. The recessional velocity (calculated from the Doppler shifts of the H and K lines) appears below each spectrum. Note that the fainter—and thus more distant—a galaxy is, the greater is its redshift. *(Carnegie Observatories)*

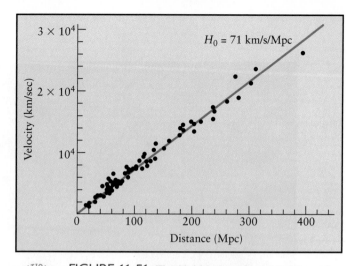

FIGURE 11-51 The Hubble Law The distances and recessional velocities of distant galaxies are plotted on this graph. The straight line is the "best fit" for the data. This linear relationship between distance and speed is called the Hubble law. For historical reasons, distances between galaxies, clusters of galaxies, and superclusters of galaxies are usually given in megaparsecs, Mpc, rather than millions of light-years.

When Hubble plotted the redshift data on a graph of distance versus velocity, he found that the points lie nearly along a straight line. Figure 11-51 shows the relationship between the distances to galaxies and their recessional motion, including many whose distances were measured using the apparent magnitudes of Type Ia supernovae observed in them. The information in this figure can be stated as a formula, called the **Hubble law:**

$$\text{Recessional velocity} = H_0 \times \text{distance}$$

This is simply the equation of a straight line, with the slope of the line denoted by the constant H_0, commonly called the **Hubble constant.**

The relationship between the distances to galaxies and their redshifts was one of the most important astronomical discoveries of the twentieth century. It tells us that we are living in an expanding universe, and the Hubble law reveals the speed of that expansion. It also shows why we are not in the center of the universe, even though the other superclusters are all receding from us, as discussed in Appendix G-3.

The discovery that the universe is expanding raised the question of whether the universe will expand forever or whether the gravitational attraction between all of its parts is enough to cause it someday to stop and recollapse. This is like trying to escape forever from Earth. A normal jump might raise you up ⅓ m or so, but if you had a sufficiently powerful cannon to shoot you upward fast enough, you could escape Earth's gravitational attraction completely and enter interplanetary space. To determine whether the superclusters are expanding away from one another fast enough to escape their mutual gravitation forever, astronomers have worked hard to determine the exact value of H_0. We explore that here and pick up the question of the fate of the universe in Chapter 12.

If the Hubble constant were 3 times its present value, how much slower or faster would superclusters be moving apart?

Power in Numbers Much of the power of science comes from expressing systematic data in mathematical form. By themselves, the spectra of galaxies give a qualitative picture of galactic motion, and the pattern of their recessional velocities suggests that a large-scale property of the universe is in effect here. When the data are converted into the equation, called the Hubble law, it becomes clear that the universe is expanding, and we can begin to look for the reason behind this expansion.

11-20 Different techniques determine the expansion of the universe at different distances from Earth

To determine the Hubble constant, astronomers must measure the redshifts and distances to many galaxies. Although redshift measurements from spectra can be quite precise, it is very difficult to measure accurately the distances to remote galaxies. We cannot use the method of stellar parallax at the distances of galaxies. Recall from Section 8-1 that only the distances to the stars in our Galaxy within about 150 pc (500 ly) can be determined very precisely this way.

Distances to other galaxies are determined from observations of the apparent magnitudes of objects in them or apparent magnitudes of entire galaxies. These observations are compared to known absolute magnitudes for similar objects. Having the apparent magnitude and the absolute magnitude enables astronomers to calculate distances. Astronomers use the term **standard candle** to denote any object whose absolute magnitude is known.

Spectroscopic parallax (see Section 8-1) provides distances to stars up to 10 kpc (33 kly, where kly denotes 1000 light-years). This distance is still within the Milky Way. To determine distances to other galaxies, we need sources in them that become very bright compared to the luminosity of normal stars, and whose absolute magnitudes are well known. Using the brightnesses of RR Lyrae variable stars (see *Variable Stars*, Chapter 9), astronomers can measure distances to the Large and Small Magellanic Clouds or about 100 kpc (330 kly). To measure the Hubble flow, we must determine distances to much more distant galaxies.

Cepheid variable stars (see Section 9-12 and Figure 11-4) can now be seen out to 30 Mpc (100 Mly, where Mly denotes a million light-years) from Earth. The distances to galaxies in this nearby volume of space can thus be determined from the Cepheid period–luminosity law. Plotting the distances to galaxies versus their recessional velocity for the small volume of space extending out 30 Mpc from Earth (shown on the left side of the graph in Figure 11-51) enabled astronomers to get a first estimate for H_0 from the slope of the line through these data. This gave $H_0 \approx 75$ km/s/Mpc. Errors in both distance and recessional velocities led to inaccuracies in the slope of this curve.

Beyond 30 Mpc, even the brightest Cepheid variables, which have absolute magnitudes of about –6, are not visible with current technology. In the 1970s, the astronomers Brent Tully and Richard Fisher developed another method for determining distances. They discovered that the width of the hydrogen 21-cm emission line of a spiral galaxy is related to the galaxy's absolute magnitude: *The broader the line, the brighter the galaxy.* The connection goes like this: Interstellar gas orbits in galaxies, as we have seen. Different hydrogen gas clouds in different places in the same galaxy have different speeds toward or away from us, and because of their different Doppler shifts we observe a variety of slightly different wavelengths from this gas, centered around 21 cm. This combination of emissions makes the 21-cm emission line appear "broad."

The **Tully–Fisher relation** says that the greater the galaxy's mass, the greater its luminosity and the more rapidly the stars and gas orbit in it. The faster it rotates, the greater the range of speeds and hence the greater the Doppler shifts we see from it. Measuring the width of the 21-cm line, therefore, tells us the galaxy's mass. Assuming that stars with different masses, and hence luminosities, occur with the same frequencies in each galaxy, the galaxy's mass tells us its *absolute* magnitude. Combining this with observations of the galaxy's *apparent* magnitude allows us to calculate its distance from us using the distance–magnitude relationship. Because line widths can be measured quite accurately, astronomers can use the Tully–Fisher relation to determine the luminosities (absolute magnitudes) of many spiral galaxies and thus their distances.

The key method for determining the distances to the most remote galaxies is to measure the brightness of Type Ia supernova explosions within them. These supernovae all reach an absolute magnitude of –19 at the peak of their outbursts. Using their observed apparent magnitudes and known absolute magnitudes, their distances can be calculated. With this method, astronomers in 2001 measured a supernova nearly 3 billion parsecs (10 billion light-years) away, and more distant Type Ia supernovae have been measured since then. Measurements of distances to galaxies using the Tully–Fisher relation and Type Ia supernovae added data to Figure 11-51 and yield a more accurate Hubble constant of 71 km/s/Mpc.

Figure 11-52 summarizes the distances for which all of the measuring techniques described here are valid. Including data from all of the methods yields a distance–velocity graph, as shown in Figure 11-51. Using this diagram, astronomers have calculated that the Hubble constant is

$$H_0 = 71 \pm 4 \text{ km/s/Mpc}$$

The plus or minus 4 km/s/Mpc indicates the possible range for H_0, when taking into account all of the errors that may still remain in the observations and calculations. A significant source of error is the effect of intergalactic gas that causes the standard candles to appear dimmer than they would if the gas were not present. Astronomers must determine the amounts of both intergalactic and interstellar gas and take their effects into account, something that is often hard to do.

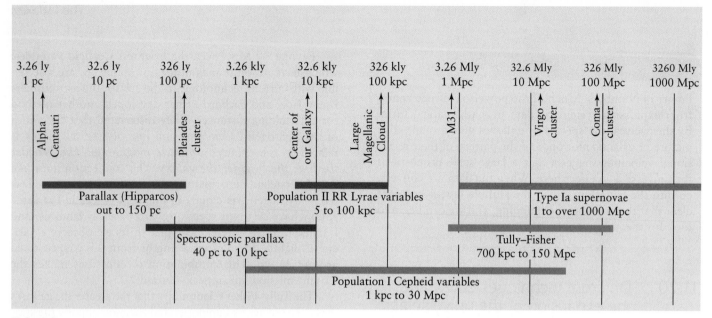

FIGURE 11-52 Techniques for Measuring Cosmological Distances Astronomers use different methods to determine different distances in the universe. All of the methods shown here are discussed in the text.

What are a few of the techniques that we use to measure distances on Earth?

11-21 Astronomers are looking back to a time when galaxies were first forming

The Hubble Space Telescope has been looking deeper and deeper into the universe. In 1998, working in concert with the Keck I telescope, it observed galaxies 8 Bly (8 billion light-years) away, many of them in the process of merging (Figure 11-53). That same year, it saw galaxies nearly 12 Bly away (Figure 11-53b). In 2004, galaxies up to 13.2 Bly away were observed. We are seeing these latter objects as they were less than a billion years after the universe came into existence. Although many of the early galaxies are not as well formed as nearby galaxies, the existence of galaxy-sized collections of stars so far back in time is an important clue to the processes of star and galaxy formation that occurred in the early universe.

The observational data now being accumulated from the period of time when the universe was less than a few billion years old are helping astronomers develop and test theories of the evolution of the early universe. We will explore this point further in Chapter 12.

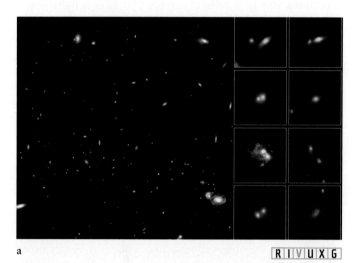

a

R I V U X G

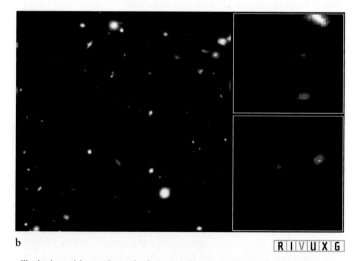

b

R I V U X G

FIGURE 11-53 Distant Galaxies (a) The young cluster of galaxies MS1054-03, shown on the left, contains many orbiting pairs of galaxies, as well as remnants of recent galaxy collisions. Several of these systems are shown at the right. This cluster is located 8 billion light-years away from Earth. (b) This image of more than 300 spiral, elliptical, and irregular galaxies contains several galaxies that are an estimated 12 billion light-years from Earth. Two of the most distant galaxies are shown in the images on the right, in red, at the centers of the pictures. *(a, b: P. Van Dokkum, Uner of Granengen, ESA, and*

QUASARS

We end this chapter by looking at active galaxies—energy sources of almost unimaginable power. While the short-lived outputs of supernovae boggle the mind, they represent miniscule amounts of energy compared to quasars, Seyfert galaxies, radio galaxies, and BL Lac objects. Quasars, for example, emit more energy each second than the Sun does in 200 years, and they continue to do so for millions of years. Evidence shows that most galaxies undergo periods of similar activity. Some truly remarkable activity must occur deep within quasars and other ultrahigh energy sources to make them so luminous.

Our knowledge of quasars began in an amateur astronomer's backyard. Grote Reber built the first radio telescope in 1936 behind his home in Illinois, opening the realm of nonvisual astronomy. By 1944, Reber had detected strong radio emissions from sources in the constellations Sagittarius, Cassiopeia, and Cygnus. Two of these sources, Sagittarius A (Sgr A) and Cassiopeia A (Cas A), are in our Galaxy. The first is the galactic nucleus (see Section 11-4), and the second is a supernova remnant (see Chapter 10). However, Reber's third source, called Cygnus A (Cyg A), proved hard to categorize (Figure 11-54). Others quickly refined Reber's observations, but the mystery only deepened in 1954, when Walter Baade and Rudolph Minkowski, using the 200-in optical telescope on Mount Palomar, discovered a strange-looking galaxy at the position of Cygnus A (Figure 11-54 inset).

The galaxy associated with Cyg A is very dim. Nevertheless, Baade and Minkowski managed to photograph its spectrum. They detected a redshift that corresponds to a speed of 14,000 km/s. According to the Hubble law, this speed indicates that Cyg A lies 635 million ly (194 Mpc) from Earth.

Because Cyg A is one of the brightest radio sources in the sky, its enormous distance intrigued astronomers. Although barely visible through the giant optical telescope at Palomar, Cyg A's radio waves can be picked up by amateur astronomers with backyard equipment. Its radio energy output must therefore be colossal. In fact, Cyg A shines with a radio luminosity 10^7 times as bright as that of an entire ordinary galaxy, such as Andromeda. The object that creates the Cyg A radio emissions has to be something extraordinary.

In what way are telescopes time machines?

11-22   Quasars look like stars but have huge redshifts

Cygnus A is not the only powerful radio source in the far-distant sky. Starting in the 1950s, radio astronomers created long lists of radio sources. One of the most famous lists, the *Third Cambridge Catalogue,* was published in 1959. (The first two catalogues were filled with inaccuracies.) Even today, astronomers often refer to its 471 radio sources by their "3C numbers." Cyg A, for example, is designated 3C 405, because it is the 405th source in the Cambridge list. Because of the extraordinary luminosity of Cyg A, astronomers were eager to learn whether any other sources in the 3C catalog had similar properties.

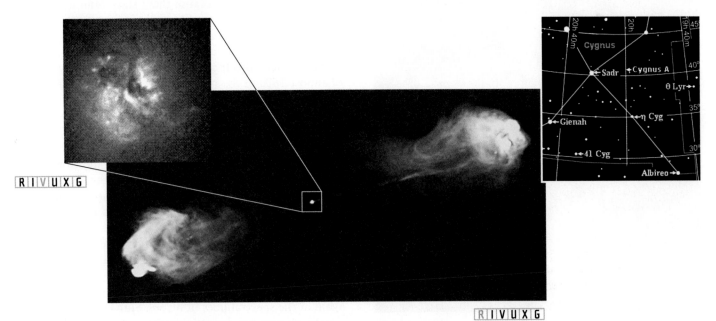

R I V U X G

R I V U X G

FIGURE 11-54 Cygnus A (3C 405) Radio image produced from observations made at the Very Large Array. Most of the radio emissions from Cygnus A come from the radio lobes located on either side of the peculiar galaxy seen in the inset, a Hubble Space Telescope image. Each of the two radio lobes extend about 160,000 light-years from the optical galaxy and contain a brilliant, condensed region of radio emission. **Inset:** At the heart of this system of gas lies a strange-looking galaxy that has a redshift that corresponds to a recessional speed of 5% of the speed of light. According to the Hubble law, Cygnus A is therefore 635 million light-years from Earth. Because Cygnus A is one of the brightest radio sources in the sky, this remote galaxy's energy output must be enormous. *(R. A. Perley, J. W. Dreher, J. J. Cowan, NRAO; inset: William C. Keel, Robert Fosbury)*

One interesting case was 3C 48. In 1960, Allan Sandage used the Palomar telescope to discover a "star" at the location of this radio source (Figure 11-55). Recall that stars are blackbodies whose peak intensity is typically visible light and whose radio emission is much less intense. Because ordinary stars are not strong sources of radio emission, 3C 48 had to be something unusual. Indeed, its spectrum showed a series of emission lines that, initially, no one could identify. Although 3C 48 was clearly an oddball, many astronomers thought it was just another strange star in our Galaxy.

Another such "star," called 3C 273, was discovered in 1962. Like 3C 48, this object (Figure 11-56) emits a series of bright spectral emission lines that no one could then identify. These spectral lines are brighter than the background radiation at other wavelengths, called the *continuum* (recall Kirchhoff's laws in Section 3-20).

A breakthrough finally came in 1963, when Maarten Schmidt at the California Institute of Technology identified four of the brightest spectral lines of 3C 273 as four spectral lines of hydrogen. However, these emission lines from 3C 273 are found at much longer wavelengths than the usual wavelengths of these lines. Schmidt concluded that the hydrogen lines are subjected to a substantial redshift. Furthermore, the intense emission lines mean that something unusual is heating the gas.

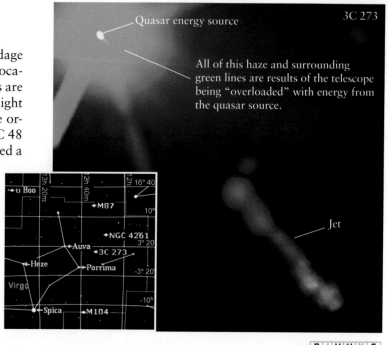

FIGURE 11-56 **Quasar 3C 273** This combined X-ray and infrared view shows the starlike object associated with the radio source 3C 273 and the luminous jet it has created. The jet is also visible in the radio and visible parts of the spectrum. By 1963, astronomers determined that the redshift of this quasar is so great that, according to the Hubble law, it is nearly 2 billion light-years from Earth. *(S. Jester, D. E. Harris, H. L. Marshall, K. Meisenheimer, H. J. Roser, and R. Perley)*

FIGURE 11-55 **Quasar 3C 48** For several years, astronomers erroneously believed that this object is simply a peculiar, nearby star that happens to emit radio waves. Actually, the redshift of this starlike object is so great that, according to the Hubble law, it must be roughly 4 billion light-years away. *(Alex G. Smith, Rosemary Hill Observatory, University of Florida)*

Spectra for stars in our Galaxy exhibit comparatively small Doppler shifts, because these stars cannot move extremely fast relative to the Sun without soon escaping from the Galaxy. Schmidt thus concluded that 3C 273 is not a nearby star after all. Pursuing this conclusion, he promptly found that its redshift corresponds to a speed away from us of almost 16% of the speed of light. According to the Hubble law, this huge redshift implies an impressive distance to 3C 273 of roughly 2 billion light-years.

Figure 11-57 shows the spectrum of 3C 273, which looks nothing like a star's blackbody spectrum with absorption lines. Remember from Chapter 3 that a spectrograph records energy intensity at different wavelengths. The emission lines appear as peaks, and absorption lines appear as valleys. The emission lines are caused by excited gas atoms that emit radiation at specific wavelengths. Figure 11-58 shows the spectrum of an object so far away that the expansion of the universe (see Section 11-19) gives it a redshift of 92% of the speed of light.

Inspired by Schmidt's success, astronomers looked again at the spectral lines of 3C 48. Sure enough, the emission lines corresponded to hydrogen but with a redshift that corresponds to a velocity of nearly one-third the speed of light. Therefore, 3C 48 must be nearly twice as far away as 3C 273, or about 4 billion light-years from Earth, assuming that the Hubble constant, H_0, is 71 km/s/Mpc.

Because of their starlike appearances and strong radio emissions, 3C 48 and 3C 273 were dubbed **quasi-stellar**

FIGURE 11-68 **Sombrero Galaxy (M104)** This spiral galaxy in Virgo is nearly edge-on to our Earth-based view. Spectroscopic observations indicate that a billion-solar-mass black hole is located at the galaxy's center. You can see the bright region in the galaxy's center created by stars and gas that orbit the black hole. *(NASA and the Hubble Heritage Team [STScI/AURA])*

11-26 Jets of protons and electrons ejected from around black holes help explain active galaxies

The gravitational energy associated with supermassive black holes at the centers of galaxies creates huge jets of gas comprised of protons and electrons. To see how this works, consider a black hole at the center of a young galaxy filled with gas and dust. Because the centers of galaxies are

congested places, a black hole there will capture a massive accretion disk of this debris (Figure 11-69a). According to Kepler's third law, the material in the inner regions of such a disk orbits the most rapidly. The inner matter, therefore, constantly interacts with the more slowly moving gases in the outer regions, and the friction between them heats all of the gas. As energized gases spiral toward the black hole, they are compressed and heated to millions of degrees. This temperature creates great pressure, causing some of the hot gas to expand and eventually squirt out where the resistance to its expansion is lowest—namely, perpendicular to the accretion disk.

In 1995, the Hubble Space Telescope took a picture of the giant elliptical galaxy NGC 4261 (Figure 11-69b) that astronomers interpret as showing this set of events. A disk of gas and dust about 800 light-years in diameter is seen orbiting a supermassive black hole. The speed of the material in this disk indicates that it is held in orbit by a 1.2-billion-solar-mass object. The inset in Figure 11-69b is a Hubble image of the galaxy's nucleus. The dark, horizontal oval in the picture is our oblique view of the accretion disk, and ground-based radio and optical observations show a double-lobed structure (Figure 11-69b) that is bisected by the oval.

What confines the ejected gas to narrow jets? At first, the gas still falling toward the black hole prevents the jets from spreading. To see why, consider air spraying out of the nozzle of a hose. The stream of air broadens and fans out through a wide angle in the air. This would be the case around these black holes if the infalling gas were not there. If the nozzle is placed in a swimming pool, however, the stream of air will not fan out as much at first (Figure 11-70). Similarly, as the two jets of hot gas leave the vicinity of the black hole, they must blast their way through the gas that is still crowding

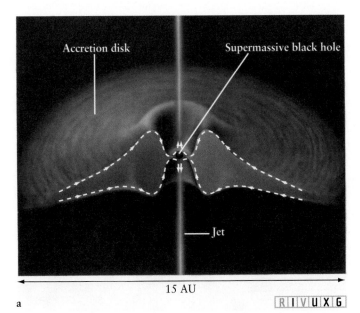

a

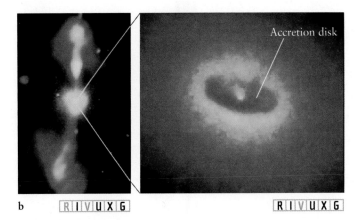

b

FIGURE 11-69 **Supermassive Black Holes as Engines for Galactic Activity** (a) In the accretion disk around a supermassive black hole, in-swirling gas heats and expands. Pulled inward, compressed, and heated further, some of it is eventually expelled perpendicular to the disk in two jets. (b) The giant elliptical galaxy NGC 4261 is a double radio source located in the Virgo cluster, about 100 million light-years from Earth. An optical photograph of the galaxy (white) is combined with a radio image (orange and yellow) to show both the visible galaxy, which does not emit much radio energy, and its jets, which do. **Inset:** This Hubble Space Telescope image of the nucleus of NGC 4261 shows a disk of gas and dust about 800 ly (250 pc) in diameter, orbiting a supermassive black hole. *(b: NASA; inset: ESA)*

inward. Passage through this material causes the jets to become narrow, concentrated beams. Then, magnetic fields created in the hot, swirling accretion disk spiral around the jets of gas and help keep the ejecta in columns (Figure 11-71).

After traveling for hundreds of thousands of years or more, the jets of gas interact with enough preexisting interstellar and intergalactic gas to be stopped, thereby forming the lobes and other features. In 2003, astronomers using the Chandra X-ray Observatory observed gas jetting out from a supermassive black hole in a galaxy at the center of the Perseus cluster of galaxies and striking gas that resides between galaxies. This *intracluster* gas became compressed by the jet and generated concentric rings or shells of sound waves 35,000 light-years in wavelength. Put another way, the sound created by the jet is a B-flat that is 57 octaves below middle C.

This "rocket nozzle" model explains not only double radio sources but also the jets and beams we see protruding from other active galaxies. We can now tie together all of the various objects discussed in this chapter. *The major difference between the different types of active galaxies is the angle at which we view the central engine.* As Figure 11-72 shows, an observer sees a radio galaxy or a double radio source when the accretion disk is viewed nearly edge-on, because the jets are nearly in the plane of the sky. At a steeper angle, a quasar is seen. If one of the jets is aimed almost directly at Earth, the galaxy is a BL Lac object. The Seyferts, active spiral galaxies, are distinguished by whether we can see the bright core (Type 1) or not (Type 2). Their placements in Figure 11-72 are still under investigation.

There are many active galaxies that we *should* see but do not. These have such thick and dusty accretion disks that little visible light or X rays get through them. We cannot see the energy sources, for example, of quasars, using visible light or X-ray telescopes in such situations. However, infrared radiation should pass through the gas and dust. Thus, in 2005, astronomers used the Spitzer Space Telescope to study regions of galaxies in which quasars *should* exist. In just one tiny patch of sky they found 21. Extrapolating to the total volume of space that contains quasars, they expect "stealth" active galaxies to exist by the millions or more.

a

b

FIGURE 11-70 Focusing Jets by Pressure (a) If a high-speed jet of gas or liquid encounters little pressure (from the surrounding air, in this image), then it will spread out. (b) If the jet encounters high pressure, such as occurs when it enters water, then it will maintain longer its shape as a column. *(a: Comstock Images/Getty Images; b: Fundamental Photographs, New York)*

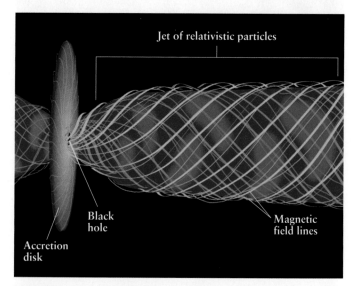

FIGURE 11-71 Focusing Jets by Magnetic Fields The hot, ionized accretion disk (red-yellow) around the black hole rotates and creates a magnetic field that is twisted into spring-shaped spirals above and below the disk. Some of the accretion disk's gas falling toward the black hole is overheated and squirted at high speeds into the two tubes created by the magnetic fields. The fields keep the gas traveling directly outward from above and below the disk, thus creating the two jets.

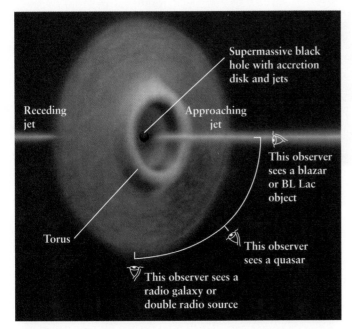

FIGURE 11-72 **Orientation of the Central Engine and Its Jets** BL Lacertae objects, quasars, and double radio sources appear to be the same type of object viewed from different directions. If one of the jets is aimed almost directly at Earth, we see a BL Lac object. If the jet is somewhat tilted to our line of sight, we see a quasar. Tilted farther and we see an active galaxy. If the jets are nearly perpendicular to our line of sight, we see a double radio source. The central region of the system is shown in Figures 11-69a and 11-71.

The other major factor in the observed properties of active galaxies and their evolutionary history is the amount of material available to make the accretion disk. Overlap in the brightness levels of different types of active galaxies occurs because central black holes and their accretion disks have different masses. Their emissions also evolve with time.

When supermassive black holes are young, they are surrounded by vast amounts of gas, which create massive accretion disks. Over time, much of this gas either enters the black hole or jets away from the disk, as discussed previously. When an accretion disk is depleted, that active galaxy becomes more and more quiet. Because the galaxies we observe close to us are about the same age as the Milky Way, over 13 billion years old, most have small or no accretion disks around the massive black holes in their nuclei. This is why we do not see many active galaxies close to us. However, when galaxies collide, gas can be available to enable quiet galaxies to once again become active.

INSIGHT INTO SCIENCE

Occam's Razor Revisited It is possible to create *separate* theories for quasars, BL Lac objects, and other active galaxies. However, scientists prefer a comprehensive single theory that explains as many things as possible, because it incorporates common properties of the various phenomena. This search for simple, more powerful explanations is the central engine of progress in science.

11-27   Gravity focuses light from quasars

The radiation emitted by all quasars is subject to the gravitational lensing described by Einstein's general theory of relativity (see Sections 10-19 and 11-18). Like the light from stars behind the Sun, when the light from quasars travels toward us, it is deflected as it moves past other galaxies or intergalactic gas and dark matter. As we saw in Section 11-18, this gravitational lensing can lead to our receiving several images of objects. Figure 11-73 shows four images of a quasar in a configuration called an **Einstein cross**. If the background object is exactly behind an intervening quasar, the light is actually focused as a ring, called an **Einstein ring** (see Figure 11-49b), rather than as several separate images.

As you know, light travels at a finite speed. For example, light from the Sun takes about 8 min to reach Earth. We see the Sun as it was 8 min ago. The farther an object is from us, the longer it takes light and other radiation from it to reach Earth. We see a galaxy 1 billion light-years away as it was a billion years ago. When we look at quasars, we are seeing objects as they appeared when the universe was younger still. In the last few years, astronomers using the Sloan Digital Sky Survey have discovered quasars at a distance corresponding to when the universe was only about 800 million years old.

11-28   Frontiers yet to be discovered

Despite our being part of it, our knowledge of the Milky Way is limited by the difficulty in making observations through its gas and dust. We have yet to locate all of the globular clusters. Even more challenging will be locating and categorizing all of the stars and clouds of gas and dust in the disk of the Galaxy. Are there any more small galaxies that the Milky Way is consuming? We also have to determine fully and accurately the structure and extent of the spiral arms and the bar of gas, dust, and stars running through the central bulge.

Microlensing is similar to the physics of which: reflecting telescopes or refracting telescopes?

Our observations of the Galaxy's nucleus hint at remarkable activity that has yet to be explained. What is heating the 100-million-K gas? What is creating the positrons there, whose resulting gamma rays we observe? In its farthest reaches, what else is in the Galaxy's halo that creates all of the gravitational force we detect, but whose source is not yet observed? In other words, what is the dark matter? Another major unsolved mystery is the formation process of our Galaxy. Did stars form first? Did the supermassive black hole form and then pull gas and dust into orbit, which then formed stars? Did smaller galaxies form and then coalesce into the Milky Way as it is today? Astronomers are presently working to answer all of these questions.

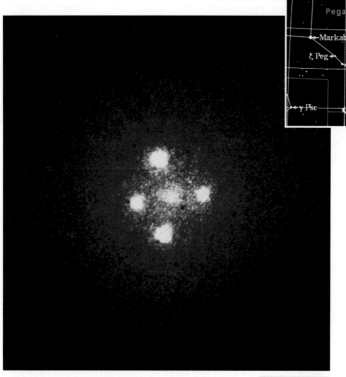

R I V U X G

 FIGURE 11-73 Gravitational Lensing of Quasars
Image from the Hubble Space Telescope that shows the gravitational lensing of a quasar in the constellation of Pegasus. The quasar, about 8 billion light-years from Earth, is seen as four separate images that surround a galaxy that is only 400 million light-years away. This pattern is called an Einstein cross. The diffuse image at the center of the Einstein cross is the core of the intervening galaxy. The physical effect that creates these multiple images is the same as that seen for galaxies, as depicted in Figure 11-49. *(NASA/ESO)*

The farther out in the universe that we explore, the more there is to discover. In the realm of the galaxies, we have yet to understand fully how the different types of galaxies form. Are central bulges formed as a result of collisions between galaxies? Are globular clusters the remnants of previous galaxies that have been consumed by bigger galaxies? How long does it take interacting galaxies to combine? Do supermassive black holes form before, simultaneously with, or after galaxies come into existence? Moving into the realm of the global structure of the universe, what is the overall distribution of clusters and superclusters of galaxies in the universe? Are most of them really found on spongelike or soap bubble–like surfaces and, if so, what is that structure telling us about the underlying distribution of dark matter and other properties of the cosmos?

What astronomical objects other than black holes have jets of gas rushing out from them?

Quasars, active galaxies, BL Lac objects, and double radio sources are rich areas for observational and theoreti-cal research. Why do different quasars emit most strongly in different parts of the spectrum? Are quasars an early phase of most galaxies, including the Milky Way, as many astronomers are coming to believe? What causes quasar variability? Do we have the proper explanations that blazars, quasars, and radio galaxies are caused by supermassive black holes seen from different orientations? How did supermassive black holes form so quickly in the life of the universe? New discoveries in this realm are being made nearly every week.

What other astronomical objects have been observed after their light passed through a gravitational lens?

SUMMARY OF KEY IDEAS

Defining the Milky Way

• A century ago, astronomers were divided on whether all the stars and nebulae are part of the Milky Way Galaxy.

• The Shapley–Curtis debate was the first major public discussion between astronomers as to whether the Milky Way contains all the stars in the universe.

• Cepheid variable stars are important in determining the distance to other galaxies.

• Edwin Hubble proved that there are other galaxies far outside of the Milky Way.

The Structure of Our Galaxy

• Our Galaxy has a disk about 100,000 light-years in diameter and about 2000 light-years thick, with a high concentration of interstellar dust and gas. It contains around 200 billion stars.

• Interstellar dust obscures our view into the plane of the galactic disk at visual wavelengths. However, hydrogen clouds can be detected beyond this dust by the 21-cm radio waves emitted by changes in the relative spins of electrons and protons in the clouds, as well by other non-visible emissions.

• The center, or galactic nucleus, has been studied at gamma-ray, X-ray, infrared, and radio wavelengths, which pass readily through intervening interstellar dust and H II regions that illuminate the spiral arms. These observations have revealed the dynamic nature of the galactic nucleus, but much about it remains unexplained.

• A supermassive black hole of about $4 \times 10^6 \ M_\odot$ exists in the galactic nucleus.

• The galactic nucleus of the Milky Way is surrounded by a flattened sphere of stars, called the *central bulge*, through which a bar of stars and gas extend. The entire Galaxy is surrounded by a halo of matter that includes a spherical distribution of globular clusters and field stars, as well as large amounts of dark matter.

• A disk with two bright and several dimmer arms of stars, gas, and dust spirals out from the ends of the bar in the galactic central bulge.

• Young OB associations, H II regions, and molecular clouds in the galactic disk outline huge spiral arms where stars are forming.

• The Sun is located about 26,000 light-years from the galactic nucleus, between two major spiral arms. The Sun moves in its orbit at a speed of about 828,000 km/h and takes about 230 million years to complete one orbit around the center of the Galaxy.

Mysteries at the Galactic Fringes
• From studies of the rotation of the Galaxy, astronomers estimate that its total mass is about 1×10^{12} $M_\odot$. Much of this mass is still undetectable.

Types of Galaxies
• The Hubble classification system groups galaxies into four major types: spiral, barred spiral, elliptical, and irregular.

• The arms of spiral and barred spiral galaxies are sites of active star formation.

• According to the theory of self-propagating star formation, spiral arms of flocculent galaxies are caused by the births and deaths of stars over extended regions of a galaxy. Differential rotation of a galaxy stretches the star-forming regions into elongated arches of stars and nebulae that we see as spiral arms.

• According to the spiral density wave theory, spiral arms of grand-design galaxies are caused by spiral density waves. The gravitational field of a spiral density wave compresses the interstellar clouds that pass through it, thereby triggering the formation of stars, including OB associations, which highlight the arms.

• Elliptical galaxies contain much less interstellar gas and dust than do spiral galaxies; relatively little star formation occurs in elliptical galaxies.

Clusters and Superclusters
• Galaxies group into clusters rather than being randomly scattered through the universe.

• A rich cluster contains at least a thousand galaxies; a poor cluster may contain only a few dozen up to a thousand galaxies. A regular cluster has a nearly spherical shape with a central concentration of galaxies; in an irregular cluster, the distribution of galaxies is asymmetrical.

• Our Galaxy is a member of a poor, irregular cluster called the Local Group.

• Rich, regular clusters contain mostly elliptical and lenticular galaxies; irregular clusters contain more spiral and irregular galaxies. Giant elliptical galaxies are often found near the centers of rich clusters.

• No cluster of galaxies has an observable mass large enough to account for the observed motions of its galaxies; a large amount of unobserved mass must be present between the galaxies.

• Hot intergalactic gases emit X rays in rich clusters.

• When two galaxies collide, their stars initially pass each other, but their interstellar gas and dust collide violently, either stripping the gas and dust from the galaxies or triggering prolific star formation. The gravitational effects of a galactic collision can cast stars out of their galaxies into intergalactic space.

• Galactic mergers occur, sometimes producing giant elliptical galaxies. Also, a large galaxy in a rich cluster may grow steadily through galactic cannibalism.

Superclusters in Motion
• A simple linear relationship exists between the distance from Earth to galaxies in other superclusters and the redshifts of those galaxies (a measure of the speed at which they are receding from us). This relationship is the Hubble law: Recessional velocity $= H_0 \times$ distance, where H_0 is the Hubble constant.

• Astronomers use standard candles—Cepheid variables, the brightest supergiants, globular clusters, H II regions, supernovae in a galaxy, and the Tully–Fisher relation—to calculate intergalactic distances. Because of difficulties in measuring the distances to remote galaxies, the value of the Hubble constant, H_0, is not known with complete certainty.

• The development of radio astronomy in the late 1940s led to the discovery of very powerful and extremely distant energy sources.

Quasars and Other Active Galaxies
• An active galaxy is an extremely luminous galaxy that has one or more unusual features: an unusually bright, star-like nucleus; strong emission lines in its spectrum; rapid variations in luminosity; and jets or beams of radiation that emanate from its core. Active galaxies include quasars, Seyfert galaxies, radio galaxies, double radio sources, and BL Lacertae objects.

• A quasar, or quasi-stellar radio source, is an object that looks like a star but has a huge redshift. This redshift corresponds to a distance of billions of light-years from Earth, according to the Hubble law.

• To be seen from Earth, a quasar must be very luminous, typically about 100 times brighter than an ordinary galaxy. Relatively rapid fluctuations in the brightness levels of some quasars indicate that they cannot be much larger than the diameter of our solar system.

• An active spiral galaxy with a bright, starlike nucleus and strong emission lines in its spectrum is categorized as a Seyfert galaxy.

• An active elliptical galaxy is called a radio galaxy. It has a bright nucleus and a pair of radio-bright jets that stream out in opposite directions.

• BL Lacertae (BL Lac) objects (some of which are called blazars) have bright nuclei whose cores show relatively rapid variations in luminosity.

• Double radio sources contain active galactic nuclei located between two characteristic radio lobes. A head–tail radio source shows evidence of jets of high-speed particles that emerge from an active galaxy.

Supermassive Engines

• Many galaxies contain huge concentrations of matter at their centers.

• Some matter that spirals in toward a supermassive black hole is squeezed into two oppositely-directed beams that carry particles and energy into intergalactic space.

• The energy sources from quasars, Seyfert galaxies, BL Lac objects, radio galaxies, and double radio sources are probably matter ejected from the accretion disks that surround supermassive black holes at the centers of galaxies.

WHAT DID YOU THINK?

1 *What is the shape of the Milky Way Galaxy?* The Milky Way is a barred spiral galaxy. A bar of stars, gas, and dust runs through its central region. It has two bright, and several dim, spiral arms and is surrounded by a spherical halo of stars and dark matter.

2 *Where is our solar system located in the Milky Way Galaxy?* The solar system is between the Sagittarius and Perseus spiral arms, about 26,000 light-years from the center of the Galaxy.

3 *Is the Sun moving through the Milky Way Galaxy and, if so, about how fast?* Yes. The Sun orbits the center of the Milky Way Galaxy at a speed of 828,000 km/h.

4 *Are most of the stars in spiral galaxies located in their spiral arms?* No. The spiral arms contain only 5% more stars than the regions between the arms.

5 *Do all galaxies have spiral arms?* No. Galaxies may be either spiral, barred spiral, elliptical, or irregular. Only spirals and barred spirals have arms.

6 *Are galaxies isolated objects?* No. Galaxies are grouped in clusters, and clusters are grouped in superclusters.

7 *Is the universe contracting, unchanging in size, or expanding?* The universe is expanding.

Review Questions

The answers to computational problems, which are preceded by an asterisk (*), appear at the back of the book.

1. Where in the Galaxy is the solar system located? **a.** in the nucleus, **b.** in the halo, **c.** in a spiral arm, **d.** between two spiral arms, **e.** in the central bulge

2. What is located in the nucleus of the Galaxy? **a.** a globular cluster, **b.** a spiral arm, **c.** a black hole, **d.** the solar system, **e.** a MACHO

3. Which statement about the Milky Way Galaxy is correct? **a.** Our Galaxy is but one of many galaxies. **b.** Our Galaxy contains all the stars in the universe. **c.** All the stars in our Galaxy take the same time to complete one orbit. **d.** Most of the stars in our Galaxy are in the central bulge. **e.** None of the stars in our Galaxy move.

4. What was the Shapley–Curtis debate all about? Was a winner declared at the end of the debate? Whose ideas turned out to be correct?

5. How did Edwin Hubble prove that M31 is not a nebula in our Milky Way Galaxy?

6. As seen in the northern hemisphere, why is the Milky Way far more prominent in July than in December? Planetarium software, such as *Starry Night Enthusiast™*, may be useful in answering this question.

7. What observations led Harlow Shapley to conclude that we are not at the center of the Galaxy?

8. Explain why globular clusters spend more time in the galactic halo than in the plane of the disk, even though their eccentric orbits take them across the disk of the Galaxy.

9. How do hydrogen atoms generate 21-cm radiation? What do astronomers learn about our Galaxy from observations of that radiation?

10. Why do astronomers believe that vast quantities of dark matter surround our Galaxy?

11. What is synchrotron radiation?

12. Why are there no massive O and B stars in globular clusters?

13. What evidence indicates that a supermassive black hole is located at the center of our Galaxy?

*14. Approximately how many times has the solar system orbited the center of the Galaxy since the Sun and planets were formed?

15. What are the most massive galaxies in the universe? **a.** normal spirals, **b.** barred spirals, **c.** giant ellipticals **d.** dwarf ellipticals, **e.** irregulars

16. Spiral density waves are directly responsible for which of the following? **a.** flocculent spirals, **b.** grand-design

spirals, **c.** supernovae, **d.** the collisions between galaxies, **e.** galactic cannibalism

17. Which of the following statements about the motion of galaxies is correct? **a.** All galaxies are moving apart. **b.** Superclusters of galaxies are all moving apart. **c.** Superclusters of galaxies are all moving toward each other. **d.** The Milky Way Galaxy is at the center of the universe. **e.** All clusters of galaxies in each supercluster are moving toward each other.

18. What is the common name for the sonic boom created by lightning?

19. What spectral classes of stars are the 5% only found in spiral arms of spiral galaxies?

20. What is the Hubble classification scheme? Which category includes the biggest galaxies? Into which category do the smallest galaxies fall? Which type of galaxy is the most common?

21. In which Hubble types of galaxies are new stars most commonly forming? Describe the observational evidence that supports your answer.

22. What is the difference between a flocculent spiral galaxy and a grand-design spiral galaxy?

23. Briefly describe how the theory of self-propagating star formation accounts for the existence of spiral arms in some spiral galaxies.

24. Briefly describe how the spiral density wave theory accounts for the existence of spiral arms in some spiral galaxies.

25. How is it possible that galaxies in our Local Group still remain to be discovered? In what part of the sky are these galaxies located? What sorts of observations might reveal these galaxies?

26. Can any galaxies besides our own be seen with the naked eye? If so, which?

27. What is the difference between a rich cluster of galaxies and a poor one? What is the difference between a regular cluster of galaxies and an irregular one?

28. How can a collision between galaxies produce a starburst galaxy?

29. Why do astronomers believe that considerable quantities of dark matter must exist in clusters of galaxies?

30. Explain why the dark matter in galaxy clusters cannot be neutral hydrogen.

31. What is the Hubble law?

32. Some galaxies in the Local Group exhibit blueshifted spectral lines. Why are these blueshifts not violations of the Hubble law?

33. What is a standard candle? Why are standard candles important to astronomers trying to measure the Hubble constant?

34. What kinds of stars would you expect to find populating space between galaxies in a cluster?

35. What is a double radio source seen along one axis of a jet called? **a.** a BL Lac object, **b.** a quasar, **c.** an active galaxy, **d.** a double radio source, **e.** a pulsar

36. What two things does the engine of a quasar contain? **a.** a supermassive black hole and a binary companion, **b.** a stellar-mass black hole and a binary companion, **c.** a supermassive black hole and an accretion disk, **d.** a stellar-mass black hole and an accretion disk, **e.** a supergiant star and an accretion disk

37. Suppose you suspected a certain object in the sky to be a quasar. What sort of observations would you perform to confirm your hypothesis?

38. Explain how the rate of variability of a source of light can be used to place an upper limit on the size of the source.

39. What is an active galaxy? List the different kinds of active galaxies. How do they differ from one another?

40. Why do astronomers believe that the energy-producing region of a quasar is very small?

41. How is synchrotron radiation produced?

42. What evidence indicates that quasars are extremely distant active galaxies?

43. What is a double radio source?

44. What is a supermassive black hole? What observational evidence suggests that supermassive black holes might be located at the centers of many galaxies?

45. Why do many astronomers believe that the engine at the center of a quasar is a supermassive black hole surrounded by an accretion disk?

46. How does the orientation of the jets that emanate from the center of a galaxy relative to our line of sight relate to the type of active galaxy that we observe?

Observing Projects

47. **Clusters of Galaxies** contain different numbers and distributions of galaxies. In this exercise you can use *Starry Night Enthusiast*™ to compare a few of these groupings. Select **Favourites > Deep Space > Galaxies > Virgo Cluster** from the menu. Position the cursor over one of the prominent galaxies in the center of the view and open the object contextual menu (right-click on a PC, Ctrl-click+C on a Mac) and select **Highlight "GA Virgo Cluster" Filament**. Use the **Increase current elevation** (∧) button in the

toolbar to step away from the cluster until the complete cluster, highlighted in yellow, is visible in the view. Use the **Location Scroller** to examine the cluster from various viewpoints. What is the general shape of the Virgo cluster? If you see other groupings of galaxies as you move the screen, make a sketch of the screen and circle what you believe are other groups on your sketch. To see how astronomers have grouped these other galaxies, open the object **Contextual Menu** over a galaxy within a suspected group and select the **Highlight** option to see the extent of the group to which this galaxy belongs. Choose three of these clusters, center and zoom in on them, as necessary, and use the **Location Scroller** to study them. Describe their distributions of galaxies compared to the Virgo cluster. For example, what are their shapes and relative sizes compared to Virgo and to each other?

 48. Use the **Search** tab in WWT to observe and classify each of the galaxies listed. For each galaxy in the list, indicate its classification as a spiral, barred spiral, irregular, or elliptical galaxy.

a. NGC 43 14
b. M95
c. M87
d. Large Magellanic Cloud
e. M100
f. M105
g. M49
h. Small Magellanic Cloud

49. Use WWT to explore **M82**, the **Cigar Galaxy**. Click the **Guided Tour** tab, then click on followed by the thumbnail of the **Galaxies** folder. Click on the thumbnail labeled **"M82 Cigar Galaxy"** and watch the tour. At the conclusion of the tour, you may wish to click the **Watch Again** button to answer the following questions: **a.** In which constellation is M82? **b.** What causes the bright X-ray emission of M82 as seen by the Chandra X-ray Observatory? **c.** What do astronomers suggest was the cause of the starburst activity in M82?

12 Cosmology

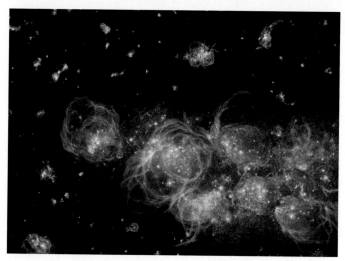

Bursts of Star Formation in the Early Universe *(Depiction by science artist Adolf Schaller, STScI/NASA/K. Lanzetta, SUNY)*

WHAT DO YOU THINK?

1 What is the universe?

2 Did the universe have a beginning?

3 Into what is the universe expanding?

4 How strong is gravity compared to the other forces in nature?

5 Will the universe last forever?

Answers to these questions appear in the text beside the corresponding numbers in the margins and at the end of the chapter.

For millennia, our ancestors speculated about whether the universe has existed forever or whether it began some finite time in the past. Did God create the universe, or was its creation an event that required no divine intervention? Science has yet to provide an answer to this question. But astronomers and other scientists *have* accumulated a great deal of knowledge about what has happened to the universe beginning a tiny fraction of a second after it began and what its fate will be. Although the overall theory of cosmic evolution is not complete, its major elements are consistent with observations.

In this chapter we will examine the distribution of matter on the largest scales in the universe and how this matter is moving and changing. Using this information, we will explore the Big Bang model of how the universe has evolved from its earliest moments, when it was smaller than an atom. We will see how it grew and created matter as we know it, forming and evolving large-scale systems, such as galaxies and groupings of galaxies. Finally, we will consider the fate of the universe.

In this chapter you will discover

- cosmology, which seeks to explain how the universe began, how it evolves, and its fate

- the best theory that we have for the evolution of the universe—the Big Bang

- how astronomers trace the emergence of matter and the formation of galaxies

- how astronomers explain the overall structure of the universe

- our understanding of the fate of the universe

THE BIG BANG

The **universe** consists of all matter, energy, and spacetime that we can ever detect or that will ever be able to affect us. (We use this definition because there may be matter and energy in other dimensions or matter and energy that are moving away from us so quickly that their influence will never reach us. These ideas are still very preliminary and theoretical, and for this book we will continue to deal with just concepts for which science has experimental evidence.) So far, in this text, we have explored matter on size scales from atoms to superclusters of galaxies. We also learned in Section 11-19 that the superclusters are all moving away from one another, implying that the universe is expanding. In this chapter we take the observational evidence from the rest of the book and use it to explore the Big Bang theory of cosmology. **Cosmology** is the study of the large-scale structure and evolution of the universe.

12-1 General relativity predicts an expanding (or contracting) universe

Modern cosmology almost began in 1915, when Einstein published his theory of general relativity. To his surprise and dismay, the equations predicted that the universe is not static: They indicated that it should be either expanding (which it is) or contracting. But Einstein was not ready for what the equations were telling him.

The prediction of a changing universe flew in the face of the widely accepted belief in an infinite, static universe, a concept promoted by Isaac Newton more than two centuries earlier. Newton believed that each star is fixed in place and held under the influence of a uniform gravitational pull from every part of the cosmos. If the stars were not uniformly distributed, he argued, one region would have more mass than another. The denser region's gravity would then attract other stars, causing them to further clump together. Because he did not observe this clumping, Newton concluded that the mass of the universe must be distributed uniformly over an infinite space. At the time that Einstein published the theory of general relativity, the existence of galaxies and clusters and superclusters of galaxies had not yet been established, and the relative motion of superclusters of galaxies had not been observed.

The apparently static universe and the prevailing philosophy that the universe had existed forever made Einstein doubt the implications of his own theory, so he missed the opportunity to propose that we live in a changing universe. Instead, he adjusted his elegant equations to yield a static cosmos. He did this by adding a repulsive (outward-pushing) term, called the **cosmological constant**, to his equations so that gravity's normal attractive force would be counterbalanced and the universe would be static. After observations revealed that the universe is expanding, Einstein said that adding the cosmological constant was the biggest blunder of his career.

Although the value of the cosmological constant that Einstein inserted was wrong, the concept of such a constant may be correct. Observations since 1997 indicate that the universe is not just expanding but actually *accelerating* outward. This means that there must be an outward pressure that more than counteracts the effects of normal gravitation, which is trying to slow the universe's expansion. One of the two current theories that can explain that acceleration is the presence of a cosmological constant that creates the outward pressure. We discuss these two theories further in Section 12-15.

12-2 The expansion of the universe creates a Dopplerlike redshift

Edwin Hubble is credited with discovering that we live in an **expanding universe** (see Section 11-19). The redshifts of clusters and superclusters of galaxies that Hubble found moving away from us appear to come from the Doppler effect, but they actually do not. Recall that the normal Doppler shift is caused by an object moving toward or away from us through *fixed* spacetime (see Section 3-2). However, using Einstein's theory of general relativity, we find that spacetime, the fabric of the universe, is *not fixed* but is actually expanding. This expansion is what

is carrying the superclusters away from one another, and, in many cases, it carries clusters in a given supercluster away from each other. (The gravitational force that holds objects, such as planets and stars and entire galaxies, together is so strong that these bodies and systems are not being pulled apart. The expansion of space just acts to increase the separation between large groups of galaxies.)

The redshift that Hubble observed, caused by the expansion of the universe, is properly called the **cosmological redshift**. In other words, the photons that we observe from galaxies in other superclusters are all redshifted because space is expanding. To understand why, consider a wave drawn on a rubber band (Figure 12-1). The wave has an unstretched wavelength (see Section 3-1) of λ_0. As the rubber band stretches, the wavelength increases ($\lambda > \lambda_0$). Now imagine a photon coming toward us from a distant galaxy. As the photon travels through space, space is expanding, and, like stretching the rubber band, this expansion stretches the photon's wavelength. When the photon reaches our eyes, we see a drawn-out wavelength—the photon is redshifted. The longer the photon's journey, the more its wavelength is stretched by the expansion of the universe. Therefore, astronomers observe larger redshifts in photons from relatively distant galaxies than in photons from relatively nearby galaxies.

As we saw in Section 11-19, Hubble discovered the linear relationship between the distances to galaxies in other superclusters and the redshifts of those galaxies' spectral lines. For example, a galaxy twice as far from Earth as another galaxy has twice the cosmological redshift of the closer one. The normal Doppler shift has the same relationship.

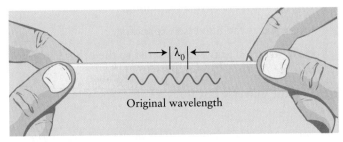

a A wave drawn on a rubber band ...

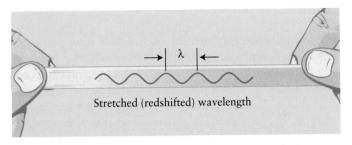

b ... increases in wavelength as the rubber band is stretched.

FIGURE 12-1 **Cosmological Redshift** Just as the waves drawn on this rubber band are stretched along with the rubber band, so too are the wavelengths of photons stretched as the universe expands.

As a result of the two effects being described by the same equations, the normal Doppler shift and the cosmological redshift predict the same relationship between redshift and motion—except for the most distant galaxies and quasars, where effects of relativity must be taken into account. Working with relatively nearby galaxies, Hubble was fully justified in using the Doppler equation to calculate the recession of galaxies.

If the cosmological constant that Einstein used had the opposite effect—that is, if it were attractive—what would happen to a universe initially at rest, as Einstein had supposed to be true?

What are the three redshifts presented in this book?

12-3 The Hubble constant is related to the age of the universe

Hubble's law gives us a way to estimate the age of the universe. Imagine watching a movie of two superclusters receding from each other. If we then run the film backward, time runs in reverse, and we observe the superclusters approaching each other. The time it would take for them to collide is the time since they were last combined as parts of the same matter. *Assuming that the universe began expanding when it came into existence and that it has always expanded at a constant rate,* we can use a simple equation to estimate the age of the universe:

$$\text{Time since the Big Bang} = \frac{\text{Separation distance}}{\text{Recessional velocity}}$$

Recall that Hubble found the relationship

$$\text{Recessional velocity} = H_0 \times \text{Separation distance}$$

which we can rewrite as

$$H_0 = \frac{\text{Recessional velocity}}{\text{Separation distance}}$$

Comparing the first and last equations here, we have that Hubble's constant is the reciprocal of the time since the universe began. Using a Hubble constant of 71 km/s/Mpc, we find:

$$1/H_0 = 1/71 \text{ km/s/Mpc} = 13.7 \text{ billion years}$$

If the universe were twice as old as it is now, how would the Hubble constant compare to the value it has today?

The universe has *not* been expanding at a constant rate, and we do not yet know the complete history of its motion. Nevertheless, calculations reveal that the age of the universe given by $1/H_0$ is within a quarter of a billion years of our best determination of the age of

the universe, and we use 13.7 billion years as the age of the universe in what follows.

12-4 Remnants of the Big Bang have been detected

In 1927, the Belgian astrophysicist and Catholic priest Georges Lemaître proposed that, as a consequence of the equations of Einstein's theory of general relativity, the universe is expanding. Using Hubble's subsequent observations of an expanding universe, Lemaître logically concluded that the superclusters must have expanded from a much smaller volume. He proposed that the universe began as an extraordinarily hot and dense *primordial atom* of energy. Just as energy created in an explosion causes the debris to expand outward, the primordial energy created at the beginning of time caused the universe to expand. This initial event is now called the **Big Bang.** Space and time (spacetime; see Chapter 10) did not exist before that moment. Rather than expanding into preexisting spacetime, the Big Bang explosion created spacetime, and that spacetime (in which we live) has been expanding ever since. The Big Bang led to the formation of all spacetime, matter, and energy.

Shortly after World War II, the astrophysicists Ralph Alpher and George Gamow proposed that seconds after the Big Bang, the universe was an incredibly hot blackbody—in excess of 10^{32} K. As heat is a measure of how many photons there are, and how energetic they are, the early universe must therefore have been filled with high-energy electromagnetic radiation. As the universe expanded, the cosmological redshift stretched the wavelength of this radiation so that most of the photons left over from this event are now radio waves. Recall from Section 3-3 that photon energies decrease with increasing wavelength. Because the early photons are being redshifted, the energy they have is decreasing. This energy is measured as heat, so the universe is cooling off as it expands.

Calculations indicated that if the universe began with a hot Big Bang, the remnants of that energy should still fill all space today, giving the entire universe a temperature only a few Kelvin above absolute zero. This radiation is called the **cosmic microwave background** because the peak of its blackbody spectrum should lie in the microwave part of the radio spectrum. In the early 1960s, Robert Dicke, P. J. E. Peebles, David Wilkinson, and their colleagues at Princeton University began designing a microwave radio telescope to detect it.

Meanwhile, just a few miles from Princeton, two physicists had already detected the cosmic background radiation. Arno Penzias and Robert Wilson of Bell Laboratories were working on a new horn antenna designed to relay telephone calls to Earth-orbiting communications satellites (Figure 12-2). However, Penzias and Wilson were mystified. No matter where in the sky they pointed their antenna, they detected a faint background noise. All efforts to eliminate this background noise failed, even the careful removal of static noise–generating pigeon droppings from the antenna.

a

b

FIGURE 12-2 **Bell Labs Horn Antenna** This Bell Laboratories horn antenna at Holmdel, New Jersey, was used by Arno Penzias (right) and Robert Wilson in 1965 to detect the cosmic microwave background. *(Lucent Technologies, Bell Laboratories)*

Thanks to a colleague, they learned of the then-theoretical cosmic microwave background and the work of Dicke, Peebles, and Wilkinson in trying to locate it. Communicating with the Princeton astronomers, Penzias and Wilson presented their finding and thus were able to claim the first detection—their annoying noise was, in fact, the remnant energy of the Big Bang. The cosmic microwave background, also commonly called the *3-degree background radiation*, is required by the Big Bang theory but is neither required nor predicted by any other competing cosmological theory. Detection of the cosmic microwave background is a principal reason why the Big Bang is accepted by astronomers as the correct cosmological theory.

What other objects that we have studied in this book have (nearly) blackbody spectra?

INSIGHT INTO SCIENCE

Consider Plausible Alternatives Until They Are Shown to Be Inconsistent with Observations Sir Fred Hoyle, who aptly named the Big Bang in 1949, was one of the very few astrophysicists who never believed that it occurred. He, Herman Bondi, and Thomas Gold proposed the *steady-state theory* in which the universe has existed forever, is continuously expanding, and, as it expands, new matter is created that takes the place of the receding matter. Where this new matter might come from is unknown, but the laws of physics do not necessarily exclude its creation. Today, astronomers accept the Big Bang theory rather than the steady-state theory because cosmic microwave background observations are consistent with the Big Bang theory but not with the steady-state theory.

FIGURE 12-3 **In Search of Primordial Photons** (a) The *Wilkinson Microwave Anisotropy Probe (WMAP)* satellite, launched in 2001, improved upon the measurements of the spectrum and angular distribution of the cosmic microwave background taken by the *COBE* satellite. (b) The balloon-carried telescope BOOMERANG orbited above Antarctica for 10 days, collecting data used to resolve the cosmic microwave background, with 10 times higher resolution than that of *COBE*. All of these experiments found local temperature variations across the sky but no overall deviation from a blackbody spectrum. *(a: NASA/WMAP Science Team; b: The BOOMERANG Group, University of California, Santa Barbara)*

Precise measurements of the cosmic microwave background were first made by the *Cosmic Background Explorer (COBE)* satellite, which operated between 1989 and 1994. The radiation has since been measured more precisely by other equipment, such as the *Wilkinson Microwave Anisotropy Probe (WMAP)* and the balloon-carried BOOMERANG (Figure 12-3). The data taken by *COBE* and shown in Figure 12-4 reveal that, as predicted, this ancient radiation has the spectrum of a blackbody with a temperature of approximately 2.73 K. It is called 3-degree radiation, for convenience. The observations by *COBE* earned the 2006 Nobel Prize in Physics for astrophysicists John C. Mather and George F. Smoot.

FIGURE 12-10 **Observable Universe Before and After Inflation** Shortly after the Big Bang, the universe expanded by a factor of about 10^{50}, due to inflation. This growth in the size of the presently observable universe occurred in a very brief time (shaded interval).

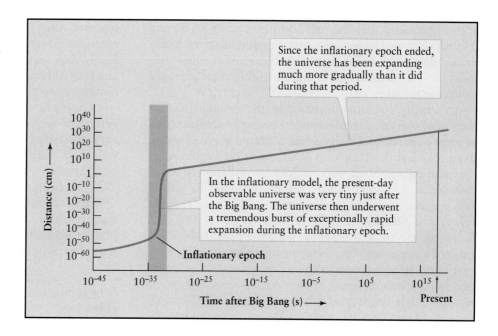

Since the inflationary epoch ended, the universe has been expanding much more gradually than it did during that period.

In the inflationary model, the present-day observable universe was very tiny just after the Big Bang. The universe then underwent a tremendous burst of exceptionally rapid expansion during the inflationary epoch.

Inflationary epoch

into space, the further back in time we are seeing. We can only see a finite distance into it, namely back to the time of decoupling, about 13.7 billion light-years away. Farther than this distance (that is, earlier in the life of the universe), the universe was opaque and so we cannot see into it. At a distance of 13.7 billion light-years, we see objects as they were just being formed (still in our nearly uniform region). Therefore, all that we can see from Earth is contained in an enormous sphere (Figure 12-11). The boundary of this sphere is called the **cosmic light horizon.** Light from all of the luminous objects in this volume has had time to reach Earth, but light from objects beyond the cosmic light horizon has not yet arrived here, so we are unaware of objects in these regions. If the speed of light were greater, we could see things farther from us.

Had the universe been perfectly uniform (everywhere homogeneous and isotropic), it would have been extremely hard for stars, galaxies, and larger groupings of matter to form, pulled together as they were by differences in mass density. Quantum theory played a vital role during the inflationary epoch in creating the clumpiness that we observe today. In 1927, British physicist Paul A. M. Dirac concluded that the laws of nature allow pairs of particles to spontaneously

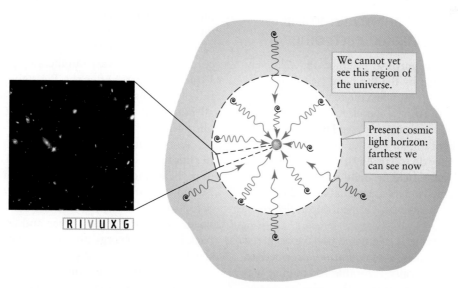

We cannot yet see this region of the universe.

Present cosmic light horizon: farthest we can see now

R I V U X G

FIGURE 12-11 **Observable Universe** This diagram shows why we only see part of the entire universe. As time passes, this volume grows, meaning that light from more distant galaxies reaches us. The farthest galaxies we see (inset) are as they were within a few hundred million years after the Big Bang. These galaxies, formed at the same time as the Milky Way, appear young because the light from their beginnings is just now reaching us. While the light from the most distant galaxies we see was traveling toward us, the universe has been growing. Therefore, objects that appear 13 billion light-years away from us are actually about three times farther away today. Put another way, if we could see it all as it is today, rather than as it was when photons from distant objects started their journeys, the visible universe would be nearly 46 billion light-years in radius. **Inset:** This image shows some of the most distant galaxies we have seen. *(Robert Williams and the Hubble Deep Field Team, STScI and NASA)*

appear, provided that they are identical except for having opposite electric charges, and that they annihilate each other and disappear within a very short period of time, typically 10^{-21} s. As discussed in Section 10-28, these pairs are called *virtual particles*.

Normally, quantum fluctuations (the spontaneous formation of pairs of virtual particles) occur on the size scales of pairs of particles. However, the quantum fluctuations on microscopic scales that were occurring at the beginning of the inflationary period were stretched by inflation to scales that spanned clusters of galaxies and beyond! To picture this, imagine a sponge compressed very tightly so that the spaces and fiber are initially pressed very close together. Released, the sponge expands, with the relationships between the spaces and the fibers remaining intact, while each part of the sponge grows to a much larger scale. The quantum fluctuations created a spongelike universe, with some regions nearly empty and others relatively full of matter. Because of this internal structure, the universe can only be considered homogeneous on size scales greater than about 1 billion light-years, so that the voids and supercluster regions average out.

Indeed, the theory predicts that tiny quantum fluctuations ultimately became the seeds for the formation of the large-scale structures of the universe that exist today. Because inflation is part of a scientific theory, its occurrence leads to predictable observations. As we will see shortly, telescopes have indeed seen the predicted distributions of matter.

12-9 During the first second, most of the matter and antimatter in the universe annihilated each other

When inflation ended, the universe resumed the relatively leisurely expansion initiated by the Big Bang (see Figure 12-10). Arriving at the time 10^{-12} s, when the temperature of the universe had dropped to 10^{15} K, the electromagnetic force separated from the weak nuclear force (see Figures 12-7 and 12-8). From that moment on, all four forces interacted with particles essentially as they do today. At 10^{-12} s, the universe's matter was primarily the individual building blocks of protons and neutrons—namely quarks, antiquarks, and the other elementary particles: neutrinos, antineutrinos, electrons, and positrons.

The next significant event is calculated to have occurred at 10^{-6} s, when the universe's temperature dropped to 10^{13} K. Just as it is energetically favorable for boiling water to turn into gas, the temperature of the universe was so high before this time that it was energetically favorable for quarks to exist as isolated particles, rather than to combine in twos and threes to create the particles that exist in the universe today. We know they were initially isolated because similar seas of quarks were recreated under equivalent conditions in the laboratory in 2001 by smashing dense atoms

together. After 10^{-6} s, a period appropriately called **confinement**, the universe was sufficiently cool so that quarks could finally stick together to form individual protons, neutrons, and their antiparticles. Figure 12-8 summarizes the connections between particle physics and cosmology.

Equations show that early in the first second, the universe was so hot that virtually all photons were gamma rays possessing incredibly high energies. These energies were high enough so that photons could create matter and antimatter, according to Einstein's equation $E = mc^2$. This process is the reverse of what we saw in Section 7-7, where we discussed how matter is converted into energy in nuclear fusion. To make a particle of mass m, you need a photon with energy E at least as great as mc^2, where c is the speed of light.

Why is inflation a necessary part of the Big Bang theory of cosmology?

The creation of matter from energy is routinely observed in laboratory experiments that involve high-energy gamma rays. A highly energetic photon colliding with an atomic nucleus can create a pair of particles (Figure 12-12a). This process, called **pair production**, always creates one ordinary particle and its antiparticle. For example, if one of the particles is an electron, the other is an antielectron, also called a positron. (Recall from Section 10-28 that similar pairs of particles are also believed to be produced near a black hole.)

A positron has the same mass as an electron but the opposite charge. When an electron and a positron meet, they annihilate each other and their energy is converted into photons (Figure 12-12b). Theory predicts that at times earlier than 1 s in the evolution of the universe, photon collisions created vast numbers of pairs of particles. Consequently, shortly after the Big Bang, all space was chock full of protons, neutrons, electrons, and their antiparticles, all immersed in a phenomenally hot bath of high-energy photons.

As the universe expanded, the temperature declined until after the first second, photons could no longer create pairs of particles and antiparticles. The remaining photons still exist and they are the cosmic microwave background that we detect today. Thereafter, particles and antiparticles often collided, annihilating each other and converting their masses back into high-energy photons, and nothing could replenish the dwindling supply of particles and antiparticles.

If all particles had annihilated their antiparticles in the early universe, no matter would be left at all and we would not be here. And yet, here we are and astronomers observe very few antiparticles in the universe. Because particles and antiparticles are always created or destroyed in pairs, why aren't they found in equal numbers in the universe today?

Physicists theorize that the symmetry or equality between the number of particles and the number of antiparticles was

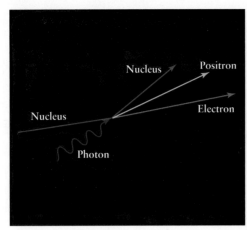

a Nucleus collides with photon and creates positron–electron pair

Time ⟶

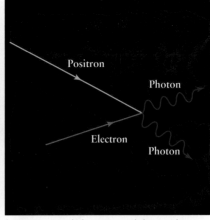

b Positron and electron annihilate each other and create two photons

Time ⟶

FIGURE 12-12 **Pair Production and Annihilation** (a) A particle and an antiparticle can be created when a high-energy photon collides with a nucleus. (b) Conversely, a particle and an antiparticle can annihilate each other and emit energy in the form of gamma rays. (The processes are more complex and are only summarized in these drawings.)

broken by the weak force—a *symmetry breaking*. This effect has been observed on Earth in high-energy particle accelerators over the past half century. This means that the number of particles was actually slightly greater than the number of antiparticles. For every billion antiprotons, perhaps a billion plus one protons formed; for every billion antielectrons, a billion plus one electrons formed. Virtually all of the antiparticles created in the first second of time annihilated normal particles shortly thereafter. The remaining particles in the universe were the slight excess of normal matter created as a result of this symmetry breaking.

12-10 The universe changed from being controlled by radiation to being controlled by matter

As the universe expanded and cooled further, the remaining protons and neutrons began colliding and fusing together, a process called **primordial nucleosynthesis.** For the most part, the nuclei with two or more particles were quickly separated again by the gamma rays in which they were bathed. However, during the first 3 min, enough helium and lithium atoms were created and not subsequently split apart to form most of the free helium and at least 10% of the trace element lithium that exists today. A lithium atom has three protons and three neutrons. Hydrogen, helium, and lithium are the three lowest-mass elements. After a few minutes, the universe was too cool to allow fusion to create more massive elements, such as carbon and oxygen. All elements other than the three lowest-mass ones formed later, due to stellar evolution, as discussed in Chapters 9 and 10.

INSIGHT INTO SCIENCE

Share Information Although the evolution of the universe is a truly large-scale event, the activity in the first 3 min was nevertheless intimately connected to the creation and properties of elementary particles, the smallest objects in the universe. Physicists who study elementary particles and astrophysicists who study cosmology now regularly hold meetings to compare notes and connect their fields.

Calculations indicate that for tens of thousands of years, photons dominated the behavior of the universe in two important ways. One effect of the photons was to contribute to the gravitational attraction in the universe. This contradicts intuition, because photons are massless and mass is what we normally associate with causing gravitational attraction. Nevertheless, the gravitational effect of photons is a tested prediction of general relativity. For about 30,000 years after the Big Bang, the energy density, and hence the gravitational effect, of photons was greater than the gravitational effect of matter. This situation was called the **radiation-dominated universe.**

The major effect of the radiation-dominated universe was that photons prevented matter from collapsing together and forming stars and galaxies. As the universe expanded, the gravitational effects of both photons and matter decreased as they became more spread out. The effect of photons decreased faster than that of matter (Figure 12-13) because the photons lost energy due to the cosmological redshift, but particles with mass did not lose energy similarly. After about 30,000 years, matter came

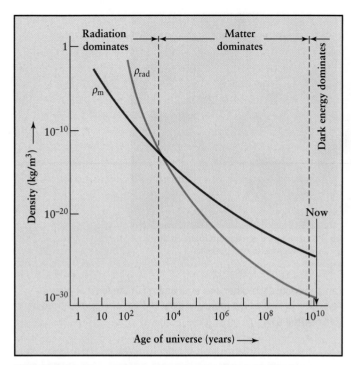

FIGURE 12-13 Evolution of Density For approximately 30,000 years after the Big Bang, the gravitational effects from photons (ρ_{rad}, shown in red) exceeded the attraction of all the matter in the universe (ρ_m, shown in blue). This early period is said to have been radiation-dominated. Later, however, continued expansion of the universe caused ρ_{rad} to become less than ρ_m, at which time the universe became matter-dominated.

to dominate the gravitational behavior of the universe, a situation that has since persisted, and thus developed the **matter-dominated universe** in which we live today. It was only after the universe became matter-dominated that gas could respond to gravitational interactions and begin clumping together where the force of gravity was sufficiently strong.

The second way that photons dominated the early universe occurred for about the first 379,000 years. During this time, photons prevented ions and electrons from connecting to form neutral atoms. During that epoch, the universe was completely filled with a shimmering expanse of high-energy photons colliding vigorously with protons and electrons. This state of matter, called a *plasma*, prevents any photons from traveling through it. In other words, in a plasma you would see a glow from local photons, but you would not be able to see photons from even a few meters or yards away (Figure 12-14a). The term **primordial fireball** describes the universe during this time.

By around 379,000 years, the universe was so large that the cosmological redshift had stretched photon wavelengths (see Figure 12-1) and therefore decreased photons' energies to the point where they could no longer ionize hydrogen. This stage is called **decoupling,** meaning that the massive particles and the photons were effectively disconnected.

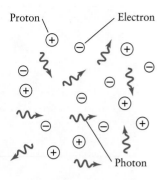

a Before recombination:
- Temperatures were so high that electrons and protons could not combine to form hydrogen atoms.
- The universe was opaque; photons underwent frequent collisions with electrons.
- Matter and radiation were at the same temperature.

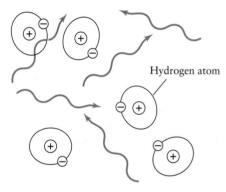

b After recombination:
- Temperatures became low enough for hydrogen atoms to form.
- The universe became transparent; collisions between photons and atoms became infrequent.
- Matter and radiation were no longer at the same temperature.

FIGURE 12-14 Era of Recombination (a) Before recombination, the energies of photons in the cosmic background were high enough to prevent protons and electrons from forming hydrogen atoms. (b) As soon as the energy of the background radiation became too low to ionize hydrogen, neutral atoms came into existence.

Electrons began orbiting nuclei, thereby creating neutral atoms (Figure 12-14b). The space between nuclei became so large that most primordial photons now passed by these particles, rather than colliding with them. Without these collisions to scatter photons, the universe ceased being hazy. It became clear, as it is today, enabling electromagnetic radiation to travel long distances unimpeded and, important from an astronomer's point of view, enabling us to see far into the universe.

How, if at all, would the universe be different if antiparticles had exceeded particles in the symmetry breaking described above?

We say that "most" of the helium in the universe was formed in primordial nucleosynthesis. Where was/is the rest formed?

This dramatic period, when the universe transformed from being opaque to transparent, took about 100,000 years to complete and is referred to as the **era of recombination,** referring to nuclei and electrons combining. There are many instances, both in space and in the laboratory, where electrons are stripped from atoms. This is the process of ionization mentioned in Section 3-21. After ionization occurs, electrons often recombine with nuclei to re-create neutral atoms. However, the gases created early in the evolution of the universe were never neutral, so the "re" in recombination is misleading. Because the universe was opaque in its first 379,000 years, we cannot see any further into the past than the era of recombination with telescopes that detect electromagnetic radiation of any wavelength.

We can use Wien's law to calculate the temperature of the cosmic background radiation at that time. The 3000 K temperature at decoupling corresponds to a peak microwave wavelength of 0.001 mm. The temperature history of the universe is graphed in Figure 12-8.

Since the primordial photons formed, the expansion of the universe has been increasing their wavelengths, so today we see these same photons as the cosmic microwave background. The universe is now about 1000 times larger than it was at decoupling. Although the universe is dominated by matter today, the number of photons left over from the Big Bang is still immense. From the physics of blackbody radiation, astronomers calculate that there are now 400 million cosmic background microwave photons in every cubic meter of space. In contrast, if all visible matter in the universe were uniformly spread throughout space, there would be roughly one hydrogen atom in every 3 cubic meters of space. In other words, photons continue to outnumber atoms by more than a billion to one. The microwave background contains the most ancient photons we expect ever to be able to observe. This microwave background is a ghostly relic of the former dazzling splendor of the universe.

12-11   Galaxies formed from huge clouds of primordial gas

The Big Bang theory explains the evolution of spacetime, matter, and energy since the Planck era. It accounts for the hydrogen, most of the helium, and some of the lithium that exist today. But how do we explain the existence of superclusters of galaxies, clusters of galaxies, and individual galaxies? As noted earlier, these must have formed after the beginning of the matter-dominated era of the universe.

To explain large-scale structures in the early matter-dominated universe, recall the quantum fluctuations that were stretched by inflation. These clumps throughout the inflated universe were initially extremely small differences in density with a wide range of sizes in the microwave background. We know this because the *COBE* and *WMAP* satellites observed the cosmic microwave background and discovered these variations in density back when the universe was only 379,000 years old. Their data show that these variations were only about 1 part in 100,000 denser than the average density of matter in the early universe (Figure 12-15). But these small variations on various size scales are all that is needed to explain the present large-scale structure of the universe.

When the universe was radiation-dominated, the particles in it were moving too fast to clump together and form structure. Then the universe became matter-dominated and particles cooled down sufficiently to clump together. The extra gravitational attraction in regions where the expanded quantum fluctuations created concentrations of matter slightly greater than average caused the particles there to draw together and ultimately form galaxies, clusters, and superclusters.

In the first stage of the clustering process, the contracting matter collapsed inward, heated up, and rebounded outward due to the increased pressure between particles. Gravity drew the particles back inward and the process repeated itself over many cycles. This contracting and re-expanding of vast quantities of gas produced oscillations analogous to sound waves in our air. The gas transferred some of the oscillating energy to the photons that are today the cosmic microwave background, creating regions of slightly hotter and slightly cooler gas. These oscillations in the microwave background have been observed. As we will see, these observations have an important bearing on our understanding of the overall shape of the universe.

Virtually all matter in the early universe consisted of clouds of hydrogen and helium. Between the time of decoupling and the first burst of star formation was a period of little starlight, called the **dark ages.** Note, however, that the dark ages were actually bright, due to light from energy created in the Big Bang. There is growing observational evidence that star formation (see Section 9-14) began within 200 million years of the Big Bang (Figure 12-16a). The earliest generation of stars are called *Population III* stars. By 600 million years after the Big Bang, galaxies were beginning to coalesce (Figure 12-16b). We also have observational evidence that galaxies were in clusters within 2 billion years of the Big Bang (Figure 12-16c).

Computer simulations indicate that the first stars were typically 100–500 $M_\odot$, a million times or more brighter than the Sun, and they lived for no more than a few million years before exploding. These stars provided the first metals in the universe (other than the primordial lithium). During the first billion years, it appears that clumps of gas that contained millions or billions of solar masses also collapsed to create supermassive black holes. These would have attracted other matter into orbit around them and thereby served as the seeds for the growth of some galaxies. By observing remote galaxies, astronomers have discovered that most galaxies initially emitted energy we associate with quasars and

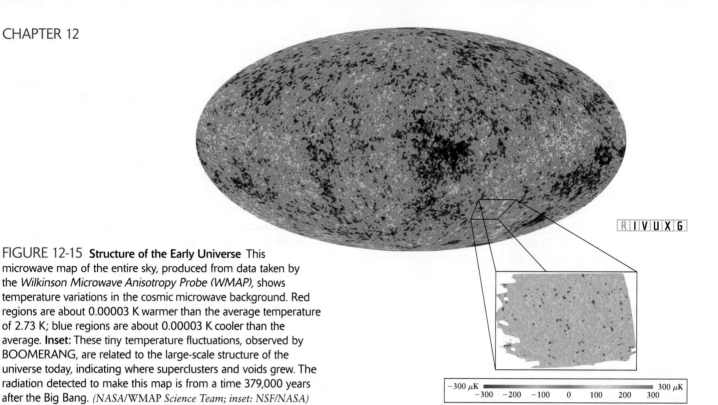

RIVUXG

FIGURE 12-15 Structure of the Early Universe This microwave map of the entire sky, produced from data taken by the *Wilkinson Microwave Anisotropy Probe (WMAP)*, shows temperature variations in the cosmic microwave background. Red regions are about 0.00003 K warmer than the average temperature of 2.73 K; blue regions are about 0.00003 K cooler than the average. **Inset:** These tiny temperature fluctuations, observed by BOOMERANG, are related to the large-scale structure of the universe today, indicating where superclusters and voids grew. The radiation detected to make this map is from a time 379,000 years after the Big Bang. *(NASA/WMAP Science Team; inset: NSF/NASA)*

−300 μK ▬▬▬▬▬▬▬▬ 300 μK
−300 −200 −100 0 100 200 300

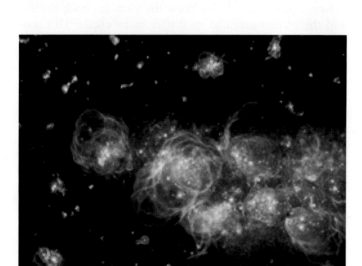

a Early bursts of star formation

b Arrows indicate galaxies beginning to form 13.4 billion years ago

RIVUXG

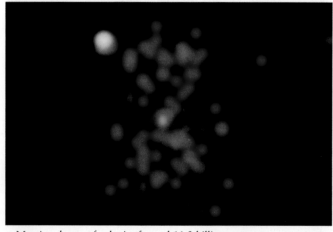

c Massive cluster of galaxies formed 11.2 billion years ago

RIVUXG

FIGURE 12-16 Galaxies Forming by Combining Smaller Units (a) This painting indicates how astronomers visualize the burst of star formation that occurred within a few hundred million years after the Big Bang. The arcs and irregular circles represent interstellar gas illuminated by supernovae. (b) Using the Hubble and Keck telescopes, astronomers discovered two groups of stars (arrows) 13.4 billion light-years away that are believed to be protogalaxies, from which bigger galaxies grew. These protogalaxies were discovered because they were enlarged by the gravitational lensing of an intervening cluster of galaxies. (c) The Chandra X-ray telescope imaged gravitationally bound gas around the distant galaxy 3C 294. The X-ray emission from this gas is the signature of an extremely massive cluster of galaxies, in this case, at a distance of about 11.2 billion light-years from us. *(a: Adolf Schaller, STScI/NASA/K. Lanzetta, SUNY; b: Richard Ellis (Caltech) and Jean-Paul Kneib (Observatorie Midi-Pyrenees, France), NASA, ESA; c: NASA)*

other active galaxies. This implies that most young galaxies have supermassive black holes at their centers. Recent observations reveal that massive and supermassive black holes typically grow until they have about 0.2% of the mass of a galaxy's nuclear bulge.

From about 600 million years to about 6 billion years after it formed, the universe underwent heavy star formation. Galaxies grew in size during this period, some continuously, others suddenly and quickly. Star formation decreased as the amount of interstellar hydrogen diminished. One observed galaxy, with about 8 times the mass in stars as the Milky Way has, reached its full star-forming potential when the universe was only 800 million years old.

Observations also reveal that galaxies were bluer and brighter in the past than they are today. These changes in color and brightness suggest a high abundance of young, bright, hot, massive stars in newly formed galaxies (Figure 12-17a). As galaxies age, these blue O and B stars become supergiants and eventually die. Therefore, galaxies grow somewhat redder and dimmer. This is especially true of those elliptical galaxies that formed early on. They appear to form nearly all of their stars in one vigorous burst of activity that lasts for about a billion years (Figure 12-17b).

If the first elements were hydrogen, helium, and lithium, where did Population II stars in globular clusters and elsewhere get their other metals?

In contrast, spiral galaxies have been forming stars for at least the past 10 billion years, although at a gradually decreasing rate. There is still plenty of interstellar hydrogen in the disks of spiral galaxies, like our Milky Way, to fuel

star formation today. That is why O and B stars still highlight their spiral arms. Figure 12-17b compares the rates at which spiral and elliptical galaxies form stars.

12-12　Star formation activity determines a galaxy's initial structure

Imagine a developing galaxy, called a *protogalaxy*, forming from a cloud of gas. Theory proposes that the rate of star formation determines whether this protogalaxy becomes a spiral or an elliptical galaxy. If stars form slowly enough, then the gas surrounding them has plenty of time to settle by collision with other infalling gas into a flattened disk, just like the early solar system. Star formation continues because the protogalactic disk contains an ample supply of hydrogen, and a spiral or lenticular (disk-shaped but without spiral arms) galaxy is created. If, however, the initial stellar birthrate is high in the protogalaxy, the theory predicts that virtually all pregalactic gas is used up in the creation of stars before a disk can form. In this case, an elliptical galaxy is created. Figure 12-18 depicts these contrasting sequences of events. Figure 12-19 summarizes the major evolutionary activities in the early universe.

Not all galaxies maintain their initial structure over the evolution of the universe. As we discussed in Section 11-17, some galaxies collide, and these collisions can change a galaxy's structure from spiral, for example, to elliptical. Indeed, astronomers who use extremely long exposures, called Hubble Deep Field images (see, for example, the inset on Figure 12-11), have observed elliptical galaxies during the first few billion years of the universe's existence that are far less uniform in color than are closer ellipticals. They

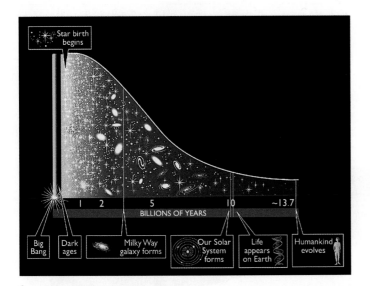

a

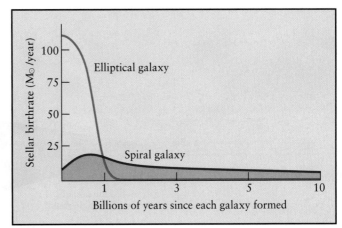

b

FIGURE 12-17 **Stellar Birth Rates** (a) This figure shows that star formation started quickly and has been tapering off ever since. (b) Most of the stars in an elliptical galaxy are created in a brief burst

of star formation when the galaxy is very young. In spiral galaxies, stars form at a more leisurely pace that extends over billions of years.

Sequence of events ——→

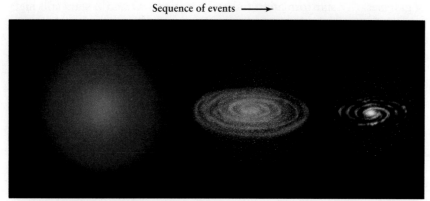

a Formation of a disk galaxy

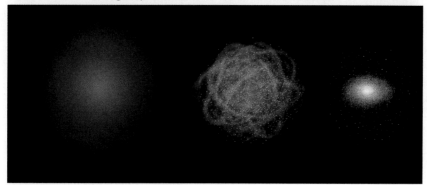

b Formation of an elliptical galaxy

FIGURE 12-18 **Creation of Spiral and Elliptical Galaxies** A galaxy begins as a huge cloud of primordial gas that collapses gravitationally. (a) If the rate of star birth is low, then much of the gas collapses to form a disk, and a spiral galaxy is created. (b) If the rate of star birth is high, then the gas is converted into stars before a disk can form, resulting in an elliptical galaxy.

observed blue stars in the young ellipticals consistent with the merger of spirals and a resulting burst of star formation, as well as lots of lower-mass (yellow and red) stars. Our understanding of galactic formation and evolution is far from complete.

Determining the location and nature of the dark matter in the universe (see Sections 11-7 and 11-18) is still of paramount importance in understanding the cosmos. The observable stars, gas, and dust in a galaxy or cluster of galaxies account for only about 10 to 20% of each object's mass.

(Recall that this observable matter does not have enough mass to hold galaxies or clusters of galaxies together.) Other than massive neutrinos, we have very little idea of what the remaining 80 to 90% of each galaxy or cluster of galaxies is composed of. However, some progress in understanding the distribution of this dark matter is being made. In 2002, astronomers who map large numbers of galaxies and use computer simulations of the effects of dark matter determined that on the scales of clusters of galaxies, the locations of the galaxies often coincide with the concentrations of dark matter, while voids between clusters coincide with voids of dark matter. It is likely that the gravitational force of the dark matter caused gas to concentrate in the same regions, thereby stimulating formation of supermassive black holes and galaxies. In 2007, using images of galaxies that were focused by a gravitational lens of dark matter, astronomers were able to plot the locations of this dark matter (Figure 12-20).

If dark matter did not exist, would any gravitational lensing occur in the universe?

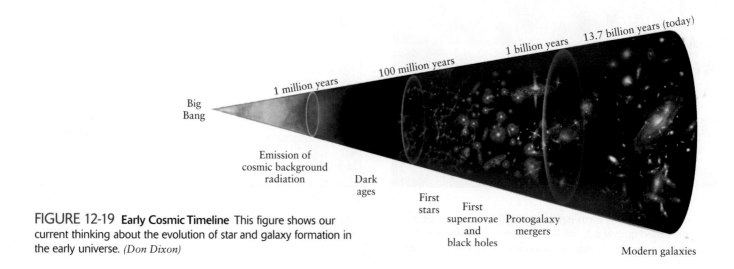

FIGURE 12-19 **Early Cosmic Timeline** This figure shows our current thinking about the evolution of star and galaxy formation in the early universe. *(Don Dixon)*

1 million years

100 million years

1 billion years

13.7 billion years (today)

Big Bang

Emission of cosmic background radiation

Dark ages

First stars

First supernovae and black holes

Protogalaxy mergers

Modern galaxies

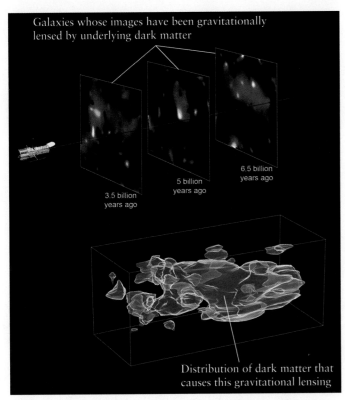

Galaxies whose images have been gravitationally lensed by underlying dark matter

6.5 billion years ago

5 billion years ago

3.5 billion years ago

Distribution of dark matter that causes this gravitational lensing

FIGURE 12-20 Mapping Dark Matter (top) The Hubble Space Telescope observed that galaxies in the same direction, but at different distances from Earth, undergo different amounts of gravitational lensing. (bottom) Much of this effect is due to dark matter. By subtracting out the lensing effects of intervening galaxies, the distorted shapes of the galaxies at various distances enable astronomers to determine the distribution of dark matter. *(NASA, ESA, and R. Massey, CIT)*

THE FATE OF THE UNIVERSE

We now turn to the future. Will the universe last forever? Or will it someday stop expanding and collapse?

12-13 The average density of matter is one factor that determines the future of the universe

As superclusters move apart in the expanding universe, their mutual gravitational attractions act on one another and thereby slow the rate at which they separate. If that were all that mattered in determining the fate of the universe, you could simply locate and add up all of the visible and (presently) dark matter in existence and see if its gravitational force is enough to eventually stop the expansion.

This is analogous to what engineers do in calculating, for example, how high a cannonball shot upward from the surface of Earth will travel. Earth's gravitational force slows the ball's ascent. If the cannonball's speed upward is less than the escape velocity from Earth's gravity (about 11 km/s

straight up), it will fall back to Earth. If the ball's speed exceeds the escape velocity, it will leave Earth and continue outward forever, despite the relentless pull of Earth's gravity. On the boundary between these two scenarios is the situation when the cannonball's speed equals the escape velocity. In that case, the cannonball will just barely escape falling back to Earth, slowing forever, and come to rest an infinite distance away.

By analogy with the cannonball fired from Earth, it would seem that if the universe is moving outward too slowly to overcome the mutual gravitational attraction of its parts, it should stop expanding and someday collapse. If it is moving outward at exactly its escape velocity, it should expand until coming to a stop an infinite time in the future. Or, if it is moving outward fast enough, it should continue to expand forever. However, none of these options is correct!

The laws of physics pertaining to the evolution of the universe, as spelled out in the theory of general relativity and in recent observations of distant objects, reveal that reality is more complex than this simple analogy. Just when astronomers were getting comfortable with the idea that the gravitational force is always attractive, general relativity once again showed that reality ignores our common sense by demonstrating that gravity can be repulsive.

We now know two additional effects that must be considered: The effect of matter on the shape of the universe and the presence of a repulsive gravitational force, called **dark energy** *(not to be confused with dark matter)*. We begin by considering the effect that matter and energy have on the shape of the universe.

12-14 The overall shape of spacetime affects the future of the universe

During the 1920s, Alexandre Friedmann in Russia, Georges Lemaître in Belgium, Willem de Sitter in the Netherlands, and Einstein himself applied the theory of general relativity to the expanding universe. General relativity predicts that the presence of matter curves the fabric of spacetime, as we saw in Chapters 10 and 11 in the form of gravitational lensing and the distortion of spacetime around black holes. Similarly, the presence of energy also curves spacetime—recall that matter and energy are related by $E = mc^2$.

Only three possibilities exist for the overall shape of the universe. They are determined by the amount of mass and energy the universe contains and how fast it is expanding. For example, imagine shining two powerful laser beams out into space. Suppose that we can align these two beams so that they are perfectly parallel as they leave Earth. Suppose, further, that nothing gets in the way of these two beams. We follow them across the spacetime whose shape we wish to determine. The light beams will begin to diverge due to the expanding universe carrying them apart. This effect happens regardless of any properties of the spacetime, and, because we are not interested in the effect of expansion right now, we compensate for it (that is, we ignore it) in what follows.

The three possibilities for the paths of the laser beams, due to the actual curvature of the universe, are

1. Two beams of light, starting out parallel, gradually get closer and closer together as they move across the universe, eventually intersecting at some enormous distance from Earth (Figure 12-21a). In this case, we say that space is *positively curved*. Analogously, the lines of constant longitude on Earth's surface are parallel at the equator but intersect at the poles. Thus, if the universe has this effect on light, the shape of the universe is like the surface of a sphere: Space is spherical and the universe has positive curvature. This is what would happen if the universe had such high mass and energy densities that the mass and energy literally curved space back in on itself. In the absence of any outward-pushing force, such as would be supplied by a sufficiently large cosmological constant (see Section 12-1), such a universe does not have enough energy to keep expanding forever. It would someday collapse.

a Parallel light beams converge

b Parallel light beams remain parallel

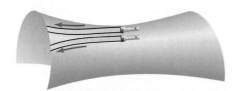

c Parallel light beams diverge

FIGURE 12-21 **Possible Geometries of the Universe** The shape of space (represented here as two-dimensional for ease of visualization) is determined by the matter and energy contained in the universe. The curvature is either (a) positive, (b) zero, or (c) negative, depending on whether the average matter and energy density throughout space is greater than, equal to, or less than a critical value. The lines on each curve are initially parallel. They converge, remain parallel, or diverge, depending on the curvature of space.

It is easiest to understand such a universe by visualizing a two-dimensional version of it as being the surface of a balloon on which you must remain. In such a universe, you could, in principle, keep going in a straight line (which is a curve on the balloon's surface) and eventually end up back where you started. Like the surface of a balloon, a three-dimensional universe with positive curvature has no outside or center and is said to be a **closed universe.**

2. Two beams of light remain parallel regardless of how far they travel (Figure 12-21b). In this case, space is not curved. That is, space is flat and it can extend without limit. This is the structure that meets our commonsense belief.

3. Two initially parallel beams of light gradually diverge farther and farther apart as they move across the universe (Figure 12-21c). This is what would happen if the energy of the Big Bang was sufficiently great to assure that the universe is going to expand forever. In this case, the universe has negative curvature. A horse's saddle is a good example of a negatively curved or *hyperbolic* surface. Initially parallel lines drawn on a saddle always diverge. Thus, in a negatively curved universe, we would describe space as hyperbolic. A hyperbolic universe extends without limit and is called an **open universe.** Both the flat and the hyperbolic universes are open and will therefore grow larger forever.

Most astronomers have a distinct preference for the flat universe. The reason for this expectation that the universe is flat stems from the observations that the universe is homogeneous and isotropic. The only mechanism we have at present to explain these properties of the matter and energy distribution in space is inflation, as discussed earlier. However, the equations predict that, if inflation occurred, the universe must be very nearly flat because the period of inflation stretched the volume of the universe and the matter and energy in it so much.

Telescopic observations strongly support the belief that the universe is flat. This conclusion comes from examining the sizes of the regions of slightly higher and lower temperature in the cosmic microwave background (see Figure 12-15) and comparing them to the sizes predicted for a flat universe. The observations are consistent with the theoretical variations. Furthermore, if the universe had positive curvature, light from the hot regions of the early universe would be curved and thereby focused, creating bigger, brighter images than are observed. Figure 12-22 summarizes the effects of space curvature on observations of the cosmic microwave background.

What, if any, objects have we given enough velocity to escape forever from Earth?

FIGURE 12-22 **Cosmic Microwave Background and the Curvature of Space** Temperature variations in the early universe appear as "hot spots" in the cosmic microwave background. The apparent sizes of these spots depend on the curvature of space. (a) In a closed universe with positive curvature, light rays from opposite sides of a hot spot bend toward each other. Hence, the hot spot appears larger than it actually is (dashed lines). (b) The light rays do not bend in a flat universe. (c) In an open universe, light rays bend apart. The dashed lines show that a hot spot would appear smaller than its actual size. *(The BOOMERANG Group, University of California, Santa Barbara)*

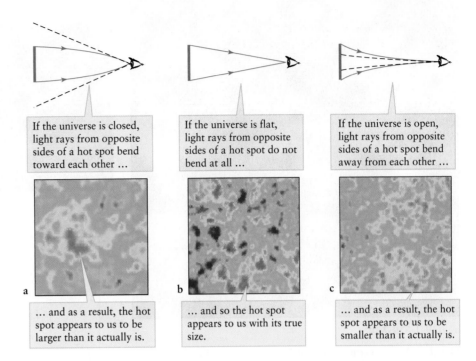

If the universe is closed, light rays from opposite sides of a hot spot bend toward each other …

… and as a result, the hot spot appears to us to be larger than it actually is.

If the universe is flat, light rays from opposite sides of a hot spot do not bend at all …

… and so the hot spot appears to us with its true size.

If the universe is open, light rays from opposite sides of a hot spot bend away from each other …

… and as a result, the hot spot appears to us to be smaller than it actually is.

a b c

12-15 Dark energy is causing the universe to accelerate outward

Until the past few years, there was a major inconsistency between the distribution of observed matter and energy in the universe and the flatness of space. Just as there is not enough visible matter to account for galaxies and clusters of galaxies that remain as bound systems, there is not enough observed matter and energy to account for a flat universe. Visible matter accounts for only 4% of the required mass. The cosmic microwave background photons add only 0.005% of the required gravitational effects needed for flatness, and calculations of the mass necessary to keep galaxies and clusters bound, combined with gravitational lensing by dark matter of distant galaxies and quasars, reveal that the dark matter known to exist accounts for only about 23% of the required mass. Therefore, the universe has only about 27% of the required mass and energy necessary to make it flat. Allowing for the errors that still exist in all of these observations, the possible range of mass and energy from all matter and photons in the universe is still between only 20% and 40% of that required for flatness. Yet flat it certainly appears to be.

By the mid-1990s, combined evidence of the microwave background and large-scale structure had forced astronomers to the conclusion that the remaining energy required to make the universe flat must be a form of dark energy with the remarkable feature that it causes the universe to accelerate outward. Corroborating evidence for the existence of such dark energy was found in the recent observations of Type Ia supernovae in extremely distant galaxies (Figure 12-23).

The light curves (see Section 10-4) for Type Ia supernovae (explosions of white dwarfs in binary star systems) are very well known. Studies show that these supernovae in nearby galaxies behave similarly to those observed in the Milky Way, regardless of their environment. Because they are so bright, such supernovae are excellent standard candles for determining distances to objects billions of light-years away. Assuming that supernovae in distant galaxies also behave similarly to those in our Galaxy, observations first made in 1998 revealed that the supernovae in distant galaxies appear dimmer than they would if the universe had been continually decelerating or even expanding at a constant speed. In other words, the universe is now expanding faster than it was, so the distant galaxies are farther away than they otherwise have been. Because they are farther away, the supernovae in them appear dimmer than in a continually slowing universe. The universe is *accelerating* outward.

There must be some kind of repulsive force acting to increase the rate at which superclusters separate today. Astronomers hypothesize that this is due to some kind of dark energy that has a repulsive gravitational effect, as mentioned earlier. Confirmation that dark energy exists came in 2003, when astronomers observed light from very distant galaxies passing through more nearby clusters of galaxies on its way toward Earth. As the distant light moved toward a cluster on its way to us, that light was gravitationally blueshifted, meaning the light gained energy as it was pulled toward the cluster. If there were no dark energy in a flat universe, then as that light left the vicinity of the cluster, it would have been redshifted (lost energy) by exactly the same amount and come to us with the same wavelengths it would have had if it had never passed through the cluster.

a R I V U X G

FIGURE 12-23 **Dimmer Distant Supernova** (a) These Hubble Space Telescope images show the galaxy in which the supernova SN 1997ff occurred. This supernova, more than 10 billion light-years away, was dimmer than expected, indicating that the distance to it is greater than the distance it would have if the universe had been continually slowing down since the Big Bang. This supports the notion that an outward (cosmological) force is acting over vast distances in the universe. The arrow on the first inset shows the galaxy in which the supernova was discovered. The bright spot on the second inset shows the supernova by subtracting the constant light emitted by all the other nearby objects. (b) The distances and brightnesses of many very distant supernovae are plotted on this diagram. The locations of the most distant supernovae in the upper region strongly indicate that the universe has been accelerating outward for the past 6 billion years. *(a: Adam Riess, Space Telescope Science Institute, NASA)*

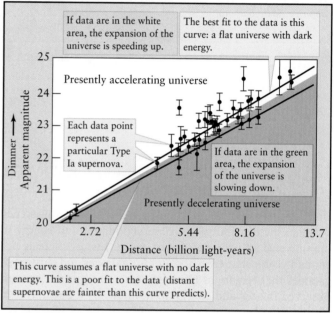

b

However, if dark energy exists, then the universe is accelerating outward (moving outward faster and faster). So the universe would have expanded more during the time that the light was leaving the cluster than when that light was first moving toward the cluster. Under those conditions, the light would have gained more energy from the cluster's gravitational attraction while traveling toward the cluster than the light would have lost to that gravitational attraction as it was moving away from the cluster. In other words, the amount of blueshift that the light underwent falling into the cluster would be greater than the amount of redshift that it underwent leaving the vicinity of the cluster. This was precisely what astronomers discovered: The light from distant galaxies undergoes a net blueshift when it passes through clusters of galaxies on its way to us—the universe is accelerating outward.

Cosmological Constant There are two theoretical explanations for dark energy. Let us first consider what would happen if the cosmological constant that Einstein introduced and then rejected actually did exist.

As Einstein had intended, it would provide a repulsive force to nature. The energy associated with that force would appear in the vacuum of space. Recall from Sections 10-28 and 12-8 that, even in a vacuum, particles are constantly appearing and disappearing. They add energy to the cosmos, so even empty space has energy. The cosmological constant form of energy has the property of contributing a repulsive gravitational force that competes with the attractive gravitational force from matter and radiation, which slows down the expansion. Which gravitational force wins depends on whether there is more vacuum energy or more matter and radiation.

As the universe expands, the average density of matter and energy decreases. In other words, on average there is less matter and energy in each cubic meter of space every second than there was the second before. However, the density of energy created by the cosmological constant is unchanged—as the universe expands, the amount of vacuum energy in each cubic meter of space remains constant. If the universe is still expanding when the vacuum energy exceeds the matter and energy per unit volume, then the repulsive gravitational force would dominate and force the universe to start accelerating outward, as is observed.

The concern that some astronomers express about the cosmological constant is that the repulsive force it creates has just the right strength to have allowed the universe to slow down for several billion years and then to slowly cause it to accelerate outward, as seen today. To explain this apparent coincidence would require fine-tuning the ratio of the vacuum energy to the matter and radiation to a remarkably high degree of precision. Scientists do not yet have a good reason why the repulsion created by a cosmological constant is not, say, a hundred times greater, in which case structure throughout the universe, such as stars and galaxies, would not have formed yet, or a billion times less, in which case the repulsive effects of the cosmological constant would be negligible today.

Quintessence Despite these concerns, observations of distant Type Ia supernovae suggest that the outward force acting on the universe is nearly constant, consistent with the dark energy being created by the physics we characterized with the cosmological constant. However, these observations do not yet rule out the major competing theory of dark energy, called **quintessence**. Scientists are exploring a variety of mathematical descriptions of quintessence. It differs from the vacuum energy associated with the cosmological constant in that the energy of quintessence is not constant but changes slowly with the expansion of the universe and can also be changed by the flow of energy and matter through the universe. Because it can change, quintessence need not have been fine-tuned to a strength that makes the universe flat and creates just the slight acceleration we are seeing. Instead, it could have started out with some arbitrary strength and then been adjusted by physical properties of the universe to match the growth of the universe.

As the universe grew, the energy in quintessence would eventually come to dominate the gravitational energy from matter. If this took place a few billion years ago, quintessence would have begun exerting a negative gravitational force and so the universe would have begun accelerating.

The unexpected discovery that the universe is accelerating outward completely alters astronomers' perspectives about the fate of the cosmos. It tells us that *the universe will expand forever.* At the same time, it neatly solves the mystery of why the universe is flat, as required by the occurrence of inflation in the early universe. Dark energy provides the remaining 73% of the energy needed to account for the universe being flat. The contributions of all of its major components to the mass and energy in the universe are listed in Appendix C-12.

We have seen that our present understanding of the laws of nature is incomplete, especially in that our theories do not explain the nature of the universe during the Planck era (or why the universe began in the first place). Driven by these challenges, scientists have begun to develop superstring theories that may resolve these issues.

12-16 Superstring Theory

To combine gravitation and the other three fundamental forces in nature into one comprehensive Theory of Everything, scientists have had to consider a universe that contains more than the four dimensions we know about today (three of space and one of time). The theories that mathematically describe this new formulation of the universe are called *superstring theories.* There are basically five such theories, which allow all of the particles we have been studying—such as quarks, protons, neutrons, electrons, and photons—to exist. Of course, only one of the superstring theories can be correct, but which, if any, remains to be seen.

Spacetime in superstring theories has 10 dimensions, of which 6 are everywhere rolled up into such tiny volumes that we cannot detect them directly. The other 4 dimensions are our normal spacetime. Superstring's more general spacetime carries with it properties that allow scientists to combine all four forces in nature into one set of equations.

The difficulty in reconciling quantum mechanics (describing the weak, strong, and electromagnetic forces) and general relativity (describing gravitation) is that the three forces in quantum mechanics are quantized, whereas general relativity is not. In other words, the weak, strong, and electromagnetic forces are transmitted by particles. For example, the quanta of electromagnetism are photons. Gravity, as described by general relativity, is based on a smooth and continuous, rather than quantized, force. Specifically, the distortion of spacetime by matter and energy creates the gravitational force.

Superstring theories begin with a different assumption about all particles and their interactions than do either quantum mechanics or general relativity. The new theories assert that each particle is actually a tiny vibrating string, with different types of particles, each vibrating at different rates. Gravitation has its own energy-sharing particle, the graviton, analogous to photons for electromagnetism. The interactions between the strings create all of the properties of matter and energy.

The predictions made by superstring theories begin with the assumption that general relativity is the correct "classical" theory for describing the gravitational interaction between matter and energy. This may seem trivial, but, because general relativity today correctly predicts everything in its realm of validity, a more comprehensive (superstring) theory needs to keep that accuracy, or the larger theory is wrong. Some additional predictions of superstring theories include the following:

- The universe cannot have positive curvature (which it does not, as we have seen).

- Some of the clumpy structure that we see as superclusters of galaxies could have been caused by the effects of superstring activity during inflation (with the rest due to the expansion of quantum fluctuations).

- Spacetime may not be entirely smooth. It may have structural defects, like a flawed diamond or ice that has broken into abutting chunks. These defects in spacetime would appear as one-dimensional "cosmic strings" with great density. Their attraction would pull normal matter around them, creating strings of galaxies. These cosmic strings have not yet been found.

- Some of the dark matter may be particles predicted by string theory.

- Most versions of superstring theory include a cosmological constant, but there is no underlying reason yet known for the value of the cosmological constant that may exist today.

- The speed of light is the same for all photons. If different wavelengths of light from the same event arrived at different times, then superstring (and relativity) theories would be wrong.

Superstring theories are, so far, consistent with observations, but it remains to be seen if any one of them will continue to maintain consistency with future observations and to correctly predict observable things that we have not yet seen.

12-17   Frontiers yet to be discovered

Some of the biggest unanswered questions about the cosmos follow from the discoveries presented in this chapter. What caused the Big Bang? What happened during the Planck time before 10^{-43} s? Is our universe the only one, or are there others, perhaps with profoundly different physical properties than ours, that were created at the same time but that are inaccessible to us? What caused the universe to come into existence in a false vacuum? What are the details of the formation processes of galaxies? Why was gas in the early universe not eventually transformed into systems of stars a million times bigger or smaller than galaxies? Did black holes grow with the evolution of the galaxies, as is currently believed? What are the origins of the dark energy? Which, if either, of the current candidates for dark energy—the cosmological constant or quintessence—is correct? One of the things that makes this time in human existence so fascinating is that it is likely we will have answers to most of these questions within our lifetimes.

SUMMARY OF KEY IDEAS

The Big Bang
- Astronomers believe that the universe began as an exceedingly dense cosmic singularity that expanded explosively in an event called the Big Bang. The Hubble law describes the ongoing expansion of the universe and the rate at which superclusters of galaxies move apart.

- The observable universe extends about 13.7 billion light-years in every direction from Earth to what is called the cosmic light horizon. We cannot see any objects that may exist beyond the cosmic light horizon because light from these objects has not had enough time to reach us.

- According to the theory of inflation, early in its existence, the universe expanded super rapidly for a short period, spreading matter that was originally near our location throughout a volume of the universe so large that we cannot yet observe it. The observable universe today is thus a growing volume of space containing matter and radiation that was in close contact with our matter and radiation during the first instant after the Big Bang. This explains the isotropic and homogeneous appearance of the universe.

A Brief History of Spacetime, Matter, Energy, and Everything
- Four basic forces—gravity, electromagnetism, the strong nuclear force, and the weak nuclear force—explain the interactions observed in the universe.

- According to current theory, all four forces were identical just after the Big Bang. At the end of the Planck time (about 10^{-43} s after the Big Bang), gravity became a separate force. A short time later, the strong nuclear force became a distinct force. A final separation created the electromagnetic force and the weak nuclear force.

- Before the Planck time, the universe was so dense that known laws of physics did not describe the behavior of spacetime, matter, and energy back then.

- In its first 30,000 years, the universe was radiation-dominated, during which time photons prevented matter from forming clumps. Then it was matter-dominated, during which time superclusters and smaller clumps of matter formed. Today it is dark-energy-dominated. Dark energy of some sort supplies a repulsive gravitational force that causes superclusters to accelerate away from each other.

- Astronomers think that during the first 379,000 years of the universe, matter and energy formed an opaque plasma, called the primordial fireball. Cosmic microwave background radiation is the greatly redshifted remnant of the universe as it existed about 379,000 years after the Big Bang.

- By 379,000 years after the Big Bang, spacetime expansion caused the temperature of the universe to fall below 3000 K, enabling protons and electrons to combine to form hydrogen atoms. This period is called the era of recombination. The universe became transparent during the era of recombination, meaning that the microwave background radiation contains the oldest photons in the universe.

- Clusters of galaxies and individual galaxies formed from pieces of enormous hydrogen and helium clouds, each of which became a separate supercluster of galaxies.

- All of the superclusters and some of the clusters of galaxies within each supercluster are moving away from one another.

- During the matter-dominated era, structure formed in the universe. As the universe goes farther into the dark-energy-dominated era, the large-scale structure of superclusters of galaxies will fade away.

The Fate of the Universe

- The average density of matter and dark energy in the universe determines the curvature of space and the ultimate fate of the universe.

- Observations show that the universe is flat and that the cosmic microwave background is almost perfectly isotropic, resulting from a brief period of very rapid expansion (the inflationary epoch) in the very early universe.

- The universe is accelerating outward and it will expand forever.

WHAT DID YOU THINK?

1 *What is the universe?* It is all of the matter, energy, and spacetime that will ever be detectable from Earth or that will ever affect us.

2 *Did the universe have a beginning?* Yes, it occurred about 13.7 billion years ago, in an event called the Big Bang.

3 *Into what is the universe expanding?* Nothing. The Big Bang created space and time (spacetime), as well as all matter and energy in the universe. Spacetime is expanding to accommodate the expansion of the universe.

4 *How strong is gravity compared to the other forces in nature?* Gravity is by far the weakest force.

5 *Will the universe last forever?* Current observations support the belief that the universe will last forever.

Review Questions

1. Which force in nature is believed to have formed second? **a.** gravity, **b.** electromagnetic force, **c.** weak force, **d.** strong force, **e.** all formed at the same time

2. Inflation most directly explains which of the following? **a.** why the universe is homogeneous, **b.** why the universe is expanding, **c.** why the universe is going to last forever, **d.** why the universe has a background temperature, **e.** why the universe has particles

3. What does it mean when astronomers say that we live in an expanding universe?

4. Explain the difference between a Doppler redshift and a cosmological redshift.

5. In what ways are the fate of the universe, the shape of the universe, and the average density of the universe related?

6. Assuming that the universe will expand forever, what will eventually become of the microwave background radiation?

7. What does it mean to say that the universe is dark-energy-dominated? When was the universe radiation-dominated? When was it matter-dominated? How did radiation domination show itself?

8. Explain the difference between an electron and a positron.

9. Where do astronomers believe most of the photons in the cosmic microwave background originated?

10. Give examples of the actions or roles of each of the four basic physical forces in the universe.

11. What is the observational evidence for the **a.** Big Bang, **b.** inflationary epoch, and **c.** confinement of quarks?

Observing Projects

12. **View into the Distant Past** The Hubble Space Telescope has been used to explore dark regions of space where no bright stars or galaxies seem to appear. In particular, the Hubble Deep Field and Hubble Ultra Deep Field images were extremely long exposures of these dark regions and revealed that rich fields of very distant galaxies do, in fact, appear in these areas. Light emitted from these galaxies early in the life of the universe is just now reaching us. These images therefore provide us with a glimpse of the universe at a time just 1 or 2 billion years after the Big Bang. We will examine these two images. You will need a ruler for this project. In *Starry Night Enthusiast*™, open the **Find** pane, ensure that the **Query (Q)** box is empty and click on the **Down (▼)** arrow in this box to display a list of image sources. (a) Click on **Hubble Images** and double-click on **Hubble Deep Field** to center the view on this dark region of space. Note its position with respect to the Big Dipper. (If you cannot identify this region of the northern sky, click on **View > Constellations > Asterisms** and **View > Constellations > Labels**. Remove these indicators after you have identified the region.) Zoom in to a field of view about 3° wide and note that the region still appears to have no bright objects within it. Zoom in again until the **Hubble Deep Field** fills the view (about 7' × 5'). With your ruler, measure and record the visual angular diameters of five of the largest galaxies. (b) Return to the **Find** pane and the list of **Hubble Images** and click on **Hubble Ultra Deep Field** to center the view of this "dark" region of the sky. Press the space bar to go there immediately. If necessary, zoom in to the same angular size as you used to view the **Hubble Deep Field**. Measure the diameters of five of the largest galaxies in this image. Give at least two plausible reasons why the two images have different-sized largest galaxies.

WWT

13. The discovery several decades ago of microwave radiation that seemed to pervade the whole of space was quickly identified with the radiation produced at the original Big Bang at the beginning of our universe and redshifted by the expansion of space since that singular event some 13.7 billion years ago. This radiation has a precise blackbody spectrum equivalent to a temperature of 2.73 K. The *Cosmic Background Explorer (COBE)* and, more recently, the *Wilkinson Microwave Anisotropy Probe (WMAP)* spacecraft have made the best observations of this cosmic background radiation.

You can examine some of the images of the whole sky produced by these space-borne experiments using WWT.

Open WWT, open **View** to click **Off** the **Figures and Boundaries** in the **Constellation Lines** box and check that **Ecliptic** is **On** in the **Overlays** box. Ensure that **Sky** is selected in the **Look At** drop-down list.

Select **COBE DIRBE Annual Average Map (Infrared)** from the **Imagery** drop-down list to examine the whole sky in infrared radiation. Move the view around the sky to see the distribution of radiation from the Milky Way. Is the plane of the Milky Way aligned with the ecliptic? If not, estimate the angle between them.

Select **WMAP W Band (Microwave)** to see the sky at longer wavelengths. Again, note the effect of radiation from the Milky Way. This radiation must be removed from the map before the effect of the cosmic background radiation distribution can be measured and assessed. Select **WMAP Cosmic Microwave Background** to see the result of this adjustment. Survey the whole sky. Describe the structure in the radiation field. Color variations represent temperature differences, the blue regions being about 0.00003 K different from the 2.73 K radiation temperature. This structure is a reflection of the uniformity of energy and matter in the early universe. The small scale of these variations across the sky is thought to be the result of a period of very rapid expansion at an early stage of the evolution of our universe.

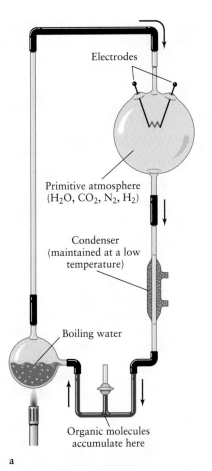

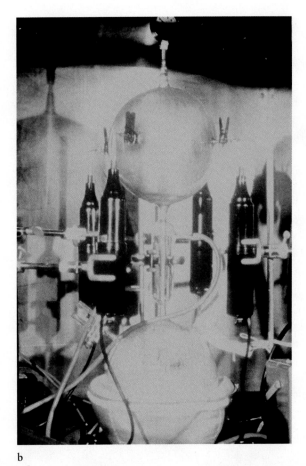

a   b

WEB LINK 13.3 **FIGURE 13-5** **Miller-Urey Experiment Updated**
(a) Modern versions of this classic experiment prove
that numerous organic compounds important to life can
be synthesized from gases that were present in Earth's primordial
atmosphere. This experiment supports the hypothesis that life
on Earth arose as a result of ordinary chemical reactions. (b) This
photograph shows a slime of organic material (in brown vials) created
in the original Miller-Urey experiment. *(b: NASA)*

then tidal forces from the star would lock the planet in synchronous rotation, making most of its surface either too hot (daytime side) or too cold (nighttime side) for life to flourish. These issues are summarized in Figure 13-6.

Synchronous rotation occurred in what other situation that we explored in this book?

13-3 Evidence is mounting that life might exist elsewhere in our solar system

Despite all of these potential limitations, scientists have strong evidence that everything does not have to be optimal for life to form. Geologists and biologists have discovered life on Earth in some incredibly challenging environments, such as on the ocean bottom, in a lake far under the Antarctic ice pack, deep inside our planet's crust, and even in hot geothermal vents (Figure 13-7). It therefore seems reasonable to believe that life could have originated off Earth

under similarly challenging conditions. Scientists consider at least four places in our solar system as possible habitats for life past or present. They are Mars and Jupiter's moons Europa, Ganymede, and Callisto.

How is water inside worlds like Europa kept liquid?

Astronomers have discovered (see Chapter 5) that Mars once had surface liquid water and that it may still have underground bodies of liquid water. We saw in Sections 5-20 through 5-22 that Europa, Ganymede, and Callisto are also likely to have liquid water under their frozen surfaces. These four bodies, then, are candidates for having evolved simple life-forms. Indeed, we saw in Chapter 5 what may be fossilized bacterial life from Mars (see Figure 5-31).

Confirmation that life evolved on any of these worlds would demonstrate that the formation of complex molecules and structures is possible off Earth. However, given the limited chemical resources available on other worlds in

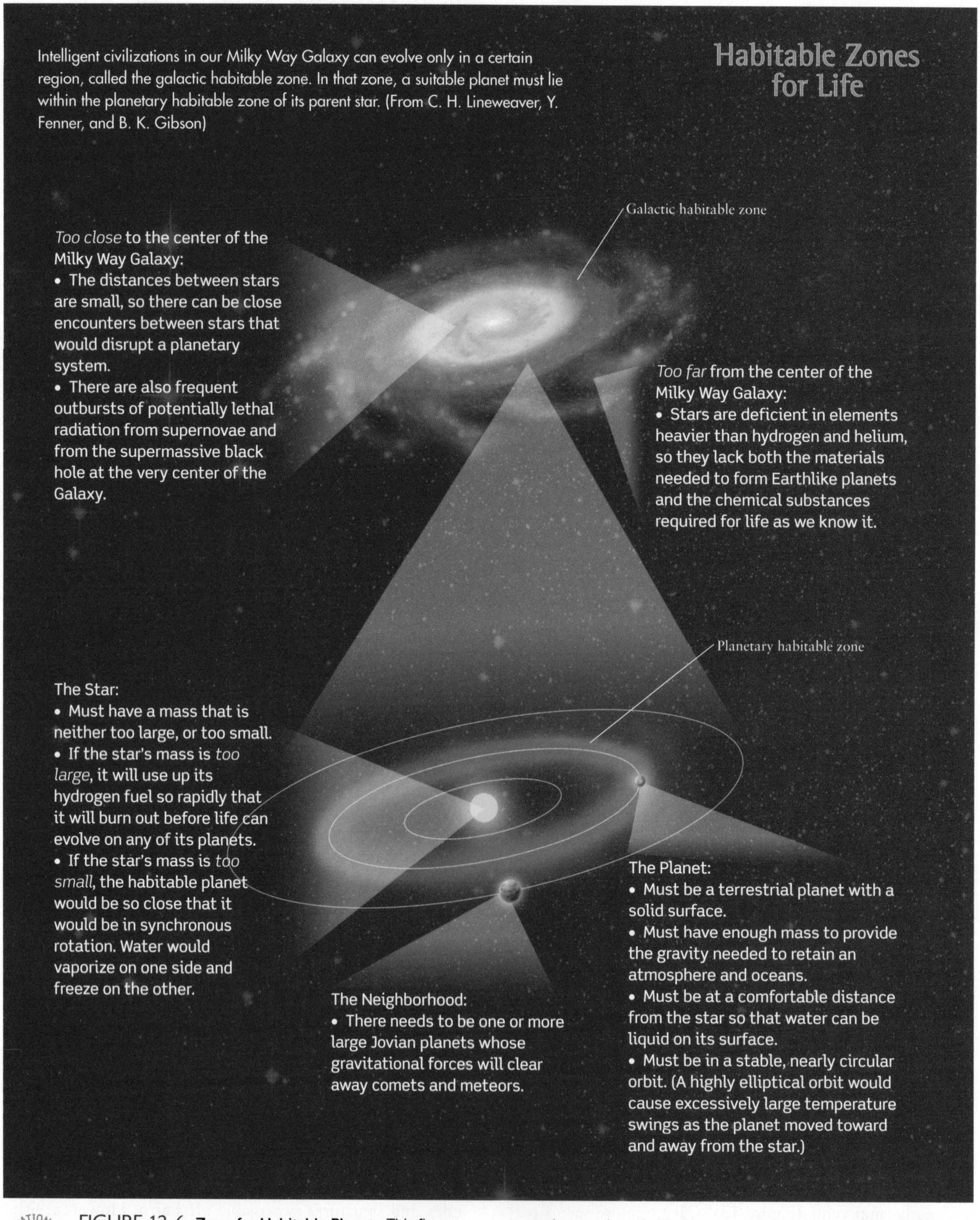

Habitable Zones for Life

Intelligent civilizations in our Milky Way Galaxy can evolve only in a certain region, called the galactic habitable zone. In that zone, a suitable planet must lie within the planetary habitable zone of its parent star. (From C. H. Lineweaver, Y. Fenner, and B. K. Gibson)

Galactic habitable zone

Too close to the center of the Milky Way Galaxy:
• The distances between stars are small, so there can be close encounters between stars that would disrupt a planetary system.
• There are also frequent outbursts of potentially lethal radiation from supernovae and from the supermassive black hole at the very center of the Galaxy.

Too far from the center of the Milky Way Galaxy:
• Stars are deficient in elements heavier than hydrogen and helium, so they lack both the materials needed to form Earthlike planets and the chemical substances required for life as we know it.

Planetary habitable zone

The Star:
• Must have a mass that is neither too large, or too small.
• If the star's mass is *too large*, it will use up its hydrogen fuel so rapidly that it will burn out before life can evolve on any of its planets.
• If the star's mass is *too small*, the habitable planet would be so close that it would be in synchronous rotation. Water would vaporize on one side and freeze on the other.

The Neighborhood:
• There needs to be one or more large Jovian planets whose gravitational forces will clear away comets and meteors.

The Planet:
• Must be a terrestrial planet with a solid surface.
• Must have enough mass to provide the gravity needed to retain an atmosphere and oceans.
• Must be at a comfortable distance from the star so that water can be liquid on its surface.
• Must be in a stable, nearly circular orbit. (A highly elliptical orbit would cause excessively large temperature swings as the planet moved toward and away from the star.)

FIGURE 13-6 **Zone for Habitable Planets** This figure summarizes the locations in the Galaxy and in orbit around stars where habitable planets might be found. Earth, of course, is in such a location.

our solar system, these locations are unsuitable for the evolution of sufficiently complex life, such as ourselves, that is self-aware and able to create advanced civilizations.

13-4   Searches for advanced civilizations try to detect their radio signals

We have not yet discovered Earthlike planets orbiting Sunlike stars. The closest case was the discovery in 2007 of a 5 Earth-mass planet orbiting the M3 star Gliese 581, which has about a third the Sun's mass. Furthermore, we saw in Section 2-14 that many stars have Jupiter-mass planets that orbit very close to them. The presence of massive planets near stars would make it very hard for those stars to support habitable planets because the gravitational tugs from the massive planets would pull Earthlike planets into highly elliptical orbits. During parts of such orbits, terrestrial worlds would be overheated and overradiated by their stars, whereas during other parts of their orbits, the planets would freeze. However, not all Sunlike stars have close-orbiting massive planets.

Billions of Sunlike stars in our Galaxy remain to be studied, and most astronomers and many other people expect that we will eventually discover habitable planets around some of them. This belief helps motivate SETI. Unlike the fossil evidence for life on Mars, the effort to locate life outside of the solar system is based on searching for high-tech evidence of advanced civilizations.

3   How might we ascertain whether extraterrestrial civilizations exist, given the tremendous distances that separate us from other stars and the huge number of stars with the potential to support life-bearing planets? Looking for individual, habitable, extrasolar planets is technically demanding and time-consuming. As we saw in Section 2-14, it requires careful observation of data available only in the past 25 years. Furthermore, the discovery of a planet does not imply that it harbors life. SETI astronomers therefore need a way to bypass the daunting search for habitable planets. Starting in the 1950s and 1960s, they identified a promising approach to searching directly for extraterrestrial intelligence—they listen for radio transmissions from distant civilizations.

As we have seen, radio waves can travel immense distances without being significantly altered by the interstellar medium. Because they penetrate gas and dust, radio waves are a logical choice for interstellar communication. Even if alien civilizations are not trying to communicate with us, they probably use radio waves in their own technology. But at what frequencies should we look to maximize the odds of detecting alien signals?

We could try random frequencies in the hopes of hearing something—anything. But the number of directions in which to look and frequencies to check are both daunting. We need to find some way to improve the odds. It turns out that some radio wavelengths travel farther without being absorbed by interstellar gas and dust, or without competing with noise from other sources, than do other radio wave-

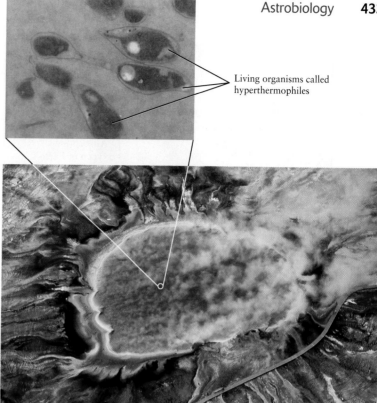

Living organisms called hyperthermophiles

a Grand prismatic spring, Yellowstone National Park

b   R I V U X G

FIGURE 13-7 Hyperthermophiles (a) These microscopic thermophiles (heat-loving organisms) live in water that is between 80°C and 100°C (85°F–140°F). (b) Tube worms (light-green tubes) with hemoglobin-rich red plumes reside around black smokers—vents in the ocean bottom that are in the same temperature range as the hot springs shown in (a). These vents are over 3 km (2 mi) under water. *(inset: NASA; a: Jim Peaco; July 2001 Yellowstone National Park Image by NPS Photo; b: Fisheries and Oceans of Canada)*

lengths. The SETI pioneer Bernard Oliver pointed out that a range of relatively clear frequencies exists in the neighborhood of the microwave emission lines of hydrogen and the hydroxyl radical (OH) (Figure 13-8). This region of the microwave part of the radio spectrum is called the **water hole,**

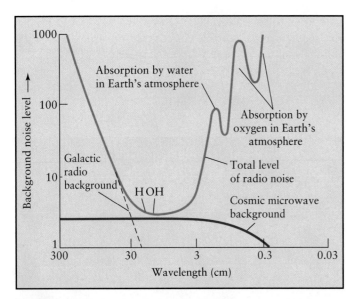

FIGURE 13-8 Water Hole The so-called water hole is a range of radio wavelengths from about 3 to 30 cm that happens to have relatively little cosmic noise. Some scientists suggest that this noise-free region would be well-suited for interstellar communication.

a humorous reference to the H and OH lines being so close together. If extraterrestrial beings are purposefully sending messages into space (to communicate with spacecraft or other worlds), it seems reasonable that they would choose to transmit on these frequencies.

Even if only a few alien civilizations are scattered across the Galaxy, we have the technology to detect radio transmissions from them. Since Project Ozma was carried out, the number of organized searches for signals from extraterrestrials approaches 100. SETI searches use radio telescopes (Figure 13-9) along with sophisticated electronic equipment and powerful computers to conduct both a targeted search and an all-sky survey. Although NASA has created astrobiology centers to study issues regarding life in the solar system and Galaxy, most of the SETI programs in operation today are privately funded. Several are actively listening and analyzing enormous volumes of data.

Indeed, so much data are available to analyze that one SETI project, called SETI@home (setiathome.berkeley.edu), has enlisted the help of home-computer users. These SETI researchers provide actual data from radio telescopes to be analyzed, along with a data-analysis program. While the computer's screensaver is on, the program runs, the data are analyzed, and the results are sent back to the researchers to be combined with data from other home-computer users. The processed data are replaced with fresh data to be analyzed. Another SETI program uses an array of small, inexpensive TV satellite antennas to look for signals from space.

4 Figure 13-10 shows the region of the Galaxy from which radio signals similar to the kinds of transmissions we humans make could be detected. Occasionally, a SETI

project does detect powerful or unusual signals from space. However, because none of these signals are ever repeated, SETI researchers do not yet believe they have discovered any extraterrestrial civilization.

The detection of a message from an alien civilization would be one of the greatest events in human history. Communicating with aliens could dramatically change the course of civilization on Earth through the sharing of scientific information or an awakening of social or humanistic enlightenment. In only a few years, our industry and social structure could advance centuries into the future. The effects of such change would touch every person on Earth.

If we detect incidental, rather than directed, signals from alien civilizations, what information might they contain?

13-5 The Drake equation: How many civilizations are likely to exist in the Milky Way?

Just how many planets throughout our Galaxy are likely to harbor complex life? The first person to tackle this question quantitatively was Frank Drake. He proposed that the number of technologically advanced civilizations in the Galaxy (designated by the letter N) could be estimated by what is now called the **Drake equation:**

$$N = R^* f_p n_e f_l f_i f_c L,$$

FIGURE 13-9 Radio Telescope Used for SETI The Arecibo Observatory's radio telescope, with a diameter of 305 m (1000 ft) is the largest in the world. It is located in Arecibo, Puerto Rico. In the past decade, it was used in an all-sky survey, along with an antenna located in the Mojave Desert in California, to search for extraterrestrial intelligence. In 1996, an antenna in Canberra, Australia, joined the network. *(Courtesy of the NAIC-Arecibo Observatory, a facility of the NSF)*

FIGURE 13-10 Region of the Search for Extraterrestrial Intelligence The red circle, centered on the solar system, shows the region of the Milky Way in which SETI searches can reasonably expect to detect radio emissions from alien civilizations. This assumes that the signals are similar in nature to the kinds of radio emissions we generate on Earth. *(NASA/Space Telescope Science Institute)*

in which

R^* = rate at which solar-type stars form in the Galaxy

f_p = fraction of solar-type stars that have planets

n_e = number of planets per solar-type star system suitable for life

f_l = fraction of those habitable planets on which life actually arises

f_i = fraction of those life-forms that evolve into intelligent species

f_c = fraction of those species that develop adequate technology and then choose to send messages out into space

L = lifetime of that technologically advanced civilization

The Drake equation expresses quantitatively the number of extraterrestrial civilizations as a product of terms, some of which can be estimated from what we know about stars and stellar evolution. For instance, thanks to new evidence for extrasolar planets, astronomers can now hope to determine the first two terms, R^* and f_p, by observation. We should probably exclude stars larger than about 1.5 $M_\odot$, because they have main-sequence lifetimes shorter than the time it took for intelligent life to develop on Earth—some 3.8 to 4.0 billion years. If that is typical of the time needed to evolve higher life-forms, then a massive star becomes a giant or even explodes as a supernova before self-aware creatures can evolve on any of its planets.

Although low-mass stars have much longer lifetimes, they are less suited for supporting life on their planets be-cause they are so cool. As noted earlier, only planets very near a low-mass star would be sufficiently warm for water to be a liquid. But, as also mentioned earlier, a planet that close would become tidally coupled to the star, developing synchronous rotation. One side would have continuous daylight, leading to the evaporation of oceans, while the other side would be in perpetual, frigid darkness. The only place that life could survive on such planets is in the narrow ring at the boundary between day and night. Such a small fraction of the land available greatly reduces the likelihood that life would be able to evolve to the complexity necessary for technological civilizations to develop.

As "ideal" life-supporting stars, this leaves main-sequence stars with masses near those of the Sun. These are stars in spectral types between F5 and M0. Based on statistical studies of star formation in the Milky Way, astronomers calculate that roughly one of these Sunlike stars forms in the Galaxy each year, yielding a value of $R^* = 1$ per year.

We learned in Sections 2-10 and 2-11 that the planets in our solar system formed in conjunction with the birth of the Sun. We have also seen evidence that similar processes of planetary formation may be commonplace around isolated stars. Many astronomers, therefore, give f_p a value of 1, meaning they believe it is likely that most Sunlike stars have planets.

Unfortunately, the rest of the factors in the Drake equation are very hypothetical. The chances that a planetary system has an Earthlike world are not known. Were we to consider our own solar system as representative, we could put n_e at 1. Let us be more conservative, however, and suppose that 1 in 10 solar-type stars is orbited by a habitable planet, making $n_e = 0.1$.

From what we know about the evolution of life on Earth, we might assume that, given appropriate conditions, the development of life is a certainty, which would make $f_l = 1$. This is, of course, an area of intense interest to biologists. For the sake of argument, we might also assume that evolution naturally leads to the development of intelligence (a conjecture that is hotly debated) and also make $f_i = 1$. It is anyone's guess as to whether these intelligent extraterrestrial beings would attempt communication with other civilizations in the Galaxy, but, if we assume they all would, f_c would also be put at 1.

The last variable, L, the longevity of technological civilization, is the most uncertain of all—it cannot be tested! Looking at our own example, we see a planet whose atmosphere and oceans are increasingly polluted, potentially destroying the food chain. When we add in how close we have come to destroying ourselves with weapons of mass destruction, it may be that we humans are among the lucky few technological civilizations to squeak through its first years. In other words, L may be as short as 100 years. Putting all of these numbers together, we arrive at

$$N = 1 \times 1 \times 0.1 \times 1 \times 1 \times 1 \times 100 = 10$$

Therefore, out of the hundreds of billions of stars in the Galaxy, there may be only 10 civilizations technologically advanced enough to communicate with us. Of course, the numbers used here are just estimates. A wide range of values has been proposed for the terms in the Drake equation, and these various numbers produce vastly different estimates of N. Increasing the average lifetime of advanced civilizations significantly increases the total number, N. Some scientists argue that there is exactly one advanced civilization in the Galaxy and that we are it. Others speculate that there may be tens of millions of planets inhabited by intelligent creatures. Although we do not know yet, science enables us to home in on the number.

13-6   Humans have been sending signals into space for more than a century

We have been doing more than just listening passively for voices from the cosmos. Astronomers have intentionally broadcast well-focused signals through radio telescopes toward star systems likely to harbor advanced life, including a 1974 "hello" message aimed at the globular cluster M13 (Figure 13-11a) and a 2008 broadcast of the Beatle's "Across the Universe." Unintentional broadcasts into space have occurred for longer than that. Since 1895, with the first radio transmission by the Italian engineer Guglielmo Marconi, we have been sending messages about ourselves

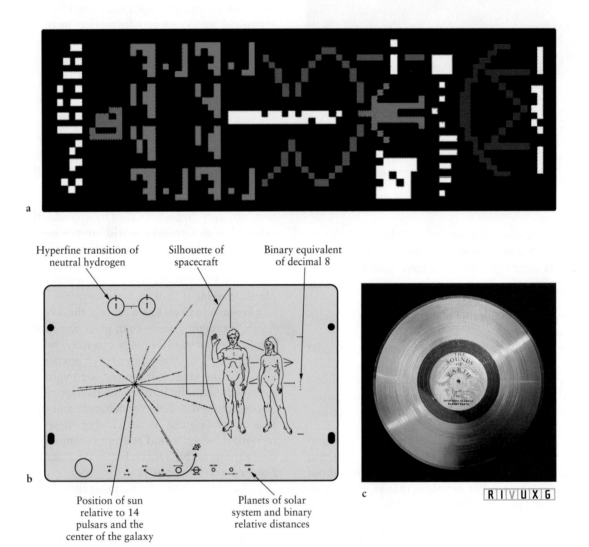

Hyperfine transition of neutral hydrogen

Silhouette of spacecraft

Binary equivalent of decimal 8

Position of sun relative to 14 pulsars and the center of the galaxy

Planets of solar system and binary relative distances

R I V U X G

FIGURE 13-11 **Human Memorabilia in Space** (a) Humans have beamed radio signals into space, hoping that the message will someday be intercepted by an alien civilization. This is a visual version of the signal sent in 1974 from the Arecibo radio telescope toward the globular cluster M13. The *Pioneer* and *Voyager* spacecraft, now in interstellar space, also carry messages from Earth. (b) The plaques on *Pioneer 10* and *Pioneer 11* provide information about where we are, what we look like, and some of the science we know.

(c) Images and sounds sent on *Voyager 1* and *Voyager 2* were stored on phonographic records, long before DVDs were even a twinkle in an engineer's eye. There are also instructions for playing the record, which contains information about our biology, our technology, and our knowledge base. Each record also contains the sounds of children's voices. It is remotely possible that another race might someday discover the spacecraft. *(a: Frank Drake/UCSC, et al., Arecibo Observatory/Cornell/NAIC: b, c: NASA)*

into space. The spherical region within some 115 light-years of the solar system is now filled with signals from radio and television shows. If other advanced civilizations fall within that sphere, they could very well be listening to radio programs or watching television shows from decades past, trying to make sense of our species.

With what two numbers used in the Drake equation calculation do you disagree most? Using your numbers, how many civilizations do you estimate exist?

Four spacecraft also carry information about the human race. Launched in 1972 and 1973, respectively, the *Pioneer 10* and *Pioneer 11* spacecraft carry plaques about us (Figure 13-11b). They are now making their way into interstellar space. In 1977, the two *Voyager* spacecraft were sent toward the outer solar system. They carry old-fashioned phonograph records encoded with human images and voices. Because space is so vast and the spacecraft are so small, the likelihood of their being discovered is truly remote. And yet millions, perhaps billions, of years from now, one of the spacecraft just might reach another race of intelligent creatures. From its path and the information on board, those creatures might be able to determine where the *Pioneer* and *Voyager* spacecraft came from. If their travel budgets are sufficiently large, they might even decide to come and visit us. At the least, they could send us radio messages. Will our descendants be here to receive them?

13-7   Frontiers yet to be discovered

We still have much to learn about astrobiology. Myriad steps in the formation of life on Earth remain to be understood. Although the evidence for liquid water once having existed on Mars is overwhelming, we still need to confirm whether life ever existed there. Does life exist on Mars, Europa, Ganymede, or Callisto today? Astronomers are still working to refine their ideas of how organic compounds formed in space. A paramount frontier is detecting life elsewhere in the universe.

What negative consequence could arise from an alien civilization discovering the *Pioneer 10* or *11* spacecraft?

SUMMARY OF KEY IDEAS

• The chemical building blocks of life exist throughout the Milky Way Galaxy.

• Organic molecules and water have been discovered in interstellar clouds, in some meteorites, in comets, and in newly forming star and planet systems.

• The Drake equation is used to estimate the number of technologically advanced civilizations in the Galaxy whose radio transmissions we might discover. Estimates of this number vary from 1 to millions.

• Astronomers are using radio telescopes to search for signals from other self-aware life in the Galaxy. This effort is called the search for extraterrestrial intelligence, or SETI. SETI is primarily done at frequencies where radio waves pass most easily through the interstellar medium. So far, these searches have not detected any life outside Earth.

• Everyday radio and television transmissions from Earth, along with intentional broadcasts into space, may be detected by other life-forms.

WHAT DID YOU THINK?

1 *Why is water so important to the formation of life?* Water allows many interactions between atoms and molecules dissolved in it.

2 *What element is uniquely suited to be the foundation of life as we know it, and why?* Carbon is the only atom that can bond flexibly (but not too flexibly) with three or more other atoms.

3 *How do astronomers search for extraterrestrial intelligence?* They search for radio signals from other advanced civilizations.

4 *Have astronomers located any extraterrestrial civilizations?* No extraterrestrial civilizations have yet been discovered.

5 *If advanced alien civilizations exist, is there any way they might know of our existence?* Yes, they might detect our intentional radio broadcasts into space, or our everyday radio and television broadcasts, or, possibly, one of our spacecraft traveling in interstellar space.

Discussion Questions

1. Which stellar spectral type is most likely to have a planet on which advanced life exists? **a.** O, **b.** B, **c.** A, **d.** G, **e.** M

2. Which element serves as the foundation or "backbone" of life? **a.** oxygen, **b.** carbon, **c.** silicon, **d.** hydrogen, **e.** nitrogen

3. What arguments could you make against sending messages via radio or spacecraft into interstellar space?

4. Try using the Drake equation with values that you find reasonable. How many civilizations do you estimate there are in our Galaxy?

5. Discuss the possible biological effects of our probes that visit other life-sustaining worlds.

6. What social effects might probes from other worlds have on us?

7. List some of the pros and cons in the argument that alien spacecraft have visited Earth.

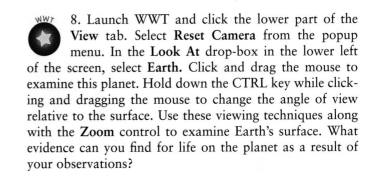 8. Launch WWT and click the lower part of the View tab. Select **Reset Camera** from the popup menu. In the **Look At** drop-box in the lower left of the screen, select **Earth.** Click and drag the mouse to examine this planet. Hold down the CTRL key while clicking and dragging the mouse to change the angle of view relative to the surface. Use these viewing techniques along with the **Zoom** control to examine Earth's surface. What evidence can you find for life on the planet as a result of your observations?

Why ET Hasn't Called

The lifetime of civilizations in the Drake equation for estimating extraterrestrial intelligences is greatly exaggerated   By MICHAEL SHERMER

Michael Shermer, "Why ET Hasn't Called," *Scientific American,* July 2002, 33.

In science there is arguably no more suppositional formula than that proposed in 1961 by radio astronomer Frank Drake for estimating the number of technological civilizations that reside in our galaxy: $N = R\, f_p\, n_e\, f_l\, f_i\, f_c\, L$

In this equation, N is the number of communicative civilizations, R is the rate of formation of suitable stars, f_p is the fraction of those stars with planets, n_e is the number of Earth-like planets per solar system, f_l is the fraction of planets with life, f_i is the fraction of planets with intelligent life, f_c is the fraction of planets with communicating technology, and L is the lifetime of communicating civilizations.

Species may simply not be equipped to survive for long periods in large populations.

Although we have a fairly good idea of the rate of stellar formation, a dearth of data for the other components means that calculations are often reduced to the creative speculations of quixotic astronomers.

Most SETI (search for extraterrestrial intelligence) scientists are realistic about the limitations of their field; still, I was puzzled to encounter numerous caveats about $L,$ such as this one from SETI Institute astronomer Seth Shostak: "The lack of precision in determining these parameters pales in comparison with our ignorance of L." Similarly, Mars Society president Robert Zubrin says that "the biggest uncertainty revolves around the value of L; we have very little data to estimate this number, and the value we pick for it strongly influences the results of the calculation." Estimates of L reflect this uncertainty, ranging from 10 years to 10 million years, with a mean of about 50,000 years.

Using a conservative Drake equation calculation, where L = 50,000 years (and $R = 10$, $f_p = 0.5$, $n_e = 0.2$, $f_l = 0.2$, $f_i = 0.2$, $f_c = 0.2$), then $N = 400$ civilizations, or one per 4,300 light-years. Using Zubrin's optimistic (and modified) Drake equation, where $L = 50,000$ years, then $N = $ five million galactic civilizations, or one per 185 light-years. (Zubrin's calculation assumes that 10 percent of all 400 billion stars are suitable G- and K-type stars that are not part of multiples, with almost all having planets, that 10 percent of these contain an active biosphere and that 50 percent of those are as old as Earth.) Estimates of N range wild-

ly between these figures, from Planetary Society scientist Thomas R. McDonough's 4,000 to Carl Sagan's one million.

I find this inconsistency in the estimation of L perplexing because it is the one component in the Drake equation for which we have copious empirical data from the history of civilization on Earth. To compute my own value of $L,$ I compiled the durations of 60 civilizations (years from inception to demise or the present), including Sumeria, Mesopotamia, Babylonia, the eight dynasties of Egypt, the six civilizations of Greece, the Roman Republic and Empire, and others in the ancient world, plus various civilizations since the fall of Rome, such as the nine dynasties (and two republics) of China, four in Africa, three in India, two in Japan, six in Central and South America, and six modern states of Europe and America.

The 60 civilizations in my database endured a total of 25,234 years, so $L = 420.6$ years. For more modern and technological societies, L became shorter, with the 28 civilizations since the fall of Rome averaging only 304.5 years. Plugging these figures into the Drake equation goes a long way toward explaining why ET has yet to drop by or phone in. Where $L = 420.6$ years, $N = 3.36$ civilizations in our galaxy; where $L = 304.5$ years, $N = 2.44$ civilizations in our galaxy. No wonder the galactic airways have been so quiet!

I am an unalloyed enthusiast for the SETI program, but history tells us that civilizations may rise and fall in cycles too brief to allow enough to flourish at any one time to traverse (or communicate across) the vast and empty expanses between the stars. We evolved in small hunter-gatherer communities of 100 to 200 individuals; it may be that our species, and perhaps extraterrestrial species as well (assuming evolution operates in a like manner elsewhere), is simply not well equipped to survive for long periods in large populations.

Whatever the quantity of $L,$ and whether N is less than 10 or more than 10 million, we must ensure L does not fall to zero on our planet, the only source of civilization we have known. **SA**

Michael Shermer is publisher of Skeptic *magazine (www.skeptic.com) and author of* In Darwin's Shadow: The Life and Science of Alfred Russel Wallace.

APPENDIX A

Powers-of-Ten Notation

Astronomy is a science of extremes. As we examine various cosmic environments, we find an astonishing range of conditions—from the incredibly hot, dense centers of stars to the frigid, near-perfect vacuum of interstellar space. To describe such divergent conditions accurately, we need a wide range of both large and small numbers. Astronomers avoid such confusing terms as "a million billion billion" (1,000,000,000,000,000,000,000,000) by using a standard shorthand system. All the cumbersome zeros that accompany such a large number are consolidated into one term consisting of 10 followed by an exponent, which is written as a superscript and called the **power of ten.** The exponent merely indicates how many zeros you would need to write out the long form of the number. Thus,

$$10^0 = 1$$
$$10^1 = 10$$
$$10^2 = 100$$
$$10^3 = 1000$$
$$10^4 = 10,000$$

and so forth. Equivalently, the exponent tells you how many tens must be multiplied together to yield the desired number. For example, ten thousand can be written as 10^4 ("ten to the fourth") because $10^4 = 10 \times 10 \times 10 \times 10 = 10,000$. Similarly, 273,000,000 can be written as 2.73×10^8.

In scientific notation, numbers are written as a figure between 1 and 10 multiplied by the appropriate power of 10. The distance between Earth and the Sun, for example, can be written as 1.5×10^8 km. Once you get used to it, you will find this notation more convenient than writing "150,000,000 km" or "one hundred and fifty million kilometers."

This powers-of-ten system can also be applied to numbers that are less than 1 by using a minus sign in front of the exponent. A negative exponent tells you that the location of the decimal point is as follows:

$$10^0 = 1.0$$
$$10^{-1} = 0.1$$
$$10^{-2} = 0.01$$
$$10^{-3} = 0.001$$
$$10^{-4} = 0.0001$$

and so forth. For example, the diameter of a hydrogen atom is approximately 1.1×10^{-8} cm. That notation is more convenient than stating "0.000000011 cm" or "11 billionths of a centimeter." Similarly, 0.000728 equals 7.28×10^{-4}.

Using the powers-of-ten shorthand, one can write large or small numbers like these compactly:

$$3,416,000 = 3.416 \times 10^6$$
$$0.000000807 = 8.07 \times 10^{-7}$$

Because powers-of-ten notation bypasses all the cumbersome zeros, a wide range of circumstances can be numerically described conveniently:

$$\text{one thousand} = 10^3$$
$$\text{one million} = 10^6$$
$$\text{one billion} = 10^9$$
$$\text{one trillion} = 10^{12}$$

and also

$$\text{one thousandth} = 10^{-3} = 0.001$$
$$\text{one millionth} = 10^{-6} = 0.000001$$
$$\text{one billionth} = 10^{-9} = 0.000000001$$
$$\text{one trillionth} = 10^{-12} = 0.000000000001$$

Temperature Scales

Three temperature scales are in common use. Throughout most of the world, temperatures are expressed in degrees Celsius (°C), named in honor of the Swedish astronomer Anders Celsius, who proposed it in 1742. The **Celsius temperature scale** (also known as the "centigrade scale") is based on the behavior of water, which freezes at 0°C and boils at 100°C at sea level on Earth.

Scientists usually prefer the **Kelvin scale,** named after the British physicist Lord Kelvin (William Thomson), who made many important contributions to our knowledge about heat and temperature. On the Kelvin temperature scale, water freezes at 273 K and boils at 373 K. Note that we do not use the degree symbol with the Kelvin temperature scale.

Because water must be heated by 100 K or 100°C to go from its freezing point to its boiling point, you can see that the size of a Kelvin is the same as the size of a degree Celsius. When considering temperature *changes,* measurements in Kelvins and in degrees Celsius lead to the same number.

A temperature expressed in Kelvins is always equal to the temperature in degrees Celsius plus 273. Scientists prefer the Kelvin scale because it is closely related to the physical meaning of temperature. All substances are made of atoms, which are very tiny (a typical atom has a diameter of about 10^{-10} m) and constantly in motion. The temperature of a substance is directly related to the average speed of its atoms. If something is hot, its atoms are moving at high speeds. If a substance is cold, its atoms are moving much more slowly.

The coldest possible temperature is the temperature at which atoms move as slowly as possible (they can never quite stop completely). This minimum possible temperature, called *absolute zero,* is the starting point for the Kelvin scale. Absolute zero is 0 K, or −273°C. Because it is impossible for anything to be colder than 0 K, there are no negative temperatures on the Kelvin scale.

In the United States, many people still use the now-archaic Fahrenheit scale, which expresses temperature in degrees Fahrenheit (°F). When the German physicist Gabriel Fahrenheit introduced this scale in the early 1700s, he intended 0°F to represent the coldest temperature then achievable (with a mixture of ice and saltwater) and 100°F to represent the temperature of a healthy human body. On the Fahrenheit scale, water freezes at 32°F and boils at 212°F. Because there are 180 degrees Fahrenheit between the freezing and boiling points of water, a degree Fahrenheit is only 100/180 (= 5/9) the size of the other scales.

The following equation converts from degrees Fahrenheit to degrees Celsius:

$$T_C = 5/9 \, (T_F - 32)$$

To convert from Celsius to Fahrenheit, a simple rearrangement of terms gives the relationship

$$T_F = 9/5 \, T_C + 32$$

where T_F is the temperature in degrees Fahrenheit and T_C is the temperature in degrees Celsius.

For example, consider a typical room temperature of 68°F. Using the first equation, we can convert this measurement to the Celsius scale as follows:

$$T_C = 5/9 \, (68 - 32) = 20°C$$

To arrive at the Kelvin scale, we simply add 273 degrees to the value in degrees Celsius. Thus, 68°F = 20°C = 293 K. The accompanying figure displays the relationships among these three temperature scales.

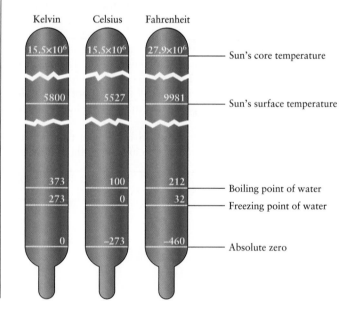

Kelvin	Celsius	Fahrenheit	
15.5×10^6	15.5×10^6	27.9×10^6	Sun's core temperature
5800	5527	9981	Sun's surface temperature
373	100	212	Boiling point of water
273	0	32	Freezing point of water
0	−273	−460	Absolute zero

Data Tables

TABLE C-1 The Planets: Orbital Data

Planet	Semimajor axis (AU)	Semimajor axis (10⁶ km)	Sidereal period (year)	Sidereal period (day)	Synodic period (day)	Mean orbital speed (km/s)	Orbital eccentricity	Inclination of orbit to ecliptic (°)
Mercury	0.3871	57.9	0.2408	87.97	115.88	47.9	0.206	7.00
Venus	0.7233	108.2	0.6152	224.70	583.92	35.0	0.007	3.39
Earth	1.0000	149.6	1.0000	365.26	—	29.8	0.017	0.00
Mars	1.5237	227.9	1.8808	686.98	779.94	24.1	0.093	1.85
Jupiter	5.2034	778.6	11.862	4332.6	398.9	13.1	0.048	1.31
Saturn	9.5371	1433.5	29.457	10,759	378.1	9.7	0.054	2.48
Uranus	19.1913	2872.5	84.01	30,685	369.7	6.8	0.047	0.77
Neptune	30.0690	4495.1	164.79	60,189	367.5	5.4	0.009	1.77

TABLE C-2 The Planets: Physical Data

Planet	Equatorial diameter (km)	Equatorial diameter (Earth = 1)	Mass (kg)	Mass (Earth = 1)	Mean density (kg/m³)	Rotation period* (days)	Inclination of equator to orbit (°)	Surface gravity (Earth = 1)	Albedo	Escape speed (km/s)
Mercury	4879	0.383	3.302×10^{23}	0.055	5427	58.6	0.0	0.38	0.12	4.3
Venus	12,104	0.949	4.869×10^{24}	0.815	5243	243.02^{R}	177.4	0.91	0.59	10.4
Earth	12,756	1.000	5.974×10^{24}	1.000	5515	0.996	23.45	1.000	0.37	11.2
Mars	6794	0.533	6.419×10^{23}	0.107	3933	1.025	25.19	0.38	0.16	5.0
Jupiter	142,984	11.209	1.899×10^{27}	317.83	1326	0.413	3.13	2.5	0.52	59.5
Saturn	120,536	9.449	5.685×10^{26}	95.16	687	0.445	26.73	1.1	0.47	35.5
Uranus	51,118	4.007	8.683×10^{25}	14.54	1270	0.717^{R}	97.77	0.91	0.56	21.3
Neptune	49,528	3.883	1.024×10^{26}	17.147	1638	0.671	28.32	1.1	0.4	23.5

For Jupiter, Saturn, Uranus, and Neptune, the internal rotation period is given. A superscript R means that the rotation is retrograde (opposite the planet's orbital motion).

TABLE C-3 Satellites of the Planets

Planet	Satellite	Discoverers of satellite	Average distance from center (km)	Orbital (sidereal) period* (days)	Orbital eccentricity	Diameter of satellite (km)	Mass (kg)
EARTH	Moon	—	384,400	27.322	0.0549	3476	7.349×10^{22}
MARS	Phobos	Hall (1877)	9378	0.319	0.02	$28 \times 23 \times 2$	1.1×10^{16}
	Deimos	Hall (1877)	23,459	1.262	0.00	$16 \times 12 \times 10$	2.4×10^{15}
JUPITER	Metis	Synott (1980)	127,960	0.2948	0.00	44	1×10^{17}
	Adrastea	Jewitt et al. (1979)	128,980	0.2983	0.00	$24 \times 16 \times 20$	1.9×10^{16}
	Amalthea	Barnard (1892)	181,300	0.4982	0.00	$270 \times 170 \times 150$	7.5×10^{18}
	Thebe	Synott (1979)	221,900	0.6745	0.02	98	8×10^{17}
	Io	Galileo (1610)	421,600	1.769	0.00	3643	8.93×10^{22}
	Europa	Galileo (1610)	670,900	3.551	0.01	3138	4.80×10^{22}
	Ganymede	Galileo (1610)	1,070,000	7.155	0.00	5268	1.48×10^{23}
	Callisto	Galileo (1610)	1,883,000	16.689	0.01	4821	1.08×10^{23}
	Themisto	Kowal (1975)	7,435,000	130.02	0.24	8	(?)
	Leda	Kowal (1974)	11,094,000	238.72	0.15	18	6×10^{15}
	Himalia	Perrine (1904)	11,480,000	250.57	0.16	170	9.5×10^{18}
	Lysithea	Nicholson (1938)	11,720,000	259.22	0.11	38	8×10^{16}
	Elara	Perrine (1905)	11,737,000	259.65	0.21	80	8×10^{17}
	S/2000/J11	Sheppard et al. (2000)	12,654,000	287	0.25	4	(?)
	Euporie	Sheppard et al. (2001)	19,017,000	553.1^R	0.16	2	(?)
	Chaldene	Sheppard et al. (2000)	20,375,000	723.8^R	0.24	4	(?)
	Iocaste	Sheppard et al. (2000)	20,733,000	632^R	0.22	5	(?)
	Kale	Sheppard et al. (2001)	20,804,000	721^R	0.48	2	(?)
	Orthosie	Sheppard et al. (2001)	20,876,000	623^R	0.27	2	(?)
	Thyone	Sheppard et al. (2001)	20,876,000	632^R	0.30	4	(?)
	Euanthe	Sheppard et al. (2001)	20,947,000	620^R	0.18	3	(?)
	Harpalyke	Sheppard et al. (2000)	21,019,000	623^R	0.23	4	(?)
	Praxidike	Sheppard et al. (2000)	21,162,000	625^R	0.22	7	(?)
	Ananke	Nicholson (1951)	21,200,000	631^R	0.17	28	4×10^{16}
	Hermippe	Sheppard et al. (2001)	21,376,000	632^R	0.25	4	(?)
	Taygete	Sheppard et al. (2000)	21,734,000	732^R	0.25	5	(?)
	Erinome	Sheppard et al. (2000)	21,948,000	728^R	0.27	3	(?)
	Carme	Nicholson (1938)	22,600,000	692^R	0.21	46	9×10^{16}
	Isonoe	Sheppard et al. (2000)	22,806,000	726	0.26	4	(?)
	Pasithee	Sheppard et al. (2001)	22,949,000	715^R	0.29	2	(?)
	Eurydome	Sheppard et al. (2001)	23,378,000	721^R	0.35	3	(?)
	Aitne	Sheppard et al. (2001)	23,449,000	741^R	0.29	3	(?)
	Pasiphae	Melotte (1908)	23,500,000	735^R	0.38	36	2×10^{17}
	Megaclite	Sheppard et al. (2000)	23,521,000	753^R	0.43	5	(?)
	Sponde	Sheppard et al. (2001)	23,592,000	749^R	0.45	2	(?)
	Sinope	Nicholson (1914)	23,700,000	758^R	0.28	28	8×10^{16}
	Autonoe	Sheppard et al. (2001)	23,979,000	765^R	0.42	4	(?)
	Kalyke	Sheppard et al. (2000)	24,164,000	743^R	0.24	5	(?)
	Callirrhoe	Scotti et al. (1999)	24,200,000	759^R	0.28	8	(?)

Reading Graphs

The graphs you will encounter in this book are compact ways of displaying patterns of information relating two variables, such as the temperature and luminosity of stars or the temperature at different altitudes in an atmosphere. The relevant values of one of the variables are presented along the horizontal or *x* axis and the relevant values of the other variable are presented along the vertical or *y* axis. The word "relevant" here indicates that often graphs do not start at 0. Consider three examples. First is data presented in Chapter 5 concerning the temperature of Venus's atmosphere at different altitudes (Figure D-1).

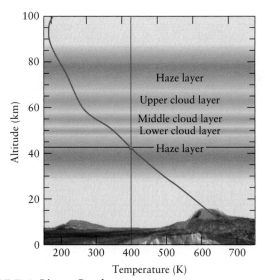

FIGURE D-1 Linear Graph

The horizontal axis of the graph indicates the temperature in degrees Kelvin (K). The vertical axis denotes the altitude above Venus's surface in kilometers (km). Note that both axes are always labeled with a name (temperature or altitude, here) and units (K or km, respectively). Bear in mind that some variables such as the temperature and altitude in Figure D-1 increase to the right or upward (these are more common), but some variables will increase to the left or down.

To read a graph, note that a value on the horizontal axis is then transferred directly upward through the graph. For example, all points on the blue line in Figure D-1 (which extends upward from 400 K) have a temperature of 400 K. Equivalently, the value given on the vertical axis is transferred to all points horizontally across from this value. All the points on the black line in Figure D-1 are at an altitude of about 43 km above Venus's surface. We have interpolated between 40 and 45 to get this value (Figure D-2).

A curve or a set of points on the graph presents the *relationship* between the variable represented on the horizontal axis and the variable on the vertical axis. Each point on a curve or each separate point relates the two variables. Choose a point, say the dot on Figure D-1. It represents the temperature at a certain altitude above Venus's surface. To find the temperature at that point, you slide directly down (along the blue line in this example) from

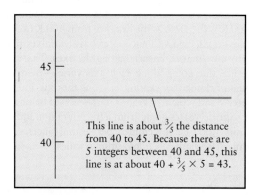

This line is about ⅗ the distance from 40 to 45. Because there are 5 integers between 40 and 45, this line is at about 40 + ⅗ × 5 = 43.

FIGURE D-2 Interpolation

the point and read the value of the horizontal variable under it. To find the altitude for that point, you slide directly over to the side (along the black line) and read the value of the vertical variable there. In our example, sliding along the blue line leads to 400 K on the temperature line. Therefore, this point corresponds to a temperature of 400 K. Moving horizontally from the point, you encounter, by interpolation, the 43-km indicator. Combining the data, you conclude that *the temperature of Venus's atmosphere 43 km above its surface is 400 K.*

The purple curve in Figure D-1 provides the relationship between altitude and temperature for a wide range of locations above Venus in a representation that is much more informative than a table of heights and temperatures. Specifically, this curve shows you the trend of temperature with altitude. Note that variables are occasionally put on the right side and on the top of graphs.

Sometimes, the known information is not a curve, but rather a set of points, as in our second example (Figure D-3) from Chapter 10. In this case, the graph connects the apparent magnitude of a nova (how bright it appears to be as seen from Earth regardless of its distance or other factors) and time. Each dot indicates how bright the nova (an explosion on the surface of certain stars)

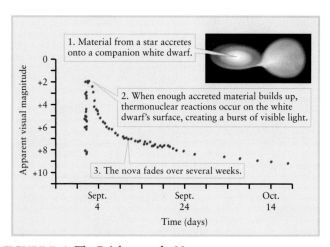

1. Material from a star accretes onto a companion white dwarf.

2. When enough accreted material builds up, thermonuclear reactions occur on the white dwarf's surface, creating a burst of visible light.

3. The nova fades over several weeks.

FIGURE D-3 The Brightness of a Nova

was at different times. For example, the peak brightness of the nova was an apparent magnitude of about +2 and it occurred on September 2. Noting that time increases to the right, you can immediately see that the trend of the nova's brightness is to increase rapidly and decrease more slowly.

The graphs so far have shown variables that change uniformly along the axes. For example, the distance on Figure D-1 from 300 K to 400 K is the same as the distance from 400 K to 500 K, and so on. Many graphs you will encounter have variables that do not change uniformly (that is, linearly) along the axes. That means that the change in value going along each axis varies—there is not the same amount of change per centimeter along the axis. In Figure D-4, from Chapter 9 and typical of graphs in the second half of the book, the temperature decreases going from left to right nonuniformly and the luminosity (total energy emitted per second) increases upward nonuniformly. These are called logarithmic scales.

The purpose of logarithmic and other nonlinear axes is to present in compact form data that varies very widely in values. For example, the dimmest star represented in Figure D-4 is just less than 0.1 times as luminous as the Sun, while the brightest star is nearly 1000 times as luminous. The process of getting information from logarithmic graphs is the same as linear graphs. You must just be careful not to think of values as doubling or tripling when you go over or up two or three intervals. Figure D-5 shows how a logarithmic scale varies over one decade of values. The same numbering intervals apply for any decade of values, for example, 1 to 10 or 10^5 to 10^6. As you can see, the numbers bunch up near the highest value, so you need to interpolate these graphs more carefully than linear graphs.

Referring to Figure D-4, the luminosity of a star with surface temperature of 4000 K is about 0.1 $L_\odot$. Note also that some graphs are linear on one axis and logarithmic on the other axis.

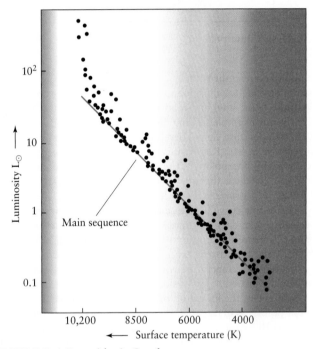

FIGURE D-4 Logarithmic Graph

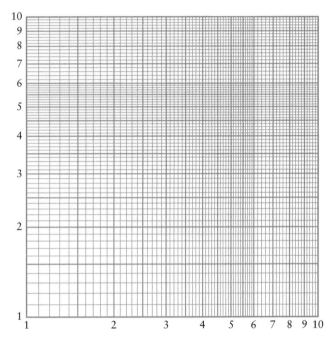

FIGURE D-5 Logarithmic Scale

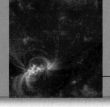

Periodic Table of the Elements

1 H Hydrogen																	2 He Helium
3 Li Lithium	4 Be Beryllium											5 B Boron	6 C Carbon	7 N Nitrogen	8 O Oxygen	9 F Fluorine	10 Ne Neon
11 Na Sodium	12 Mg Magnesium											13 Al Aluminum	14 Si Silicon	15 P Phosphorus	16 S Sulfur	17 Cl Chlorine	18 Ar Argon
19 K Potassium	20 Ca Calcium	21 Sc Scandium	22 Ti Titanium	23 V Vanadium	24 Cr Chromium	25 Mn Manganese	26 Fe Iron	27 Co Cobalt	28 Ni Nickel	29 Cu Copper	30 Zn Zinc	31 Ga Gallium	32 Ge Germanium	33 As Arsenic	34 Se Selenium	35 Br Bromine	36 Kr Krypton
37 Rb Rubidium	38 Sr Strontium	39 Y Yttrium	40 Zr Zirconium	41 Nb Niobium	42 Mo Molybdenum	43 Tc Technetium	44 Ru Ruthenium	45 Rh Rhodium	46 Pd Palladium	47 Ag Silver	48 Cd Cadmium	49 In Indium	50 Sn Tin	51 Sb Antimony	52 Te Tellurium	53 I Iodine	54 Xe Xenon
55 Cs Cesium	56 Ba Barium	57 La Lanthanum	72 Hf Hafnium	73 Ta Tantaium	74 W Tungsten	75 Re Rhenium	76 Os Osmium	77 Ir Iridium	78 Pt Platinum	79 Au Gold	80 Hg Mercury	81 Tl Thallium	82 Pb Lead	83 Bi Bismuth	84 Po Polonium	85 At Astatine	86 Rn Radon
87 Fr Francium	88 Ra Radium	89 Ac Actinium	104 Rf Rutherfordium	105 Db Dubnium	106 Sg Seaborgium	107 Bh Bohrium	108 Hs Hassium	109 Mt Meitnerium	110 Ds Darmstadium	111 Rg Roentgenium							

58 Ce Cerium	59 Pr Praseodymium	60 Nd Neodymium	61 Pm Promethium	62 Sm Samarium	63 Eu Europium	64 Gd Gadolinium	65 Tb Terbium	66 Dy Dysprosium	67 Ho Holmium	68 Er Erbium	69 Tm Thulium	70 Yb Ytterbium	71 Lu Lutetium
90 Th Thorium	91 Pa Protactinium	92 U Uranium	93 Np Neptunium	94 Pu Plutonium	95 Am Americium	96 Cm Curium	97 Bk Berkelium	98 Cf Californium	99 Es Einsteinium	100 Fm Fermium	101 Md Mendelevium	102 No Nobelium	103 Lr Lawrencium

Changing Pluto's Status as a Planet

Pluto has been removed from the pantheon of planets and relegated to the status of a "dwarf planet," leaving eight surviving planets. Although this change requires everyone to relearn the number of planets, it places Pluto into a new category that will eventually help astronomers better understand all the types of bodies that orbit the Sun.

Changing categories in astronomy is not new. Imagine that the year is 1780. Six planets are known: Mercury, Venus, Earth, Mars, Jupiter, and Saturn. Now move ahead a year: Uranus has just been discovered, requiring people to learn that there are seven planets, not six. Fast forward 19 years. . . . It is New Year's Day, 1801, and the Swiss-Italian amateur astronomer Giuseppe Piazzi discovers the eighth planet, which he names *Ceres* (Figure F-1). It is located between Mars and Jupiter. During the following year, German astronomer H. Wilhelm Olbers discovers the ninth planet, Pallas, in the same region of the solar system. The discovery of Pallas is followed within 5 years by the discovery of the tenth and eleventh planets—Juno by German astronomer Karl Harding and Vesta by Olbers (Figure F-2). These bodies (Ceres, Pallas, Juno, and Vesta) also orbit the Sun between Mars and Jupiter.

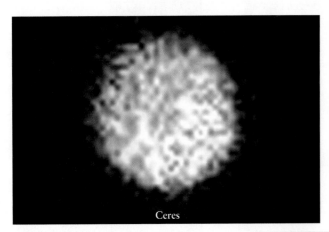

Ceres

RIVUXG

FIGURE F-1 Ceres, the Largest Asteroid in Orbit between Mars and Jupiter *(NASA, ESA, J. Parker [Southwest Research Institute], P. Thomas [Cornell University], L. McFadden [University of Maryland, College Park], and M. Mutchler and Z. Levay [STScI])*

The fifth such body, Astraea, located between Mars and Jupiter was discovered in 1845 by amateur German astronomer Karl Hencke. It is significantly smaller than any previously discovered planet, just like all but one[1] of the next 18 planets discovered by 1852. Furthermore, all these smaller bodies are located in the same general region of the solar system—between Mars and Jupiter. Seeing a physically meaningful pattern emerging, some astronomers began listing in journals the bodies smaller than Vesta (namely, Astraea and all the subsequently discovered small bod-

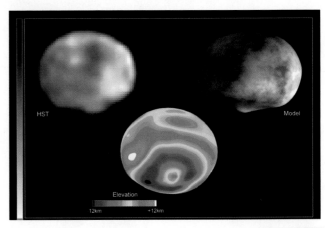

RIVUXG

FIGURE F-2 Asteroid Vesta, between Mars and Jupiter *(Ben Zellner [Georgia Southern Univ.], Peter Thomas [Cornell Univ.], and NASA)*

ies) as *asteroids*, or equivalently, *minor planets* or *planetoids*. The number of planets plunged from more than two dozen down to 12. The terms *asteroids* and *minor planets* are still in common use today.

Prior to 1855, the sizes of Ceres, Pallas, Juno, and Vesta had been grossly overestimated. During that year, another measurement scheme was used that actually underestimated their sizes. Based on these smaller sizes, these four bodies were added to the list of asteroids, which resulted in classifying eight bodies as planets: Mercury, Venus, Earth, Mars, Jupiter, Saturn, Uranus, and Neptune. This categorization supports the idea that the smaller objects—the asteroids—are fundamentally different than the eight planets.[2]

Modern observational techniques have finally established the sizes of the four largest asteroids as being smaller than the sizes originally measured and larger than the sizes measured in 1855. Nevertheless, these asteroids are all very small compared to even the smallest of the eight planets, Mercury. Indeed, the largest of the asteroids, Ceres, is only a quarter the diameter of our Moon.

Zoom ahead to 1930. During that year, Pluto is discovered by American astronomer Clyde Tombaugh. Pluto was considered to be at the edge of the known solar system and has an orbit that sometimes takes it closer to the Sun than Neptune. Indeed, Pluto's orbit is much more elliptical than the orbits of the eight planets, but similar to the very elliptical orbits of many asteroids. However, unlike the asteroids, Pluto is not located between Mars and

---

[1] The one exception is giant Neptune, discovered in 1846 out beyond Uranus.

[2] We honestly do not yet know all the differences between planets and asteroids or between the various asteroids.

Jupiter. Pluto's size is decidedly unusual, too. With a diameter 18% that of Earth, just less than half that of Mercury, just two-thirds that of our Moon, and just 4 times that of Ceres, Pluto doesn't fit well as a planet or as an asteroid. It was grouped with the eight planets, despite debate about whether it was more like them or like the ever-growing number of known asteroids.

Although asteroids continued to be discovered throughout the twentieth century, the status of Pluto as a planet was unthreatened until 1978, when Pluto was discovered to have a moon, Charon, which is almost as large as Pluto (Figure F-3). (Pluto has a diameter of 2300 km while Charon has a diameter of 1190 km.) For comparison, Earth and our Moon, which are the pair of planet and moon that is most similar to Pluto and Charon, have diameters of 12,800 km and 3500 km, respectively. Pluto and Charon are a pair of nearly equal mass bodies orbiting each other, which is unique in the solar system. The status of Pluto as a planet was further questioned as similar-sized bodies, such as Sedna and Eris, were discovered in the same remote area of the solar system.

R I V U X G

FIGURE F-3 **Pluto and Its Moon Charon** *(Dr. R. Albrecht, ESA/ ESO Space Telescope European Coordinating Facility; NASA)*

More than 100,000 asteroids and bodies orbiting beyond Neptune have now been identified. Their sizes and locations suggest that they are all quite different from the eight planets, and similar to Pluto. Pluto's size is more similar to the larger asteroids than to any planet, and its orbit is more similar to those of other bodies orbiting the Sun in space beyond Neptune than to the orbit of any planet. Pluto has finally been reclassified to better reflect these connections. These classifications were done during the summer of 2006 at a meeting of the International Astronomical Union, the group charged with classifying and naming objects in space. That meeting was filled with tension, as numerous competing definitions of planets and other objects in the solar system were proposed and rejected. In the end, the following definitions were adopted.

A **planet** is a celestial body that (a) is in orbit around the Sun, (b) has sufficient mass for its own gravity to overcome rigid body forces so that it assumes a *hydrostatic equilibrium* (in other words, the object is nearly spherical), and (c) it has cleared the neighborhood around its orbit (of smaller bodies).

Although (a) is self-explanatory and Pluto indeed orbits the Sun, the conditions (b) and (c) deserve some explanation. If an object has its shape because of the bonds between its atoms, then it can have essentially any shape. Rocks look like rocks, potatoes look like potatoes, and you and I look as we do because of such bonding. However, if an object has enough mass, then the gravitational attraction between its particles are strong enough to reshape the object—pulling down high places and pushing up low ones—until the object becomes nearly spherical. The resulting balance between the force of gravity pulling inward and the pressure created by the matter pressing on itself resulting in the matter pushing outward is called hydrostatic equilibrium. There are about three dozen objects in the solar system that meet this criterion, including Mercury, Venus, Earth, Mars, Jupiter, Saturn, Uranus, Neptune, Pluto, the asteroid Ceres, our Moon, several moons of each of the large planets, Pluto's moon Charon, and several of the Pluto-like objects orbiting beyond Neptune that have been discovered, including Sedna, Makemake, Haumea, and Eris. Conversely, there are well over a hundred thousand objects in the solar system that look more like potatoes than anything else. These bodies include most of the moons and asteroids and all of the comets.

The third condition to be considered a planet, (c) above, means that the object must also have enough gravitational attraction to pull onto itself, or to fling far away, the myriad smaller pieces of debris that orbit in its neighborhood. The inner eight planets satisfy this condition, but observations reveal numerous objects still in Pluto's vicinity. Pluto does not have enough gravitational attraction to clear its surrounding area; so, by using the new criteria for determining planets, Pluto is not a planet.

Nevertheless, Pluto and numerous other objects are still a part of the solar system, so they need to be classified. Because we know very little about the chemical compositions of these bodies (including Pluto), we do not yet know if they are a single class of objects or several distinct classes of objects. For example, some of them may be composed mostly of rock and metal, while others may be mostly rock and ice; some may have sheaths of ice, while others have rocky surfaces. Until such details are known, all the other spherical objects orbiting the Sun that have *not* cleared their neighborhoods in space of debris and that are not satellites[3] of other bodies are classified as **dwarf planets**. Dwarf planets that orbit farther from the Sun, on average, than Neptune are called **Plutoids**. Pluto is both a dwarf planet and a Plutoid, but its moon Charon is *neither* because Charon orbits the slightly larger body, Pluto. The asteroid Ceres is now classified as a dwarf planet. Plutoids Eris, Makemake (pronounced MAH-kay MAH-kay), and Haumea are also dwarf planets.

Finally, the panoply of nonsatellite, nonspherical objects in the solar system are called **small solar-system bodies (SSSBs)**. Neither this name nor the name "dwarf planets" is very satisfying. The four inner planets, Mercury, Venus, Earth, and Mars, are much smaller than the four outer planets, so it is easy to confuse the inner "terrestrial planets" with "dwarf planets." Just as the name "planetoid" was dropped in favor of "asteroid," the two names—dwarf planets and small solar-system bodies—may eventually be changed to better reflect the objects they represent.

---

[3] In other words, moons.

FIGURE F-4 Scales of the Solar System All these bodies are presented to scale. *(International Astronomical Union/NASA)*

As has occurred several times since the eighteenth century, people now have to go through the uncomfortable process of learning new names for old astronomical objects. Clearly, changing Pluto's designation to "dwarf planet (134340) Pluto" does not change it physical properties. Indeed, astronomers hope that the new designations will *better* represent Pluto's properties, along with the properties of the other small objects orbiting the Sun. Although the renaming will be confusing for a while, it will eventually help clarify our perception of the solar system (Figure F-4).

The Earth-Centered Universe

As we move through the twenty-first century, most of us find it hard to understand why anyone would believe that the Sun, planets, and stars orbit Earth. After all, we *know* that Earth spins on its axis. We *know* that the gravitational force from the Sun holds the planets in orbit, just as Earth's gravitational force holds the Moon in orbit. These facts have become part of our understanding of the motions of the heavenly bodies, and we are taught these things from the time we are children.

Psychologists call this background information that we use to help explain things a *conceptual framework*. Any conceptual framework contains all of the information we take for granted. For example, when the Sun rises, moves across the sky, and sets, we take for granted that it is Earth's rotation that causes the Sun's apparent motion.

Our ancestors possessed a different conceptual framework for understanding the cosmos. They did not know that Earth rotates. They did not know that the then-mysterious force that held them to the ground is the same force that attracts Earth to the Sun and the Moon to Earth. They did not know that the Sun is a star, just like the fixed points of light in the sky, and they did not know any of the other laws of motion that we take for granted.

Because they did not feel Earth move under their feet, or see any other indication that Earth is in motion, our forebears sensed nothing to support the belief that we are in motion. The obvious conclusion for one who has a prescientific conceptual framework, even today, is that Earth stays put while objects in the heavens move around it.

This prescientific conceptual framework for understanding the motions of the heavenly bodies was based on the senses and on common sense. That is, people observed motions and drew "obvious," commonsense conclusions. Today, we incorporate the known and tested laws of physics in our understanding of the natural world. Many of these realities are utterly counterintuitive, and, therefore, the conceptual frameworks that we possess are less consistent with common sense than those held in the past. Studying science helps us develop intuition that is consistent with the actual workings of nature.

Geocentric Explanation of the Planets' Retrograde Motions

The early Greeks developed many theories to account for the occasional retrograde motion of the planets and the resulting loops that the planets trace out against the background stars. One of the most successful ideas was expounded by the last of the great ancient Greek astronomers, Ptolemy, who lived in Alexandria, Egypt, 1900 years ago. His basic concepts are sketched in the accompanying figure. Each planet is assumed to move in a small circle called an *epicycle,* the center of which moves in a larger circle called a *deferent,* whose center is offset from Earth. As viewed from Earth, the epicycle moves eastward along the deferent, and both it and the planet on it revolve in the same direction (counterclockwise).

Most of the time, the motion of the planet on its epicycle adds to the eastward motion of the epicycle on the deferent. Thus, the planet is seen from Earth to be in direct motion (to the left or eastward) against the background stars throughout most of the year (Figure G-1a). However, when the planet is on the part of its epicycle nearest Earth, its motion along the epicycle subtracts from the motion of the epicycle along the deferent. The planet thus appears to slow and then halt its usual movement to the left (eastward motion) among the constellations, and then seems to move to the right (westward) among the stars for a few weeks or months (Figure G-1b). This concept of epicycles and deferents enabled Greek astronomers to explain the retrograde loops of the planets.

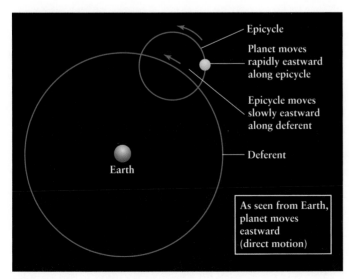

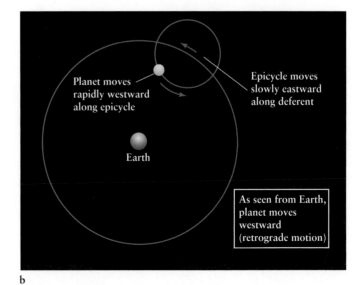

a

b

FIGURE G-1 A Geocentric Explanation of Planetary Motion Each planet revolves around an epicycle, which, in turn, revolves around a deferent centered approximately on Earth. As seen from Earth, the speed of the planet on the epicycle alternately (a) adds to or (b) subtracts from the speed of the epicycle on the deferent, thus producing alternating periods of direct and retrograde motions.

Using the wealth of astronomical data in the library at Alexandria, including records of planetary positions covering hundreds of years, Ptolemy deduced the sizes of the epicycles and deferents and the rates of revolution needed to produce the recorded paths of the planets. After years of arduous work, Ptolemy assembled his calculations in the *Almagest,* in which the positions and paths of the Sun, Moon, and planets were described with unprecedented accuracy. In fact, the *Almagest* was so successful that it became the astronomer's bible. For more than 1000 years, Ptolemy's cosmology endured as a useful description of the workings of the heavens.

Eventually, however, the commonsense explanation of Earth-centered cosmology began to go awry. Errors and inaccuracies that were unnoticeable in Ptolemy's day compounded and multiplied over the years, especially errors due to precession, the slow change in the direction of Earth's axis of rotation. Fifteenth-century astronomers made some cosmetic adjustments to the Ptolemaic system. However, the system became less and less satisfactory as more fanciful and arbitrary details were added to keep it consistent with the observed motions of the planets. After Newton's time, scientists knew that orbital motion required a force to be acting on the body. However, nothing in Ptolemy's epicycle theory produced such a force.

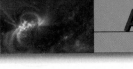

Energy and Momentum

Scientists identify two types of energy that are available to any object. The first, called **kinetic energy**, is associated with the object's motion. For speeds much less than the speed of light, we can write the amount of kinetic energy, KE, of an object as

$$KE = \frac{1}{2}mv^2$$

where m is the object's mass and v is its speed. Kinetic energy is a measure of how much work the object can do on the outside world or equivalently, how much work the outside world has done to give the object this speed.

Work is also a rigorously defined concept that often is at odds with our intuition. It is defined as the product of the force, F, acting on an object and the distance, d, over which the object moves in the direction of the force:

$$W = Fd$$

For example, if I exert a horizontal force of 50 N (N is the unit newtons and is the metric unit of force) and thereby move an object 10 m in that direction, then I have done 50 N × 10 m = 500 J of work. (I have used the relationship that 1 newton × 1 meter = 1 joule.)

The second type of energy is called **potential energy**. It represents how much energy is available to an object as a result of its location in space. For example, if you hold a pencil above the ground, the pencil has potential energy that can be converted into kinetic energy by Earth's gravitational force. How does that conversion get underway? Just let go of the pencil.

There are various kinds of potential energy, such as the potential energy stored in a battery and the potential energy stored in objects under the influence of gravity. We will focus on *gravitational potential energy*. Far from extremely massive objects, like stars, or extremely dense objects, like black holes, gravitational potential energy, PE, can be written as

$$PE = \frac{GmM}{r}$$

where the constant $G = 6.6683 \times 10^{-11}$ N m²/kg², m is the mass of the object whose gravitational potential energy you are measuring, M is the mass of the object generating the gravitational attraction, and r is the distance between the centers of mass of these two objects.

Near the surface of Earth, this equation simplifies to

$$PE = mgh$$

where $g = 9.8$ m/s² (32 ft/s²) is the gravitational acceleration at Earth's surface, and h is the height of the object above Earth's surface.

Potential energy can be converted into kinetic energy and vice versa. By dropping the pencil, its gravitational potential energy begins to decrease while its kinetic energy begins to increase at the same rate. The pencil's total energy is conserved. Conversely, if you throw a pencil up in the air, the kinetic energy you give it will immediately begin to decrease, while its potential energy increases at the same rate.

Related to the motion of an object, and hence to its kinetic energy, are the concepts of *linear momentum,* usually just called **momentum**, and **angular momentum**. Momentum, p, is described by the equation

$$\boldsymbol{p} = m\boldsymbol{v}$$

where $\boldsymbol{v}$ is the velocity of the object. Both $\boldsymbol{p}$ and $\boldsymbol{v}$ are in boldface to indicate that they both represent motion in some direction or another, as well as a numeric value. Simple algebra reveals that kinetic energy and momentum are related by

$$KE = \frac{p^2}{2m}$$

Linear momentum, then, indicates how much energy is available to an object because of its motion in a straight line (linear motion).

Angular momentum, L, can be expressed mathematically as

$$L = I\omega$$

where I is the **moment of inertia** of an object, and $\boldsymbol{\omega}$ (lowercase Greek omega) is the angular speed and direction of the rotating object. Just as an object's mass indicates how hard it is to change an object's straight-line motion, the moment of inertia indicates how hard it is to change the rate at which an object rotates or revolves. The moment of inertia depends on an object's mass and shape. Kinetic energy due to angular motion can be written as

$$KE = \frac{L^2}{2I}$$

Newton's first law can also be expressed in terms of **conservation of linear momentum:** *A body maintains its linear momentum unless acted upon by a net external force.*

Equivalently, for angular motion we can write the **conservation of angular momentum:** *A body maintains its angular momentum unless acted upon by a net external torque. Torques* are created when a force acts on an object in some direction other than toward the center of the object's angular motion, as shown in the accompanying figure.

Earth has angular momentum and keeps spinning on its rotational axis and orbiting the Sun. Likewise, the Moon has angular momentum and keeps spinning on its rotation axis and orbiting Earth. Virtually all objects in astronomy have angular momentum, and it is probably fair to say that conservation of angular momentum is among the most important laws in the cosmos. After all, conservation of angular momentum is what keeps the planets in orbit around the Sun, the moons in orbit around the planets, and astronomical bodies rotating at relatively constant rates, as well as causing many other rotation-related effects that we will encounter throughout this book.

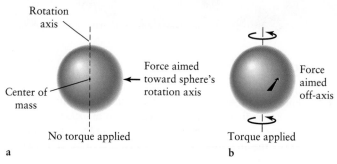

Angular Momentum and Torque (a) When a force acts through an object's rotation axis or toward its center of mass, the force does not exert a torque on the object. (b) When a force acts in some other direction, then it exerts a torque, causing the body's angular momentum to change. If the object can spin around a fixed axis, like a globe, then the rotation axis is the rod running through it. If the object is not held in place, then the rotation axis is in a line through a point called the object's *center of mass*. The center of mass of any object is the point that follows a smooth, elliptical path as the object moves in response to a gravitational field. All other points in the spinning object wobble as it moves.

Radioactivity and the Ages of Objects

The isotopes of many elements are radioactive, meaning that the elements spontaneously transform into other elements. Each radioactive isotope has a distinctive *half-life,* which is the time that it takes half of the initial concentration of the isotope to transform into another element. After two half-lives, a radioactive isotope is reduced to $\frac{1}{2} \times \frac{1}{2}$ or $\frac{1}{4}$ of its initial concentration (Figure H-1). Among the most important radioactive elements for determining the age of objects in astronomy is the isotope of uranium with 146 neutrons, ^{238}U ("U two-thirty-eight"). The half-life of ^{238}U decaying into lead is 4.5 billion years.

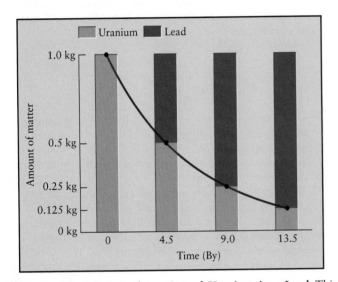

FIGURE H-1 **The Transformation of Uranium into Lead** This figure shows the rate that 1 kg of uranium decays into lead, as described in the text. It contains $\frac{1}{8}$ kg of uranium after 13.5 billion years.

To determine the time since an object, such as a piece of space debris discovered on Earth, solidified, scientists estimate how much lead it had when it formed. Then they measure the amounts of uranium and lead that it contains now. Subtracting the amount of original lead, they use the amount of uranium and lead, together with the graph of radioactive decay, to determine the object's age.

Example: Suppose a piece of space debris discovered on Earth was determined to have equal amounts of lead and uranium. How long ago did this debris form? Assuming that it originally had no lead, we see from the chart that a 1-to-1 mix of lead to uranium occurs 4.5 billion years after the object formed. This is one half-life of uranium.

Compare! This process works with any radioactive isotope. However, some isotopes have such short half-lives that they are not useful in astronomy. For example, the well-known carbon dating used to determine the ages of ancient artifacts on Earth is of little use in astronomy because ^{14}C has a half-life of only 5730 years. Because so much of the carbon in a sample has decayed away by then, carbon dating is useful only for time intervals shorter than 100,000 years, usually a period over which little of astronomical importance occurs.

Thermonuclear Fusion

What drives the thermonuclear fusion that powers the Sun? For nuclei to fuse, they must be brought together at incredibly high temperatures and pressures. That is exactly what occurs in the Sun's core, where the entire mass of the Sun compresses inward. The core's temperature is 15.5 million K, its pressure is about 3.4×10^{11} atm, and its density is 160 times greater than that of water.

Normally, nuclei cannot contact one another because the positive electric charge on each proton prevents nearby protons from coming too closely. (Remember that like charges repel each other.) But in the extreme heat and pressure of the Sun's center, the protons move so fast in such close proximity that they can stick, or fuse, together.

The nuclear transformations inside the Sun follow several routes, but each begins with the simplest atom, hydrogen (H). Most hydrogen nuclei consist of a single proton. The final outcome of fusion is creation of the nucleus of the next simplest atom, helium (He), consisting of two protons and two neutrons. The fusion of hydrogen into helium takes several steps.

Figure H-2 shows the most common path for hydrogen fusion in the Sun. This particular sequence is called the *proton–proton,* or *PP, chain.*

Note that proton–proton fusion releases positively charged electrons, e^+, called **positrons,** and neutral, nearly massless particles, called **neutrinos,** v. When these positrons encounter regular electrons in the Sun's core, both particles are annihilated, and their mass is converted into energy in the form of gamma-ray photons. The final fusion, in which the helium forms, also returns two protons to the core, which are then available to fuse again.

Because the PP chain produces neutrinos and the Sun's energy, we can summarize hydrogen fusion this way:

$$4\,^1\text{H} \rightarrow 1\,^4\text{He} + \text{neutrinos} + \text{energy}$$

Example: From our summary equation, we can calculate the energy released during a fusion reaction. We simply look at how much mass is converted into energy:

$$\begin{aligned}
\text{Mass of 4 hydrogen atoms} &= 6.693 \times 10^{-27} \text{ kg} \\
- \text{ Mass of 1 helium atom} &= 6.645 \times 10^{-27} \text{ kg} \\
\text{Mass lost} &= 0.048 \times 10^{-27} \text{ kg}
\end{aligned}$$

Thus, a small fraction (0.7%) of the mass of the hydrogen going into the nuclear reactions does not show up in the mass of the helium. Ignoring the relatively small mass of the neutrino, this lost mass is converted into energy, as predicted by Einstein's famous equation:

$$\begin{aligned}
E &= mc^2 \\
&= (0.048 \times 10^{-27} \text{ kg}) \times (3 \times 10^8 \text{ m/s})^2 \\
&= 4.3 \times 10^{-12} \text{ J}
\end{aligned}$$

Compare! The energy released from the formation of a single helium atom would light a 10-watt lightbulb for almost one-half a trillionth of a second.

Now let us add up the energy emitted from the entire Sun. As noted in Chapter 7, the Sun's mass, usually designated 1 $M_\odot$, is equal to 333,000 Earth masses. Its total energy output per second, called the **solar luminosity** and denoted 1 $L_\odot$, is 3.9×10^{26} W. To produce this luminosity, the Sun converts 600 million metric tons of hydrogen into helium within its core each second. This is twice the weight of the Empire State Building in New York City. This enormous rate is possible because the Sun contains a vast supply of hydrogen—enough to continue the present rate of energy output for another 5 billion years.

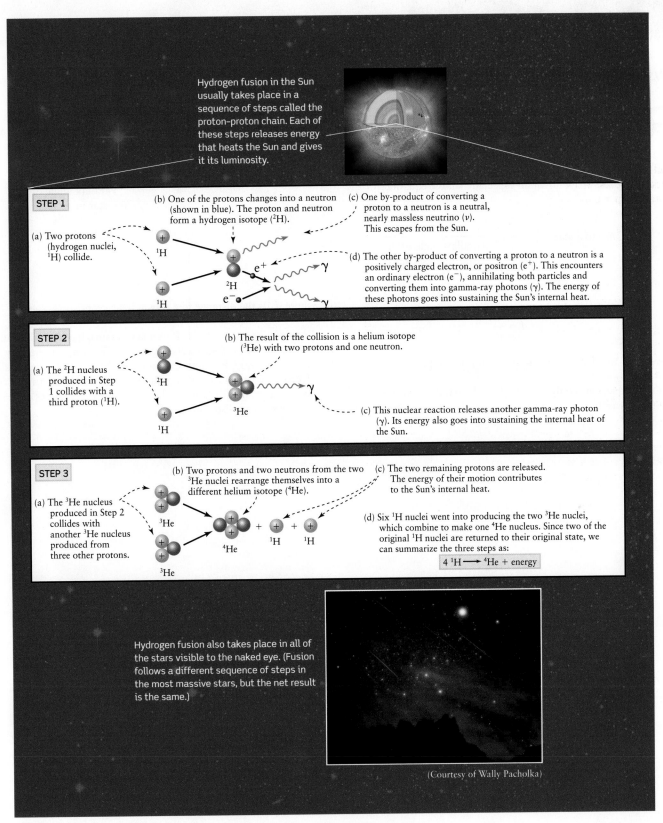

Hydrogen fusion in the Sun usually takes place in a sequence of steps called the proton–proton chain. Each of these steps releases energy that heats the Sun and gives it its luminosity.

STEP 1

(a) Two protons (hydrogen nuclei, ^1H) collide.

(b) One of the protons changes into a neutron (shown in blue). The proton and neutron form a hydrogen isotope (^2H).

(c) One by-product of converting a proton to a neutron is a neutral, nearly massless neutrino (ν). This escapes from the Sun.

(d) The other by-product of converting a proton to a neutron is a positively charged electron, or positron (e^+). This encounters an ordinary electron (e^-), annihilating both particles and converting them into gamma-ray photons (γ). The energy of these photons goes into sustaining the Sun's internal heat.

STEP 2

(a) The ^2H nucleus produced in Step 1 collides with a third proton (^1H).

(b) The result of the collision is a helium isotope (^3He) with two protons and one neutron.

(c) This nuclear reaction releases another gamma-ray photon (γ). Its energy also goes into sustaining the internal heat of the Sun.

STEP 3

(a) The ^3He nucleus produced in Step 2 collides with another ^3He nucleus produced from three other protons.

(b) Two protons and two neutrons from the two ^3He nuclei rearrange themselves into a different helium isotope (^4He).

(c) The two remaining protons are released. The energy of their motion contributes to the Sun's internal heat.

(d) Six ^1H nuclei went into producing the two ^3He nuclei, which combine to make one ^4He nucleus. Since two of the original ^1H nuclei are returned to their original state, we can summarize the three steps as:

$$4 \,^1H \longrightarrow \,^4He + energy$$

Hydrogen fusion also takes place in all of the stars visible to the naked eye. (Fusion follows a different sequence of steps in the most massive stars, but the net result is the same.)

(Courtesy of Wally Pacholka)

FIGURE H-2 Steps to Fuse Hydrogen into Helium by the Proton–Proton Chain

Distances to Nearby Stars

Recall from Appendix H-1: Units of Astronomical Distance that 1 parsec (1 pc) is the distance at which two objects 1 AU apart appear 1 arcsec apart. This distance is 3.09×10^{13} km, or 206,265 AU. The word *parsec* (from *parallax second*) originated in the use of parallax to measure distance. Using parsecs, we can write down an especially simple equation for the distances to stars:

$$\text{Distance to a star in parsecs} = \frac{1}{\text{Parallax angle of that star in arcseconds}}$$

or

$$d = \frac{1}{p}$$

where d is the distance to the star and p is the parallax angle of that star.

The equation is only this simple in these units, which is one of the main reasons why many astronomers discuss cosmic distances in parsecs rather than light-years. We will primarily use light-years (ly) throughout this book, however, as they are more intuitive. In light-years, this same equation becomes approximately

$$\text{Distance to a star in light-years} \approx \frac{3.26}{\text{Parallax angle of that star in arcseconds}}$$

or

$$d_{\text{ly}} = \frac{3.26}{p}$$

where d_{ly} is the distance to a star in light-years.

Example: The nearest star, Proxima Centauri, has a parallax angle of 0.77 arcsec, and so its distance is 1/0.77, or approximately 1.3 pc. Equivalently, Proxima Centauri is 4.24 ly away. The parallax of Proxima Centauri is comparable to the angular diameter of a dime seen from a distance of 3 km. The parallax angles of the 25 nearest stars are listed in Table C–5.

The Distance–Magnitude Relationship

The closer a star, the brighter it appears. The inverse-square law leads to a simple equation for absolute magnitude, M. Suppose a star's apparent magnitude is m and its distance from Earth is d (measured in parsecs). Then

$$M = m - 5 \log (d/10)$$

where log stands for the base-10 logarithm. This distance–magnitude relation can be rewritten as

$$m - M = 5 \log d - 5$$

Example: Consider Proxima Centauri, the nearest star to Earth, other than the Sun. By measuring its parallax angle, we know this star is at a distance from Earth of $d = 1.3$ pc. Its apparent magnitude is $m = +11.1$. Therefore, its absolute magnitude is

$$M = 11.1 - 5 \log (1.3/10) = 11.1 - (-4.4) = +15.5$$

Compare! The Sun is an average star with M = +4.8, so Proxima Centauri is an intrinsically dim star. If you know any two of d, m, and M, you can calculate the third variable. For example, if we know a star's absolute and apparent magnitudes, the equation can be used to determine its distance.

Gravitational Force

From Newton's law of gravitation, if two objects that have masses, m_1 and m_2, are separated by a distance, r, then the gravitational force, F, between them is

$$F = \frac{Gm_1m_2}{r^2}$$

In this formula, G is the **universal constant of gravitation**, whose value has been determined from laboratory experiments:

$$G = 6.668 \times 10^{-11} \text{ N m}^2 \text{ kg}^{-2}$$

where N is the unit of force, a newton.

The equation $F = G(m_1m_2/r^2)$ gives, for example, the force from the Sun on Earth and, equivalently, from Earth on the Sun. If m_1 is the mass of Earth (6.0×10^{24} kg), m_2 is the mass of the Sun (2.0×10^{30} kg), and r is the distance from the center of Earth to the center of the Sun (1.5×10^{11} m):

$$F = 3.6 \times 10^{22} \text{ N}$$

This number can then be used in Newton's second law, $F = ma$, to find the acceleration of Earth due to the Sun. This yields

$$a_{\text{Earth}} = F/m_1 = 6.0 \times 10^{-3} \text{ m/s}^2$$

Newton's third law says that Earth exerts the same force on the Sun, so the Sun's acceleration due to Earth's gravitational force is

$$a_{\text{Sun}} = F/m_2 = 1.8 \times 10^{-8} \text{ m/s}^2$$

In other words, Earth pulls on the Sun, causing the Sun to move toward it. Because of the Sun's greater mass, however, the amount that the Sun accelerates Earth is more than 300,000 times greater than the amount that Earth accelerates the Sun.

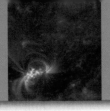

CHAPTER 1

QUESTION: The planets move through constellations, while the stars do not. What two possible properties of the planets could allow them to do that?

ANSWER: Two explanations for why the planets move through the constellations, but the stars do not are 1. The planets are closer than the stars and the difference in distances allows us to see the planets move (correct explanation). 2. The planets are about the same distances as the stars, but the planets are moving faster across the sky (incorrect explanation).

QUESTION: Explain why Figure 1-14 must have been taken facing west. Hint: Examine Figure 1-18.

ANSWER: In the northern hemisphere, stars move southward (to the right) as they rise and northward (to the right) as they set. If this figure were taken in the morning, the stars would be moving in the wrong direction.

CHAPTER 2

QUESTION: We saw in Chapter 1 that the Moon's orbit around Earth is not circular. Where in its orbit is the Moon moving fastest and where is it moving slowest?

ANSWER: By Kepler's second law, the Moon is moving fastest where it is closest to Earth (perigee) and slowest where it is farthest away (apogee).

QUESTION: Sitting in a moving car, how can you experimentally verify that your body has inertia?

ANSWER: To show that your body has inertia, drive a car and then rapidly put on the brakes. You will feel yourself forced forward against the seatbelt. Without inertia, you would slow down at the same rate as the car does without having to be restrained by the seatbelt.

QUESTION: Which elements on Earth may have been unchanged since the universe began?

ANSWER: Much of the hydrogen and helium, and some of the lithium, on Earth formed shortly after the universe began. All other elements on Earth (and elsewhere in the universe) were definitely formed from these building blocks since then.

QUESTION: Why is Earth's albedo continually changing?

ANSWER: The surface and atmospheric features of Earth, including cloud cover, ocean surface activity, snow, ice, and vegetation cover change, thereby changing the amount of light scattered back into space (the albedo).

CHAPTER 3

QUESTION: The speed of sound is about 0.34 km/s (0.21 mi/s). How can this value and the information in this section be used to determine a person's distance from a lightning strike?

ANSWER: To determine the distance to a lightning strike, measure the time interval between the lightning and subsequent thunder. Since you see the lightning virtually instantaneously after it strikes, while the thunder travels at the much slower "speed of sound," the distance to the strike is just the speed of sound times the time between the lightning and thunderclap.

QUESTION: Where on Figure 3-10c does the light from the primary mirror actually come to a focal point?

ANSWER: The focal point of the primary mirror on Figure 3-10c is where the two light rays reflecting off the primary mirror meet to the left of the secondary mirror.

QUESTION: Why does the human eye/brain clear the images it receives many times a second?

ANSWER: The brain clears images it receives in order to detect motion. Without clearing what we see very often, the brain would get saturated with light and would be unable to detect changes (in other words, motion).

QUESTION: Where in a typical house would infrared detectors indicate most activity?

ANSWER: At home, infrared detectors are useful for detecting things in the kitchen, in the furnace room, in fireplaces, and wherever else there are heat sources.

QUESTION: From what we have just discussed, what do you think causes fireworks to have their distinctive colors?

ANSWER: Transitions of electrons in the elements and molecules used in the fireworks emit specific wavelengths, giving fireworks their distinctive colors.

QUESTION: Which Balmer line in Figure 3-49 is H_α?

ANSWER: H_α on Figure 3-49 is the transition from $n = 3$ to $n = 2$. This transition's energy is the least (the resulting photon has the longest wavelength) of all Balmer transitions.

QUESTION: Does a police siren approaching you sound higher or lower in pitch than the siren at rest relative to you?

ANSWER: An approaching siren has a higher pitch than does the same siren at rest relative to you.

3. 9 times more

CHAPTER 4

QUESTION: An increase in the vegetation on Earth will have what effect on the carbon dioxide level in the atmosphere?

ANSWER: Increased vegetation will decrease the carbon dioxide level in the atmosphere by converting some of the CO_2 (along with water) into oxygen and glucose.

QUESTION: What happens to the Van Allen belts when Earth's magnetic field is flipping?

ANSWER: When the Van Allen belts are flipping, Earth's magnetic field and the resulting Van Allen belts vanish, allowing more high energy particles to enter Earth's atmosphere than occurs today.

QUESTION: Why haven't the ocean tides on Earth put our planet into synchronous rotation with respect to the Moon?

ANSWER: Earth's angular momentum is too great for it to have been slowed down enough to be in synchronous rotation with the Moon today.

CHAPTER 5

QUESTION: If Mercury were struck by a large planetesimal, why might that planet not have had a moon form, as Earth did when it was struck?
ANSWER: Two possible reasons the large impact on Mercury didn't create a moon there are that the impacting body hit head on, putting debris in all directions, but not in orbit, and that the impacting body was composed of low-density material that didn't have enough mass to cause an ejection.

QUESTION: Why isn't Mercury in synchronous rotation with respect to the Sun?
ANSWER: Mercury is not in synchronous orbit around the Sun because the planet's orbit is too elliptical to keep the same face toward the Sun.

QUESTION: Why does little cratering occur on Venus today, even compared to the present low cratering rate on our Moon?
ANSWER: Venus's thick atmosphere vaporizes virtually every object heading toward the planet's surface, preventing cratering from occurring there today.

QUESTION: Why do you think it is easier to stand up in a dust devil on Mars than in the equivalent-speed event here on Earth?
ANSWER: The air on Mars is less dense than the air on Earth, so the air pressure on Mars is lower and winds of equivalent speed have less impact there than on Earth.

QUESTION: Would our Moon have to be closer or farther away to orbit in the same direction, but rise in the west?
ANSWER: By Kepler's third law, our Moon would have to be closer to Earth for it to rise in the west.

QUESTION: Why is the terrestrial body inside Jupiter so small compared to what it would be in orbit without the outer layers around it?
ANSWER: The outer layers of Jupiter tremendously compress the terrestrial body in its core.

QUESTION: What creates most of the heat inside Io that causes it to have volcanoes and geysers?
ANSWER: Friction created by tidal distortion (rubbing of rock on rock) in Io creates the heat necessary to form volcanoes and geysers there.

QUESTION: What two factors cause Saturn's belt and zone system?
ANSWER: Convection of Saturn's outer layers and the planet's rotation cause that planet to have a belt and zone system.

QUESTION: What evidence do astronomers have that Enceladus has liquid water in its interior?
ANSWER: Ice floes on the surface and water vapor in its atmosphere strongly indicate that there is liquid water inside Enceladus.

QUESTION: Why do Uranus's rings remain in orbit?
ANSWER: Uranus's rings are held in orbit by two shepherd moons, Cordelia and Ophelia.

QUESTION: Is our Moon inside or outside Earth's Roche limit?
ANSWER: Our Moon is outside Earth's Roche limit. Otherwise, the Moon would have been pulled apart and made into a ring system around Earth.

CHAPTER 6

QUESTION: What objects are today classified as planets? Dwarf planets?
ANSWER: Planets are Mercury, Venus, Earth, Mars, Jupiter, Saturn, Uranus, and Neptune. Dwarf planets are Pluto, Ceres, Eris, Makemake, and Haumea.

QUESTION: Why do astronomers doubt that the asteroid belt was once made up of a single planet?
ANSWER: The total mass of all the objects in the asteroid belt combined is much less than that of any planet.

QUESTION: What evidence do we have that Comet Shoemaker-Levy 9 was not one solid chunk of rock surrounded by ice?
ANSWER: Shoemaker-Levy 9 broke into numerous pieces under the relatively weak tidal influence of Jupiter. A solid comet would not have been so easily pulled apart.

QUESTION: Why are stony-iron meteorites so rare compared to stony or iron meteorites?
ANSWER: Stony-iron meteorites come from the boundaries between the rocky and metal parts of asteroids. These are very thin layers; hence, the meteorites from them are rare.

CHAPTER 7

QUESTION: Which of the Sun's three atmospheric layers is coolest? Which is densest?
ANSWER: A portion of the Sun's chromosphere is the coolest atmospheric layer, while its photosphere is the densest part.

QUESTION: Why does the Sun not collapse under the influence of its own enormous gravitational attraction?
ANSWER: The Sun doesn't collapse because some of the energy it generates in its core pushes its gases outward, counteracting the inward force of gravity.

QUESTION: Why did the earlier neutrino detectors not detect the predicted number of neutrinos from the Sun?
ANSWER: The earlier neutrino detectors were only sensitive to one type of neutrino. Since some of the Sun's neutrinos transform from one type to another before reaching Earth, the early detectors did not detect all of them.

5. Next maximum in 2005, next minimum in 2012

CHAPTER 8

QUESTION: For which object—the Moon, Mars, or the star Sirius—is the parallax angle from Earth smallest?
ANSWER: The parallax angle for Sirius is smaller than that of Mars and our Moon because Sirius is farthest away of the three.

QUESTION: A main-sequence star of which spectral type—F5, A8, or K0—is largest?
ANSWER: Hotter main-sequence stars are larger than cooler main-sequence stars. Therefore, the A8 main-sequence stars are larger than F5 or K0 main-sequence stars.

QUESTION: For what types of stars does the mass–luminosity relationship not apply?
ANSWER: The mass–luminosity relationship does not apply for any type of star other than those on the main sequence.

QUESTION: What do the spectral lines of a spectroscopic binary do as observed from Earth?
ANSWER: The spectral lines of a spectroscopic binary Doppler shift in opposite directions as seen from Earth, unless the two stars orbit perpendicular to our line of sight. In the latter case, the spectral lines do not change at all.

15. 25 times brighter

CHAPTER 9
QUESTION: How are T Tauri stars different from main-sequence stars like the Sun?
ANSWER: Unlike solar-mass main-sequence stars, T Tauri stars are emitting large amounts of gas and dust, and they are changing brightness significantly.

QUESTION: Which star arrives on the main sequence first, one that is 0.5 $M_\odot$ or one that is 2 $M_\odot$?
ANSWER: Higher mass stars arrive on the main sequence more quickly than lower mass stars, so a 2 solar mass star would arrive there more rapidly than a 0.5 solar mass star.

QUESTION: Will the Sun undergo a helium flash? Why or why not?
ANSWER: The Sun will undergo a helium flash because its core will remain degenerate as it expands into the giant phase.

QUESTION: How many helium atoms does it take to make one oxygen atom?
ANSWER: It takes four helium atoms to make one oxygen atom.

QUESTION: How do astronomers observe that Cepheids are changing size?
ANSWER: Their cyclic changes in brightness and Doppler shift indicate that Cepheids are changing size.

CHAPTER 10
QUESTION: Why do type Ia supernovae not have any hydrogen lines in their spectra?
ANSWER: Type Ia supernovae lack hydrogen lines because the star that is exploding was composed primarily of carbon and oxygen.

QUESTION: Why were neutrinos from SN 1987A observed before the light from this event?
ANSWER: Neutrinos were observed from SN 1987A before light from this event because the neutrinos were released by the dying star before the visible light was generated by the blast.

QUESTION: If the lighthouse model of pulsars is correct, do we see all the nearby pulsars? Why or why not?
ANSWER: We do not see the nearby pulsars whose beams do not sweep in our direction.

QUESTION: Why do glitches change the rotation rates of pulsars?
ANSWER: Glitches change the rotation rates of a pulsar by redistributing its mass. To conserve its angular momentum, the pulsar's rotation rate must also change.

QUESTION: How much mass, m, would have to be destroyed in order to create an amount of energy, E?
ANSWER: By Einstein's equation $E = mc^2$, to create energy, E, requires mass E/c^2.

QUESTION: How would a 1 $M_\odot$ black hole 1 AU from Earth affect our planet?
ANSWER: Since the black holes we have discussed so far have more mass than the Sun, such a black hole 1 AU from Earth would have a greater pull on our planet than does the Sun.

QUESTION: Why does at least one particle in the Hawking process always fall into the black hole?
ANSWER: At least one particle must enter the black hole in the Hawking process to conserve the linear momentum of the pair of particles that the process creates.

CHAPTER 11
QUESTION: Referring to Figure 11-4, how would the brightness of a Cepheid variable with peak luminosity of 1000 $L_\odot$ change if it were observed every 5 days?
ANSWER: A Cepheid variable with peak luminosity of 1000 $L_\odot$ has a period of 5 days, so if it were observed every 5 days, its luminosity would appear unchanged from one observation to the next.

QUESTION: The presence of supernova remnants at the center of our Galaxy implies what other activity is occurring in that region?
ANSWER: Since stars that become supernovae, and the remnants they leave, have short lives in astronomical terms, supernova remnants at the center of our Galaxy imply that new star formation must be occurring there.

QUESTION: Comparing it to the galaxies in Figure 11-20, what is the spiral classification of M 74, shown in Figure 11-23?
ANSWER: M 74 is an Sc galaxy.

QUESTION: The Large Magellanic Cloud is visible to the naked eye in the southern hemisphere. From Figure 11-32, what two things do you think it could be mistaken for?
ANSWER: The Large Magellanic Cloud is sometimes mistaken for part of the Milky Way or for a cloud in the night sky.

QUESTION: If the dark matter in Figure 11-49c suddenly vanished, what would we see of the clusters of galaxies behind it?
ANSWER: If the dark matter between us and a cluster of galaxies suddenly vanished, the shapes we would see of many of those galaxies would change (since the dark matter lenses the light from them).

QUESTION: If the Hubble constant were 3 times its present value, how much slower or faster would superclusters be moving apart?
ANSWER: If the Hubble constant were 3 times its present value, superclusters would be moving apart 3 times faster.

QUESTION: In what way are telescopes time machines?
ANSWER: The light entering telescopes travels at a finite speed, so what we see through them are events and objects as they were some time in the past.

QUESTION: How are the spectra of quasars different from the spectra of stars?
ANSWER: Quasar spectra have emission lines, while stellar spectra have absorption lines.

ANS-4 ANSWERS TO COMPUTATIONAL QUESTIONS AND SOME CHAPTER QUESTIONS

QUESTION: What other astronomical objects have been observed after their light passed through a gravitational lens?
ANSWER: Besides quasars, astronomers have observed galaxies that have been gravitationally lensed.

14. 20 times

CHAPTER 12

QUESTION: If the cosmological constant that Einstein used had the opposite effect—that is, if it were attractive—what would happen to a universe initially at rest, as Einstein had supposed to be true?
ANSWER: If the cosmological were attractive, then a universe initially at rest would collapse—the objects in it would fall together.

QUESTION: If the universe were twice as old as it is now, how would the Hubble constant compare to the value it has today?
ANSWER: Since the inverse of the Hubble constant is its age, if the universe were twice as old, the Hubble constant would be half as large as it is today.

QUESTION: If dark matter did not exist, would any gravitational lensing occur in the universe?
ANSWER: Even without dark matter, gravitational lensing would occur due to the gravitational effects of superclusters of galaxies, clusters of galaxies, individual galaxies, and other mass distributions.

CHAPTER 13

QUESTION: Synchronous rotation occurred in what other situation that we explored in this book?
ANSWER: Examples of synchronous rotation that we have considered are the orbits of most of the moons in the solar system around their respective planets, including our Moon orbiting Earth, and the synchronous rotation of Pluto with respect to its moon Charon.

QUESTION: What negative consequence could arise from an alien civilization discovering the Pioneer 10 or Pioneer 11 spacecraft?
ANSWER: If an alien civilization discovered Pioneer 10 or 11, they could trace them back to Earth. Visits from them might introduce diseases for which we have no defense, or they might lead to interplanetary war, if we can't get along with our new neighbors.

GLOSSARY

A ring The outermost ring of Saturn visible from Earth; it is located just beyond Cassini's division.

absolute magnitude The apparent magnitude that a star would have if it were 10 parsecs from Earth.

absorption line A dark line in a continuous spectrum created when photons of a certain energy are absorbed by atoms or molecules.

absorption line spectrum Dark lines superimposed on a continuous spectrum.

acceleration A change in the direction or magnitude of a velocity.

accretion The gradual accumulation of matter by an astronomical body, usually caused by gravity.

accretion disk An orbiting disk of matter spiraling in toward a star or black hole.

achromatic lens A compound lens designed to minimize the effect of chromatic aberration.

active galactic nuclei (AGN) Supermassive black holes in the cores of some galaxies that emit particles and radiation that, when viewed from different angles, create Seyfert galaxies, radio galaxies, double radio sources, BL Lacertae objects, and quasars.

active galaxy A very luminous galaxy, often containing an active galactic nucleus.

active optics A system that adjusts a reflecting telescope in response to changes in temperature and shape of the mount; it helps optimize an image.

adaptive optics Primary telescope mirrors that are continuously and automatically adjusted to compensate for the distortion of starlight due to the motion of Earth's atmosphere.

AGB star *See* **asymptotic giant branch (AGB) star.**

albedo The fraction of sunlight that a planet, asteroid, or satellite scatters directly back into space.

amino acids A class of chemical compounds that are the building blocks of proteins.

angle The opening between two straight lines that meet at a point.

angular diameter (angular size) The arc angle across an object.

angular momentum A measure of how much energy an object has stored in its rotation and/or revolution.

angular resolution The angular size of the smallest detail of an astronomical object that can be distinguished with a telescope.

annular eclipse An eclipse of the Sun in which the Moon is too distant to cover the Sun completely so that a ring of sunlight is seen around the Moon at mideclipse.

anorthosite A light-colored rock found throughout the lunar highlands and in some very old mountains on Earth.

aphelion The point in its orbit where a planet or other solar system body is farthest from the Sun.

Apollo asteroid An asteroid that is sometimes closer to the Sun than Earth is.

apparent magnitude A measure of the brightness of light from a star or other object as seen from Earth.

arc angle The measurement of the angle between two objects or two parts of the same object.

arcminute (arcmin) One-sixtieth of a degree of arc.

arcsecond (arcsec) One-sixtieth of a minute of arc.

asteroid (minor planet) Any of the rocky objects larger than a few hundred meters in diameter (and not classified as a planet or moon) that orbits the Sun.

asteroid belt A 1½-astronomical-unit-wide region between the orbits of Mars and Jupiter in which most of the asteroids are found.

astrobiology The study of life in the universe.

astronomical unit (AU) The average distance between Earth and the Sun: 1.5×10^8 km = 93 million mi.

astronomy The branch of science dealing with objects and phenomena that lie beyond Earth's atmosphere.

astrophysics That part of astronomy dealing with the physics of astronomical objects and phenomena.

asymptotic giant branch (AGB) star A red giant star that has completed core helium fusion and has reexpanded for a second time.

atom The smallest particle of an element that has the properties characterizing that element.

atomic number The number of protons in the nucleus of an atom.

aurora (*plural* aurorae) Light radiated by atoms and ions formed by the solar wind in Earth's upper atmosphere; seen most commonly in the polar regions.

autumnal equinox The intersection of the ecliptic and the celestial equator where the Sun crosses the equator moving from north to south. The beginning of autumn (around September 23).

average density The mass of an object divided by its volume.

B ring The brightest of the three rings of Saturn visible from Earth; it lies just inside the Cassini division.

barred spiral galaxy A spiral galaxy in which the spiral arms begin from the ends of a bar running through the central bulge.

belt asteroid An asteroid whose orbit lies in the asteroid belt.

belts (of Jupiter) Dark, reddish bands in Jupiter's cloud cover.

Big Bang An explosion that took place roughly 15 billion years ago, creating all space, time, matter, and energy in which the universe emerged.

binary star Two stars revolving about each other; a double star.

birth line   A line on the Hertzsprung-Russell diagram corresponding to where stars with different masses transform from protostars to pre–main-sequence stars.

BL Lacertae (BL Lac) object   A type of active galaxy; a blazar.

black hole   An object whose gravity is so strong that the escape velocity from it exceeds the speed of light.

blackbody   A hypothetical perfect radiator that absorbs and reemits all radiation falling upon it.

blackbody curve   The curve obtained when the intensity of radiation from a blackbody at a particular temperature is plotted against wavelength.

blackbody radiation   Electromagnetic radiation emitted by a blackbody.

blazar   A BL Lacertae object.

blueshift   A shift of all spectral features toward shorter wavelengths; the Doppler shift of light from an approaching source.

Bohr atom   A model of the atom, described by Niels Bohr, in which electrons revolve about the nucleus in various circular orbits.

Bok globule   A small, roundish, dark nebula in which stars are forming.

brown dwarf   Any of the planetlike bodies with less than 0.08 $M_\odot$ and more than about 13 $M_{Jupiter}$; such bodies do not have enough mass to sustain fusion in their cores.

C ring   The faint, inner portion of Saturn's main ring system.

caldera   The crater at the summit of a volcano.

capture theory   The idea that the Moon was created at a different location in the solar system and subsequently captured by Earth's gravity.

carbon fusion   The thermonuclear fusion of carbon nuclei to produce oxygen and neon.

carbonaceous chondrites   A class of extremely ancient, carbon-rich meteorites.

Cassegrain focus   An optical arrangement in a reflecting telescope in which light rays are reflected by a secondary mirror through a hole in the primary mirror.

Cassini division   A prominent gap between Saturn's A and B rings discovered in 1675 by J. D. Cassini.

celestial equator   A great circle on the celestial sphere 90° from the celestial poles.

celestial poles   Points about which the celestial sphere appears to rotate.

celestial sphere   A hypothetical sphere of very large radius centered on the observer; the apparent sphere of the night sky.

Celsius temperature scale   *See* **temperature (Celsius)**.

center of mass   The point around which a rigid system is perfectly balanced in a gravitational field; also, the point in space around which mutually orbiting bodies have elliptical orbits.

central bulge   A distribution of stars in the shape of a flattened sphere that surrounds the nucleus of a spiral galaxy like the Milky Way.

Cepheid variable star   One of two types of yellow, supergiant, pulsating stars.

Cerenkov radiation   Radiation produced by particles traveling through a substance faster than light can.

Ceres   The largest known asteroid and the first to be discovered.

Chandrasekhar limit   The maximum mass of a white dwarf, about 1.4 $M_\odot$.

charge-coupled device (CCD)   A type of solid-state silicon wafer designed to detect photons.

chromatic aberration   An optical property whereby different colors of light passing through a lens are focused at different distances from it.

chromosphere   The layer in the solar atmosphere between the photosphere and the corona.

circumpolar stars   All the stars that never set at a given latitude; all the stars between Polaris and the northern horizon.

close binary   A binary star whose members are separated by a few stellar diameters.

closed universe   A universe that contains enough matter to cause it to recollapse. It is finite in extent and has no "outside."

cluster (of galaxies)   A collection of a few hundred to a few thousand galaxies bound by gravity.

co-creation theory   The theory that the Moon formed simultaneously with Earth and in orbit around it.

collision-ejection theory   The theory that the Moon was created by the impact of a planet-sized object with Earth; presently considered the most plausible theory of the Moon's formation.

color–magnitude diagram   A plot of the surface temperatures (colors) versus the absolute magnitudes of stars.

coma (of a comet)   The nearly spherical, diffuse gas surrounding the nucleus of a comet near the Sun.

comet   A small body of ice and dust in orbit about the Sun. While passing near the Sun, a comet's vaporized ices give rise to a coma, tails, and a hydrogen envelope.

conduction (thermal)   The transfer of heat by passing energy directly from atom to atom.

configuration (of a planet)   A particular geometric arrangement of Earth, a planet, and the Sun.

confinement   The moment shortly after the Big Bang when quarks bound together to form particles like protons and neutrons.

conic section   The curve of intersection between a circular cone and a plane. This curve can be a circle, ellipse, parabola, or hyperbola.

conjunction   The alignment of two bodies in the solar system so that they appear in the same part of the sky as seen from Earth.

conservation of angular momentum   The law of physics stating that the total amount of angular momentum in an isolated system remains constant.

conservation of linear momentum   If the sum of the external forces on a system remains zero, the total linear momentum of the system remains constant.

constellation   Any of the 88 contiguous regions that cover the entire celestial sphere, including all the objects in each region; also, a configuration of stars often named after an object, a person, or an animal.

contact binary   A close binary system in which both stars fill or overflow their Roche lobes.

continental drift   The gradual movement of the continents over the surface of Earth due to plate tectonics.

continuous spectrum (continuum)   A spectrum of light over a range of wavelengths without any spectral lines.

convection   The transfer of energy by moving currents of fluid or gas containing that energy.

convective zone   A layer in a star where energy is transported outward by means of convection; also known as the *convective envelope* or *convection zone*.

core   The central portion of any astronomical object.

core-accretion model   The traditional theory of giant planet formation that begins with terrestrial planets slowly pulling abundant amounts of water, hydrogen, and helium onto themselves.

core helium fusion   The fusion of helium to form carbon and oxygen at the center of a star.

core hydrogen fusion   The fusion of hydrogen to form helium at the center of a star.

corona   The Sun's outer atmosphere.

coronagraph   A specially designed telescope with a baffle that blocks out the solar disk so that the corona can be photographed.

coronal hole   A dark region of the Sun's inner corona as seen at X-ray wavelengths.

coronal mass ejection   Large volumes of high energy gas released from the Sun's corona.

cosmic censorship   The belief that the only connection between a black hole and the universe is the black hole's event horizon.

cosmic light horizon   A sphere, centered on Earth, whose radius equals the distance traveled by light since the Big Bang.

cosmic microwave background   Photons from every part of the sky with a blackbody spectrum at 2.73 K; the cooled-off radiation from the primordial fireball that originally filled all space.

cosmic ray (primary cosmic ray)   High-speed particles traveling through space.

cosmic ray shower (secondary cosmic rays)   Groups of particles from Earth's atmosphere propelled Earthward by the impact of a cosmic ray.

cosmological constant   A number sometimes inserted in the equations of general relativity that represents a pressure that opposes gravity throughout the universe.

cosmological redshift   An increase in wavelength from distant galaxies and quasars caused by the expansion of the universe.

cosmology   The study of the formation, organization, and evolution of the universe.

coudé focus   A reflecting telescope in which a series of mirrors direct light to a remote focus away from the moving parts of the telescope.

crater   A circular depression on a celestial body caused by the impact of a meteoroid, asteroid, or comet, or by a volcano.

crescent Moon   A lunar phase during which the Moon appears less than half full.

crust   The solid surface layer of some astronomical bodies, including the terrestrial planets, the moons, the asteroids, and some stellar remnants.

dark ages   The age of the universe between the time of decoupling and the first burst of star formation.

dark energy   A repulsive gravitational effect that is causing the universe to accelerate outward.

dark matter (missing mass)   The as-yet-undetected matter in the universe that is underluminous and probably quite different from ordinary matter.

dark nebula   A cloud of interstellar gas and dust that obscures the light of more distant stars.

declination (dec)   The coordinate on the celestial sphere exactly analogous to latitude on Earth; measured north and south of the celestial equator.

decoupling   The epoch in the early universe when electrons and ions first combined to create stable atoms; the time when electromagnetic radiation ceased to dominate over matter.

deferent   A fixed circle in the Earth-centered universe along which a smaller circle (an epicycle) moves carrying a planet, the Sun, or the Moon.

degree (°)   A unit of angular measure or a temperature measure.

dense core   Any of the regions of interstellar gas clouds that are slightly denser than normal and destined to collapse to form one or a few stars.

density   The ratio of the mass of an object to its volume.

density wave   A spiral-shaped compression of the gas and dust in a spiral galaxy.

density-wave theory   An explanation of spiral arms in galaxies elaborated by C. C. Lin and F. Shu. (*See* **spiral density wave**.)

detached binary   A binary system in which the surfaces of both stars are inside their Roche lobes.

differential rotation   The rotation of a nonrigid object in which parts at different latitudes or different radial distances move at different speeds.

differentiation   *See* **planetary differentiation**.

diffraction grating   An optical device consisting of closely spaced lines ruled on a piece of glass that is used like a prism to disperse light into a spectrum.

direct motion   The gradual, eastward apparent motion of a planet against the background stars as seen from Earth.

disk (of a galaxy)   A flattened assemblage of stars, gas, and dust in a spiral galaxy like the Milky Way.

distance modulus   The difference between the apparent and absolute magnitudes of an object.

diurnal motion   Cyclic motion with a 1-day period.

Doppler effect (or Doppler shift)   The change in wavelength of radiation due to relative motion between the source and the observer along the line of sight.

double-line spectroscopic binary   A spectroscopic binary whose spectrum exhibits spectral lines of both stars.

double radio source   An extragalactic radio source characterized by two large lobes of radio emission, often located on either side of an active galaxy.

Drake equation   A mathematical equation used to estimate the number of extraterrestrial civilizations that may exist in our Galaxy.

dust devil   Whirlwind found in dry or desert areas on both Earth and Mars.

dust tail   A comet tail caused by dust particles escaping from the comet's nucleus.

dwarf elliptical galaxy   A small elliptical galaxy with far fewer stars than a typical galaxy.

dwarf planet   A celestial body that is in orbit around the Sun and has sufficient mass for its self-gravity to pull the body into a nearly spherical shape, but does not have enough gravity to clear its orbital neighborhood of all the small debris orbiting there.

dwarf star   Any star smaller than a giant, such as a main-sequence star or a white dwarf.

dynamo theory   The generation of a magnetic field by circulating electric charges.

eccentricity   *See* **orbital eccentricity**.

eclipse   The blocking of part or all of the light from the Moon by Earth (lunar eclipse) or from the Sun by the Moon (solar eclipse).

eclipse path   The track of the tip of the Moon's shadow along Earth's surface during a total or annular solar eclipse.

eclipsing binary   A double star system in which stars periodically pass in front of each other as seen from Earth.

ecliptic   The annual path of the Sun on the celestial sphere; the plane of Earth's orbit around the Sun.

Einstein cross   The appearance of four images of the same galaxy or quasar due to gravitational lensing by an intervening galaxy.

Einstein ring   The circular or arc-shaped image of a distant galaxy or quasar created by gravitational lensing by an intervening galaxy.

ejecta blanket   The ring of material surrounding a crater that was ejected during the crater-forming impact.

electromagnetic force   The interaction between charged particles, the second of four fundamental forces in nature.

electromagnetic radiation   Radiation consisting of oscillating electric and magnetic fields, namely gamma rays, X rays, visible light, ultraviolet and infrared radiation, and radio waves.

electromagnetic spectrum   The entire array of electromagnetic radiation.

electron   A negatively charged subatomic particle usually found in orbit about the nucleus of an atom.

electron degeneracy pressure   A powerful pressure produced by repulsion of closely packed (degenerate) electrons.

element   A substance that cannot be decomposed by chemical means into simpler substances. Every atom of the same element contains the same number of protons.

ellipse   A closed curve obtained by cutting completely through a circular cone with a plane; the shape of planetary orbits.

elliptical galaxy   A galaxy with an elliptical shape, little interstellar matter, and no spiral arms.

elongation   The angle between a planet and the Sun as seen from Earth.

emission line   A bright line of electromagnetic radiation.

emission line spectrum   A spectrum that contains only bright emission lines.

emission nebula   A glowing gaseous nebula whose light comes from fluorescence caused by a nearby star.

Encke division   A thin gap in Saturn's A ring, possibly first seen by J. F. Encke in 1838.

energy   The ability to do work.

energy flux   The amount of energy emitted from each square meter of an object's surface per second.

energy level (in an atom)   A particular amount of energy possessed by an electron in orbit around a nucleus.

epicycle   In the Earth-centered universe, a moving circle about which planets revolve.

equations of stellar structure   A set of relationships that describe the interactions of matter, energy, and gravity inside a star.

equinox   Either of the 2 days of the year when the Sun crosses the celestial equator and is therefore directly over Earth's equator; *see also* **autumnal equinox** *and* **vernal equinox**.

era of recombination   The time, roughly 500,000 years after the Big Bang, when the universe became transparent.

ergoregion   The region of space immediately outside the event horizon of a rotating black hole where it is impossible to remain at rest.

event horizon   The location around a black hole where the escape velocity equals the speed of light; the boundary of a black hole.

evolutionary track   On the Hertzsprung-Russell diagram, the path followed by a point representing an evolving star.

excited state   The orbit of an electron with energy greater than the lowest energy orbit (or state) available to that election.

expanding universe   The motion of the superclusters of galaxies away from each other.

eyepiece lens   A magnifying lens used to view the image produced at the focus of a telescope.

F ring   A thin ring just beyond the outer edge of Saturn's main ring system.

Fahrenheit scale   *See* **temperature (Fahrenheit)**.

filament   A dark curve seen above the Sun's photosphere that is the top view of a solar prominence.

first quarter Moon   A phase of the waxing Moon when Earth-based observers see half of the Moon's illuminated hemisphere.

fission theory   The theory that the Moon formed from matter flung off Earth because the planet was rotating extremely fast.

flare   *See* **solar flare**.

flocculent spiral galaxy   A spiral galaxy whose spiral arms are broad, fuzzy, and poorly demarcated.

focal length   The distance from a lens or concave mirror to where converging light rays meet.

focal plane   The plane at the focal length of a lens or concave mirror on which an extended object is focused.

focal point   *See* **focus**.

focus (of a lens or concave mirror)   The place at the focal length where light rays from a point object (that is, one that is too distant or tiny to resolve) are converged by a lens or concave mirror.

focus (*plural* **foci**) (of an ellipse)   The two points inside an ellipse, the sum of whose distances from any point on the ellipse is constant.

force   That which can change the momentum of an object.

frequency   The number of peaks or troughs of a wave that pass a fixed point each second. Also, the number of complete vibrations or oscillations per second.

full Moon   A phase of the Moon during which its full daylight hemisphere can be seen from Earth.

galactic cannibalism   A collision between two galaxies of unequal mass and size in which the smaller galaxy is absorbed by the larger galaxy.

galactic merger   A collision and subsequent merger of two roughly equal-sized galaxies.

galactic nucleus   The center of a galaxy; the center of the Milky Way Galaxy.

galaxy   A large assemblage of stars, gas, and dust bound together by their mutual gravitational attraction.

Galilean satellite (Galilean moon)   Any one of the four large moons of Jupiter (Callisto, Europa, Ganymede, Io) that is visible from Earth through a small telescope.

gamma ray   The most energetic form of electromagnetic radiation.

gamma-ray burst   A short burst of gamma rays; the sources of the bursts are outside our Galaxy.

gas (ion) tail   The relatively straight tail of a comet produced by the solar wind acting on ions in a comet's coma.

geocentric cosmology   The belief that Earth is at the center of the universe.

giant elliptical galaxy   A very large, extremely massive elliptical galaxy, usually located near the center of a rich cluster of galaxies.

giant molecular cloud   A large interstellar cloud of cool gas and dust in a galaxy.

giant star   A star whose diameter is roughly 10 to 100 times that of the Sun.

gibbous Moon   A phase of the Moon in which more than half, but not all, of the Moon's daylight hemisphere is visible from Earth.

glitch   A sudden speedup in the period of a pulsar.

globular cluster   A large spherical cluster of gravitationally bound stars usually found in the outlying regions of a galaxy.

grand-design spiral galaxy   A spiral galaxy whose spiral arms are thin, graceful, and well defined.

grand unified theory (GUT)   A theory that describes and explains the four physical forces.

granulation   The rice-grain–like structure of the solar photosphere due to convection of solar gases.

granules   Lightly colored convection features about 1000 km in diameter seen constantly in the solar photosphere.

gravitation (gravity)   The tendency of all matter to attract all other matter.

gravitational instability model   The theory of giant planet formation in which gases collapse together quickly, without the need for a seed terrestrial planet to form first.

gravitational lensing   The distortion of the appearance of an object by a source of gravity between it and the observer.

gravitational redshift   The redshift of photons leaving the gravitational field of any massive object, such as a star or black hole.

gravitational waves (gravitational radiation)   Ripples in the overall geometry of space produced by nonspherical moving objects.

gravity   *See* **gravitation**.

Great Dark Spot   A large, dark, oval-shaped storm that used to be in Neptune's southern hemisphere.

Great Red Spot   A large, red-orange, oval-shaped storm in Jupiter's southern hemisphere.

Great Wall   A huge arc of galaxies between two voids in the cosmos.

greatest elongation   The largest possible angle between the Sun and an inferior planet.

greenhouse effect   The trapping of infrared radiation near a planet's surface by the planet's atmosphere.

ground state   The lowest energy level of an atom.

H II region   A region of ionized hydrogen in interstellar space.

habitable zone   The region around any star wherein water can exist in liquid form and, hence, life as we know it can conceivably exist.

halo (of a galaxy)   A spherical distribution of globular clusters, isolated stars, and possibly dark matter that surrounds a galaxy.

Hawking process   The formation of real particles from virtual ones just outside a black hole's event horizon; the means by which black holes evaporate.

head–tail source   A radio galaxy whose radio emission is deflected from the galaxy.

heliocentric cosmology   A theory of the formation and evolution of the solar system with the Sun at the center.

helioseismology   The study of vibrations of the solar surface.

helium flash   The explosive ignition of helium fusion in the core of a low-mass, giant star.

helium fusion   The thermonuclear fusion of helium to produce carbon and oxygen.

helium shell flash   The explosive ignition of helium fusion in a thin shell surrounding the core of a low-mass star.

helium shell fusion   Helium fusion that occurs in a thin shell surrounding the core of a star.

Hertzsprung-Russell (H-R) diagram   A plot of the absolute magnitude or luminosity of stars versus their surface temperatures or spectral classes.

highlands   Heavily cratered, mountainous regions of the lunar surface.

homogeneity   The property of the universe being smooth or uniform as measured over suitably large distance intervals.

horizon problem (isotropy problem)   The difficulty in explaining why seemingly disconnected regions of the universe have the same temperature.

horizontal-branch stars   A group of post–helium-flash stars near the main sequence on the Hertzsprung-Russell diagram of a typical globular cluster.

hot-spot volcanism   The creation of volcanoes on a planet's surface caused by a reservoir of hot magma in the planet's mantle under a thin part of the crust.

Hubble classification   A system of classifying galaxies according to their appearance into one of four broad categories: spirals, barred spirals, ellipticals, and irregulars.

Hubble constant (H_0)   The constant of proportionality in the relation between the recessional velocities of remote galaxies and their distances; the correct value will determine the age of the universe.

Hubble flow   The recession of the galaxies caused by the expansion of the universe.

Hubble law   The relationship that states that the redshifts of remote galaxies are directly proportional to their distances from Earth.

hydrocarbon   A molecule based on hydrogen and carbon.

hydrogen envelope   An extremely large, tenuous sphere of hydrogen gas surrounding the head of a comet.

hydrogen fusion (hydrogen burning)   The thermonuclear fusion of hydrogen to produce helium.

hydrogen shell fusion   Hydrogen fusion that occurs in a thin shell surrounding the core of a star.

hydrostatic equilibrium   A balance between the weight of a layer in a star and the pressure that supports it.

hyperbola   An open curve obtained by cutting a cone with a plane.

impact breccia   A rock consisting of various fragments cemented together by the impact of a meteoroid.

impact crater   A crater on the surface of a planet or moon produced by the impact of an asteroid, meteoroid, or comet.

inferior conjunction   The configuration when Mercury or Venus is directly between the Sun and Earth.

inflation   A sudden expansion of space.

inflationary epoch   A brief period shortly after the Big Bang during which the scale of the universe increased very rapidly.

infrared radiation   Electromagnetic radiation of a wavelength longer than visible light but shorter than radio waves.

initial mass function   The numbers of stars on the main sequence at all different masses.

instability strip   A region on the Hertzsprung-Russell diagram occupied by pulsating stars.

interferometry   A method of increasing resolving power by combining electromagnetic radiation obtained by two or more telescopes.

intergalactic gas   Gas located between the galaxies within a cluster of galaxies.

intermediate-mass black hole   A black hole with a mass between a few hundred and few thousand solar masses.

interstellar dust   Microscopic solid grains of various compounds in interstellar space usually encased in ice.

interstellar extinction   The dimming of starlight as it passes through the interstellar medium.

interstellar gas   Sparse gas in interstellar space.

interstellar medium   Interstellar gas and dust.

interstellar reddening   The reddening of starlight passing through the interstellar medium resulting from the scattering of short-wavelength light more than long-wavelength light.

inverse-square law   The gravitational attraction between two objects and the apparent brightness of a light source are both inversely proportional to the square of its distance.

ion   An atom that has become electrically charged due to the loss or addition of one or more electrons.

ionization   The process by which an atom loses or gains electrons.

ionosphere (thermosphere)   Region of Earth's atmosphere, above the mesosphere, in which sunlight ionizes many atoms.

iron meteorite   A meteorite composed primarily of iron with an admixture of nickel; also called an *iron*.

irregular cluster (of galaxies)   An unevenly distributed group of galaxies bound together by their mutual gravitational attraction.

irregular galaxy   An asymmetrical galaxy having neither spiral arms nor an elliptical shape.

isotope   Atoms that all have the same number of protons (atomic number) but different numbers of neutrons. Their nuclear properties often differ greatly.

isotropy   The fact that the average number of galaxies at different distances from Earth is the same in all directions; also, the fact that the temperature of the cosmic microwave background is essentially the same in all directions.

isotropy problem (horizon problem)   The difficulty in explaining why seemingly disconnected regions of the universe have the same temperature.

Jeans instability   The condition under which gravitational forces overcome thermal forces to cause part of an interstellar cloud to collapse and form stars and planets.

Kelvin   *See* **temperature (Kelvin)**.

Kepler's laws   Three statements, formulated by Johannes Kepler, that describe the motions of the planets.

Kerr black hole Any rotating, uncharged black hole.

kiloparsec (kpc) One thousand parsecs; about 3260 light-years.

kinetic energy The energy an object has as a result of its motion.

Kirchhoff's laws Three statements formulated by Gustav Kirchhoff describing what physical conditions produce each type of spectra.

Kirkwood gaps Gaps in the spacing of asteroid orbits discovered by Daniel Kirkwood caused by gravitational attractions of planets.

Kuiper belt A doughnut-shaped ring of space around the Sun beyond Pluto that contains many frozen comet bodies, some of which are occasionally deflected toward the inner solar system.

Large Magellanic Cloud (LMC) An irregular galaxy, companion to the Milky Way about 179,000 ly away.

last quarter Moon A phase of the waning Moon when Earth-based observers see half of the Moon's illuminated hemisphere.

law of equal areas Kepler's second law.

law of inertia (Newton's first law of motion) The physical law that an object will stay at rest or move at a constant speed in a fixed direction unless acted upon by an outside force.

law of universal gravitation Newton's law of gravitation, which describes how the gravitational force between two bodies depends on their masses and separation.

lenticular galaxy A disk-shaped galaxy without spiral arms.

light Electromagnetic radiation, which travels in packets called photons.

light curve A graph that displays variations in the brightness of a star or other astronomical object over time.

light-gathering power A measure of how much light a telescope intercepts and brings to a focus.

light-year (ly) The distance that light travels in a vacuum in 1 year.

lighthouse model The explanation that a pulsar pulses by rotating and funneling energy outward via magnetic fields that are not aligned with the rotation axis.

limb (of the Sun) The apparent edge of the Sun as seen in the sky.

limb darkening The phenomenon whereby the Sun is darker near its limb than near the center of its disk.

line of nodes The line along which the plane of the Moon's orbit intersects the plane of the ecliptic.

liquid metallic hydrogen A metal-like form of hydrogen that is produced under extreme pressure.

Local Group A cluster of about 40 galaxies, of which our own Galaxy is a member.

long-period comet A comet that takes tens of thousands of years or more to orbit the Sun once.

luminosity The rate at which electromagnetic radiation is emitted from a star or other object.

luminosity class The classification of a star of a given spectral type according to its luminosity and density; the classes are supergiant, bright giant, giant, subgiant, and main sequence.

lunar Referring to the Moon.

lunar eclipse An eclipse during which Earth blocks light that would have struck the Moon.

lunar month *See* **synodic month**.

lunar phases The names given to the apparent shapes of the Moon as seen from Earth.

Lyman series A series of spectral lines of hydrogen produced by electron transitions to and from the lowest energy state of the hydrogen atom.

magnetar Especially hot, rapidly rotating neutron stars whose motion helps generate extra-strong magnetic fields.

magnetic dynamo A theory that explains phenomena of the solar cycle as a result of periodic winding and unwinding of the Sun's magnetic field in the solar atmosphere.

magnetic field A region of space near a magnetized body within which magnetic forces can be detected.

magnetosphere The region around a planet occupied by its magnetic field.

magnification (magnifying power) The number of times larger in angular diameter an object appears through a telescope than when it is seen by the naked eye.

magnitude A measure of brightness. *See* **absolute magnitude**; **apparent magnitude**.

magnitude scale The system of denoting magnitudes.

main sequence A grouping of stars on the Hertzsprung-Russell diagram extending diagonally across the graph from the hottest, brightest stars to the dimmest, coolest stars.

main-sequence star A star, fusing hydrogen to helium in its core, whose surface temperature and luminosity place it on the main sequence on the Hertzsprung-Russell diagram.

mantle (of a planet) That portion of a terrestrial planet located between its crust and core.

mare (*plural* **maria**) Latin for "sea," a large, relatively crater-free plain on the Moon.

mare basalt Dark, solidified lava that covers the lunar maria.

mascons Regions of high-density matter near the surface of the Moon.

mass A measure of the total amount of material in an object.

mass–luminosity relation The direct relationship between the masses and luminosities of main-sequence stars.

matter-dominated universe A universe in which the radiation field that fills all space is unable to prevent the existence of neutral atoms.

megaparsec (Mpc) One million parsecs.

mesosphere The layer in Earth's atmosphere above the stratosphere.

metal All elements except hydrogen and helium.

metal-poor star *See* **Population II star**.

metal-rich star *See* **Population I star**.

meteor The streak of light seen when any space debris vaporizes in Earth's atmosphere; a "shooting star."

meteor shower Frequent meteors that seem to originate from a common point in the sky.

meteorite A fragment of space debris that has survived passage through Earth's atmosphere.

meteoroid A small rock in interplanetary space.

microlensing The gravitational focusing of light from a distant star by a closer object to give a brighter image of the star.

Milky Way Galaxy The galaxy in which our solar system resides.

minor planet *See* **asteroid**.

missing mass *See* **dark matter**.

model A hypothesis that has withstood observational or experimental tests.

molecular clouds Nebulae that are often embedded in much larger bodies of gas and dust.

molecule Two or more atoms bonded together.

moment of inertia (about an axis) A measure of the inertial resistance of an object to changes in the object's rotational motion about the axis.

momentum A measure of the inertia of an object; an object's mass multiplied by its velocity.

moons (natural satellites) Bodies that orbit larger objects, which in turn orbit stars.

neap tide The least change from high to low tide during a day; it occurs during the first- and third-quarter phases of the Moon.

nebula (*plural* **nebulae**) A cloud of interstellar gas and dust.

neon fusion The thermonuclear fusion of neon nuclei.

neutrino A subatomic particle, with no electric charge and little mass, that is important in many nuclear reactions and in supernovae.

neutron A nuclear particle with no electric charge and with a mass nearly equal to that of the proton.

neutron degeneracy pressure A powerful pressure produced by degenerate neutrons.

neutron star A very compact, dense stellar remnant composed almost entirely of neutrons.

New General Catalogue (NGC) A catalog of star clusters, nebulae, and galaxies, first published in 1888.

new Moon The phase of the Moon when it is nearest the Sun in the sky.

Newtonian reflector An optical arrangement in a reflecting telescope in which a small, flat mirror reflects converging light rays to a focus on one side of the telescope tube.

Newton's laws of motion Newton's equations that describe the motion of matter as a result of forces acting on it.

nonthermal radiation *See* **synchrotron radiation**.

north celestial pole The location on the celestial sphere directly above Earth's northern rotation pole.

northern lights (aurora borealis) Light radiated by atoms and ions in Earth's upper atmosphere due to high-energy particles from the Sun and seen mostly in the northern polar regions.

northern vastness (northern lowlands) Relatively young, crater-free terrain in the northern hemisphere of Mars.

nova (*plural* **novae**) A star in a binary system that experiences a sudden outburst of radiant energy, temporarily increasing its luminosity by a factor of between 10^4 and 10^6.

nuclear Referring to the nucleus of an atom.

nuclear density The density of matter in the nucleus of an atom; about 10^{17} kilograms per cubic meter (kg/m^3).

nucleosynthesis The formation, by fusion, of higher-mass elements from lower-mass ones.

nucleus (of an atom) The massive part of an atom, composed of protons and neutrons; electrons surround a nucleus.

nucleus (of a comet) A collection of ices and dust that constitute the solid part of a comet.

nucleus (of a galaxy) *See* **galactic nucleus**.

OB association An unbound group of very young, massive stars predominantly of spectral types O and B.

OBAFGKM sequence The sequence of stellar spectral classifications from hottest to coolest stars.

objective lens The principal lens of a refracting telescope.

observable universe All space that is nearer to us than the distance traveled by light since the time of the Big Bang.

Occam's razor The principle of choosing the simplest scientific theory that correctly explains any phenomenon.

occultation The eclipsing of an astronomical object other than the Moon or Sun by another astronomical body.

Oort comet cloud A hollow spherical region of the solar system beyond the Kuiper belt where most comets are believed to spend most of their time.

open cluster A loosely bound group of young stars in the disk of the galaxy; a galactic cluster.

open universe A universe with a hyperbolic shape; lacks the mass necessary to someday stop expanding and recollapse. It will expand forever.

opposition The configuration of a planet when it is at an elongation of 180° and thus appears opposite the Sun in the sky.

optical double A pair of stars that appear to be near each other but are unbound and at very different distances from Earth.

optics The branch of physics dealing with the behavior and properties of light.

orbit The path of an object that is moving about a second object.

orbital eccentricity A measure between 0 and 1 indicating how close to circular a planet's orbit is (the eccentricity of a circular orbit is 0).

orbital inclination The tilt or angle of an object's orbital plane around the Sun compared to the ecliptic.

organic molecule A carbon-based compound.

overcontact binary   A close binary system in which the two stars share a common atmosphere.

oxygen fusion   The thermonuclear fusion of oxygen nuclei.

ozone layer   The lower stratosphere, where most of the ozone in the air exists.

pair production   The creation of a particle and an antiparticle from energetic photons.

parabola   An open curve formed by cutting a circular cone at an angle parallel to the sides of the cone.

parallax   The apparent displacement of an object relative to more distant objects caused by viewing it from different locations.

parsec (pc)   A unit of distance equal to 3.26 light-years.

partial eclipse   A lunar or solar eclipse in which the eclipsed object does not appear completely covered.

Pauli exclusion principle   A principle of quantum mechanics that says that two identical particles cannot simultaneously have the same position and momentum.

peculiar galaxy   Any Hubble class of galaxy that appears to be blowing apart.

penumbra   The portion of a shadow in which only part of the light source is covered by the shadow-making body.

penumbral eclipse   A lunar eclipse in which the Moon passes only through Earth's penumbra.

perihelion   The point in its orbit where a planet is nearest the Sun.

period   The interval of time between successive repetitions of a periodic phenomenon.

period–luminosity relation   A relationship between the period and average luminosity of a pulsating star.

periodic table   A listing of the chemical elements according to their properties; created by D. Mendeleev.

phase (of the Moon)   The appearance of the Moon at different points in its orbit of Earth.

photodisintegration   The breakup of nuclei in the core of a massive star due to the effects of energetic gamma rays.

photometry   The measurement of light intensities.

photon   A discrete unit of electromagnetic energy.

photon pressure   The force per unit area exerted by photons on stellar or interstellar gas.

photosphere   The region in the solar atmosphere from which most of the visible light escapes into space.

physics   Basic principles that govern the behavior of physical reality.

pixel   A contraction of the term "picture element"; usually refers to one square of a grid into which the light-sensitive component of a charge-coupled device is divided.

plage   A bright spot on the Sun believed to be associated with an emerging magnetic field.

Planck era   Time from the Big Bang until the Planck time (10^{-43} s).

Planck time   The brief interval of time, about 10^{-43} s, immediately after the Big Bang, when all four forces (gravity, electromagnetism, weak, strong) had the same strength.

Planck's law   A relationship between the energy carried by a photon and its wavelength.

planet   An object orbiting a star that is held together by its own gravitational force in a nearly spherical shape, that is able to clear its neighborhood of debris, and is not the moon (or satellite) of a larger orbiting body.

planetary differentiation   The process early in the life of each planet whereby denser elements sank inward and lighter ones rose.

planetary nebula   A luminous shell of gas ejected from an old, low-mass star.

planetesimal   Primordial asteroidlike object from which the planets accreted.

plasma   A hot, ionized gas.

plate tectonics   The motions of large segments (plates) of Earth's surface caused by convective motions in the underlying mantle.

polymer   A long molecule composed of many smaller molecules.

poor cluster (of galaxies)   A cluster of galaxies with only a few members.

Population I star   A star, such as the Sun, whose spectrum exhibits spectral lines of many elements heavier than helium; a metal-rich star.

Population II star   A star whose spectrum exhibits comparatively few spectral lines of elements heavier than helium; a metal-poor star.

positron   An electron with a positive rather than negative electric charge; an antielectron.

potential energy   The energy stored in an object as a result of its location in space.

powers of ten   A convenient method of writing large and small numbers that uses a number between 1 and 10 multiplied by a power of 10.

pre–main-sequence star   The stage of star formation just before the main sequence; it involves slow contraction of the young star.

precession (of Earth)   A slow, conical motion of Earth's axis of rotation caused by the gravitational pull of the Moon and Sun on Earth's equatorial bulge.

precession of the equinoxes   The slow westward motion of the equinoxes along the ecliptic because of Earth's precession.

primary mirror   The large, concave, light-gathering mirror in a reflecting telescope, analogous to the objective lens on a refracting telescope.

prime focus   The point in a reflecting telescope where the primary mirror focuses light.

primordial black hole   A relatively low-mass black hole hypothetically formed at the beginning of the universe.

primordial fireball   The extremely hot gas that filled the universe immediately following the Big Bang.

primordial nucleosynthesis   The transformation by fusion of protons and electrons into hydrogen isotopes, helium, and some lithium in the first few minutes of the existence of the universe.

prism A wedge-shaped piece of glass used to disperse white light into a spectrum.

prograde orbit An orbit of a moon or satellite around a planet that is in the same direction as the planet's rotation.

prominence Flamelike protrusion seen near the limb of the Sun and extending into the solar corona. The side view of a filament.

proper motion The change in the location of a star on the celestial sphere.

proton A heavy, positively charged nuclear particle.

protoplanet The embryonic stage of a planet when it is growing because of collisions with planetesimals.

protoplanetary disk (proplyd) A disk of material encircling a protostar or a newborn star.

protostar The earliest stage of a star's life before fusion commences and when gas is rapidly falling onto it.

protosun The Sun prior to the time when hydrogen fusion began in its core.

pulsar A pulsating source associated with a rapidly rotating neutron star with an off-axis magnetic field.

quantum mechanics The branch of physics dealing with the structure and behavior of atoms and their interactions with each other and with light.

quark A particle that is a building block of the heavy nuclear particles such as protons and neutrons.

quarter Moon A phase of the Moon when it is located 90° from the Sun in the sky; halfway between the new and full phases.

quasar (quasi-stellar radio source) A starlike object with a very large redshift.

quasi-stellar object (QSO) A quasar.

quintessence One of the explanations of the dark energy causing the universe to accelerate outward.

radial velocity That portion of an object's velocity parallel to the line of sight.

radial-velocity curve A plot showing the variation of radial velocity with time for a binary star or variable star.

radiation Electromagnetic energy; photons.

radiation-dominated universe The time at the beginning of the universe when the electromagnetic radiation prevented ions and electrons from combining to make neutral atoms.

radiation (photon) pressure The transfer of momentum carried by radiation to an object on which the radiation falls.

radiative zone A region inside a star where energy is transported outward by the movement of photons through a gas from a hot location to a cooler one.

radio astronomy The branch of astronomy dealing with observations at radio wavelengths.

radio galaxy A galaxy that emits an unusually large amount of radio waves.

radio lobes Vast regions of radio emission on opposite sides of a radio galaxy.

radio telescope A telescope designed to detect radio waves.

radio wave Long-wavelength electromagnetic radiation.

radioactive The property whereby certain atomic nuclei naturally decompose by spontaneously emitting particles.

red dwarf A low-mass main-sequence star.

red giant A large, cool star of high luminosity.

red supergiant An extremely large, cool star of high luminosity; a star in the upper right corner of the Hertzsprung-Russell diagram.

redshift The shifting to longer wavelengths of the light from remote galaxies and quasars; the Doppler shift of light from any receding source.

reflecting telescope (reflector) A telescope in which the principal light-gathering component is a concave mirror.

reflection The rebounding of light rays off a smooth surface.

reflection nebula A comparatively dense cloud of gas and dust in interstellar space that is illuminated by a star between it and Earth.

refracting telescope (refractor) A telescope in which the principal light-gathering component is a lens.

refraction The bending of light rays when they pass from one transparent medium to another.

regolith The powdery, lifeless material on the surface of a moon or planet.

regular cluster (of galaxies) An evenly distributed group of galaxies bound together by mutual gravitational attraction.

resolution The degree to which fine details in an optical image can be distinguished.

resolving power A measure of the ability of an optical system to distinguish fine details in the image it produces.

resonance The large response of an object to a small periodic gravitational tug from another object.

retrograde motion The occasional backward (that is, westward) apparent motion of a planet against the background stars as seen from Earth. Retrograde motion is an optical illusion.

retrograde orbit The orbit of a moon or satellite around a planet that is in the direction opposite to the planet's rotation.

retrograde rotation The rotation of a planet opposite to its direction of revolution around the Sun. Only Uranus and Venus have retrograde rotation.

revolution The orbit of one body about another.

rich cluster (of galaxies) A cluster of galaxies with many members.

right ascension (r.a.) The celestial coordinate analogous to longitude on Earth and measured around the celestial equator from the vernal equinox.

rille A winding crack or depression in the lunar surface caused by the collapse of a solidified lava tube.

ringlet Any one of numerous, closely spaced, thin bands of particles in planetary ring systems.

Roche limit The shortest distance from a planet or other object at which a second object can be held together by its own gravitational forces.

Roche lobe   The teardrop-shaped regions around each star in a binary star system inside of which gas is gravitationally bound to that star.

rotation   The spinning of a body about an axis passing through it.

rotation curve (of a galaxy)   A graph showing how the orbital speed of material in a galaxy depends on the distance from the galaxy's center.

RR Lyrae variable star   A type of pulsating star with a period of less than 1 day.

Sa, Sb, Sc   Categories of spiral galaxies determined by the sizes of their central bulges or how tightly wound their spiral arms are; an Sa galaxy is the most tightly wound. Spiral arms are directly attached to the central bulge.

Sagittarius A   The strong radio source associated with the nucleus of the Milky Way Galaxy.

satellite   A body that revolves about a larger one.

SBa, SBb, SBc   Categories of barred spiral galaxies determined by how tightly wound their spiral arms are; SBa galaxies are the most tightly wound.

scarp   A cliff on Mercury believed to have formed when the planet cooled and shrank.

Schmidt corrector plate   A specially shaped lens used with spherical mirrors that corrects for spherical aberration and provides an especially wide field of view.

Schwarzschild black hole   Any nonrotating, uncharged black hole.

Schwarzschild radius   The distance from the center to the event horizon in any black hole.

scientific method   The method of doing science based on observation, experimentation, and the formulation of hypotheses (theories) that can be tested.

scientific notation   The style of writing large and small numbers using powers of ten.

scientific theory (hypothesis)   An idea about the natural world that is subject to verification and refinement.

seafloor spreading   The process whereby magma upwelling along rifts in the ocean floor causes adjacent segments of Earth's crust to separate.

secondary cosmic rays   Particles from Earth's atmosphere given high speeds Earthward by cosmic rays from space.

secondary mirror   A relatively small mirror used in reflecting telescopes to guide the light out the side or bottom of the telescope.

seeing disk   The size that a star appears to have on a photographic or charge-coupled-device image as a result of the changing refraction of the starlight passing through Earth's atmosphere.

seismic waves   Vibrations traveling through or around an astronomical body usually associated with earthquakelike phenomena.

seismograph   A device used to record and measure seismic waves, such as those produced by earthquakes.

seismology   The study of earthquakes and related phenomena.

self-propagating star formation   The process whereby the deaths of stars in one part of a galaxy stimulate star formation in a neighboring region of that galaxy.

semidetached binary   A close binary system in which one star fills or is overflowing its Roche lobe.

semimajor axis (of an ellipse)   Half of the longest dimension of an ellipse.

SETI   The search for extraterrestrial intelligence.

Seyfert galaxy   A spiral galaxy with a bright nucleus whose spectrum exhibits emission lines.

Shapley–Curtis debate   An inconclusive debate between Harlow Shapley and Heber Curtis in 1920 about whether certain nebulae were beyond the Milky Way.

shepherd satellite (moon)   A small satellite whose gravitational tug is responsible for maintaining a sharply defined ring of matter around a planet such as Saturn or Uranus.

shock wave   An abrupt, localized region of compressed gas caused by an object traveling through the gas at a speed greater than the speed of sound.

shooting star   *See* **meteor.**

short-period comet   A comet that orbits the Sun in the vicinity of the planets, thereby reappearing with tails every 200 years or less.

sidereal month   The period of the Moon's revolution about Earth, measured with respect to the Moon's location among the stars; $27\frac{1}{3}$ Earth days.

sidereal period   The orbital period of one object about another, measured with respect to the stars.

single-line spectroscopic binary   A spectroscopic binary whose periodically varying spectrum exhibits the spectral lines of only one of its two stars.

singularity   A place of infinite curvature of spacetime in a black hole.

Small Magellanic Cloud (SMC)   An irregular galaxy that is a companion to the Milky Way.

small solar-system bodies (SSSBs)   All objects orbiting the Sun that are not planets, dwarf planets, or moons.

solar corona   The Sun's outer atmosphere.

solar cycle   A 22-year cycle during which the Sun's magnetic field reverses its polarity twice.

solar day   From noontime to the next noontime; for Earth it is 24 h.

solar eclipse   An eclipse during which the Moon blocks the Sun.

solar flare   A violent eruption on the Sun's surface.

solar luminosity ($L_\odot$)   The total energy emitted by the Sun each second.

solar model   A set of equations that describe the internal structure and energy generation of the Sun.

solar nebula   The cloud of gas and dust from which the Sun and the rest of the solar system formed.

solar seismology   The study of the Sun's interior from observations of vibrations of its surface.

solar system   The Sun, planets, their satellites, asteroids, comets, and related objects that orbit the Sun.

solar wind   An outward flow of particles (mostly electrons and protons) from the Sun.

solstice   Either of two points along the ecliptic at which the Sun reaches its maximum distance north or south of the celestial equator.

south celestial pole   The location on the celestial sphere directly above Earth's south rotation pole.

southern highlands   Older, cratered terrain in the Martian southern hemisphere.

southern lights (aurora australis)   Light radiated by atoms and ions in Earth's upper atmosphere due to high-energy particles from the Sun; seen mostly in the southern polar regions.

Southern Wall   A huge sheet of galaxies between two voids in the cosmos.

spacetime   The concept from special relativity that space and time are both essential in describing the position, motion, and action of any object or event.

spectral analysis   The identification of chemicals by the appearance of their spectra.

spectral lines   Dark or bright lines at specific wavelengths in a spectrum.

spectral type   A classification of stars according to the appearance of their spectra.

spectrogram   The photograph of a spectrum.

spectrograph   A device for photographing a spectrum.

spectroscope   A device for directly viewing a spectrum.

spectroscopic binary   A double star whose binary nature can be deduced from the periodic Doppler shifting of lines in its spectrum.

spectroscopic parallax   A method of determining a star's distance from Earth by measuring its surface temperature, luminosity, and apparent magnitude.

spectroscopy   The study of spectra.

spectrum (*plural* **spectra**)   The result of electromagnetic radiation passing through a prism or grating so that different wavelengths are separated.

speed   The rate at which an object moves.

spherical aberration   An optical property whereby different portions of a spherical lens or spherical, concave mirror have slightly different focal lengths, thereby producing a fuzzy image.

spicule   A narrow jet of rising gas in the solar chromosphere.

spin (of an electron or proton)   A small, well-defined amount of angular momentum possessed by electrons, protons, and other particles.

spiral arms   Lanes of interstellar gas, dust, and young stars that wind outward in a plane from the central regions of some galaxies.

spiral density wave   A spiral-shaped pressure wave that orbits the disk of a spiral galaxy and induces new star formation.

spiral galaxy   A flattened, rotating galaxy with pinwheel-like spiral arms winding outward from the galaxy's central bulge.

spoke   A moving dark region of Saturn's rings.

spring tide   The greatest daily difference between high tide and low tide, occurring when the Moon is new or full.

stable Lagrange points   Locations throughout the solar system where the gravitational forces from the Sun and a planet keep space debris trapped.

standard candle   An object whose known luminosity can be used to deduce the distance to a galaxy.

star   A self-luminous sphere of gas.

starburst galaxy   A galaxy where there is an exceptionally high rate of star formation.

Stefan-Boltzmann law   The relationship stating that an object emits energy at a rate proportional to the fourth power of its temperature, in Kelvins.

stellar evolution   The changes in size, luminosity, temperature, and chemical composition that occur as a star ages.

stellar model   The result of theoretical calculations that give details of physical conditions inside a star.

stellar parallax   The apparent shift in a nearby star's position on the celestial sphere resulting from Earth's orbit around the Sun.

stellar spectroscopy   The study of the properties of stars encoded in their spectra.

stony-iron meteorite   A meteorite composed of roughly equal amounts of rock and iron.

stony meteorite (chondrites)   A meteorite composed of rock with very little iron; also called a *stone*.

stratosphere   The second layer in Earth's atmosphere, directly above the troposphere.

strong nuclear force   The force that binds protons and neutrons together in nuclei.

subgiant   A star whose luminosity is between that of main-sequence stars and normal giants of the same spectral type.

summer solstice   The point on the ecliptic where the Sun is farthest north of the celestial equator; the day with the largest number of daylight hours in the northern hemisphere, around June 21.

Sun   The star about which Earth and the other planets revolve.

sunspot   A temporary cool region in the solar photosphere created by protruding magnetic fields.

sunspot cycle   The semiregular 11-year period with which the number and location of sunspots fluctuate.

sunspot maximum   The time during the solar cycle when the number of sunspots is greatest.

sunspot minimum   The time during the solar cycle when the number of sunspots is least.

supercluster (of galaxies)   A gravitationally bound collection of many clusters of galaxies.

supergiant   A star of very high luminosity.

supergranule   A large convective cell in the Sun's chromosphere containing many granules.

superior conjunction   The configuration when Mercury or Venus is on the opposite side of the Sun from Earth.

supermassive black hole A black hole whose mass exceeds 1000 solar masses.

supernova (*plural* **supernovae**) A stellar explosion during which a star suddenly increases its brightness roughly a millionfold.

supernova remnant A nebula left over after a supernova detonates.

superstring A set of theories that hope to describe the nature of spacetime and matter at a more fundamental level than is presently possible.

synchronous rotation The condition when a moon's rotation rate and revolution rate are equal or when a planet's rotation rate equals its moon's revolution rate.

synchrotron radiation The radiation emitted by charged particles moving through a magnetic field; nonthermal radiation.

synodic month (**lunar month**) The period of revolution of the Moon with respect to the Sun; the length of one cycle of lunar phases; 29½ Earth days.

synodic period The interval between successive occurrences of the same configuration of a planet as seen from Earth.

T Tauri stars Young, variable pre–main-sequence stars associated with interstellar matter that show erratic changes in luminosity.

tail (of a comet) Gas and dust particles from a comet's nucleus that have been swept away from the comet's nucleus by the radiation pressure of sunlight and impact of the solar wind.

telescope An instrument for viewing remote objects.

temperature (**Celsius**) Temperature measured on a scale where water freezes at 0° and boils at 100°.

temperature (**Fahrenheit**) Temperature measured on a scale where water freezes at 32° and boils at 212°.

temperature (**Kelvin**) Absolute temperature measured in Celsius degree intervals. Water freezes at 273 K and boils at 373 K.

terminator The line dividing day and night on the surface of any body orbiting the Sun; the line of sunset or sunrise.

terrestrial planet Any of the planets Mercury, Venus, Earth, or Mars; a planet with a composition and density similar to that of Earth.

theories of everything Theories under development that comprehensively explain all four fundamental forces in nature.

theory *See* **scientific theory.**

theory of general relativity A description of spacetime formulated by Einstein explaining how gravity affects the geometry of space and the flow of time.

theory of special relativity A description of mechanics and electromagnetic theory formulated by Einstein according to which measurements of distance, time, and mass are affected by the observer's motion.

thermal energy The energy associated with the motions of atoms or molecules in a substance.

thermal equilibrium A balance between the input and outflow of heat in a system.

thermonuclear fusion A reaction in which the nuclei of atoms are fused together at a high temperature.

3-to-2 spin-orbit coupling The rotation of Mercury, which makes 3 complete rotations on its axis for every 2 complete orbits around the Sun.

tidal force A gravitational force whose strength and/or direction varies over a body and thus tends to deform the body.

time zone One of 24 divisions of Earth's surface separated by 15° along lines of constant longitude (with allowances for some political boundaries).

total eclipse A solar eclipse during which the Sun is completely hidden by the Moon, or a lunar eclipse during which the Moon is completely immersed in Earth's umbra.

trailing-arm spiral A spiral arm pointing away from the direction of rotation, characteristic of all spiral galaxies.

transition (electronic) The change in energy and orbit of an electron around an atom or molecule.

transition zone Region between the Sun's chromosphere and corona where the temperature skyrockets to about 1 million K.

transverse velocity The portion of an object's velocity perpendicular to our line of sight to it.

Trojan asteroid One of several asteroids at stable Lagrange points that share Jupiter's orbit about the Sun.

troposphere The lowest level of Earth's atmosphere.

Tully–Fisher relation A correlation between the width of the 21-centimeter line of a spiral galaxy and its absolute magnitude.

turnoff point The location of the brightest main-sequence stars on the Hertzsprung-Russell diagram of a globular cluster.

21-cm radiation Radio emission from a hydrogen atom caused by the flip of the electron's spin orientation.

twinkling The apparent change in a star's brightness, position, or color due to the motion of gases in Earth's atmosphere.

Type I Cepheid A Population I Cepheid variable star found in the disks of spiral galaxies.

Type Ia supernova A supernova occurring after a white dwarf accretes enough mass from a companion star to exceed the Chandrasekhar limit.

Type II Cepheid A Population II Cepheid variable star, found in elliptical galaxies and in the halos of disk galaxies, that is 1.5 magnitudes dimmer than a Type I Cepheid.

Type II supernova A supernova occurring after a massive star's core is converted to iron.

ultraviolet (**UV**) **radiation** Electromagnetic radiation of wavelengths shorter than those of visible light but longer than those of X rays.

umbra The central, completely dark portion of a shadow.

universal constant of gravitation The constant of proportionality in Newton's law of gravitation, usually denoted G.

universe All space along with all the matter and radiation in space.

Van Allen radiation belts Two flattened, doughnut-shaped regions around Earth where many charged particles (mostly protons and electrons) are trapped by Earth's magnetic field.

variable star   A star whose luminosity varies.

velocity   A quantity that specifies both the direction and speed of an object.

vernal equinox   The point on the ecliptic where the Sun crosses the celestial equator from south to north; the beginning of spring, around March 21.

very-long-baseline interferometry (VLBI)   A method of connecting widely separated radio telescopes to make observations of very high resolution.

virtual particle   A particle and its antiparticle, created simultaneously in pairs and which quickly disappear without a trace.

visual binary star   A double star in which the two components can be resolved through a telescope.

void   A huge, roughly spherical region of the universe where exceptionally few galaxies are found.

water hole   The part of the electromagnetic spectrum at a few thousand megahertz where there is very little background noise from space.

waning   An adjective that means "decreasing," as in the "waning crescent Moon" or the "waning gibbous Moon."

wavelength   The distance between two successive peaks in a wave.

waxing   An adjective that means "increasing," as in the "waxing crescent Moon" or the "waxing gibbous Moon."

weak nuclear force   A nuclear interaction involved in certain kinds of radioactive decay.

weight   The force with which a body presses down on the surface of a world such as Earth.

white dwarf   A low-mass stellar remnant that has exhausted all its thermonuclear fuel and contracted to a size roughly equal to the size of Earth.

Widmanstätten patterns   Crystalline structure seen inside iron meteorites.

Wien's law   The relationship that the dominant wavelength of radiation emitted by a blackbody varies inversely with its temperature.

winter solstice   The point on the ecliptic where the Sun is farthest south of the celestial equator; fewest hours of daylight in the northern hemisphere, around December 22.

Wolf-Rayet stars   Rotating stars of at least 20 $M_\odot$ with strong magnetic fields and stellar winds.

work   Change in an object's energy as a result of a force being applied to it.

wormhole   A hypothetical connecting passage between black holes and other places in the universe.

X-ray burster   A neutron star in a binary star system that accretes mass, undergoes thermonuclear fusion on its surface, and therefore emits short bursts of X rays.

X rays   Electromagnetic radiation whose wavelength is between that of ultraviolet light and gamma rays.

year   The sidereal period of revolution of a planet around the Sun.

Zeeman effect   A splitting of spectral lines in the presence of a magnetic field.

zenith   The point on the celestial sphere directly overhead.

zero-age main sequence (ZAMS)   The positions of stars on the Hertzsprung-Russell diagram that have just begun to fuse hydrogen in their cores.

zodiac   A band of 13 constellations around the sky through which the Sun appears to move throughout the year.

zones (on Jupiter)   Light-colored bands in Jupiter's cloud cover.